A&P TECHN

AIRFRAME

STUDY GUIDE

JEPPESEN®

Sanderson Training Products

JS312604C

ISBN 0-89100-451-3

TABLE OF CONTENTS

SECTION 1

SECTION 2

PREFACE

The scope of the work which may be performed by the holder of a Mechanics Certificate with Airframe and Powerplant ratings issued in the United States of America is probably the greatest of any country in the world. In this age of specialization, the Airframe and Powerplant Mechanics Certificate is, in many ways, an anachronism. At any rate, the wide range of privileges permitted by these licenses require testing over many subject areas.

Persons seeking certification may come from one or more backgrounds, including military aviation experience, general aviation, or those who complete training in a FAR Part 147-approved school, with no previous aviation experience. Regardless of background, each candidate must pass the required written, oral, and practical exams. This study guide is written to help at that time.

All of the subjects mentioned on the FAA exams are included in section one of this manual. The depth includes only enough to answer the questions. Candidates who wish to study the subject further will find additional text material and pertinent FAA publications listed in appendix three of section two. No attempt has been made to create a complete textbook for all subject areas. Those books are already in print and may be obtained by contacting Jeppesen Sanderson, Inc., directly or through your local aviation book store.

Material presented here is intended to help the candidate understand the systems presented on the FAA exams. It does not necessarily represent current technology. Information on specific systems should be obtained form the manufacturer and followed in detail.

For product, service, or sales information call **1-800-621-JEPP, 303-799-9090, or FAX 303-784-4153**. If you have comments, questions, or need explanations about any component of our Maintenance Training System, we are prepared to offer assistance at any time. If your dealer does not have a Jeppesen catalog, please request one and we will promptly send it to you. Just call the above telephone number, or write:

Marketing Manager, Training Products
Jeppesen Sanderson, Inc.
55 Inverness Drive East
Englewood, CO 80112-5498

Please direct inquiries from Europe, Africa, and the Middle East to:

Jeppesen & Co., GmbH
P. O. Box 70-05-51
Walter-Kolb Strasse 13
60594 Frankfurt
GERMANY
Tel: 011-49-69-961240
Fax: 011-49-69-96124898

INTRODUCTION
How To Use This Manual

This manual is designed for the person who has qualified for the Airframe and Powerplant Mechanics rating (either by completing an FAA-certified aviation maintenance technician school program or by meeting the experience rquirements of FAR 65.77) and wants to achieve the highest possible score on the mechanic written exam.

It is very important that the student follow a few simple pointers to help make this book work. Study should be a systematic review of the material with the specific objective of preparing you for the written exam. Begin by reading the text material for a given subject. The text provided here is not meant to represent all aspects of the material being discussed. Only those items pertinent to the written exam are covered. The text is arranged in the same order as the written test and should be followed from front to back.

When you have read the material and feel that you understand it, complete the review questions at the end of each section. Check your answers with the key provided. If you missed any of the questions, go back and study the section again.

Next, complete the written exam questions on the subject. They are indicated at the completion of each subject area. Make no marks in the exam section of your book as you will be using those pages for the practice exams at the completion of your studies. Check your answers using appendix three of section two, the "answers & explanations", of this book. These answers will tell you what the correct response is and why. Also included are references for the material in this book and other texts. If you miss any questions, research the material before proceeding to the next subject.

When you have completed all subject areas, you are ready for a practice exam. The practice exams provided in this book are very similar to those used by the FAA. We have included answer sheets for each practice exam like those for the FAA tests. Mark your answers on the answer sheet and, when you have finished, check it against the answer key provided. A score of 70 is passing. If you feel that you need it or just for more practice, a second practice exam is included.

SECTION 1

A&P Technician Airframe Study Guide
Self Study Text Material

AIRCRAFT WOOD STRUCTURES

Wood was used for the first airplanes because of its favorable strength-to-weight ratio. It's use has been almost entirely superseded by all-metal construction for commercial production. The inspection and repair of wooden structures which may be encountered in antique or homebuilt aircraft is the responsibility of the airframe maintenance technician.

Advisory Circular 43.13-1A details the acceptable types of wood for use on aircraft structure, defects in wood, and repair techniques. This manual should be consulted anytime wooden structure is encountered. There are, however, a few items that the technician should be familiar with.

Wood used for aircraft structures will be in one of the following forms:

- *Solid Wood* — refers to a member consisting of one piece of wood.
- *Laminated Wood* — is an assembly of two or more layers of wood which have been glued together with the grain of all layers or laminations approximately parallel. Laminated wood spars may be substituted for solid rectangular wood spars if the same quality wood is used in both.
- *Plywood* — is an assembled product of wood and glue that is usually made of an odd number of thin plies (veneers) with the grain of each layer at an angle of 90° with the adjacent ply or plies.

Sitka spruce is the reference wood used for aircraft structures because of its excellent shock resistance qualities. Information may be found concerning acceptable species substitutions for wood materials used in aircraft repair in AC 43.13-1A.

Defects permitted in aircraft wood are:

- *Cross grain* — Spiral grain, diagonal grain or a combination of the two providing the grain variation does not exceed 1:15.
- *Wavy, curly, and interlocked grain* — if irregularities do not exceed limits for cross grain.
- *Hard knots* — sound hard knots up to 3/8" may be acceptable.
- *Pin knot clusters* — small clusters are acceptable providing they produce only a small effect on grain direction.
- *Pitch pockets* — acceptable depending on location and size.
- *Mineral streaks* — acceptable providing careful inspection fails to reveal any decay.

Defects not permitted in aircraft wood:

- *Grain defects, knots, pitch pockets and mineral streaks* — exceeding limits for acceptable defects.
- *Compression wood* — reject all material containing compression wood.
- *Compression failures* — indicated by a faint line running across the grain of a wood spar or other solid piece. Compression failures in wood structures are characterized by buckling of the fibers that appear as streaks on the surface at right angles to the grain.
- *Decay* — unlike some other wood defects, is not acceptable in any form or amount. Three forms of decay in wood structures are dote, redheart, and purplewood.

Aircraft wooden structures are joined with glue. The strength of a well-designed and properly prepared wood splice joint is provided by the glue. The strength of a glue joint must be equal to the strength of the wood. Glue deterioration in wood aircraft structure is indicated when a joint has separated and the glue surface shows only the imprint of the wood with no wood fibers clinging to the glue.

One type of wood repair is the spar splice. The wing spar is the major structural member of the wing and its integrity is required for safe flight. This is espe-

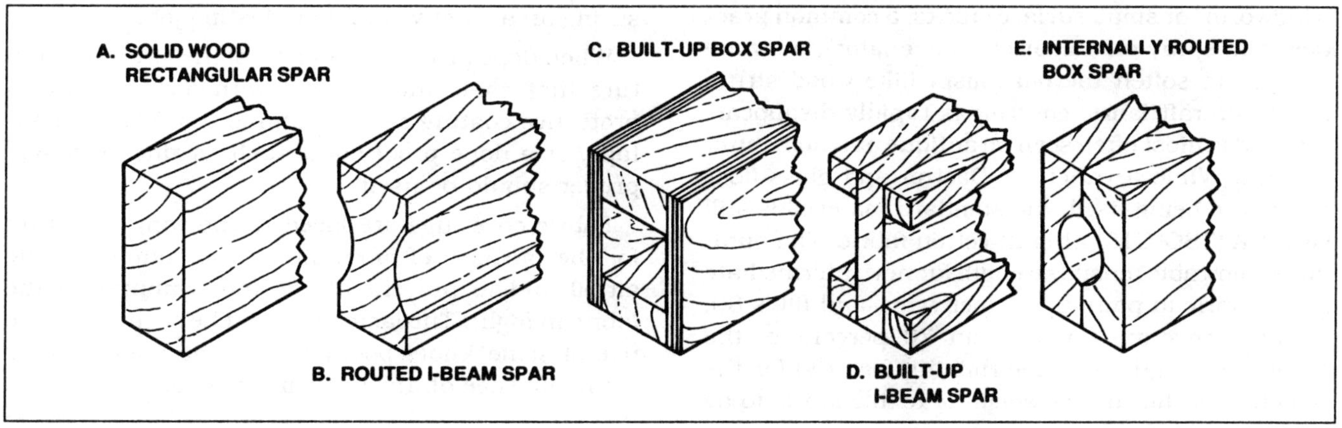

Figure 1. Types of wood wing spars.

cially true when working on a cantilever wing. The cantilever wing uses no external bracing.

Wooden spars may use any of a variety of forms. The simplest is the solid spar. The I-beam wooden spar is routed to reduce weight. Other spars may be built up of several pieces of solid and laminated wood.

Elongated bolt holes in a wood spar may be repaired by splicing in a new section of spar and drilling new holes. When making an aircraft wood repair, special care should be exercised when determining the proper grain direction because the high shearing strength parallel to the grain should be used to the fullest advantage. An optional method of repair would be to replace the spar entirely.

Stressed plywood skin may be patched with a scarf type patch. When patching a plywood skin, abrupt changes in cross-sectional areas which will develop dangerous stress concentrations should be avoided by using circular or elliptical patches.

REVIEW QUESTIONS — Aircraft Wood Structures

1. *The strength of a glue joint must be equal to what?*
2. *Where is information found concerning acceptable species substitutions for wood materials used in aircraft repairs?*
3. *Why is the I-beam wooden spar routed?*
4. *What type of wing uses no external bracing?*
5. *What type of wood is used as the standard for strength?*

Answer questions 8001 through 8014 in the FAA Test section of this book.

AIRCRAFT FABRIC COVERING

Before refinishing a doped surface, or to correct "ringworm" or small surface cracks, a common practice is to spray on a coat of "rejuvenator", a strong solvent, to soften the old finish. Like wood structures, aircraft fabric covering is rapidly disappearing, and is most often seen on antique and homebuilt aircraft. While a variety of Dacron and glass-fiber cloths are being used, the standard material is still Grade A (TSO-C15) Aircraft cotton fabric. The minimum allowable tensile strength for new intermediate grade fabric in pounds per inch warp and fill is 80. Existing covering must retain 70 percent of the original strength. The determining factor(s) for the selection of the correct weight of textile fabric to be used in covering any type of aircraft is the speed of the aircraft and the maximum wing loading.

The fabric covering is secured to the structure, and then treated with "dope" to make it airtight, offer protection from the elements, and give a decorative finish. This "dope" may be the old standard CAB (cellulose acetate butyrate) or a new epoxy finish.

It is very important that all materials and techniques are consistent with the covering system (fabric type) being used. Because one system may permit all seams to be glued, this is not possible with all types of fabrics. Consider each type of fabric to represent a "covering system" and be certain that all materials, chemicals and techniques are consistent with that "system".

There are a few terms that the technician should be familiar with when dealing with aircraft fabric. They include:

- *Warp* — the direction along the length of the fabric.
- *Fill, woof, or weft* — the direction across the width of the fabric.
- *Count* — the number of threads per inch, in warp or fill.
- *Bias* — a cut, fold, or seam made diagonally to the warp or fill threads.
- *Selvage edge* — an edge of cloth, tape, or webbing, woven to prevent raveling.
- *Pinked edge* — an edge which has been cut by machine or shears in a continuous series of V's to prevent raveling.

Machine sewed seams should be of the folded-fell or French-fell types. A plain lapped seam is satisfactory where selvage edges or pinked edges are joined.

One of the most important items in covering aircraft is proper preparation of the structure. Dope proofing, covering edges which are likely to wear the fabric, preparing plywood surfaces, and similar operations, if properly done will do much toward ensuring an attractive and long-lasting job.

When dope proofing the parts of the aircraft structure that come into contact with doped fabric, a protective coating such as aluminum foil, cellulose tape, or a dope-proof paint such as zinc chromate primer should be used.

Fabric coverings are laced to the wing structure for the purpose of maintaining the contour of the airfoil and to prevent lifting and whipping of the fabric in flight. The lacings should be secured by the use of seine knots positioned in the middle of, or along the edge of, the bottom capstrip.

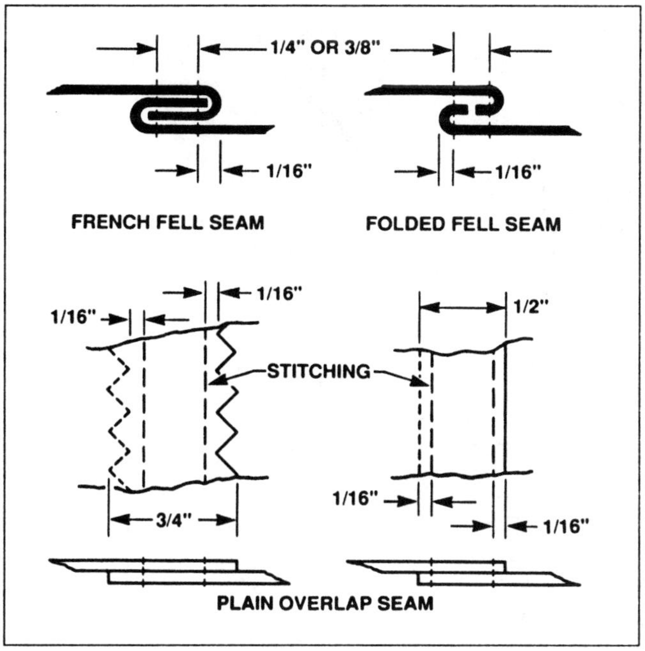

Figure 2. Types of machine-sewed seams.

Hand sewing thread and lacing cord should be waxed before using. The wax should not exceed 20% of the weight of the finished cord. A beeswax free from paraffin should be used for waxing. Dacron, linen, glass, or cotton cords are used for rib lacing cord.

Surface, or finishing, tape is installed on fabric-covered aircraft to prevent wear over lacings, seams, corners, and edges. Prior to application of surface tape to the trailing edge of a wing or control surface, the tape should be notched at intervals to prevent the entire tape from loosening in the event the tape begins to separate from the surface. The finishing tape is doped on after the first or second coat of dope.

Repair of Fabric Covers

Glass fabric used for aircraft is not affected by moisture, mildew, chemicals and acids, but even it is still vulnerable to mechanical damage from ground handling and foreign objects.

Repair fabric-covered surfaces so that the original strength and tautness are returned to the fabric. The recommended procedure for making an unsewn (doped-on) fabric patch is to clean and remove all aluminum and color coats from the surface to which the patch is to be doped. A small patch may then be applied using dope as an adhesive, and the finish restored.

Small cuts or tears are repaired by sewing the edges together and doping a pinked edge patch over the area. The baseball stitch is used in repairing tears. A sewed-in repair patch may be used if the damage is not longer than 16″ in any one direction. When the damage exceeds 16″ in any direction, a new panel should be installed.

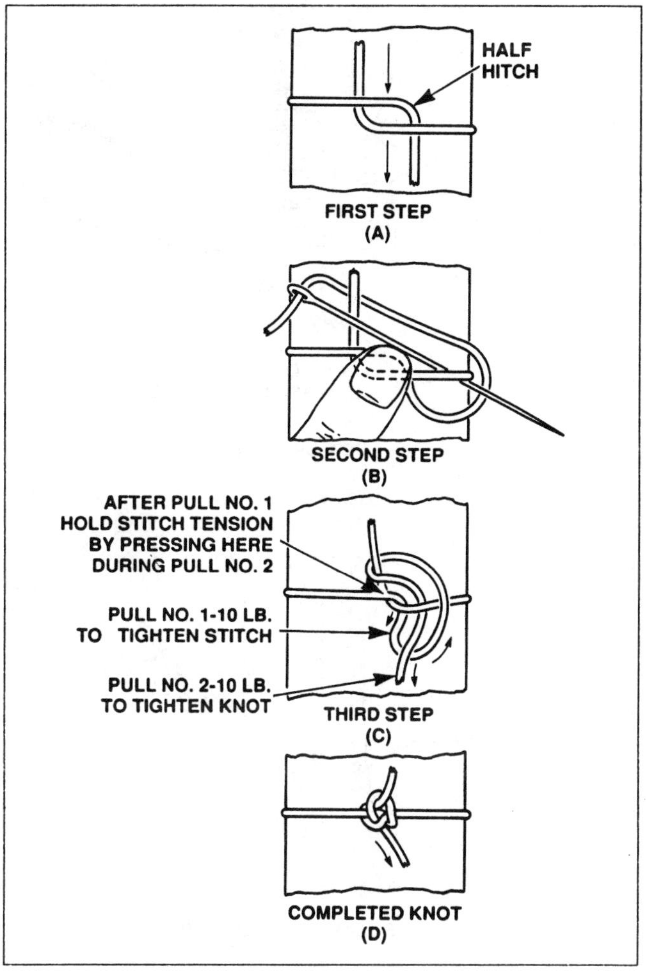

Figure 3. The modified seine knot for locking baseball stitches.

Aircraft Finishes

Whether it is new covering or a repair, aircraft fabric must be doped before it is returned to service. The initial coats may be brushed on, but the finish coats and decorative trim work is almost always accomplished by spraying.

The products used to finish a fabric covering must be those approved for the type of fabric covering. Fungicidal dopes are used in aircraft finishing as the first coat to prevent fabric rotting, and are applied extremely thin to saturate the fabric. The first coat of dope is brushed onto the fabric and not sprayed. This is to permit thorough penetration of the fabric.

When making repairs, it will be necessary to finish the surface using the same type of materials originally applied. A practical method for determining whether a fabric surface has been finished with

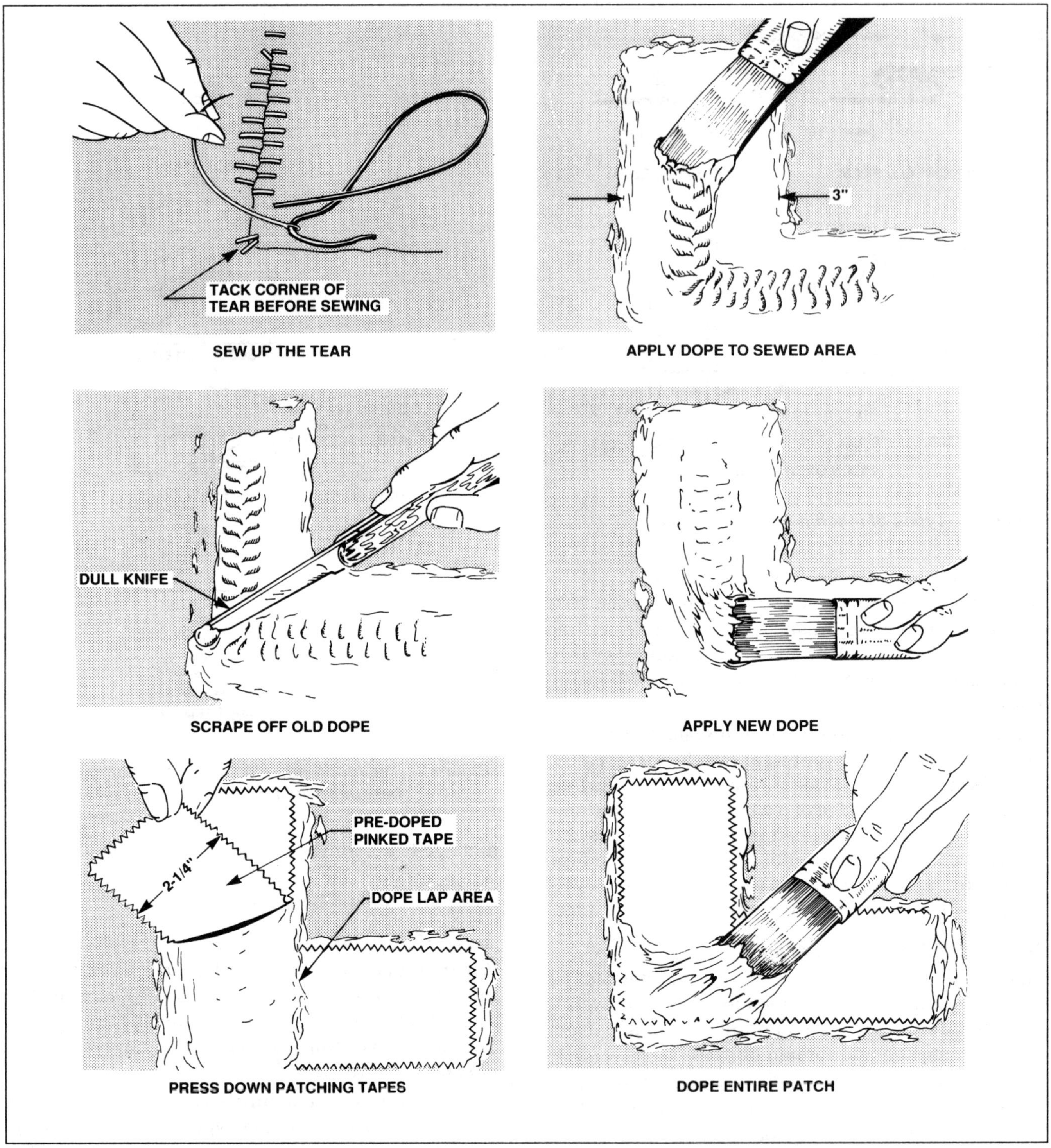

Figure 4. Repair of an L-shaped tear in a fabric surface.

nitrate (cellulose nitrate) or butyrate (cellulose acetate butyrate) dope is to rub a small section with lacquer thinner. Butyrate finish will not be affected, but nitrate finish will be softened.

When finishing a fabric surface, apply at least the number of coats necessary to result in a taut and well-filled finish job. Each coat should be thin and uniform. Too much material applied in one coat may result in runs and sags in the finish.

It is inadvisable to dope fabric on humid days because this increases the tendency of the dope to blush. Blushing may be caused by moisture in the air supply line, adverse humidity, drafts, or sudden changes in the temperature. Blushing appears as a "clouding" or "blooming" of the finish. Blushing is not acceptable in doped finishes.

Aluminum (or pigment) is normally added to the final coats of dope. The chief purpose for this is to exclude sunlight from the fabric. Ultraviolet rays from sunlight will speed deterioration of the fabric.

Whenever operations involving sanding, rubbing, or wiping of detached fabric covered surfaces, such as wings, the surface should always be grounded to prevent accumulation of static charges during the finishing process. Static discharge could ignite dust or fumes and result in fire or explosion. Do not sweep sanding residue with a dry broom. Acetone may be used to remove grease from fabric.

When finishing an aircraft, paint may be used in conjunction with pigmented dope. This will allow the metal portions of the aircraft to be matched to the doped fabric. Before spraying a paint finish on cleaned aluminum, never touch the surface with bare hands. Paint system compatibility must be observed. Old wash primer coats may be overcoated directly with epoxy finishes.

Aircraft Registration Markings

All aircraft are required to display nationality and registration marks. Aircraft registered in the United States must display the Roman capital letter "N" followed by the registration number of the aircraft. The location and size of these marks are prescribed by Part 45 of the Federal Aviation Regulations.

The letters and numbers of identification markings must be of a contrasting color and must be uniform in shape and size. The width of the characters must be ⅔ their height. When appling masking tape for trim or lettering, it should not be left on the airplane for too long, as the tape will cure to the finish and be difficult to remove.

Metal Finishing

Synthetic enamel, acrylic lacquer and polyurethane are common top coat materials. Polyurethane is the most durable. While enamel may be undercoated with zinc chromate or epoxy primers, the others require a wash primer. Wash primers consist of phosphoric acid, an alcohol based thinner, and water. Wash primers are applied in a thin coat and must be allowed at least 30 minutes to convert to a phosphate film. If sprayed in a very dry climate, sufficient moisture from the surrounding air to allow proper conversion. Without sufficient moisture, the finish coat will "sink in" the primer and not cure properly.

All primers and top coats are part of a "finishing system" and should not be intermixed.

REVIEW QUESTIONS — Aircraft Fabric Covering

6. *What is the minimum allowable tensile strength of new Grade A fabric?*
7. *What type of aircraft fabric is not affected by moisture, mildew, chemicals, and acids?*
8. *What type of stitch should be used to repair a sewn in patch?*
9. *What condition may occur when doping on humid days?*
10. *What may be used to remove grease from aircraft fabric?*
11. *What causes runs and sags in aircraft finishes?*
12. *What is the chief purpose of pigment (or aluminum) in dope?*

Answer questions 8015 through 8040 in the FAA Test section of this book.

SPECIAL RIVETS AND FASTENERS

The vast majority of aircraft structure is held together with solid rivets. Many of the modern designs use some bonded construction, and in almost all aircraft there are special fasteners; but the ability to install a solid rivet is a basic skill for all airframe maintenance technicians.

Solid Rivets

Solid rivets are used for most aircraft repairs. A well-designed rivet joint will subject the rivets to shear loads. Most field repairs are made using the 2117 rivet. These rivets are heat treated by the manufacturer and do not require heat treatment before being driven. They are most often referred to as "field rivets".

The proper diameter rivet must be used to maintain the strength desired. The general rule for finding the proper rivet diameter is three times the thickness of the thickest sheet. When making repairs, observe the rivet diameter used by the manufacturer.

For most riveted repairs your best source of information about rivet layout, edge distance, pitch, etc. will be the manufacturer's structural repair manual. If this information is not available, use the same values as observed in adjacent structure.

Rivet length must reach through the material being joined and furnish enough material to form a head that is ½ the rivet diameter thick and spread out to a diameter of 1½ times the shank diameter. To do

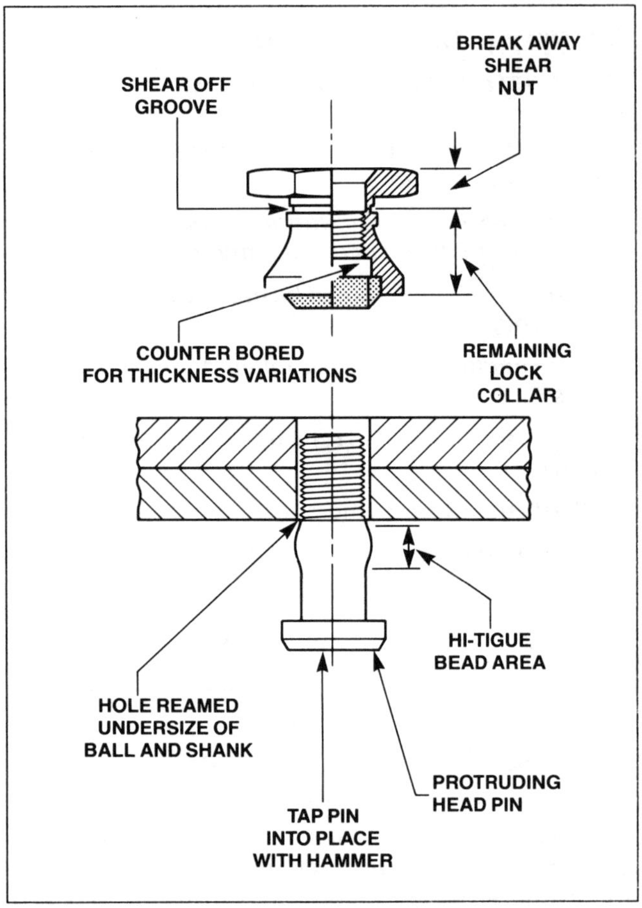

Figure 5. Many fasteners require specialized tools for installation. These tools can be quite expensive. One type of fastener that can be installed with standard tools is the Hi-Loc style.

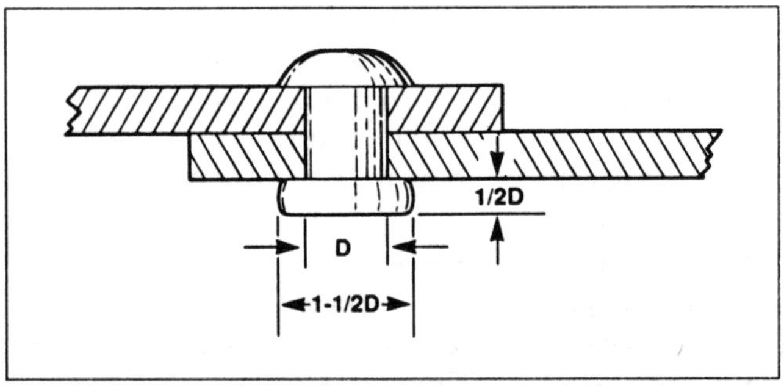

Figure 6. Dimensions of a formed, or shop, head of a solid rivet.

this, the undriven rivet must be long enough to go through the material and protrude $1\frac{1}{2}$ diameters.

Special Fasteners

There are many places on an aircraft where access to both sides of a riveted structure or structural part is impossible, or where limited space will not permit the use of a bucking bar. For use in such places, special fasteners have been designed. Because they

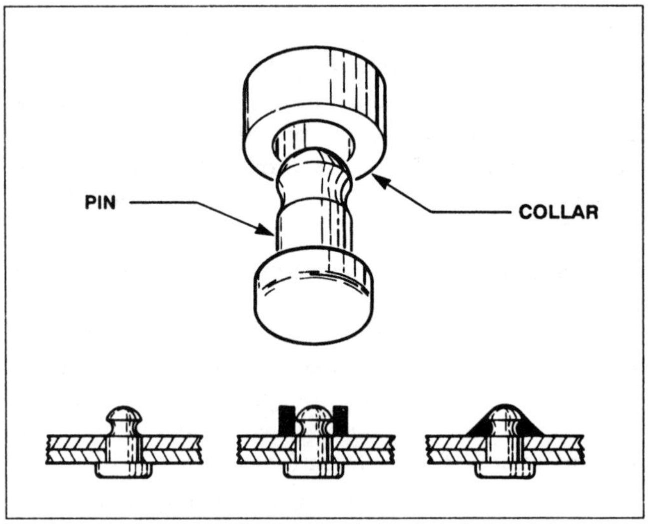

Figure 7. Hi-Shear pin rivet.

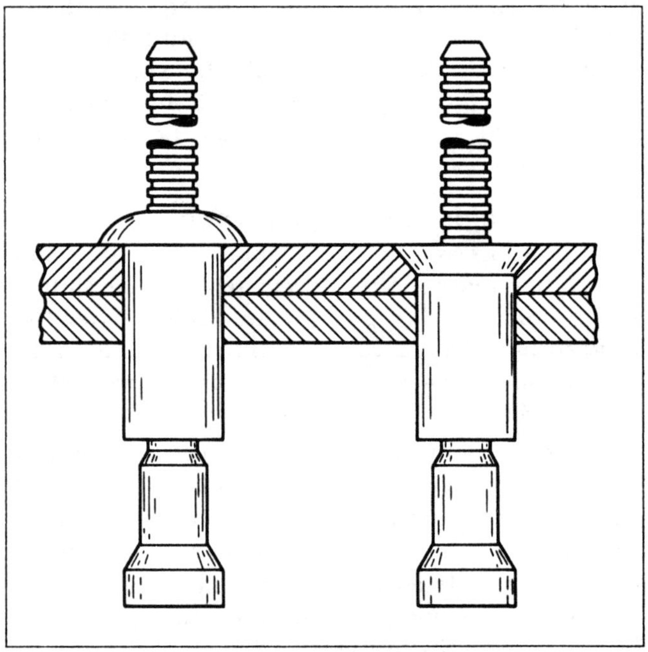

Figure 8. Self-plugging, friction-lock Cherry rivets.

are often inserted in locations where one head cannot be seen, they are also called blind fasteners.

Lockbolts and blind rivets are manufactured for various applications, both shear and tension. Most use a collar swagged over grooves in the shank to lock the pulling stem. More grooves are necessary for fasteners loaded in tension.

Pin (Hi-Shear) rivets have the same shear strength as bolts of equal diameter. These fasteners use a pin made of high-strength steel held in place by a steel or aluminum alloy collar swaged over the end of the pin. If a Hi-shear rivet is underdriven during installation, the collar will be incompletely swaged into the groove.

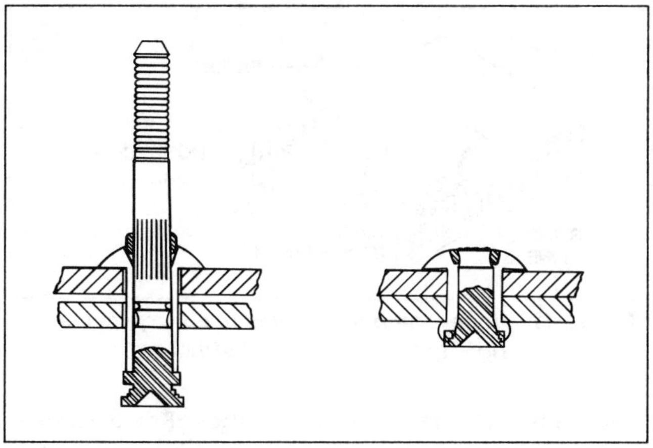

Figure 9. Mechanical-lock Cherry rivets.

Self-plugging (friction lock) Cherry rivets made by the Townsend Division of Textron may be installed when there is access to only one side of the structure. The blind head is formed by pulling the tapered stem into the hollow shank. This swells the shank and clamps the skins tightly together. Selection of the proper rivet can best be made using a gage supplied by the manufacturer. Self-plugging Cherry rivet shank diameters are measured in increments of $1/32''$.

Mechanical-lock Cherry rivets are similar to the friction-lock blind rivet except that in the process of pulling the stem through the hollow shank, a locking collar is swaged down onto the stem, and when the

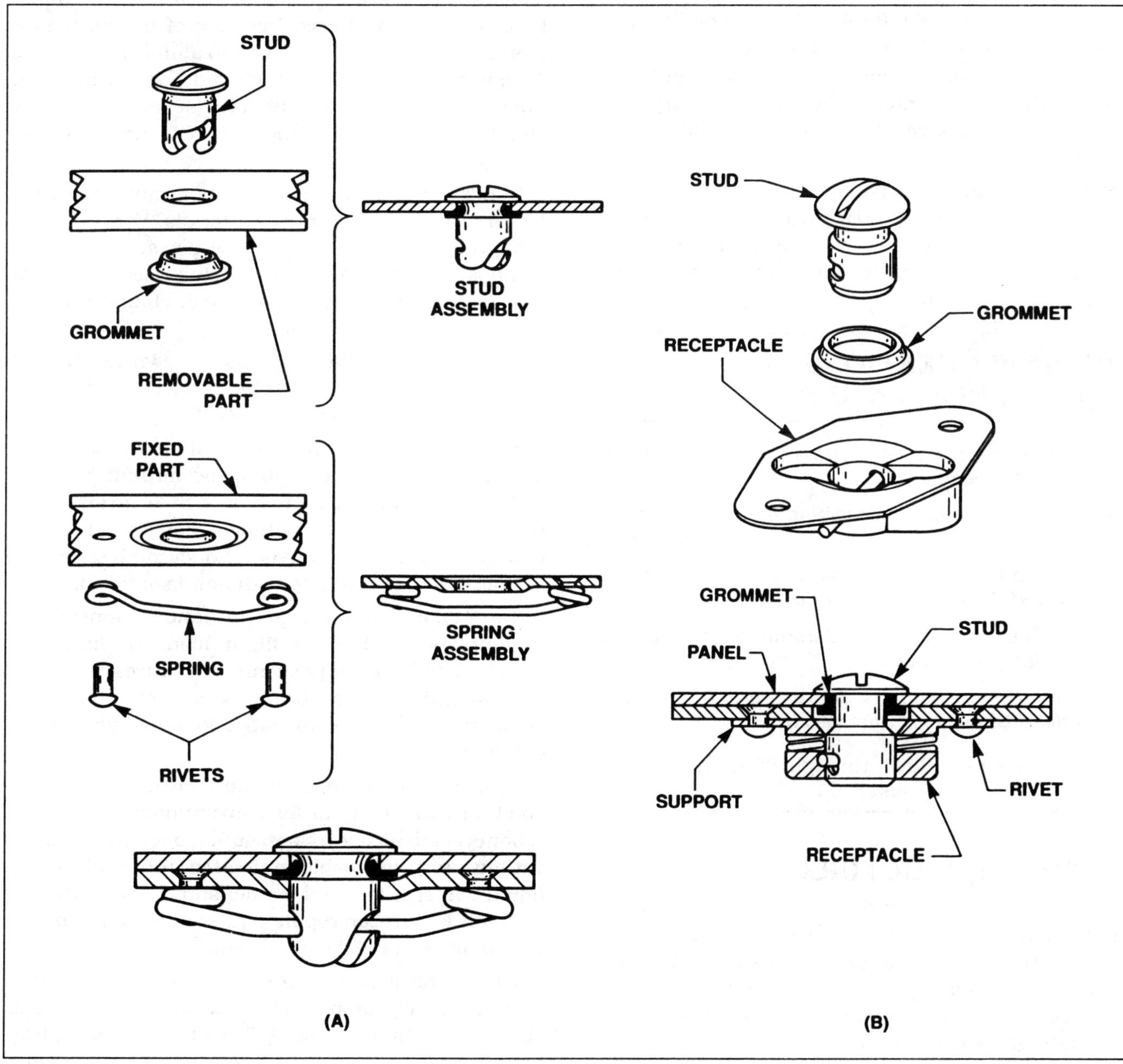

Figure 10. Dzus fastener.

stem breaks off, the portion remaining inside the shank is tightly locked in place.

Bulbed Cherrylock® rivet grip length is measured in increments of 1⁄16″.

The Dzus turnlock fastener consists of a stud, grommet, and receptacle. Figure 10 illustrates an installed Dzus fastener and the various parts. The studs are fabricated from steel and are cadmium plated. They are available in three head styles; wing, flush, and oval. Body diameter, length, and head type may be identified by the markings found on the head of the stud. The diameter is always measured in sixteenths of an inch. Stud length is the distance from the head of the stud to the bottom of the spring hole, and is measured in hundredths of an inch.

A quarter turn of the stud (clockwise) locks the fastener. The fastener may be unlocked only by turning the stud counterclockwise. A Dzus key or a specially ground screwdriver locks or unlocks the fastener.

Rivnuts, manufactured by B. F. Goodrich, are used for attaching non-structural parts or components to the airframe. On older large aircraft, they are used for attaching de-ice boots, hence their nickname, "boot nut".

REVIEW QUESTIONS — Special Rivets and Fasteners

13. *A well-designed rivet joint will subject the rivets to what type of loads?*
14. *What is the general rule for finding the proper rivet diameter?*
15. *What are the dimensions of a properly upset (bucked) head on a driven rivet?*
16. *Pin (Hi-shear) rivets have the same shear strength as what other fastener?*
17. *In what increments is the rivet grip length of a Bulbed Cherrylock rivet measured?*

Answer questions 8041 through 8052 in the FAA Test section of this book.

BONDED STRUCTURES

The introduction of bonded honeycomb (sandwich construction) members in airframe design and manufacture came as a major breakthrough in the search for a more efficient type of structure. The preferred method to repair any composite is by laminating. This allows for a stronger repair and the smooth transition of stresses. Composite fiber ma-

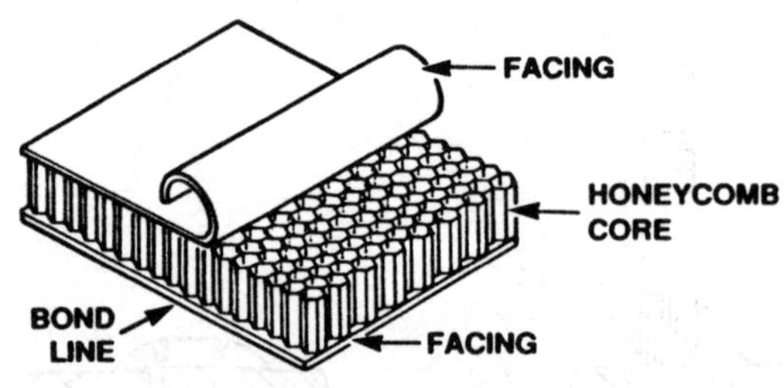

Figure 11. Honeycomb material is used to provide rigidity to a lightweight structure.

terials have different characteristics. For instance, Kevlar has a high tinsel strength and high flexibility while carbon has a high compressive strength and a high degree of stiffness. Because of the dissimilar metal, carbon fiber is corrosive to aluminum. Figure 11 illustrates a section of bonded honeycomb. The honeycomb stands on end and separates facings which are bonded to the core by means of an adhesive or resin.

Bonded honeycomb construction may be manufactured using a variety of materials. Various combinations of metal, wood, fiberglass, and exotic composite materials are now being used. Composites exhibit low-thermal expansion characteristics, and are gaining popularity.

Each different material has its own method of construction and repair. Methods used for fiberglass may prove unsatisfactory for Kevlar.

Sandwich panels made of metal honeycomb construction are used in high-speed aircraft because this type of construction has a high strength-to-weight ratio and greater stiffness than a single sheet. One of the main advantages of honeycomb structure is that it is better able to withstand sonic vibration.

The majority of damages to bonded honeycomb assemblies result from flight loads or improper ground handling. Honeycomb structures may also be damaged by sonic vibrations. Exposure of honeycomb structures to sonic vibrations usually causes delaminations.

Inspection for damage is more critical for honeycomb assemblies than for conventional structures. A honeycomb structure can suffer extensive damage without any observable indication. Sonic vibration, liquid leakage, internal condensation, or a mistake in manufacture or repair can cause, or result in, varied amounts of delamination.

The simplest test in checking for a delaminated area on a honeycomb structure would be the metallic ring test. When a coin (25-cent piece) is lightly bounced against a solid structure, a clear metallic ring should be heard. If delamination is present, a dull thud will be heard. A clear metallic sound indicates the facing and core are sound.

A similar technique may be used on fiberglass panels. When inspecting a fiberglass panel using the "tapping" test procedure, a dull thud from the tapping tool indicates separation of the laminates used in the lay-up. One method of inspecting a laminated fiberglass structure that has been subjected to damage is to strip the damaged area of all paint and shine a strong light through the structure. Ultra sonic and acoustic emission inspection procedures that have been developed for corrosion inspection in metal structures have also been adapted to composite inspection.

Other testing methods may be employed when inspecting for certain types of damage or failure. Radiography (x-ray) inspection procedures may be used to check for water in the honeycomb of a composite structure.

After inspections of bonded honeycomb structures are completed, any damage found must be evaluated to determine the type of repair needed. Damage evaluation charts for honeycomb structures of a large aircraft can be found in the applicable section of its structural repair manual. Once the type of repair is determined, procedures outlined in the structural repair manual should be rigidly followed.

Bonded honeycomb structure damages up to 1″ in diameter may be repaired by a hole-filling technique, using approved materials. The repair method is commonly known as the "potted compound repair". It is the easiest and fastest method of repairing a damaged area of a bonded honeycomb structure.

When repairing puncture-type damage of a metal faced laminated honeycomb panel, a doubler should be prepared to cover the puncture, and applied with the potting compound. The edges of the doubler should be tapered 100 times the thickness of the metal.

More severe damage repair will probably involve the removal and replacement of a section of the core material. When removing damage from one face sheet of a laminated honeycomb panel, it is important to avoid damaging the opposite face sheet. The most effective tool for removing a damaged area 1″ diameter or less is a router and template. A typical repair used to retain 100% strength to a damaged area of bonded honeycomb, 1″ in diameter, is an overlap type metal facing patch.

The material used to replace the core may be balsa wood or honeycomb. When balsa wood is used to replace a damaged honeycomb core, the plug should be cut so that the grain is perpendicular to the skin.

Honeycomb core material used to replace damaged aluminum cores of honeycomb bonded structure is usually fiberglass. When replacing a metal honeycomb core section, the ribbon direction of the repair should align with the original core material. Repairs inside a metal honeycomb structure should be treated to prevent corrosion by priming the repair with a corrosion inhibitor and sealed from the atmosphere.

Special fasteners, designed for composite structures, must be used for attachment of structure and components. Standard "hole filling" lockbolts and rivets will further damage a composite structure by causing delamination and cracking. Carbon-graphite structures must be resistant to possible corrosion that can be caused by contact with metal fasteners.

Some bonded honeycomb panels are constructed of a doubler separating an upper and lower skin and sectioned into bays of honeycomb core material. The edges of a bay area in which the honeycomb structure joins the laminated area of the panel or section is known as the transition area. Effective repairs to the transition areas are particularly essential because of the local transferring of the stresses. In a transition area bonded honeycomb panel repair, the circular shape repair should not exceed 2″ in diameter.

Fiberglass must be identified as to it's material makeup before it can be repaired. Two common classifications are S-glass (high tensile strength) and E-glass (high di-electric). E-glass is the most common. Fiberglass laminate repair uses products and techniques similar to those used for repair of bonded honeycomb structures. When making repairs to fiberglass structures, cleaning of the area to be repaired is essential for a good bond. The final cleaning should be made using methyl ethyl ketone (MEK).

Superficial scars, scratches, surface abrasions, or rain erosion on fiberglass laminates can generally be repaired by applying one or more coats of suitable resin (room temperature catalyzed) to the surface.

Fiberglass laminate damage not exceeding the first layer (ply) can be repaired by filling with a putty consisting of a compatible resin and clean, short glass fibers. Fiberglass laminate damage that extends completely through one facing and into the core requires the replacement of the damaged core and facing. Fiberglass damage that extends completely through a laminated sandwich structure can also be repaired, using approved methods and materials. Stress concentrations can be eliminated when making scarf repairs in composite structure by avoiding abrupt changes in cross-sectional area. Fiberglass and other fabrics are strongest in the warp direction.

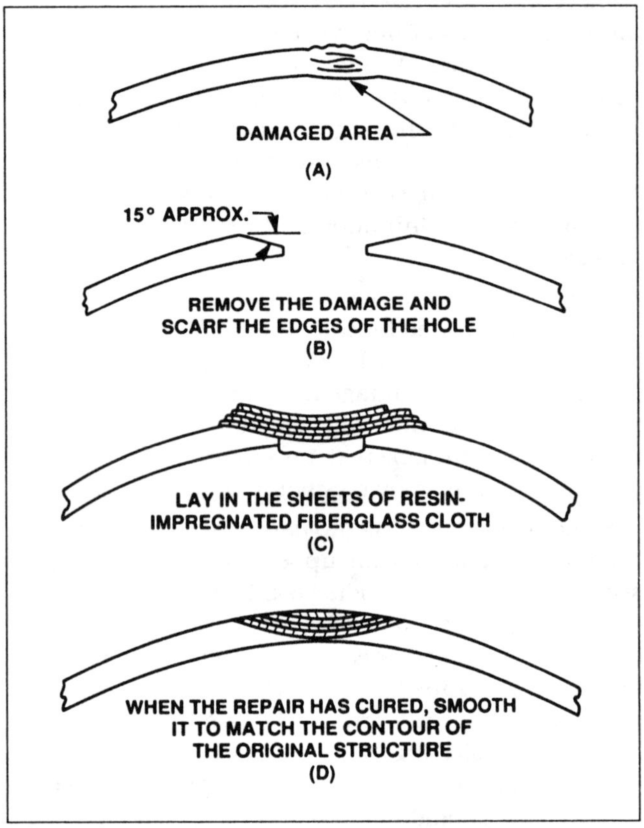

Figure 12. Scarfed repair to a laminated fiberglass structure.

A tool, called a "warp compass", is used to determine the proper ply orientation. Strength and stiffness depend on proper ply orientation.

Pressure must be applied to most repairs. The most effective is "vacuum bagging". This is a system whereby the repair is sealed in a bag and excess resin is drawn off into a "bleeder" by applying vacuum to the bag and forcing the repair into intimate contact with the damaged part.

The repairs we have been discussing all involve the use of certain resins and catalysts. The airframe maintenance technician should be familiar with these resins, their mixing, handing, and use as mixed materials have a definite working, or pot, life. Polyester resin in its liquid state is of little use, but when the molecules are joined into solidly-linked chains we have the rigidity necessary to combine with the strength of other materials and produce composite materials for structural applications.

Polyester resin is cured by adding a catalyst which joins the molecules. The amounts of materials mixed must be carefully controlled, as this will affect the ultimate strength of the product. The most effective method for determining if the proper ratio of resin-to-catalyst was used is to mix enough to have a sample for testing after curing. The resin to catalyst ratio is important. A 50/50 mix is good, but in advanced composites, a 60/40 mix is better. Some repair systems use pre-impregnated fabric. These already have the resin and fabric contained in a sealed package. The package needs only to be mixed and used.

When repairing large, flat surfaces with polyester resins, warping of the surface is likely to occur. One method of reducing the amount of warpage is to use short strips of fiberglass in the bonded repair.

Composite materials are easily contaminated. Never use a cutting fluid for lubrication. Should a lubrication be necessary, use clean water.

Because some plastic resins may be so thin as to run off of a vertical surface it may be necessary to thicken the resin to create a suitable product. Thixotropic agents may be added to thicken the resin to a paste-like consistency. Among the thixotropic agents most familiar in aviation are microballoons. These hollow glass or phenolic balls range in diameter from about 10 to 300 microns, and resemble fine sand. The benefits of using microballoons when making repairs are that they lower the density and give greater flexibility, thus lowering stress concentrations in the repair area.

Aircraft Windows and Interior Furnishings

Transparent plastics may be used in single sheets (monolithic) or in laminated assemblies. The two most common types are Cellulose Acetate and Acrylic plastics (Plexiglas). Cellulose acetate was used in the past, but since it is dimensionally unstable and turns yellow after it has been installed for a time it has just about passed from the scene. Plexiglas is also a heat forming, or thermoplastic, material.

Before repairs can begin the technician must determine which type is being used. Zinc chloride may be used to determine the type of plastic. When zinc chloride is applied to cellulose acetate plastic, the plastic will turn milky. Zinc chloride has no effect on acrylic plastic.

Aircraft windows may be cleaned by first flushing with plenty of water. Wash with mild soap and water. If no scratches are visible after transparent plastic enclosure materials have been cleaned, their surfaces should be covered with a thin coat of wax.

If an aircraft's transparent plastic enclosures exhibit fine cracks which may extend in a network over or under the surface or through the plastic, the plastic is said to be crazing. There are certain new products on the market which may polish out these

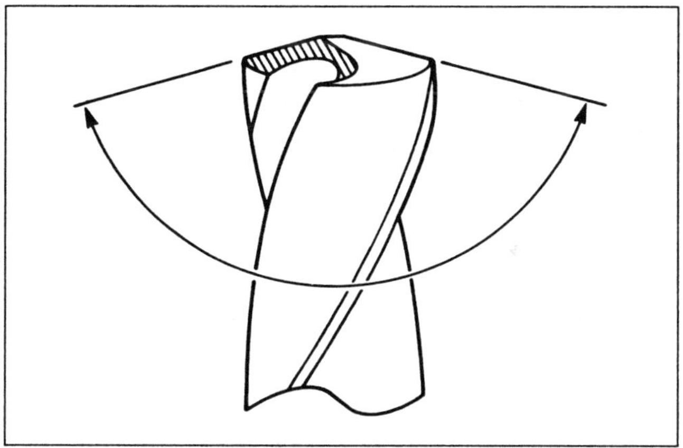

Figure 13. Drill acrylic plastic with a drill having an included angle of approximately 150°.

small cracks and restore clear vision. Generally, windows exhibiting crazing must be replaced.

Before replacing windows, their method of mounting must be examined to assure a proper fit for the replacement panel. Plastics expand and contract considerably more than the metal channels in which they are mounted. To accommodate this, windshields are mounted to a sufficient depth in the channel to prevent them from falling out when the panel contracts at low temperatures, or deforms under load.

Windows and windshields may be secured with rivets or bolts extending through the plastic material. When holes are drilled completely through Plexiglass, a specially modified twist drill should be used. When rivets are used to secure panels, adequate spacers or other satisfactory means to prevent excessive tightening of the frame to the plastic should be provided. When installing panels which are retained by bolts extending through the plastic material and elastic stop nuts, the stop nuts should be tightened to a firm fit then backed off one full turn.

The most common method of cementing transparent plastics is the "soak method". This consists of dipping one of the two pieces to be cemented into the cement until a sufficient cushion is formed. When this is pressed against the opposite dry surface, the excess cement forms a second cushion—shallow, but enough to permit thorough intermingling of the two surfaces.

Cabin upholstery materials installed in current standard category airplanes must be at least flame resistant (FAR 91.613).

If a new safety belt is to be installed in an aircraft, the new belt must conform to the strength requirement of Technical Standard Order (TSO) C22.

REVIEW QUESTIONS — Bonded Structures

18. *What is the simplest test for delamination of bonded honeycomb structure?*
19. *When balsa wood is used to replace a damaged honeycomb core, the plug should be cut so that the grain lies in what direction to the skin?*
20. *What effect will zinc chloride have on acrylic plastics?*
21. *A potted compound repair on honeycomb can usually be made on damages less that what size?*
22. *Exposure of honeycomb structures to sonic vibrations usually causes what condition?*
23. *The final cleaning of fiberglass structures before repairs are made should be made with what chemical?*
24. *If an aircraft window exhibits fine cracks which may extend in a network over or under the surface, the plastic is said to have what condition?*
25. *Seat belts used in aircraft must conform to what TSO?*
26. *What is the most common method of cementing transparent plastics?*

Answer questions 8053 through 8098 in the FAA Test section of this book.

SHEET METAL STRUCTURES

Most modern aircraft are constructed primarily of sheet aluminum. It is necessary for the airframe maintenance technician to understand the basic repair techniques associated with this material. The goal of any repair is to restore the damaged area to its original strength. To do this, the selection of the proper material, rivets, tools, and repair scheme is essential.

There are two basic types of sheet metal structures used for aircraft: monocoque and semi-monocoque. They differ in the amount of internal structure they use. Monocoque fuselages derive their principle strength from the actual covering, metal or plywood. This is referred to as "stressed skin".

In the semi-monocoque fuselage, primary bending loads are taken by the longerons. The longerons are supplemented by other longitudinal members, called stringers. Stringers are more numerous and lighter in weight than longerons. The longerons and stringers prevent tension and compression from bending the fuselage.

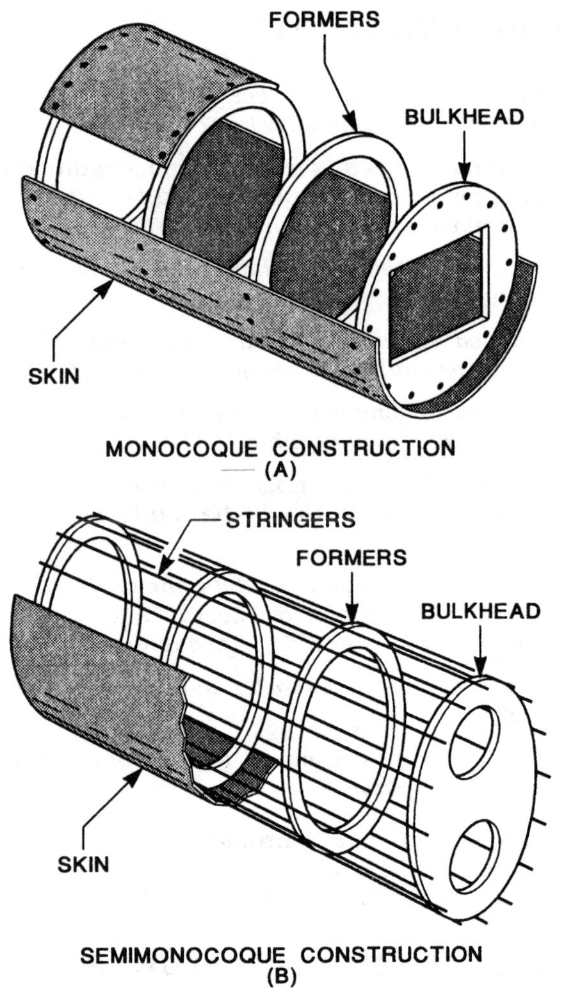

Figure 14. Semi-monocoque construction.

Sheet aluminum used on aircraft is identified by a 4-digit identification number followed by a heat-treat code. This is clearly marked on new sheets, and may often still be seen on the inside of aircraft structure. Before beginning any repair, determine the type of material used. In making wing skin repairs the rivets selected should be the same alloy as the material being riveted.

The first digit of the aluminum identification code indicates the primary alloying agent. A large percentage of aircraft aluminum alloy has as its primary alloying agent, copper. Alloys in which copper is the primary alloying agent will begin with the number 2 (for example: 2024-T36).

Clad aluminum alloys are used in aircraft because they are less subject to corrosion than uncoated alloys. Clad alloys, or "Alclad", is an aluminum alloy sheet to which a thin coat of pure aluminum has

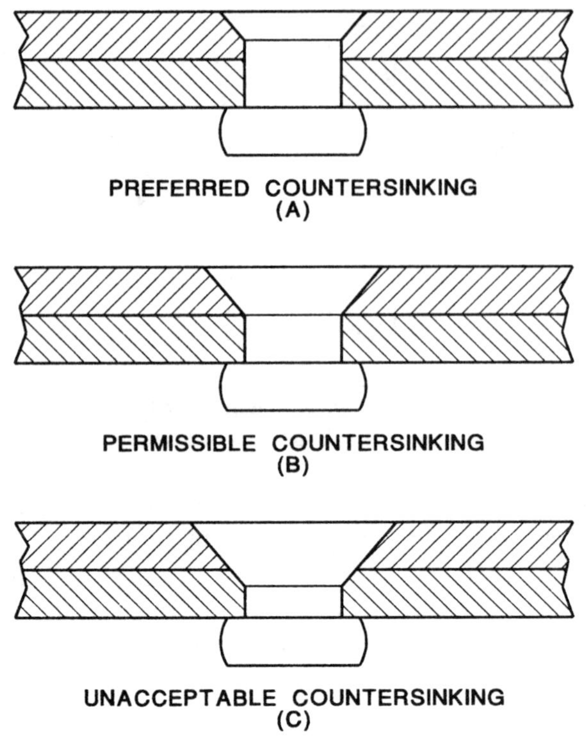

Figure 15. General rules for countersinking rivets.

been clad to both sides. Pure aluminum is quite resistant to corrosion because it forms a thin uniform film of oxide that prevents further corrosive damage.

The vast majority of aircraft structure is held together with solid rivets. The ability to properly install a rivet, to evaluate the condition of a riveted joint, and to properly remove a solid rivet are basic skills required of airframe maintenance technicians.

Rivet Head Shapes

In the past there have been a large number of rivet head shapes, but in most repairs only two head styles are seen. The universal head AN470 (MS20470) is the most popular rivet style and may be used to replace any protruding head rivet.

Where smooth skin is important, flush rivets such as the AN426 (MS20426) are used. This rivet has a 100° countersunk head. The skin is prepared for this type of rivet by either cutting the hole with a countersink or by dimpling it so that the head will be flush with the surface of the skin.

An improperly made countersink reduces the strength of a flush-riveted joint and may even cause failure of the sheet or the rivet head. As a general rule, use the countersink method when the thickness of the material is greater than the thickness of

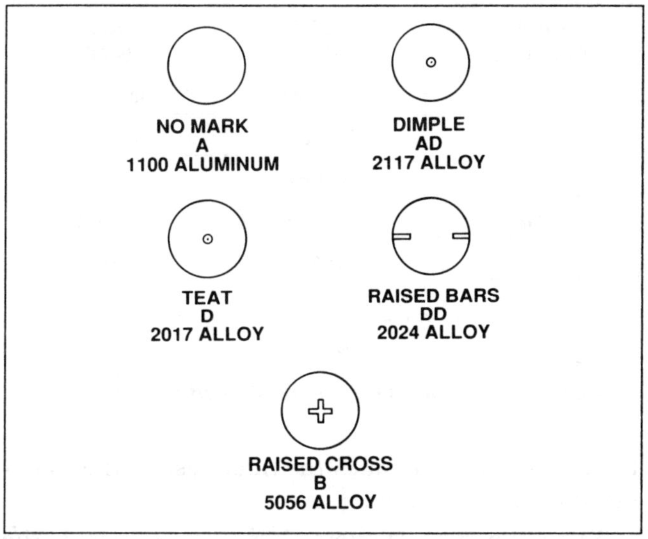

Figure 16. Alloy designations and identification for solid rivets.

the rivet head, and use the dimpling method on thinner material.

Figure 15 illustrates general rules for countersinking. Note in view (A) that the material is quite thick and the head of the countersunk rivet extends only about halfway through the upper layer of metal. Countersinking will leave plenty of material for gripping.

In view (B), the countersunk head reaches completely through the upper layer. This condition should be avoided.

View (C) has the rivet head extending well into the second layer of material. This indicates that the material is thin and most of it would be ground away by drill countersinking; therefore, dimpling is preferred.

Rivet Alloys

Solid rivets are available in a great number of metal alloys, but for most aircraft construction we use only five. The identifying marks on the heads of aluminum alloy rivets indicate the specific alloy used in the manufacture of the rivets. Figure 16 illustrates the commonly used alloys and their head markings.

For nonstructural applications we can use a rivet that is made of commercially pure aluminum (1100), known as a type A rivet. Type A rivets carry no identifying marks on the head. Under certain conditions, type A rivets are not used because of their low strength characteristics.

The most common rivet used in aircraft construction and repair is the AD rivet. These are manufactured from 2117 aluminum alloy. These rivets are heat treated by the manufacturer and can be driven in the condition in which they are received. AD rivets are identified by a small dimple in the center of their head.

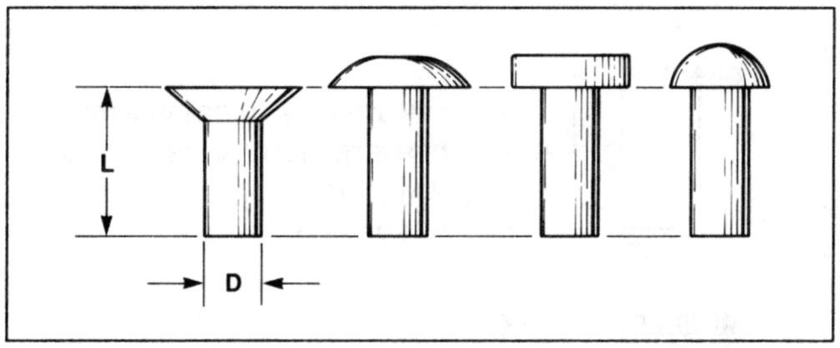

Figure 17. Dimension measurement for solid rivets.

When more strength is required that can be furnished with type AD rivets, we may find "icebox" rivets used. These are D rivets made of 2017 alloy and identified by a small raised teat in the center of the head, and DD rivets made of 2024 alloy which have two raised dashes on the head. These are called icebox rivets because they are heat treated and then stored in a sub-freezing icebox until they are to be driven. The heat treatment softens the rivet to facilitate riveting. The purpose of refrigeration is to retard age hardening.

After driving, the shear strength is approximately 75% of its ultimate shear strength. The remaining 25% is obtained by standing at ordinary room temperature for about four days (age harden). Heat-treated rivets in the D and DD series that are not driven within the prescribed time after heat treatment or removal from refrigeration must be reheat treated before use.

Magnesium is highly susceptible to dissimilar metal corrosion, and rivets used to hold sheets of magnesium together should contain no copper. Type B rivets made of alloy 5056 are used to rivet magnesium. B rivets are identified by a raised cross on the head.

When riveting dissimilar metals together, certain precautions must be exercised in order to prevent electrolytic action. The best procedure to follow is to place a thin gasket of aluminum tape, cellophane, or properly impregnated fabric between the two parts.

While not as common, other rivet types may be encountered. Mild steel rivets may be used for riveting steel parts. Monel rivets may be used for riveting nickel-steel alloys.

Rivet Identification

Aircraft rivet diameter is measured in increments of $1/32''$, and the length in increments of $1/16''$. The length is measured from the end of the shank to the portion of the head that will be flush with the surface of the metal. The flush rivet, AN426 (MS20426), is measured from the top of its head, and protruding head rivets are measured from the bottom of their head.

The complete identification of a rivet includes its head style, the material of which it is made, and its

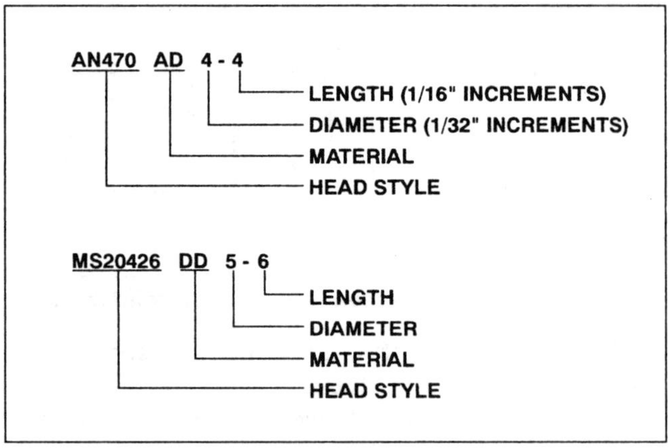

Figure 18. Solid rivet identification.

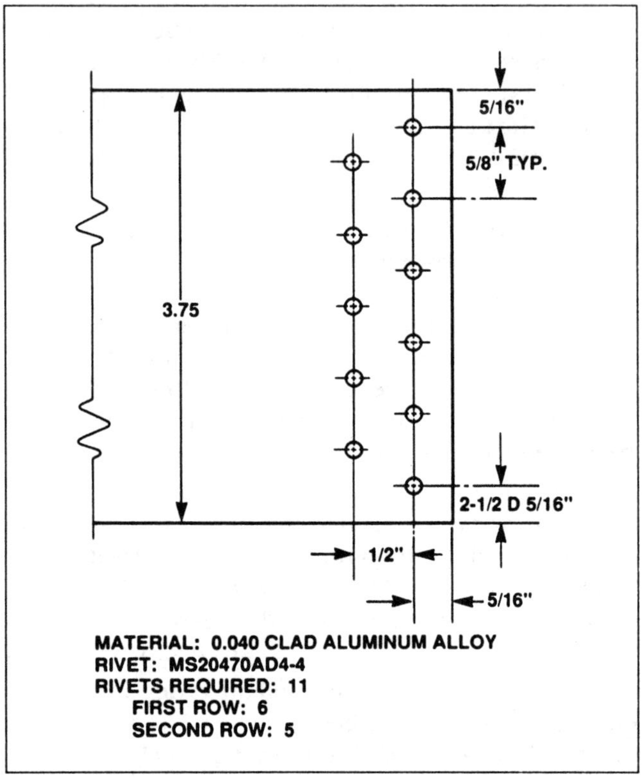

Figure 19. Rivet layout for a two-row splice.

diameter and length. For example, an MS20470AD4-4 rivet has a universal head (MS20470), is made of alloy 2117 (AD), is 1/8 of an inch in diameter (4/32"), and is 1/4 of an inch long (4/16").

Rivet Layout

When a repair is made to an aircraft structure the type, size, and spacing of rivets are an important consideration to assure a repair that is as close to the strength of the original structure as possible.

A commonly used rule of thumb for selecting the rivet size (diameter) is to choose a rivet whose diameter is at least three times the thickness of the thickest sheet being joined.

RIVET DIAMETER	PILOT SIZE	FINAL SIZE
3/32	3/32 (.0937)	40 (.098)
1/8	1/8 (.125)	30 (.1285)
5/32	1/8 (.1562)	21 (.159)
3/16	3/16 (.1875)	11 (.191)
1/4	1/4 (.250)	F (.2557)
5/16	5/16 (.3125)	O (.316)
3/8	3/8 (.375)	V (.377)

Figure 20. Drill sizes for the most used aircraft rivets.

When the rivet diameter has been selected, the length must be determined. For a rivet to furnish enough material to form a proper shop (bucked) head, the rivet must be long enough to go through the material and protrude 1½ times the shank diameter.

It is important when installing a row of rivets that they be placed neither too close to the edge of the sheet, nor too far away. The minimum edge distance for aircraft rivets is two times the diameter of the rivet shank.

Rivet pitch is the distance between the centers of adjacent rivets in the same row. Transverse pitch is the distance between rivet centers in adjacent rows. The minimum pitch for a single row of aircraft rivets is three times the diameter of the rivet shank (3D), and maximum pitch is 12D.

Rivet Installation

In figure 20, we have the twist drill sizes for rivet holes. You will notice that the holes made by these drills are usually 3 or 4 thousandths of an inch larger than the diameter of the rivet. The allows the rivet to be slipped into place without forcing it and scraping the protective oxide coating on the shank.

Most rivet holes in aircraft structure are drilled with a hand-held drill motor turning a carbon steel or high-speed twist drill. Twist drills may have from one to four spiral flutes. The function of the flute section of the drill is to form the cutting edge of the drill point.

The angle at which the drill point is ground is the lip clearance angle. On standard drills used for hard metal (steel and cast iron), the included angle should be 118°. The included angle of a twist drill used for soft metals should be 90°.

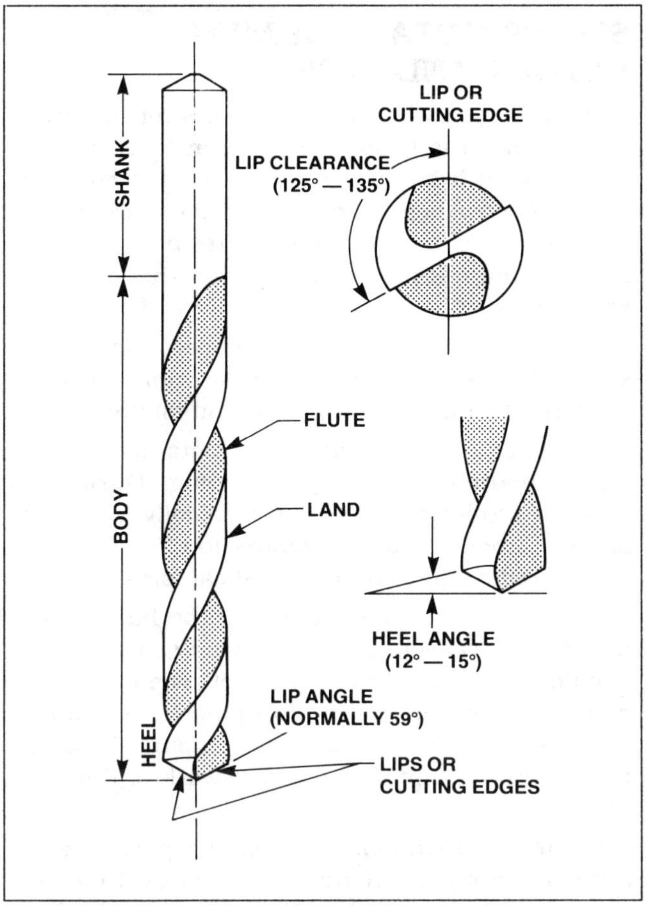

Figure 21. Twist drill, showing how the cutting edges are formed by the ground angle of the flutes.

When comparing the machining techniques for stainless steel sheet material to those for aluminum alloy sheet, it is normally considered good practice to drill the stainless steel at a lower speed with more pressure applied to the drill. When drilling stainless steel, the drill used should have an included angle of 140° and turn at a low speed.

Some holes may require a very accurate dimension. These may be drilled to a size very near the requirement and finished with a reamer. When using a reamer to finish a drilled hole to the correct size, turn the reamer only in the cutting direction.

Gun Riveting

Most of the rivets used in aircraft repairs are driven by a reciprocating air hammer, more commonly called a rivet gun. There must be a rivet set for each rivet size and each rivet head style. The set is placed against the rivet head and a polished steel bucking bar is held against the rivet shank.

It is important when selecting a rivet set for the job that you use the correct size. A rivet set used to drive AN470 or MS20470 rivets should have a slightly greater radius than the rivet head. If the set is too large, it will produce small indentations in the skin around the rivet head. If the set is too small, it will produce a similar mark on the head of the rivet. In either case, the marks indicate improper riveting and the rivet must be removed and replaced.

Rivet removal is accomplished by drilling through the head and removing it with a chisel or a pin punch. The rivet shank is then punched out with a pin punch. A drill one size smaller than the rivet shank diameter is used to remove the head.

Sheet Metal Repairs

Aircraft structural units, such as spars, engine supports, etc., which have been built up from sheet metal, are normally repairable using approved methods.

Evaluation of damage and determining the proper repair scheme are skills that come largely with experience. In all cases, first consult the manufacturer's structural repair manual, if one exists. Should one not exist, consult AC 43.13-1A. If a similar repair is not found there, consult with FAA engineering. Below are a few tips and suggestions to situations that commonly occur:

- Shallow scratches in sheet metal may be repaired by burnishing.
- Repairs or splices involving stringers on the lower surface of stressed skin metal wings are usually permitted, but are normally more critical in reference to strength in tension than similar repairs to the upper surface.
- A gusset or gusset plate is used in the construction and repair of aircraft structure to join and reinforce intersecting structural members.
- When straightening members made of 2024-T4 aluminum, you should straighten cold and reinforce. This material is factory heat treated and no attempt to alter that in the field should be made.
- When fabricating parts from alclad 2024-T3 aluminum sheet stock, all scratches, kinks, tool marks, nicks, etc., must be held to a minimum.
- When selecting a file to remove nicks, the terms "double-cut" and "second-cut" have the same meaning.
- Relief holes may be drilled in fabricated parts to relieve stress concentrations and prevent cracking.
- If a stressed skin metal wing assembly is known to have been critically loaded, genuine rivet tipping will be indicated by groups of consecutive rivets heads tipped in the same direction. Shear loads will cause

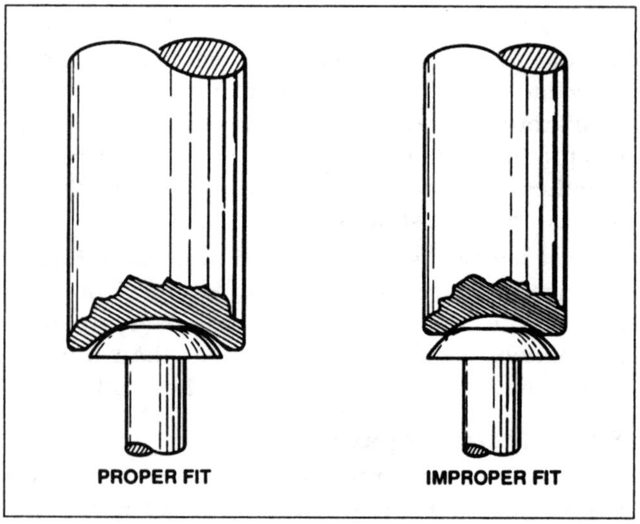

Figure 22. The radius of the cup of the rivet set must be slightly larger than the radius of the rivet head.

the most rivet failures. Joggles in the removed rivet shanks would indicate partial shear failure.

- When repairing a small hole in a metal stressed skin, the major consideration in the design of the patch should be the shear strength of the riveted joint.

REVIEW QUESTIONS — Sheet Metal Structures

27. *What should be the included angle of a twist drill for soft metals?*
28. *What should be the included angle of a twist drill for stainless steel?*
29. *Why are clad aluminum alloys used in aircraft structures?*
30. *What type of wing uses no external bracing?*
31. *What is the purpose of a relief hole?*
32. *What are the dimensions of an AN430AD4-8 rivet?*
33. *What head markings do most rivets used in aircraft construction bear?*
34. *What drill size is used to remove a rivet?*
35. *What rivet is used for riveting magnesium alloy structures?*
36. *What rivet is used for riveting nickel-steel alloys?*

Answer questions 8099 through 8178 in the FAA Test section of this book.

SHEET METAL FORMING, LAYOUT AND BENDING

The fabrication of sheet metal parts for an aircraft require that the technician have at least a fundamental knowledge of the metal being used, and a working knowledge of applied geometry. We must be able to lay a part out as a flat pattern so that when it is cut and folded into the desired shape, the size will be correct.

The methods used in forming operations include such sheetmetal work processes as shrinking, stretching, bumping, crimping, and folding.

Shaping or forming malleable metal by hammering or pounding is called bumping. During this process the metal is supported by a dolly, a sandbag, or a die. Each contains a depression into which the hammered portions of the metal can sink.

If a streamline cover plate is to be hand formed using a form block, a piece of dead soft aluminum should first be placed over the hollow portion of the mold and securely fastened in place. The bumping operation should be started by tapping the aluminum lightly around the edges and working down into the center.

Bumping operations are usually performed on pure aluminum, as many alloys are heat treated, or will work harden quickly. You can distinguish between aluminum and aluminum alloy by testing with a 10% solution of caustic soda. Alloys will turn black because of the copper content.

Making bends in sheets, plates, or leaves is called folding. Folds are usually thought of as sharp, angular bends; they are generally made on folding machines (sheet metal brake).

A piece of sheet metal is bent to a certain radius. The curvature of the bend is referred to as the bend radius. Unless otherwise specified, the radius of a bend is the inside radius of the metal being formed.

When forming straight bends, the thickness of the material, its alloy composition, and its temper condition must be considered. Generally speaking, the thinner the material, the sharper it can be bent (the

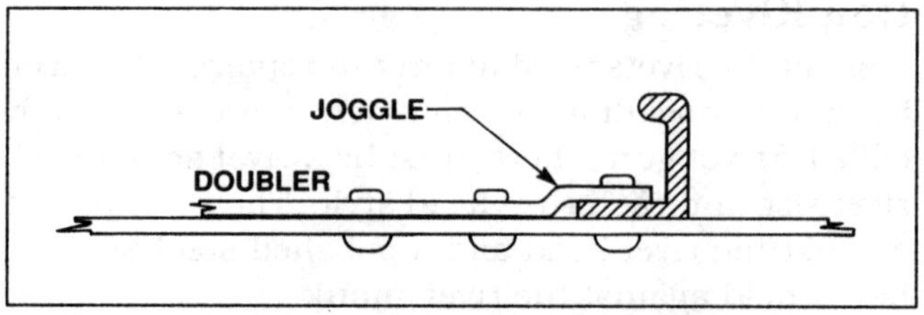

Figure 23. Joggle a doubler where it must fit over a bulb angle, rather than attempting to pull the doubler against the skin with the rivets.

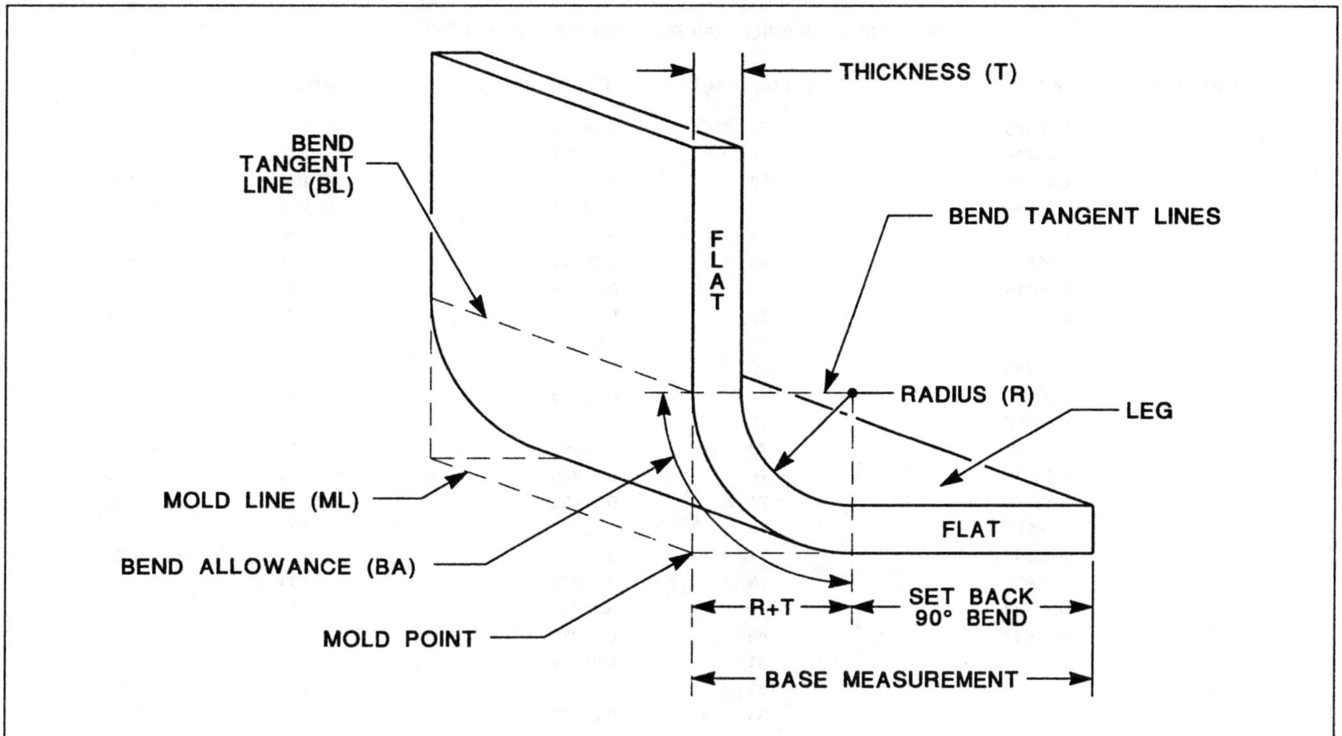

Figure 24. Bend allowance terms.

smaller the radius of bend), and the softer the material, the sharper the bend. The sharpest bend that can be placed in a piece of metal without critically weakening the part is called the minimum radius of bend. Given all other factors remaining the same, the smaller the bend radius, the greater the amount of material needed to make the bend.

A simple combination of two shallow bends, called a joggle, may be used to allow clearance for a sheet or extrusion.

The sheet metal used for aircraft construction and repair was formed by a rolling process to reduce its thickness. In the process of rolling, the metal assumes a grain structure which is easy to see in a piece of aluminum alloy. When a piece of aluminum alloy is to be bent using a minimum radius for the type and thickness of material, the layout should be made so that the bend will be 90° to the grain of the sheet.

The aluminum alloys used in aircraft construction are usually hardened by heat treatment. This heat treatment must be taken into account when selecting materials for fabricating replacement parts. Many alloys are too brittle to bend with normal shop tools. The heat treating process gives the material more strength, but also makes it more prone to cracking. Holes are usually drilled at the intersection of two bends to prevent the metal from cracking. These are referred to as relief holes.

When making a bend in sheet metal, the bend allowance must be calculated. Bend allowance is the length of material required for the bend. This amount of metal must be added to the overall length of the layout pattern to assure adequate metal for the bend.

Familiarity with the following terms is necessary for an understanding of bend allowance and its application to an actual bending job.

- *Leg* — The longer part of a formed angle.
- *Flange* — The shorter part of a formed angle—the opposite of leg.
- *Neutral line* — When bending metal, the material on the outside of the curve stretches while the material on the inside of the curve compresses. That part of the material which is not affected by either stress is the neutral line.
- *Mold Line (ML)* — The line formed by extending the outside surfaces of the leg and flange. An imaginary point from which the real base measurements are provided on drawings.
- *Bend Tangent Line (BL)* — The line at which the metal starts to bend and the line at which the metal stops curving. All the space between the bend tangent lines is the bend allowance.
- *Setback (SB)* — The setback is the distance from the bend tangent line to the mold point. In a 90° bend SB = R + T (radius of the bend plus the thickness of

For bend other than 90° Setback = K (R + T)

Ang [deg]	K value	Ang [deg]	K value	Ang [deg]	K value
1	0.00873	61	0.58904	122	1.8040
2	0.01745	62	0.60086	123	1.8418
3	0.02618	63	0.61280	124	1.8807
4	0.03492	64	0.63487	125	1.9210
5	0.04366	65	0.63707	126	1.9626
6	0.05241	66	0.64941	127	2.0057
7	0.06116	67	0.66188	128	2.0503
8	0.06993	68	0.67451	129	2.0965
9	0.07870	69	0.68728	130	2.1445
10	0.08749	70	0.70021	131	2.1943
11	0.09629	71	0.71329	132	2.2460
12	0.10510	72	0.72654	133	2.2998
13	0.11393	73	0.73996	134	2.3558
14	0.12278	74	0.75355	135	2.4142
15	0.13165	75	0.76733	136	2.4751
16	0.14054	76	0.78128	137	2.5386
17	0.14945	77	0.79543	138	2.6051
18	0.15838	78	0.80978	139	2.6746
19	0.16734	79	0.82434	140	2.7475
20	0.17633	80	0.83910	141	2.8239
21	0.18534	81	0.85408	142	2.9042
22	0.19438	82	0.86929	143	2.9887
23	0.20345	83	0.88472	144	3.0777
24	0.21256	84	0.90040	145	3.1716
25	0.22169	85	0.91633	146	3.2708
26	0.23087	86	0.93251	147	3.3759
27	0.24008	87	0.94890	148	3.4874
28	0.27933	88	0.96569	149	3.6059
29	0.25862	89	0.98270	150	3.7320
30	0.26795	90	1.0000	151	3.8667
31	0.27732	92	1.0355	152	4.0108
32	0.28674	93	1.0538	153	4.1653
33	0.29621	94	1.0724	154	4.3315
34	0.30573	95	1.0943	155	4.5107
35	0.31530	96	1.1106	156	4.7046
36	0.32492	97	1.1303	157	4.9151
37	0.33459	98	1.1504	158	5.1455
38	0.34433	99	1.1708	159	5.3995
39	0.35412	100	1.1917	160	5.6713
40	0.36397	101	1.2131	161	5.9758
41	0.37388	102	1.2349	162	6.3137
42	0.38386	103	1.2572	163	6.6911
43	0.39391	104	1.2799	164	7.1154
44	0.40403	105	1.3032	165	7.5957
45	0.41421	106	1.3270	166	8.1443
46	0.42447	107	1.3514	167	8.7769
47	0.43481	108	1.3764	168	9.5144
48	0.44523	109	1.4019	169	10.385
49	0.45573	110	1.4281	170	11.430
50	0.46631	111	1.4550	171	12.706
51	0.47697	112	1.4826	172	14.301
52	0.48773	113	1.5108	173	16.350
53	0.49858	114	1.5399	174	19.081
54	0.50952	115	1.5697	175	22.904
55	0.52057	116	1.6003	176	26.636
56	0.53171	117	1.6318	177	38.138
57	0.54295	118	1.6643	178	57.290
58	0.55431	119	1.6977	179	114.590
59	0.56577	120	1.7320	180	Infinite.
60	0.57735	121	1.7675		

Figure 25. K-chart for determining setback for bends of other than 90°.

the metal). The setback dimension must be determined prior to making the bend as it is used in determining the location of the beginning bend tangent line.

- *Bend Line* — Also called Brake or Sight Line, the bend line is set even with the nose of the brake and serves as a guide in bending the work.
- *Flat (or flat portion)* — The flat is that portion not included in the bend. It is equal to the base measurement minus the setback.
- *Base measurement* — The outside dimensions of a formed part.
- *Closed angle* — An angle that is less than 90° when measured between legs, or more than 90° when the amount of bend is measured. A piece of stock that is to be bent to a closed angle of 15° must be bent through an angle of 165°
- *Open angle* — An angle that is more than 90° when measured between legs, or less than 90° when the amount of bend is measured.
- *"K" No.* — 1 of 179 numbers on the "K" chart corresponding to one of the angles between 0 and 180° to which metal can be bent. Whenever metal is to be bent to any angle other than 90°, the corresponding "K" No. is selected from the chart and is multiplied by the sum of the radius and the thickness of the metal. The product is the amount of setback for the bend, SB = K (R + T).

Flat Pattern Layout

The most important factors needed to make a flat pattern layout are radius, thickness, and degree of bend.

When considering the location of the bend line for a fitting with only one bend, the setback must be known. On a sheet metal fitting layout with a single bend, allow for stretching by subtracting the setback from both legs.

When bending metal to exact dimensions, the length of the neutral line must be determined so that sufficient material can be allowed for the bend. To save time and calculation of the bend allowance, formulas and charts for various angles, radii of bends, material thicknesses, and other factors have been established. The bend allowance formula for a 90° bend is shown below.

$$\frac{2\pi\,(R + \tfrac{1}{2}\,T)}{4}$$

Using this formula we can determine the amount of material required to make the 90° bend in figure 26.

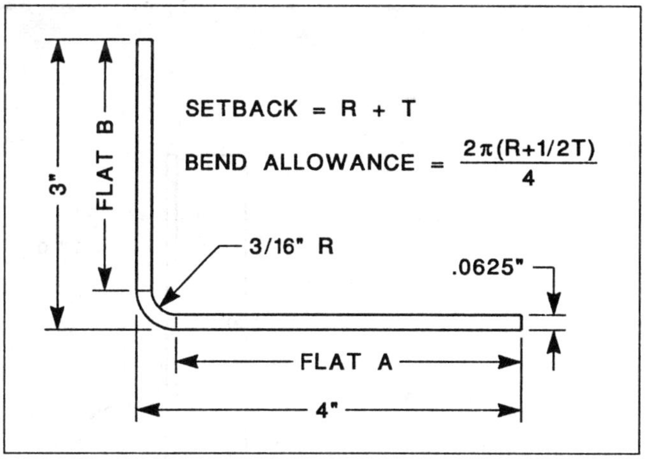

***Figure 26.** This view is used to figure the material needed for bend allowance and flat pattern layout.*

$$\frac{2 \times 3.141(\tfrac{3}{16} + \tfrac{1}{2}\,.0625)}{4}$$

$$= \frac{6.282\,(.1875 + .0312)}{4}$$

$$= \frac{1.3738}{4}$$

A much faster way to calculate bend allowance is the use of bend allowance charts. To determine bend allowance for any degree of bend by use of the chart, find the allowance per degree for the radius and gage of material used and multiply by the number of degrees of the desired bend.

If it is necessary to compute a bend allowance problem and bend allowance tables are not available, the neutral axis of the bend can be found by adding approximately one-half of the stock thickness to the bend radius.

We will go through the procedure for making a flat pattern layout for a formed channel of 0.051″ aluminum alloy. As we see in figure 27, this channel will be 4″ long, 2″ deep, and 2″ wide. A bend radius of ¼″ is used for both of the bends. We want to find the total length of the material we will need for this channel.

Begin by finding the setback, which for a 90° bend is equal to the radius plus the thickness of the metal. We will round this value to 0.30″.

The length of flat A is the distance from the end of the leg to the mold point minus the setback, or 1.70″.

The length of flat B is the distance between the two mold points, less a setback on each end. This gives us 1.40″.

The length of flat C is the same as flat A, or 1.70″.

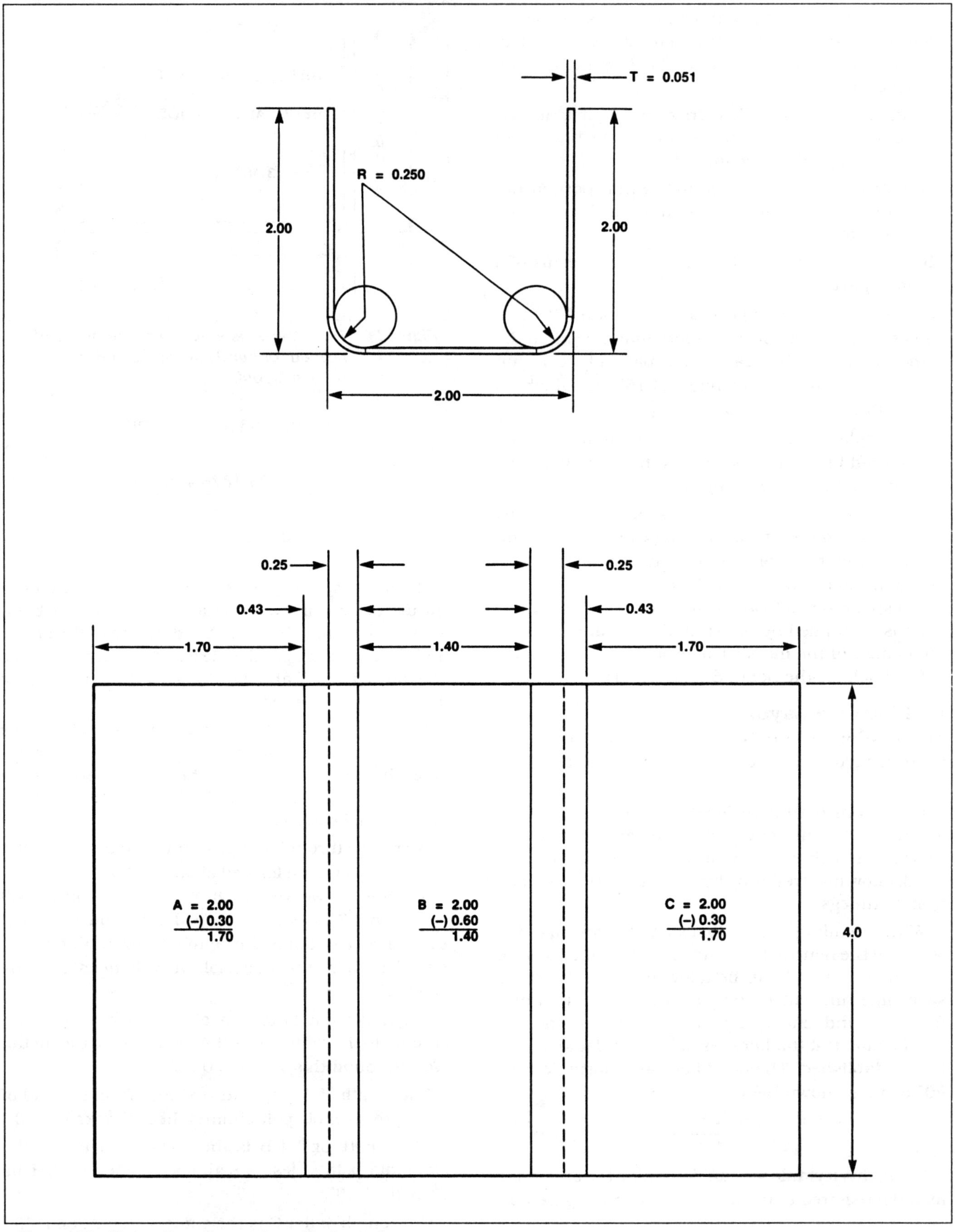

Figure 27. Flat pattern layout for a 2″ channel.

The bend allowance for 0.051″ material with a ¼″ bend radius may be calculated using the formula shown above. The bend allowance is 0.43″.

The total length of our material is the sum of the length of the three flats plus two bend allowances, or 5.66″.

REVIEW QUESTIONS — Sheet Metal Forming, Layout and Bending

37. *If a streamline cover plate is to be formed by hand bumping, the bumping operation should begin at what portion of the mold?*
38. *Should bends be made parallel or perpendicular to the grain of the aluminum?*
39. *If it is necessary to compute a bend allowance problem and bend allowance tables are not available, the neutral axis of the bend can be found by what means?*
40. *The sharpest bend that can be placed in a piece of metal without critically weakening the part is called what?*
41. *How can you distinguish between pure aluminum and aluminum alloy?*
42. *When making a bend, the part of the material which is neither compressed or stretched is known as what?*
43. *What name is give to hole drilled at the intersection of two bends to prevent the metal from cracking?*
44. *By what process are the aluminum alloys used in aircraft construction hardened?*

Answer questions 8157 through 8178 in the FAA Test section of this book.

AIRCRAFT WELDING

Metals can be joined by mechanical means (bolting or riveting) or by welding, brazing, or soldering. All of these methods are used in aircraft construction.

Welding is done by melting the edges of two pieces of metal and allowing the molten material to flow together so that the two pieces will become as one.

Brazing is similar to welding, but rather than melting the actual pieces being joined, a bonding material (brazing rod) with a much lower melting point is used.

Soldering is similar to brazing except that brazing materials normally melt at temperatures above 800°F, while solders melt at temperatures much lower than this. Soft solder is used chiefly for copper, brass, and coated iron, in combination with mechanical means.

Gas Welding

Most aircraft welded repairs are accomplished using oxyacetylene welding equipment. In this process, acetylene gas is burned with oxygen. The result is a transparent blue flame with a flame temperature of about 5,600°F. A typical portable welding rig is shown in figure 28.

Acetylene is a stable compound when it is not under pressure, but if it is stored at a pressure of more than 15 PSI, it becomes unstable. At pressures near 30 PSI it becomes so unstable that even a slight jar can cause it to explode.

To avoid this problem acetylene is supplied in steel cylinders that are filled with a porous mineral material, or an organic substance, saturated with acetone. Acetone will absorb many times its own weight of acetylene gas, and the absorbed gas can be safely stored under a pressure of more than 250 PSI.

The gaseous oxygen used for welding is stored in steel cylinders under a pressure of more than 2,000 PSI. Any compressed gas at this pressure is dangerous if the cylinder is dropped or the valve broken off.

In addition to the problems related to high pressure gases, oxygen creates a great danger of fire or

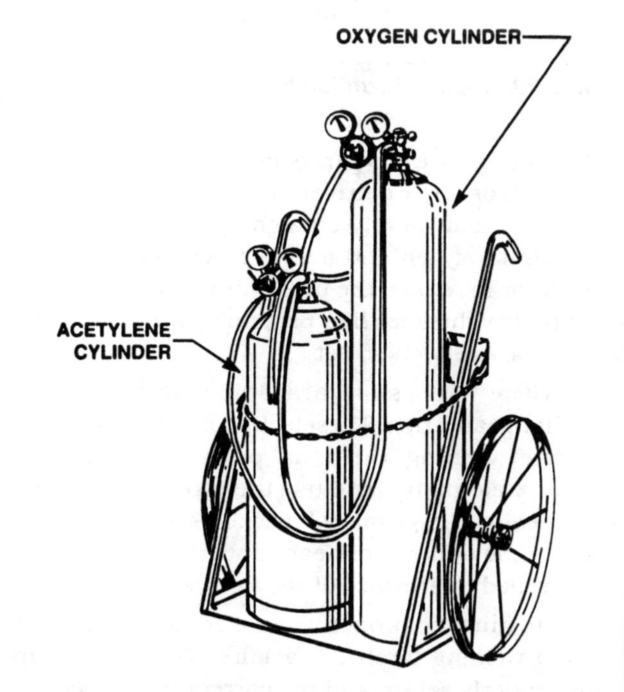

***Figure 28.** Typical portable oxyacetylene welding equipment used in aircraft maintenance shops.*

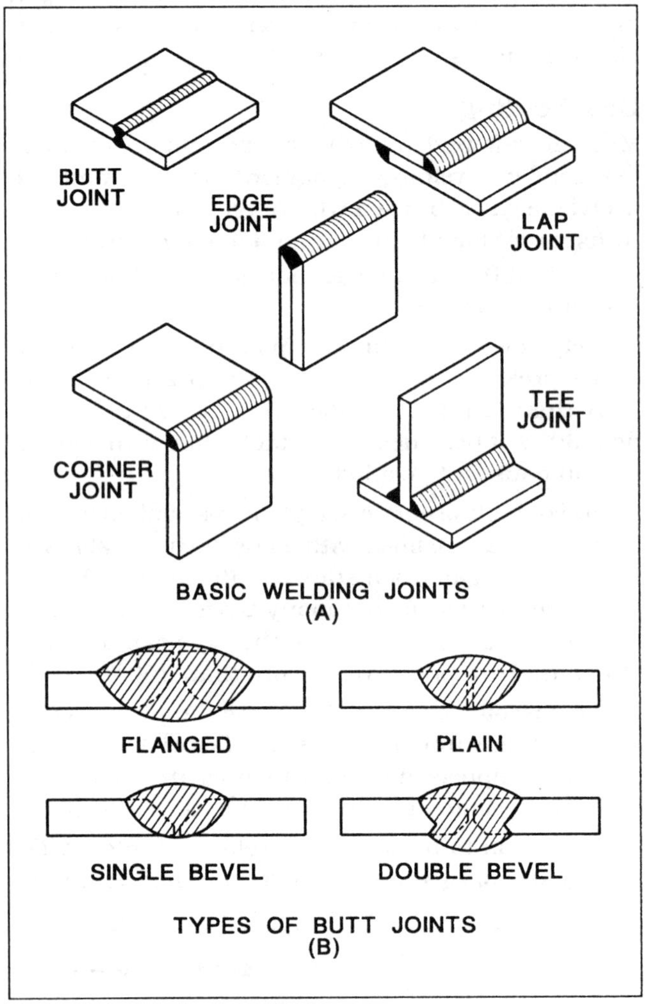

Figure 29. Types of butt joints.

explosion. If pure oxygen comes into contact with grease, oil, or other petroleum products, it will combine with them rapidly enough to ignite spontaneously. Both oxygen and acetylene cylinders require a regulator to reduce the pressure to a usable value. Acceptable values for line pressures would be: Oxygen 20 PSI, Acetylene 5 PSI.

Acetylene torches are available with a variety of interchangeable tips. Tip selection is very important. In aircraft welding, the usual practice in the selection of a welding tip is to use the type of material to be welded as the primary basis for selection. In gas welding, the amount of heat applied to the material is controlled by the size of the tip opening.

The tip must be properly installed, a loose tip may cause a welding torch to backfire. With the equipment properly set up and the correct tip screwed in, you are ready to light the torch and adjust the flame.

Open the acetylene valve on the torch about 1/6 turn, and using a flint lighter, ignite the gas coming from the torch. Open the acetylene valve until all of the smoke clears from the flame.

When the acetylene is properly adjusted, slowly open the oxygen valve on the torch until the brilliant flame turns to a transparent blue and an inner cone begins to form. As you increase the amount of oxygen, you will see a definite inner and outer cone; increase the oxygen until the outer cone (or feather) just disappears into the inner cone. This flame has a proper relationship between oxygen and acetylene and is called a neutral flame.

If the feather is pronounced, the flame is a reducing, or carburizing, flame and has an excess of acetylene. When welding stainless steel, the flame should be slightly carburizing. If too much acetylene is used in the welding of stainless steel the metal will absorb carbon and lose its resistance to corrosion. A carburizing flame should be avoided when welding steel, as it hardens the surface.

If the inner cone is sharply pointed, rather than rounded on the end, and the torch makes a hissing sound, the flame is oxidizing and there is too much oxygen for the amount of acetylene flowing.

When you are ready to extinguish the torch, it is good practice to turn off the acetylene first and allow the gas remaining in the torch tip to burn out. If the torch is not to be used again for a long period, the pressure should be turned off at the cylinder. The hose lines should then be relieved of pressure by opening the torch needle valves and the working pressure regulator, one at a time, allowing the gas to escape. Again, it is good practice to relieve the acetylene pressure first and then the oxygen pressure.

The five fundamental types of welded joints are the butt joint, tee joint, lap joint, corner joint, and edge joint. Each of these are illustrated in figure 29.

A butt joint is made by placing two pieces of material edge to edge, so that there is no overlapping, and then welding. If the metal is thicker than 1/8", it is necessary to bevel the edges so that the heat from the torch can penetrate completely through the metal.

When inspecting a butt-welded joint by visual means, the penetration should be 100% of the thickness of the base metal. A filler rod is used to add strength and reinforcement to the weld. The most important consideration when selecting a welding rod is the type of metal being welded.

A gas weld which is successfully completed should have the weld metal tapered smoothly into the base metal. Bad welds may be removed and the material welded again. Filing or grinding on a weld for appearances sake is never permitted. Welding over a previously brazed or soldered joint is not permitted.

Welding Steel Parts

Aircraft steel parts manufactured from SAE 4130 chrome/molybdenum tubing are normally considered repairable by welding. AC 43.13-1A provides detailed information on welded repairs to aircraft structures.

Engine mount members should preferably be repaired by using a larger diameter tube with fish-mouth and rosette welds.

A steel tube longeron dented at a cluster is best repaired using a welded patch plate.

Welding Aluminum Parts

Most aluminum welding is now done with a gas shielded electric arc, but there may be times when you will have to weld aluminum with an oxyacetylene torch. When welding aluminum with a torch, it is necessary to use a flux to prevent oxides from forming ahead of the torch. This is not necessary with the inert-arc welding process because it uses an inert gas to protect the weld zone from the atmosphere. The inert gases normally used when welding aluminum consist of helium or argon, or a mixture of the two.

Flux used for welding aluminum is commercially available under many trade names. The flux should be painted on the surface to be welded and applied to the welding rod. After welding aluminum, it is important that all traces of flux be removed. If left on the weld, it will corrode the metal.

Edge notching is generally recommended in butt welding above a certain thickness of aluminum because it aids in getting full penetration and prevents local distortion. Aluminum plates 1/4" or more thick are usually preheated before welding. This reduces internal stresses and assures more complete penetration.

When welding a tank fabricated from .071 sheet aluminum, a welding tip slightly larger than that used for welding steel of the same thickness should be used to obtain sufficient heat to melt the base metal. The flame used for aluminum welding should be soft and neutral.

Annealing of aluminum removes stresses caused by welding.

Welding Magnesium Parts

Repair of magnesium by welding is limited by two factors. Because magnesium used in aircraft structure is usually heat treated, the welded section can never have the strength of the original metal. Because a flux must be used in making all magnesium welds, the type of joint is limited to those that provide no possibility of trapping the flux.

When welding magnesium the filler rod should be of the same composition as the base metal.

Brazing

Brazing requires less heat than welding and can be used to join metals that are damaged by high heat. However, the strength of a brazed joint is less than that of a welded one.

A brazing flux is necessary to obtain a good union between the base metal and the filler metal.

Silver soldering is actually a type of brazing operation. The principle use of silver solder in aircraft work is the fabrication of high-pressure oxygen lines and other parts which must withstand vibration and high temperatures.

It is necessary in all silver soldering operations to use flux to chemically clean the base metal of oxide film. The oxyacetylene flame for silver soldering should be neutral.

Soft Soldering

Soft solder yields gradually under a steadily applied load and should not be used unless the transmitted loads are very low. It should never be used as a means of joining structural members.

A soldering copper (called a soldering iron if it is electrically heated) is the tool used in soldering. A new or resurfaced soldering iron cannot be used effectively until after the working face has been *tinned*. This tinning operation coats the surface with solder to aid in the transfer of heat from the soldering tip to the joint to be soldered.

When resurfacing soldering coppers, be careful to maintain the original tip design. A very thin and pointed tip on a soldering copper is undesirable because it will cool too rapidly.

Most aircraft maintenance technicians will only use soft solder on electrical connections. Keep in mind that the solder does not have the strength to secure the connection. Joints in electric wire to be soldered should be mechanically secure prior to soldering. It is the mechanical attachment that will provide strength, not the solder.

REVIEW QUESTIONS— Aircraft Welding

45. *What is the primary reason for using flux when welding aluminum?*

46. *What type of flame is used for silver soldering?*

47. *What method of repair is recommended for a steel tube longeron dented at a cluster?*

48. *Is welding over a previously brazed or soldered joint permitted?*

49. *A resurfaced soldering iron cannot be used effectively until after the working face has been _____.*

50. *What steel parts are normally considered repairable by welding?*

51. *What is an acceptable line pressure for acetylene?*

52. *What is used in acetylene cylinders to stabilize the gas?*

53. *What type of flame should be used to weld stainless steel?*

54. *The maximum working pressure of acetylene should never exceed what value?*

55. *What will happen if the flux is not removed after welding aluminum?*

Answer questions 8179 through 8211 in the FAA Test section of this book.

ROTARY WING ASSEMBLY AND RIGGING

A helicopter is an aircraft whose lift is created by a rotating wing called a rotor. The cross-section of a rotor blade is very similar to the cross section of the wing of an airplane. Lift is created by airflow over the rotor in the same manner as lift is created by airflow over a wing. Lift is a function of the angle of attack of the blade. Angle of attack is the acute angle formed by the chord line of a wing and the relative wind.

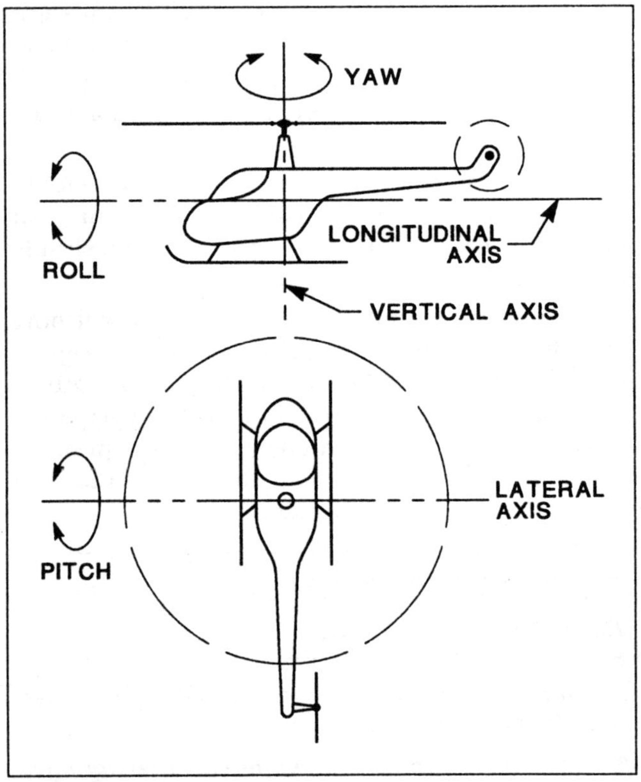

Figure 30. Axes of flight.

As a helicopter maneuvers through the air, its attitude in relation to the ground changes. These changes are described with reference to three axes of flight: (1) vertical, (2) longitudinal, and (3) lateral.

Movement about the vertical axis produces yaw, a change in direction of the nose of the helicopter to the left or right. This is controlled by the directional-control pedals. In a hovering helicopter equipped with a tail rotor, directional control is maintained by varying the pitch of the tail rotor blades.

The pedals increase or decrease tail rotor thrust. A decrease in pitch angle of the tail rotor blades on a helicopter causes the tail to pivot in the direction of torque rotation around the main rotor axis. Torque direction is the opposite of rotor blade rotation. The tail rotor of a helicopter could fail to compensate accurately for the torque of the main rotor if the tail rotor is inaccurately rigged.

Movement about the longitudinal axis is called roll. This movement is effected by moving the cyclic pitch control to the right or left. The cyclic pitch control stick is similar to the control stick of a conventional aircraft. It acts through a mechanical linkage to change pitch of each main-rotor blade during a cycle of rotation.

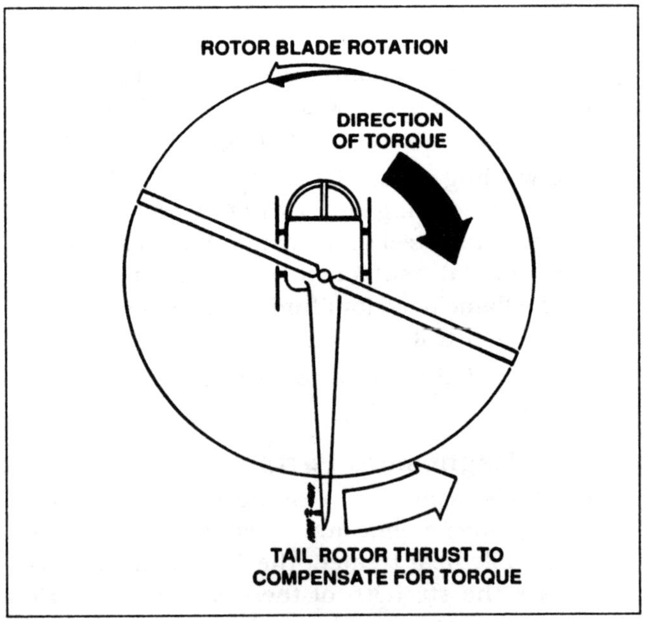

Figure 31. The use of tail rotor thrust to counteract the torque of the main rotor.

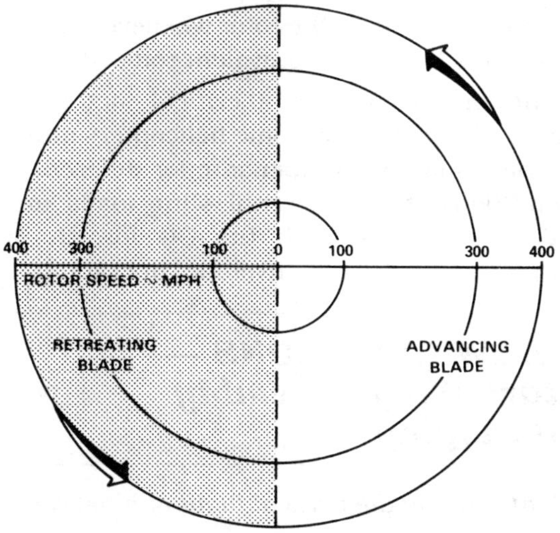

WHEN THE HELICOPTER IS HOVERING, THE ROTOR SPEED IS THE SAME ON BOTH SIDES OF THE HELICOPTER, AND THE LIFT IS THE SAME ON BOTH SIDES.
(A)

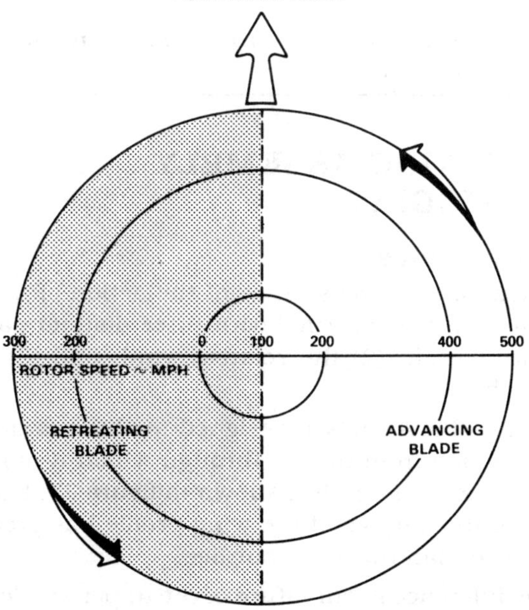

WHEN THE HELICOPTER IS MOVING THROUGH THE AIR, THE FORWARD SPEED IS ADDED TO THE ROTOR SPEED ON THE SIDE OF THE ADVANCE BLADE AND SUBTRACTED ON THE SIDE OF THE RETREATING BLADE. THE DIFFERENCE IN SPEED CAUSES A DIFFERENCE IN LIFT.
(B)

Figure 32. Creation of dissymmetry of lift.

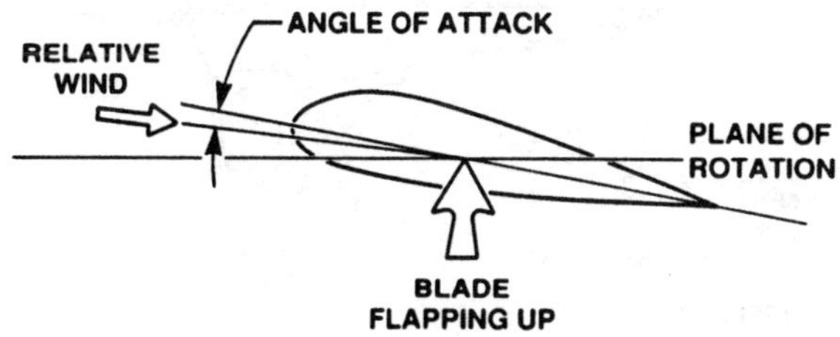

AS THE BLADE FLAPS UP, THE ANGLE OF ATTACK DECREASES AND THE LIFT DECREASES.
(A)

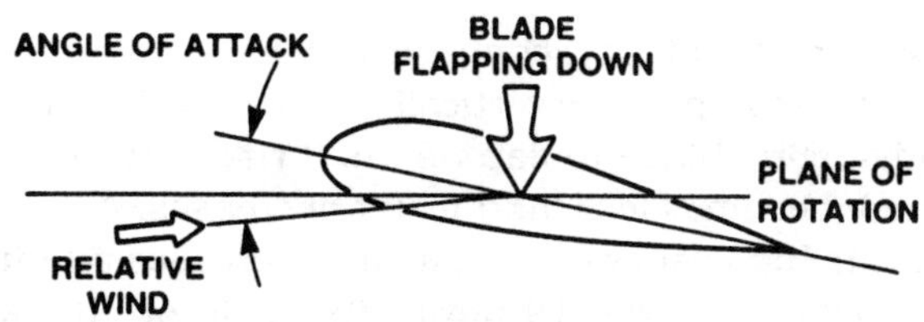

AS THE BLADE FLAPS DOWN, BOTH THE ANGLE OF ATTACK AND THE LIFT INCREASE.
(B)

Figure 33. Lift is changed by allowing the blades to flap.

Movement about the lateral axis produces a nose-up or nose-down attitude. This movement is effected by moving the cyclic pitch control fore and aft.

The collective pitch control varies the lift of the main rotor by increasing or decreasing the pitch of all of the blades at the same time. Raising the collective pitch control increases the pitch of the blades, thereby increasing the lift. Lowering the control decreases the pitch of the blades, causing a loss of lift. The vertical flight of a helicopter is controlled by increasing or decreasing collective pitch. Collective pitch is also used in coordination with cyclic pitch control to regulate the airspeed of the helicopter.

The area within the tip-path plane of the main rotor is known as the disk area or rotor disk. When hovering in still air, lift created by the rotor blades at all corresponding positions around the rotor disk is equal. A helicopter in forward flight, cruise configuration, changes direction by tilting the main rotor disk in the desired direction.

If a single-rotor helicopter is in horizontal flight, the angle of attack of the advancing blade is less than the retreating blade. This phenomenon creates dissymmetry of lift. Due to the greater lift of the advancing blade, the helicopter would roll to the left unless something is done to equalize the lift of the blades on both sides of the helicopter.

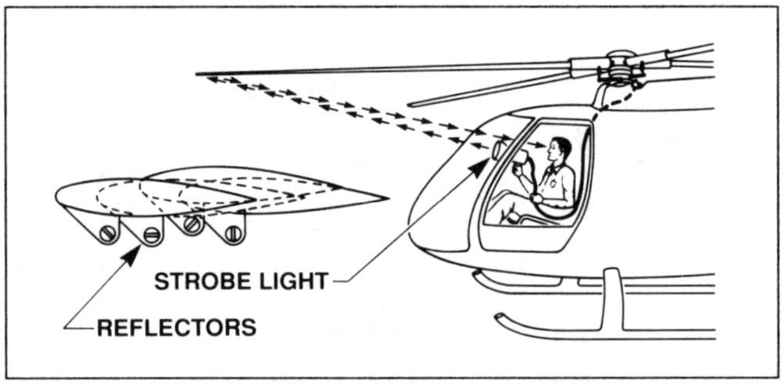

Figure 34. Checking blade track, using a strobe light, may be done either on the ground or in flight.

Dissymmetry of lift is compensated for by allowing the blades to flap (move vertically). The lift will cause the advancing blade to flap up. As it rises, its angle of attack becomes less, and its lift decreases. As the retreating blade slows down, its lift decreases and it flaps down. As it moves down, its angle of attack increases and so does its lift.

Greater downwash effect through the aft part of the main rotor disk compared to the forward part of the disk is known as transverse flow lift. This effect is corrected through design and rigging adjustments to the cyclic pitch control system.

Main rotor blades that do not cone by the same amount during rotation are said to be out of track. Tracking of rotor blades is a common rigging operation performed by helicopter technicians.

Tracking of helicopter rotor blades is normally done while the blades are rotating at a specific RPM and collective pitch angle. Most tracking is now accomplished using a strobe light to determine what adjustment must be made.

Vibration is a problem in helicopter operations, and is one often corrected by proper rigging. Vibrations can cause the helicopter to jump up and down in what is called a vertical mode, or to shake sideways which is called lateral vibration. Vertical vibrations are generally caused by some disparity in lift, and lateral vibrations are most often caused by an out-of-balance condition of the rotor.

If a severe abnormal vertical vibration developed in a helicopter equipped with a 2-blade rotor system, this would indicate the rotor blades are out of track.

Rotor system-to-engine attachments are usually through a transmission which reduces the speed of the rotor and provides a power takeoff to drive the tail rotor. One purpose of the clutch provided between the engine and the helicopter transmission is to disconnect the rotor from the engine to relieve the starting load.

A free wheeling unit may also be installed in the helicopter drive system.

This unit disconnects the rotor whenever the engine stops or slows below the equivalent rotor RPM.

Weight and balance is a critical factor in the operation of a helicopter. Some helicopters must use removable ballast to compensate for different loading conditions. Most external sling loads are attached in alignment with the CG for the same reason; to reduce "out of envelope" operations.

REVIEW QUESTIONS — Rotary Wing Assembly and Rigging

56. *What controls the vertical flight of a helicopter?*
57. *What name is given to the acute angle formed by the chord line of a wing and the relative wind?*
58. *In a hovering helicopter equipped with a tail rotor, how is directional control maintained?*
59. *Main rotor blades that do not cone by the same amount during rotation are said to be out of _____.*
60. *Movement about the longitudinal axis (roll) in a helicopter is effected by the movement of which cockpit control?*

Answer questions 8212 through 8226 in the FAA Test section of this book.

FIXED WING ASSEMBLY AND RIGGING

Aerodynamics

Aerodynamics is the science of the action of air on an object. It is concerned with three distinct parts; (1) the aircraft, (2) the relative wind, and (3) the atmosphere.

An airfoil is a surface designed to obtain a desirable reaction from the air through which it moves. The profile of an airfoil is shown in figure 35. Notice that the top surface of the wing profile has greater curvature than the lower surface.

The difference in curvature of the upper and lower surfaces of the wing builds up the lift force. A pressure differential is created between the upper and lower surfaces of the wing, forcing the wing upward in the direction of the lower pressure.

Before beginning the discussion on angle of attack and its effect on airfoils, we shall first consider the terms "chord" and "center of pressure".

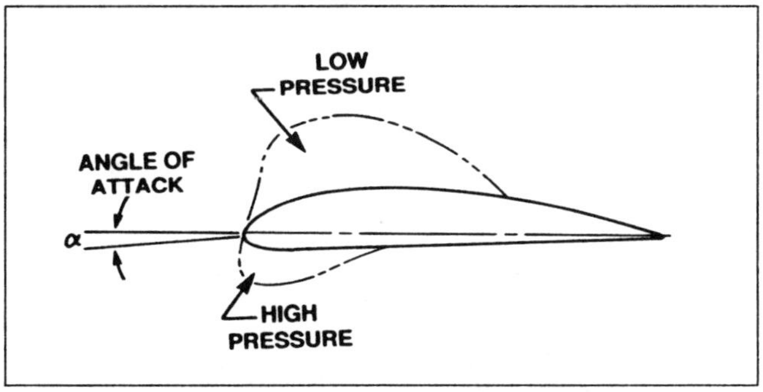

Figure 35. Air flowing over an airfoil section creates a pressure differential between the top and the bottom surfaces.

The chord of an airfoil is an imaginary straight line which passes through the section from the leading edge to the trailing edge. The chord line provides one side of an angle which ultimately forms the angle of attack. The other side of the angle is formed by a line indicating the direction of the relative airstream. Thus, angle of attack is defined as the angle between the chord line of the wing and the direction of the relative wind. This is not to be confused with the angle of incidence, which is the angle between the chord line of the wing and the longitudinal axis of the aircraft. The angle of incidence does not change in flight.

Aspect ratio describes the relationship between the chord of a wing and its span (length).

On each minute part of an airfoil, a small force is present. This force is different in magnitude and direction from any forces acting on other areas forward or rearward from this point. If all of these points are added mathematically, we can obtain a single resultant force. The point of intersection of the resultant force with the chord line of the airfoil is called the center of pressure.

The center of pressure moves along the airfoil as the angle of attack changes. As the angle of attack of an airfoil increases, the center of pressure will move toward the leading edge.

In a properly designed airfoil, the lift increases as the angle of attack is increased. When the lift of an airfoil increases, the drag will also increase. When the angle of attack increases to the angle of maximum lift, the burble point is reached. This is known as the critical angle. When the critical angle is reached, the air ceases to flow smoothly over the top surface of the airfoil and begins to burble, or eddy. When this occurs the amount of lift drops and drag becomes excessive. The force of gravity exerts itself, and the nose of the aircraft drops. This is known as a stall.

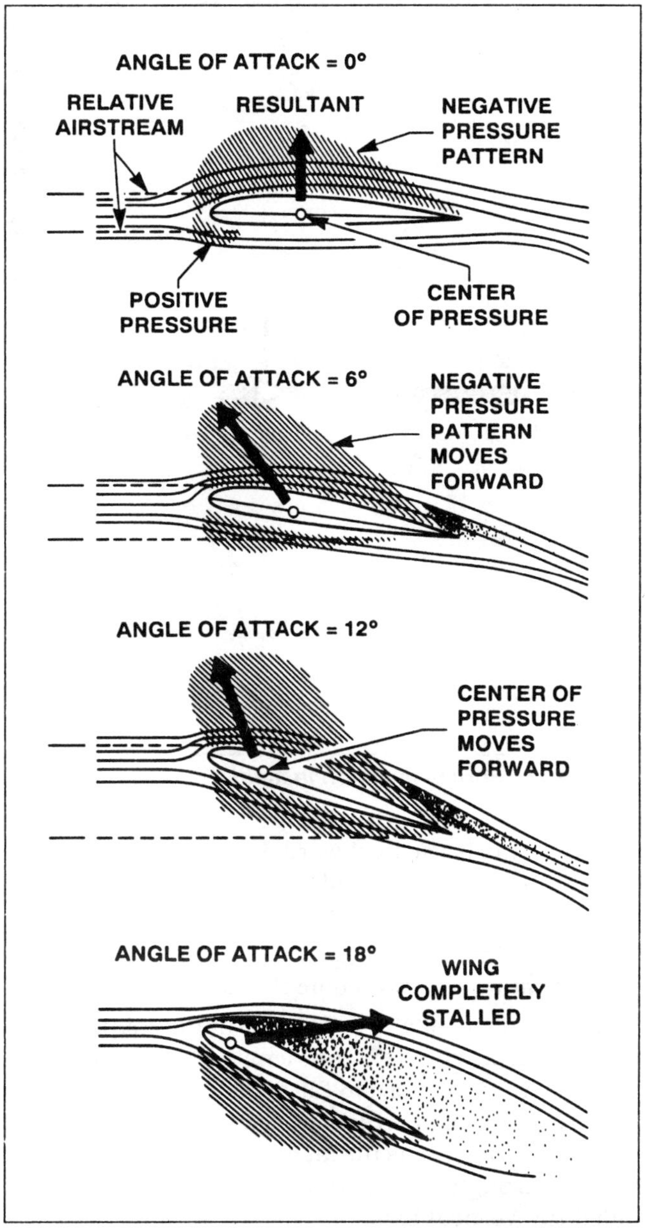

Figure 36. Effect of increasing angle of attack.

Buffeting is the intermittent application of forces to a part of an airplane. It is caused by an unsteady flow from turbulence.

Forces Acting on an Aircraft in Flight

There are four forces acting on an aircraft in flight, and only in straight and level, unaccelerated flight are they equal.

- *Weight* — This is the force of gravity acting downward upon the aircraft and everything that is in it.
- *Lift* — This is the vertical force created by the airflow over the wing. It acts to overcome weight.

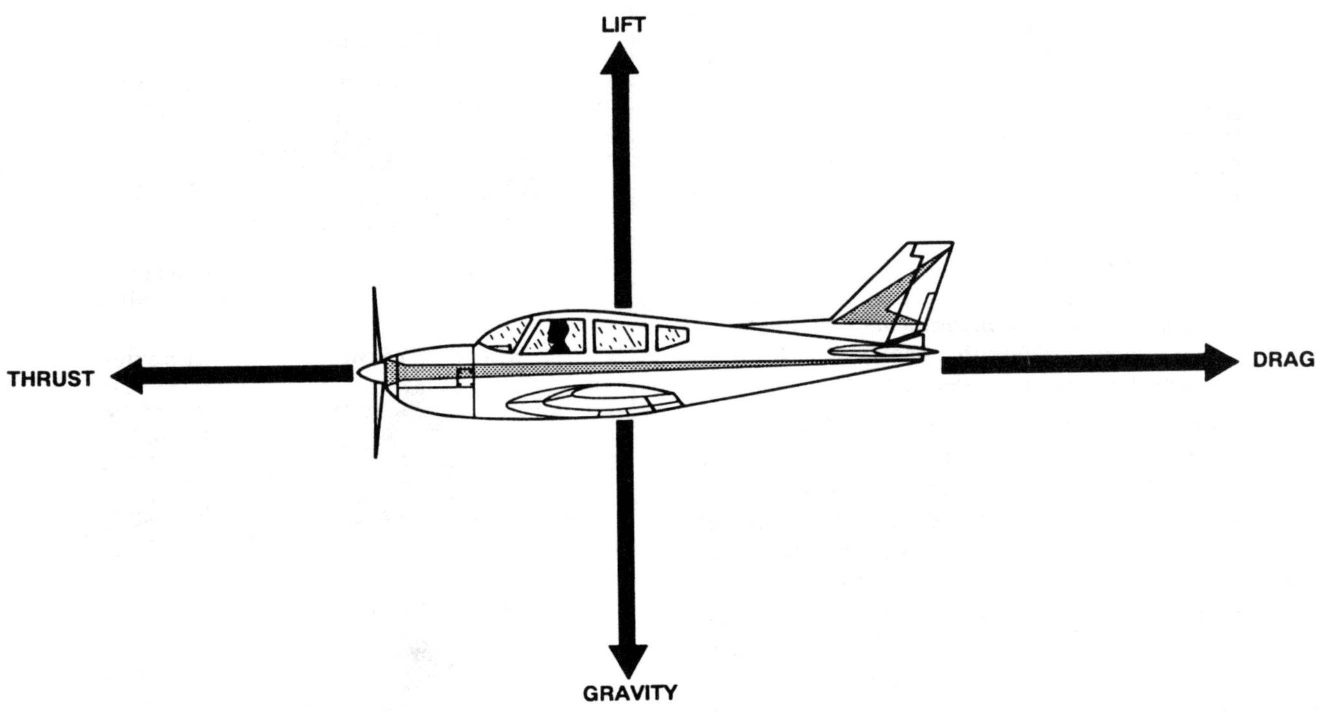

Figure 37. Forces in action in flight.

- *Drag* — This is the backward deterrent force and is caused by a disruption of the airflow by the wings, fuselage, and protruding objects.
- *Thrust* — This is the force produced by the power-plant that acts to overcome the force of drag.

An aircraft in flight is free to rotate about three axes. The axis which extends lengthwise through the fuselage from the nose to tail is called the longitudinal axis. The axis which extends crosswise, from wing tip to wing tip, is the lateral axis. The axis which passes through the center, from top to bottom, is called the vertical axis.

The motion about the longitudinal axis is called roll; motion along the lateral axis is called pitch; and an aircraft moves about its vertical axis in a motion which is called yaw. This is a horizontal movement of the nose of the aircraft.

Roll, pitch, and yaw are controlled by three control surfaces. Roll is produced by the ailerons, which are located at the trailing edges of the wings. Pitch is affected by the elevators, the rear portion of the horizontal tail assembly. Yaw is controlled by the rudder, the rear portion of the vertical tail assembly.

Stability

Three terms that appear in any discussion of aerodynamics are (1) stability, (2) maneuverability, and (3) controllability. Stability is the characteristic of an aircraft which tends to cause it to fly (hands off) in a straight and level flight path. Maneuverability is the ability of an aircraft to be directed along a desired flight path and to withstand the stresses imposed. Controllability is the quality of the response of an aircraft to the pilot's commands while maneuvering the aircraft.

Stability is very important to the safety and comfort of flight. An airplane should be rigged to fly hands-off at cruising speed.

An airplane which has good longitudinal stability should have a minimum tendency to pitch. An airplane that has a tendency to gradually increase a pitching moment that has been set into motion has poor longitudinal stability. The horizontal stabilizer is the primary surface which controls longitudinal stability.

Stability about the vertical axis is referred to as directional stability. The purpose of the vertical fin is to provide directional stability. The corkscrewing slipstream of air pushed back by the propeller strikes the vertical tail surfaces on the left side and tries to push the tail to the right. On most airplanes this effect is cancelled by offsetting the leading edge of the vertical fin slightly to the left.

Stability about the axis which runs parallel to the line of flight is referred to as lateral stability. Motion about the longitudinal axis is a lateral, or rolling,

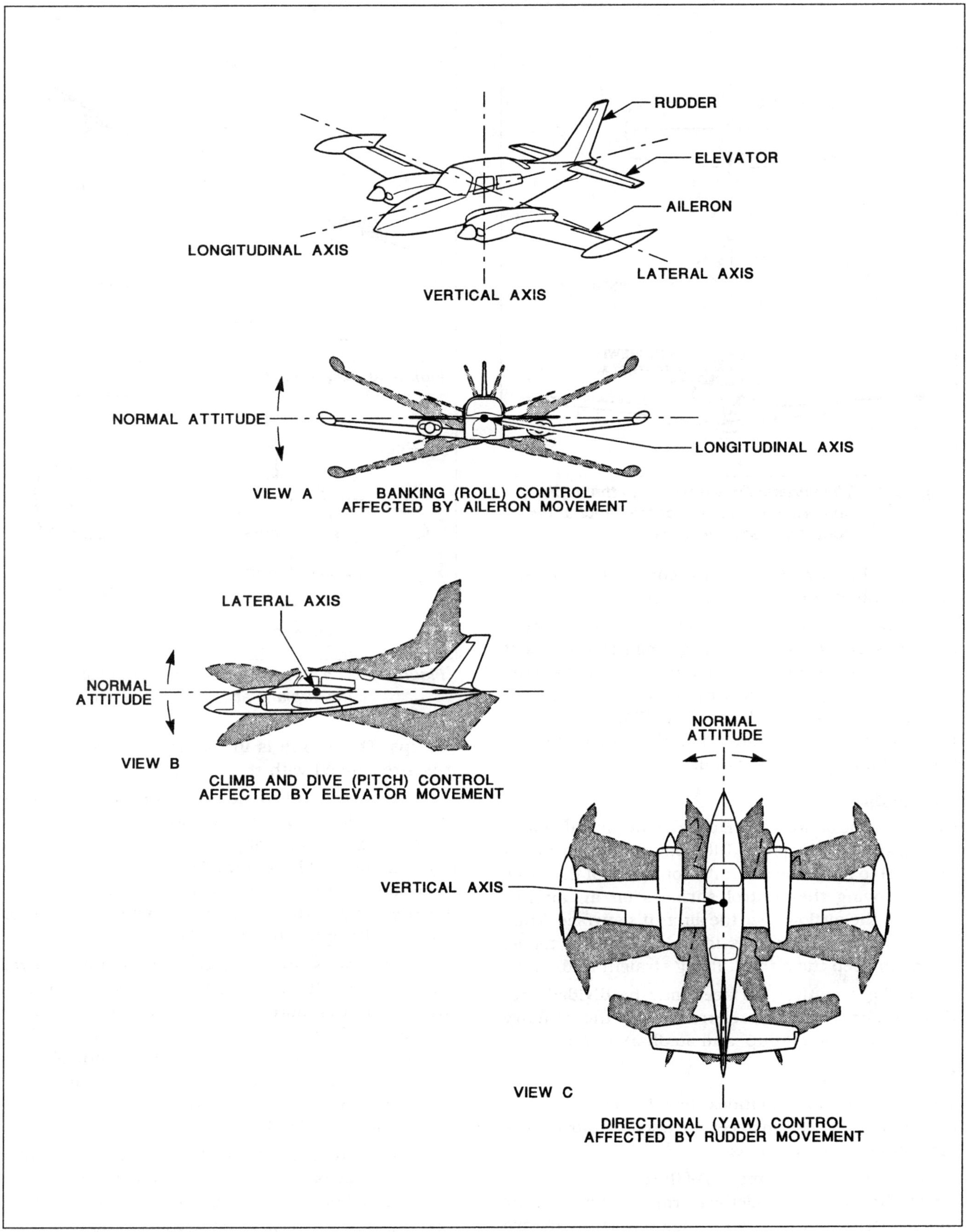

Figure 38. Motion of an aircraft about its axis.

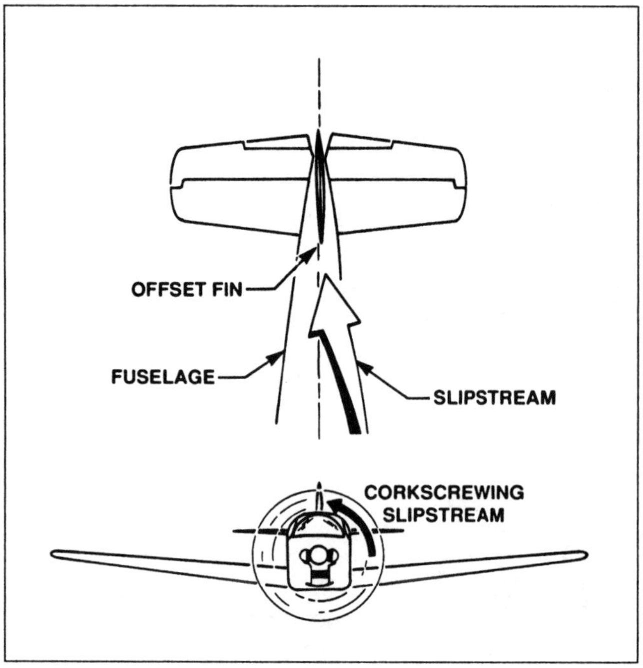

Figure 39. ***The vertical fin is offset from the airplane centerline to counteract the effect of the corkscrewing slipstream.***

motion. The principal surface contributing to the lateral stability of an airplane is the wing.

Lateral dihedral, a rigging consideration on most airplanes of conventional design, contributes most to stability of the airplane about its longitudinal axis. Dihedral is the positive acute angle between the wing and the lateral axis of the airplane. This means that the wing tips appear higher than the wing root when the airplane is on the ground.

Controls

Control is the action taken to make the aircraft follow any desired flight path. Different control surfaces are used to move the aircraft about each of the three axes. Moving the control surfaces on an aircraft changes the airflow over the aircraft's surface. This, in turn, creates changes in the balance of forces acting to keep the aircraft flying straight and level.

The flight control surfaces may be divided into three groups, usually referred to as the primary group, secondary group, and auxiliary group.

Primary Group

The primary group includes the ailerons, elevators, and rudders. These surfaces are used for moving the aircraft about its three axes.

The ailerons control motion of the aircraft about the longitudinal axis. The ailerons form a part of the wing and are located in the trailing edge of the wing toward

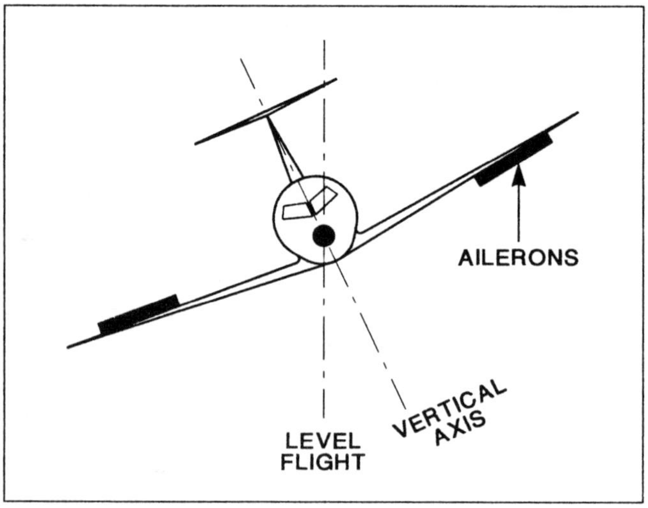

Figure 40. ***Aileron action.***

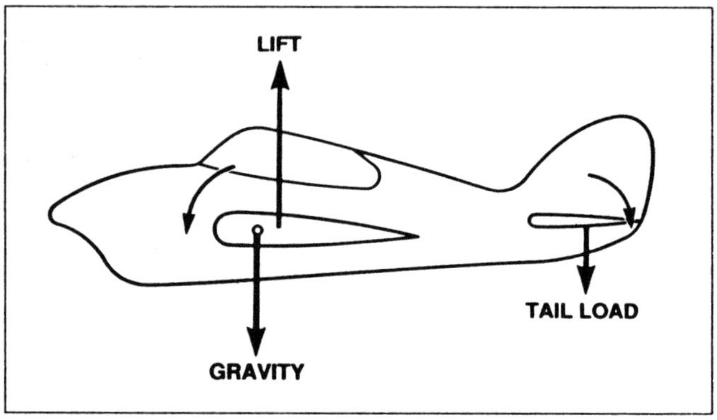

Figure 41. ***The elevators vary the tail load to rotate the airplane about its lateral axis.***

the tips. The aileron is in the neutral position when it is streamlined with the trailing edge of the wing.

Ailerons respond to side pressure applied to the control stick, or a turning motion of the control wheel. This pressure applied to the right raises the right aileron and lowers the left aileron, causing the aircraft to bank to the right. Ailerons are linked together by control cables to that when one aileron is down, the opposite aileron is up.

The aileron system utilizes differential control. Differential control on an aileron system means that the up travel is more than the down travel. During inspection of the flight control system of an airplane equipped with differential-type aileron control, side-to-side movement of the control stick will cause each aileron to have a greater up travel (from the streamlined position) than down travel.

The elevators, which are the horizontal movable control surfaces mounted on the tail of the airplane, cause the airplane to rotate about its lateral axis.

An airplane's center of lift is usually located aft of its center of gravity so that the airplane will have a

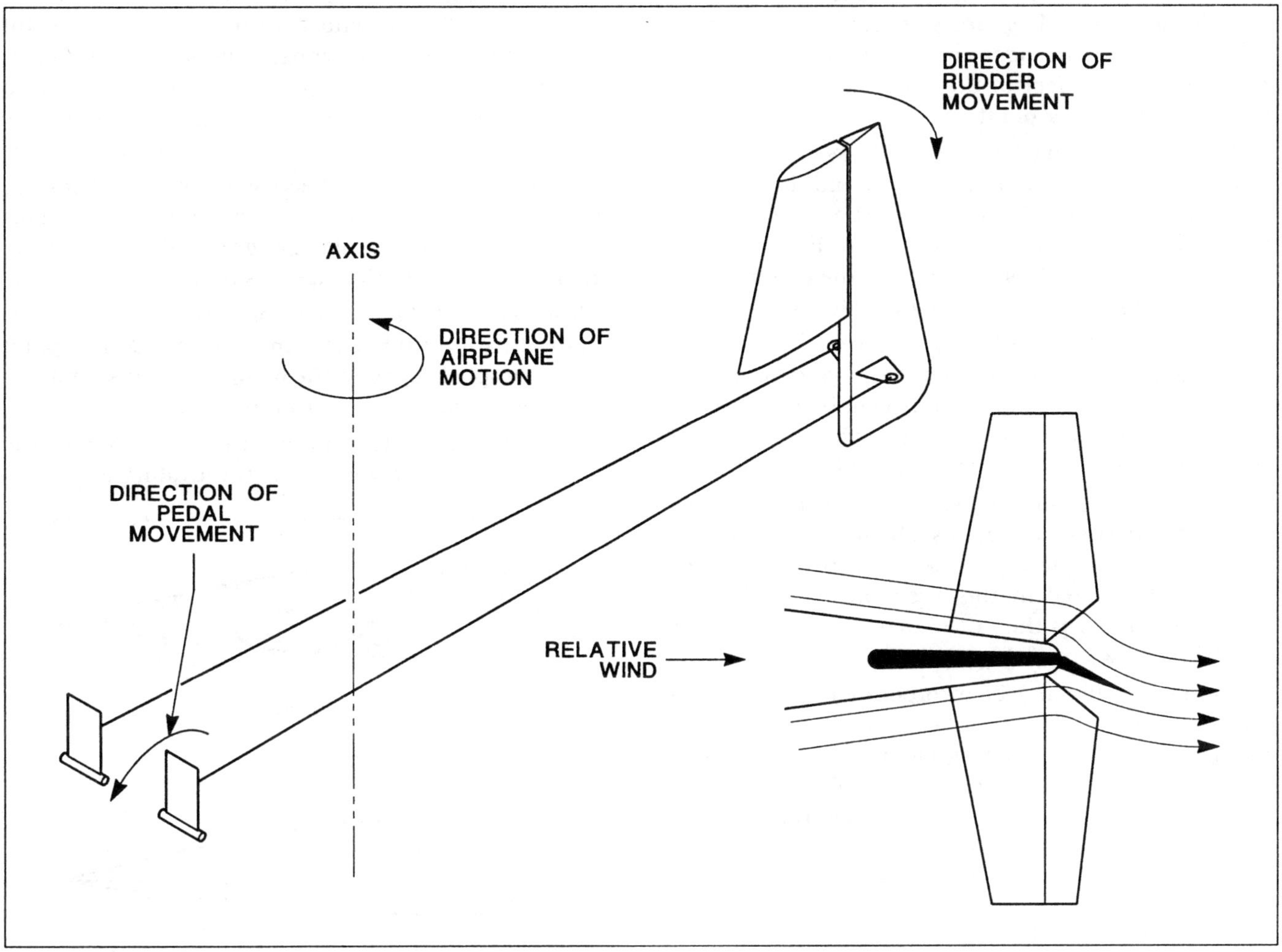

Figure 42. Rudder action.

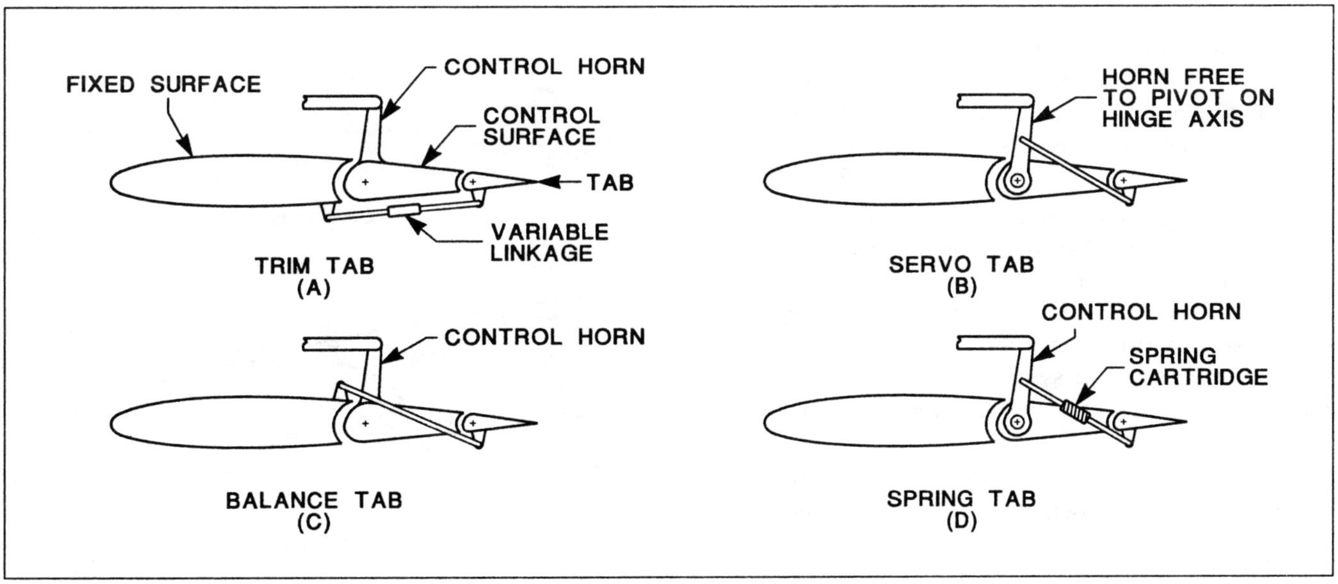

Figure 43. Flight control trim tab types.

nose-heavy tendency. In straight and level flight, the airplane is balanced by the horizontal tail load which tends to rotate the nose up, exactly counteracting the nose-down effect of the center of gravity being ahead of the center of lift.

When the control wheel is pulled back, the trailing edge of the elevator moves upward and deflects the airstream upward. This increases the down load on the tail and rotates the nose of the airplane upward. Moving the control forward deflects the trailing edge of the elevators downward and deflects the air downward. This pushes the tail up, lowering the nose.

Rotation of the aircraft about the vertical axis is called yaw. This motion is controlled by the rudder. The rudder is a movable control surface attached to the trailing edge of the vertical stabilizer.

Contrary to common belief, an airplane is not turned by use of its rudder as a boat is turned in the water. For an airplane to make a coordinated turn, it must be banked in the direction of the desired turn. The rudder is then used to overcome the effect of adverse yaw created by the roll.

Counterweights may be incorporated in the leading edge of primary control surfaces to prevent the surfaces from fluttering during flight. Control surfaces with counterweights should be statically balanced to reduce the possibility of flutter. After repairing or recovering a control surface, the surface should be rebalanced to the manufacturer's specifications.

Secondary Group

Included in the secondary group are the trim tabs and spring tabs. To offset the forces that tend to unbalance an aircraft in flight, ailerons, elevators, and rudders may be provided with auxiliary controls known as tabs. These are small, hinged control surfaces attached to the trailing edge of the primary control surface. Tabs can be moved up or down from the cockpit.

Aircraft flight control trim systems must be designed and installed so that the pilot can determine the relative position of the trim tab from the cockpit. Movement of the cockpit control toward the nose-down position during a ground operational check of the elevator trim tab system will cause the trailing edge of the trim tab to move upward regardless of elevator position.

The purpose of spring tabs, or servo-tabs, is to assist the pilot in moving the control surfaces.

Auxiliary Group

Included in the auxiliary group of flight controls surfaces are the wing flaps, spoilers, speed brakes, slats, leading edge flaps and slots.

The auxiliary groups may be divided into two sub-groups. Those whose primary purpose is lift augmenting and those whose primary purpose is lift decreasing. In the first group are the flaps, both trailing edge and leading edge (slats), and slots. The lift decreasing devices are speed brakes and spoilers.

Flaps increase the effective lift of an airfoil by increasing the camber of the airfoil. Flaps are usually mounted on the trailing edge of the wing between the inboard end of the ailerons and the fuselage.

The types of flaps in use on aircraft include: (1) plain, (2) split, (3) Fowler, and (4) slotted. The plain flap is simply hinged to the wing and forms a part of the wing surface when lowered.

The split flap gets is name from the hinge at the bottom part of the wing near the trailing edge per-

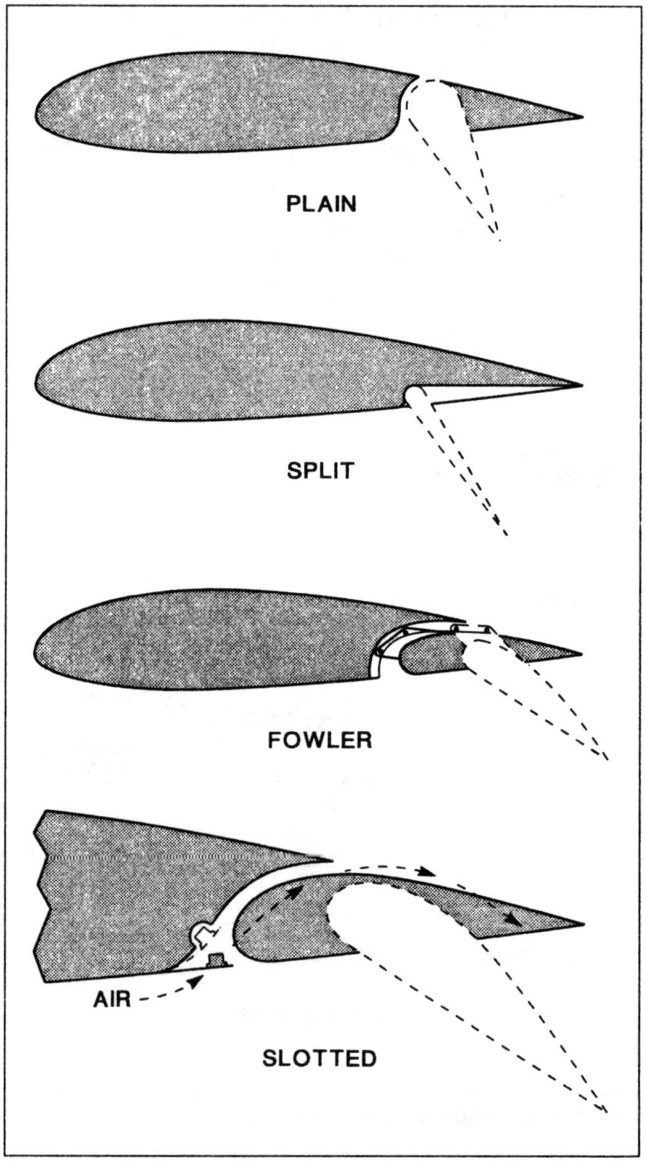

Figure 44. Types of wing flaps.

mitting it to be lowered from the fixed top surface. The Fowler flap fits into the lower part of the wing so that it is flush with the surface. When the flap is operated, it slides backward on its tracks and tilts downward at the same time. This increases wing camber, as do the other types of flaps. However, Fowler flaps also increase the wing area; thus, they provide added lift without unduly increasing drag.

The slotted flap is like the Fowler flap in operation, but in appearance it is similar to the plain flap. This flap is equipped with either tracks or hinges of a special design. During operation, the flap moves downward and rearward away from the position of the wing. The "slot" thus opened allows a flow of air over the upper surface of the flap. This effect is to streamline the airflow and to improve the efficiency of the flap.

A wing slat is a movable airfoil attached to the leading edges of high-performance airplane wings. Their purpose is to reduce stalling speed.

Some airplanes have a fixed slot in the leading edge of the wing ahead of the aileron. At high angles of attack, air flows through this slot from the high pressure area below the wing to the low-pressure area above. This air has enough energy that it will maintain smooth airflow over this portion of the wing. This will help maintain aileron effectiveness and lateral control as the aircraft approaches a stall.

It is extremely important that an airplane wing stall progressively from the root out to the tip, so the ailerons will be effective throughout the stall. If the wing does not have this stall progression characteristic, it is possible for the manufacturer to place small triangular strips of metal (stall strips) on the leading edge of the wing at the root area. The primary purpose of stall strips is to stall the inboard portions of the wing first.

Aircraft Rigging

Rigging checks may be as simple as confirming control surface travel during a routine inspection, or as complex as a complete alignment check following assembly of an aircraft. The principals observed are the same, the difference is only with the number of items checked.

The position or angle of the main structural components is related to a longitudinal datum line parallel to the aircraft center line, and a lateral datum line parallel to a line joining the wing tips. The buttock line, or buttline, of an aircraft is a width measurement left or right of, and parallel to, the vertical centerline. Fuselage stations are designated in inches aft of the zero or fixed reference line.

Certain precautions must be observed in all instances. Rigging and alignment checks should not be undertaken in the open; however, if this cannot be avoided, the aircraft should be positioned with the nose into the wind. Before checking the position or angle of the main components, the aircraft should be leveled.

The Aircraft Type Certificate Data Sheet includes the necessary information on control surface travel. The precise information needed to perform a symmetry alignment check for a particular aircraft may be found in the aircraft service or maintenance manual.

When checking alignment, a suitable sequence should be developed and followed to be certain that the checks are made at all the positions specified. The alignment checks specified usually include:

(1) Wing dihedral angle.
(2) Wing incidence angle.
(3) Engine alignment.
(4) Horizontal stabilizer incidence.
(5) Horizontal stabilizer dihedral.
(6) Verticality of the fin.
(7) A symmetry check.

The correct dihedral angle can be determined by using a dihedral board and bubble level along the front spar of each wing.

With a few exceptions, the dihedral and incidence angles of conventional modern aircraft cannot be adjusted. Some manufacturers permit adjusting the wing angle of incidence to correct for a wing-heavy condition.

If as the result of an accident, it is necessary to replace the fuselage fittings for the wing attachment, a great deal of care should be exercised in locating them in the proper position, because the angle of incidence is one of the factors to be re-established at this time.

If the right wing of a monoplane is improperly rigged to a greater angle of incidence than designated

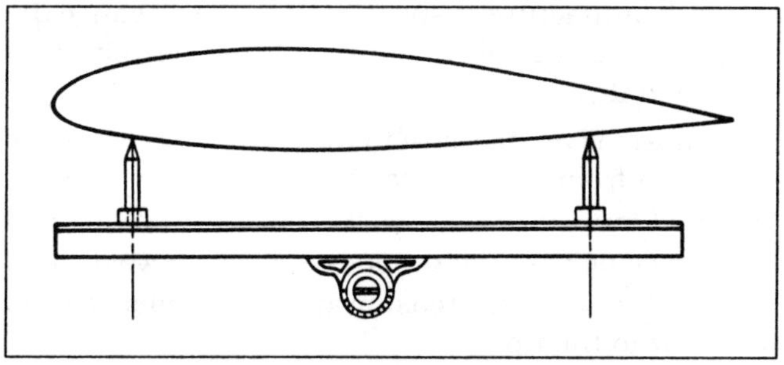

Figure 45. An incidence board is used to check for wing warpage and for the proper wash-in/wash-out.

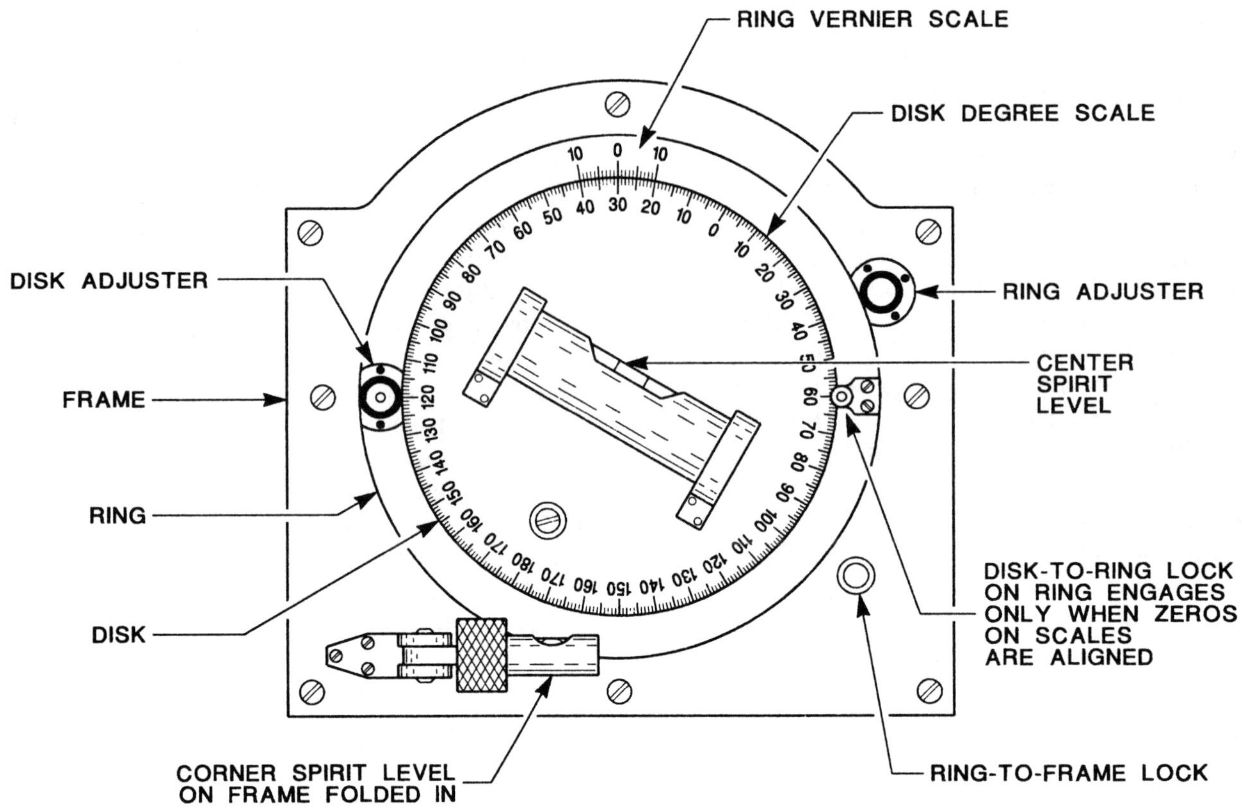

1. WITH DISK-TO-RING LOCK IN THE DEEP SLOT, TURN DISK ADJUSTER TO LOCK DISK TO RING.
2. MOVE CONTROL SURFACE TO NEUTRAL. PLACE PROTRACTOR ON CONTROL SURFACE AND TURN RING ADJUSTER TO CENTER BUBBLE IN CENTER SPIRIT LEVEL (RING MUST BE UNLOCKED FROM FRAME).
3. LOCK RING TO FRAME WITH RING-TO-FRAME LOCK.
4. MOVE CONTROL SURFACE TO EXTREME LIMIT OF MOVEMENT.
5. UNLOCK DISK FROM RING WITH DISK-TO-RING LOCK.
6. TURN DISK ADJUSTER TO CENTER BUBBLE IN CENTER SPIRIT LEVEL.
7. READ SURFACE TRAVEL IN DEGREES ON DISK AND TENTHS OF A DEGREE ON VERNIER SCALE.

Figure 46. Using the universal propeller protractor to measure control surface travel.

in the manufacturer's specifications, it will cause the airplane to be off balance both laterally and directionally.

Wash-in is an increase in the angle of incidence of the wing from the root to the tip. If a wing has an angle of incidence of 2° at the root and an angle of incidence of 3° at the outer end, it has a wash-in of 1°. Wash-out is a decreasing angle of incidence from the root to the tip.

If a pilot reports that an airplane flies left wing heavy, this condition is corrected by washing-in the left wing, or washing-out the right wing, or both. Washing-in the left wing of a monoplane, for purposes of rigging corrections after flight test, will increase both drag and lift due to the increased angle of attack.

Proper wing twist in a sheet metal constructed wing can usually be checked by utilizing a bubble level and special fixtures described by the manufacturer.

If the vertical fin of a single-engine, propeller driven airplane is rigged properly, it will generally be parallel to the vertical axis but not the longitudinal axis.

Improper rigging of the elevator trim tab system will affect the balance of the airplane about its lateral axis.

The universal propeller protractor can be used to measure degrees of flap travel. A universal propeller protractor used to measure the degrees of aileron travel should be zeroed with the aileron in the NEUTRAL position.

If the control stick of an aircraft with properly rigged flight controls is moved rearward and to the left, the right aileron will move down and the elevator will move up. If the control stick is moved forward and to the right, the left aileron will move down and the elevator will move down.

Flight Control Operating Systems

Three types of control systems commonly used are (1) the cable, (2) push-pull and (3) the torque tube system. The cable system is the most widely used, because deflections of the structure do not affect its operation.

There are three types of steel cable used for aircraft control systems: non-flexible, flexible, and extra-flexible. Non-flexible cable may be of either the 1 × 7 or 1 × 19 type. This designation means that the 1 × 7 cable is made up of 7 strands, each having only 1 wire. The 1 × 19 cable is made of 19 strands of 1 wire each. Nonflexible cable may only be used for straight runs where the cable does not pass over any pulleys.

Flexible cable is made up of seven strands, each of which has seven wires. Flexible cable may be used for straight runs or where the pulleys are large. When cables must change direction over relatively small diameter pulleys, extra-flexible cable must be used. This cable is made of seven strands, each having 19 separate wires.

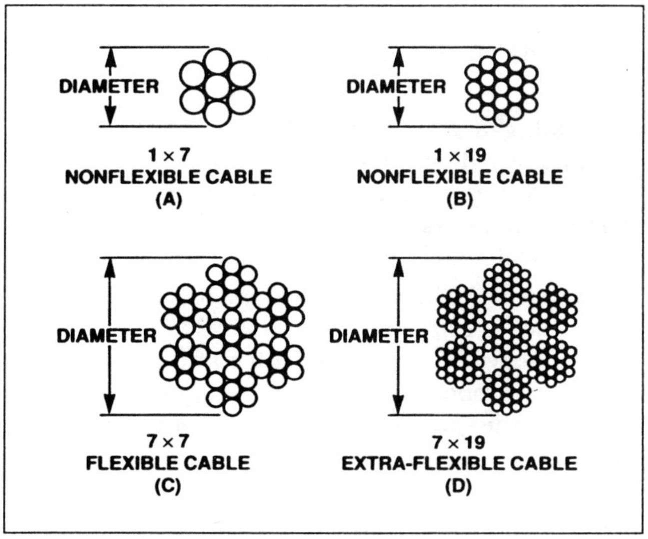

Figure 47. Types of control cable.

Cable type and size must be matched to the system requirements. The smallest size cable that may be used for aircraft primary control systems is 1/8″.

Most aircraft cables terminate with either a copper Nicopress sleeve and fitting, or a swaged terminal. The control cable terminals on most late model aircraft are swaged, and a painted band is placed around the cable adjacent to the terminal in order to detect slippage of the cable fitting.

If all instructions issued by the swaging tool manufacturer are followed when swaging a cable terminal, the resultant swaged terminal strength should be the full rated strength of the cable. An AFTER gage us used to check the diameter of the swaged portion of the terminal to ensure that the correct amount of swaging has taken place.

Since cables are a vital part of an aircraft control system, they should be carefully examined at each required inspection. A satisfactory method of inspecting control cables for broken wires is to place a piece of cloth around the cable and run it back and forth over the length of the cable.

The aircraft manufacturer routes the cables in the most direct manner possible. When a change in

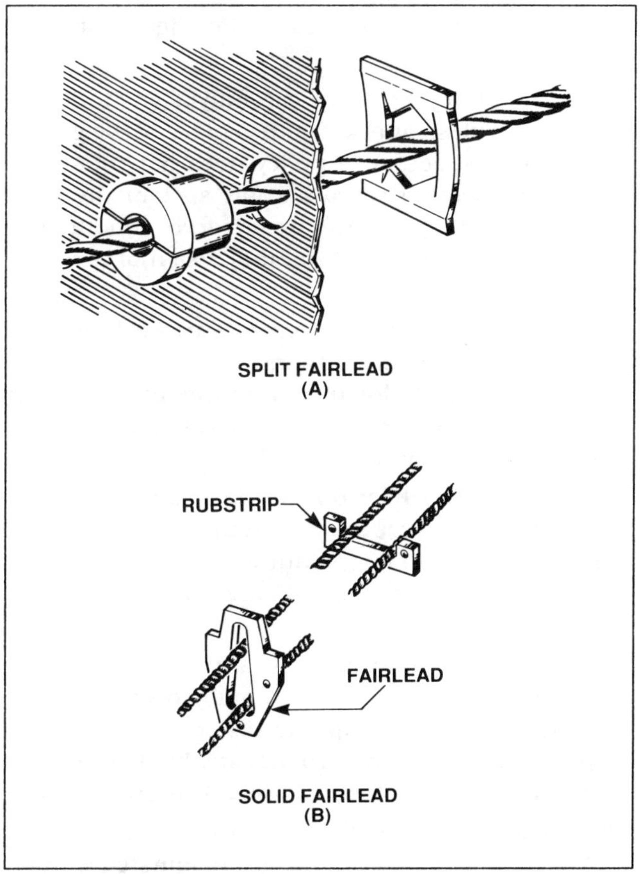

Figure 48. Fairleads.

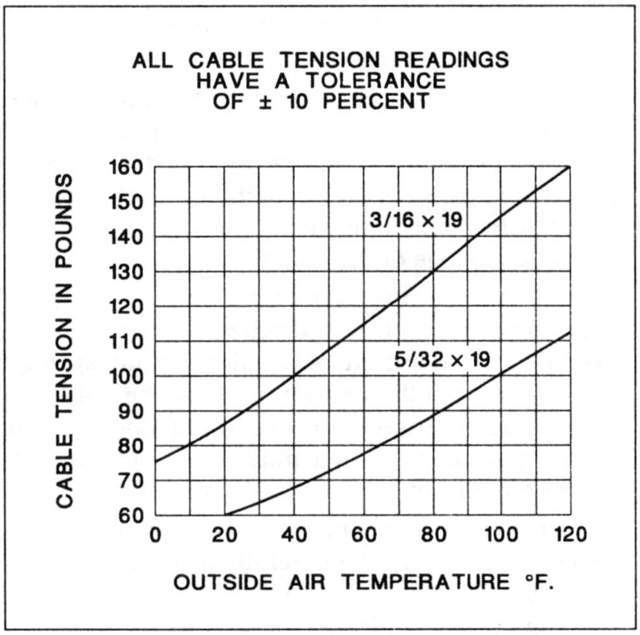

Figure 49. Cable tension temperature compensation chart.

direction is necessary, a pulley may be used. At any point in a run where there is the likelihood of the cable contacting the structure, a fairlead is used. Fairleads should never deflect the alignment of the cable more than 3°.

Control pulleys must be carefully aligned with the cable so it rides squarely in the center of the groove. If, during inspection, a pulley does not rotate and further inspection reveals a flat spot in the cable groove, the pulley should be replaced. If both sides of a pulley groove are worn, it is a sign of excessive cable tension.

In addition to proper control surface travel, the tension on the cable system must be checked at inspection or following adjustments. Tension changes may be made by adjusting turnbuckles provided for this purpose.

Large aircraft whose dimensions change appreciably with temperature must have the tension adjusted for a given ambient temperature. Temperature charts like the one seen in figure 49 are used to determine the proper values.

Some larger aircraft that are subject to large change in dimension, and normally operate at high altitudes and low temperatures, may use tension regulators. A tension regulator in the flight control cable system of a large all-metal aircraft is used primarily to retain a set tension.

After the cable tension has been adjusted with the turnbuckles, they must all be checked for the proper amount of thread showing, and then safetied. When

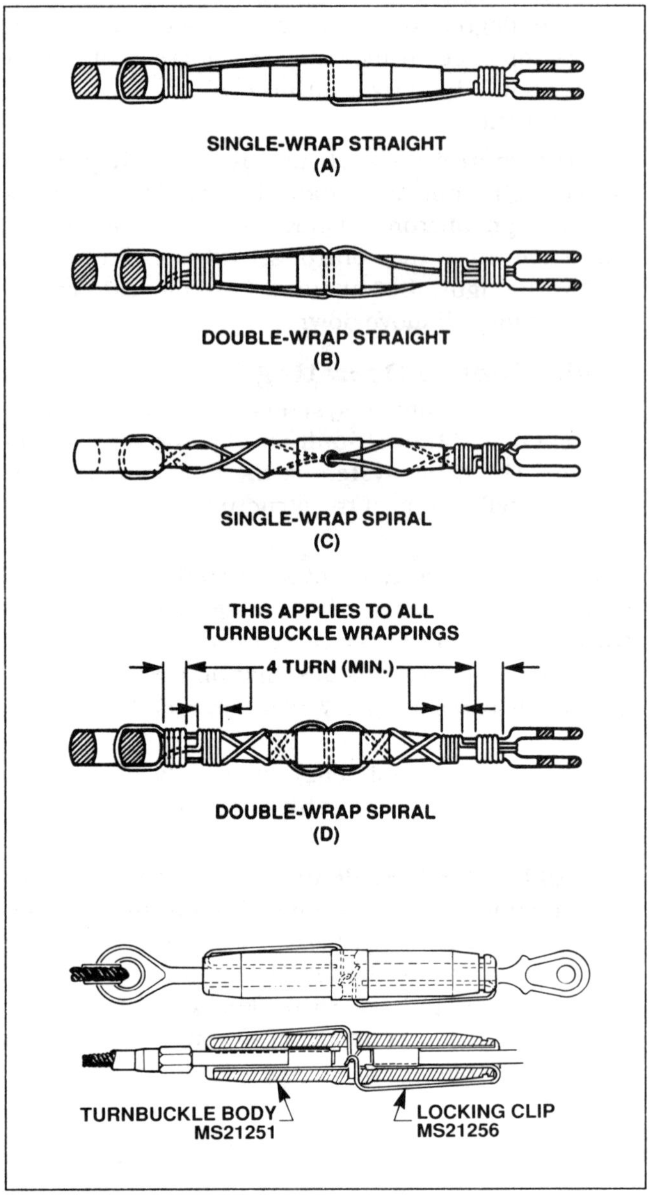

Figure 50. Turnbuckle safetying.

inspecting a control cable turnbuckle for proper installation, determine that the safety wire ends are wrapped a minimum of four turns around the terminal end shanks and no more than three threads are showing outside the turnbuckle barrel.

If the travel of an airplane's controls is correct, but the cables are rigged exceptionally tight, the airplane will be heavy on the controls. If control surfaces are adjusted properly and the control surfaces tend to vibrate, the probable cause is worn attachment fittings.

When installing a control cable, it is found that one end of a splice comes within ½" of entering a pulley groove with the loaded cable at the end of its travel. This must be corrected by adjustment or replacement. Splices must be located so that they come no closer than 2" to any pulley or fairlead.

Push-pull rods may be used instead of control cables on some aircraft and most helicopters. The tubes are usually fabricated from seamless aluminum alloy tubing with threaded rod ends riveted into both ends, and fittings screwed into these threads. To be sure that the rod ends are screwed far enough in to the fitting, each of the fittings has a small hole drilled in it. If you can pass a piece of safety wire through the hole, the rod end is not screwed in far enough. At each inspection, the threaded rod ends should be checked for the amount of thread engagement by means of the inspection hole provided.

Before a rigging check is complete all hardware should be checked for proper safety. A variety of hardware may be found, and various safety methods used. Keep in mind: (1) a cotter pin is an acceptable safety device for a castle nut when installed on secondary structures, (2) when used in close proximity to magnetic compasses, cotter pins are made of corrosion resisting steel, and (3) when a fiber stop nut can be threaded on a bolt or stud through the fiber with only the fingers, it should be rejected.

REVIEW QUESTIONS — Fixed Wing Assembly and Rigging

61. *Lateral dihedral contributes most to stability of the airplane about which axis?*
62. *An airplane which has good longitudinal stability should have a minimum tendency to _____.*
63. *What is the angle of incidence?*
64. *What controls an aircraft directionally about its vertical axis?*
65. *The elevators provide rotation about which axis?*
66. *How is the chord of a wing measured?*
67. *What two factors are involved in the aspect ratio of a wing?*
68. *What is the purpose of a wing slat?*
69. *What is the primary purpose of stall strips?*
70. *What is the purpose of the vertical fin?*
71. *Why should control surfaces with counterweights be statically balanced?*
72. *With which system is differential control associated?*
73. *What is the purpose of spring tabs or servo tabs?*
74. *What type of cable is used in primary control systems where operation over pulleys is frequent?*
75. *Why is a tension regulator used on some large aircraft?*

Answer questions 8227 through 8292 in the FAA Test section of this book.

AIRCRAFT CONFORMITY AND AIRWORTHINESS INSPECTION

Inspection programs are designed to maintain high levels of reliability and durability of the aircraft structure and systems. During normal service, each aircraft will encounter a number of types of inspections. They may include:

(1) Preflight inspection
(2) Annual inspection
(3) Progressive inspection
(4) Continuous airworthiness inspection
(5) 100-hour inspection
(6) Altimeter and static system inspection
(7) ATC Transponder checks

In addition, special inspections may be required when an overweight landing has been made, extreme turbulence encountered, or when specified by an Airworthiness Directive (AD).

On most aircraft under 12,500 lbs. gross weight, the annual and 100-hour inspection are the most common. The scope and detail of these two types of inspections is the same. They differ only in the person authorized to return the aircraft to service. The 100-hour inspection may be performed by, and the aircraft returned to service by an appropriately rated aircraft maintenance technician. The annual inspection must be performed by, and the aircraft returned to service by, a technician who also holds the Inspection Authorization (IA).

Large airplanes and turbine-powered multiengine airplanes operated under FAR Part 91, General Operating and Flight Rules, must be inspected in accordance with an inspection program authorized under FAR Part 91, Subpart E.

The operating conditions that establish the annual inspection, and that make a 100-hour inspection mandatory, are found in FAR Part 91. As the name implies, the annual inspection is required once in each twelve calendar months. This time may not be extended. An aircraft that is due an annual inspection may be operated if a Special Airworthiness Certificate has been issued for the aircraft. This permit is know commonly as a "Ferry Permit" and must be issued by the local office of the FAA.

Because an aircraft has been inspected in accordance with an annual inspection does not mean that it will always be returned to service immediately. If there are repairs required to return the aircraft to an airworthy condition, they must be performed first. In this case, the inspector may record the inspection in the aircraft's maintenance records and list the items that must be corrected. A copy of this list must be sent to the owner or operator of the aircraft.

If the items listed are minor repairs or alterations, an appropriately rated technician may repair the defects and approve the aircraft for return to service. When the defects involve major repairs or alterations an appropriately rated technician may repair the defect, and an IA may approve the aircraft for return to service.

Aircraft which are being operated for hire are required to be inspected each 100 hours. An aircraft can be operated with a 100-hour inspection overdue, if necessary to reach a place at which the inspection can be done. The maximum time a 100-hour inspection may be extended is 10 hours. This time must be deducted from the next inspection interval.

The Federal Aviation Regulations are very specific on the wording required in the maintenance records which record the approval or disapproval for return to service of an aircraft after a 100-hour or annual inspection. The required statement may be found in FAR Part 43.

An important part of any inspection routine is to check for Airworthiness Directives that may require compliance. These may include newly issued ADs and recurring ADs that are due. Compliance with an applicable AD is mandatory and must be recorded in the permanent maintenance records.

While aircraft are in the hangar for inspection it is common for other maintenance and modifications to be performed. When overhauling electrical equipment, all necessary information should be obtained from maintenance instructions published by the aircraft and/or equipment manufacturer. Radio equipment installations made in accordance with Supplemental Type Certificate data require approval using major alteration procedures.

When it becomes necessary to jack an aircraft for maintenance, inspection, or weighing purposes, it should be accomplished only after the installation of critical stress panels or plates if they are required. Aircraft being weighed should be jacked indoors to stabilize the weighing scales.

REVIEW QUESTIONS — Aircraft Conformity and Airworthiness Inspection

76. *Is compliance with ADs mandatory?*
77. *When can an aircraft be operated with a 100-hour inspection overdue?*
78. *What is the maximum time a 100-hour inspection can be extended?*
79. *When may an aircraft overdue for an annual inspection be operated?*
80. *Who is authorized to conduct a 100-hour inspection of an aircraft?*

Answer questions 8292 through 8304 in the FAA Test section of this book.

LANDING GEAR SYSTEMS

The landing gear of a fixed-wing aircraft consists of main and auxiliary units, either of which may or may not be retractable. The main landing gear forms the principal support of the aircraft on land or water and may include any combination of wheels, floats, skis, shock-absorbing equipment, brakes, retracting mechanism with controls and warning devices, cowling, fairing, and structural members to attach any of the foregoing to the primary structure. The auxiliary landing gear consists of tail or nose wheel installations, outboard pontoons, skids, etc., with necessary cowling and reinforcement.

Aircraft will generally be identified by the location of the auxiliary landing gear. Tailwheel type aircraft utilize main landing gear located ahead of the center of gravity, and an auxiliary wheel located at the tail of the aircraft. This type of arrangement is also known as "conventional" landing gear.

Almost all current production airplanes use the tricycle landing gear configuration. The two main wheels are located behind the aircraft's center of gravity and the nose of the airplane is supported by the auxiliary wheel.

Fixed or Retractable Landing Gear

Landing gear offer a considerable amount of parasitic drag. This drag increases as the speed increases. Slower aircraft lose little efficiency by using a lighter weight fixed landing gear, but faster aircraft may benefit from utilizing a retractable landing gear system. The gain from the reduced drag offsets the increased weight of the retraction mechanism.

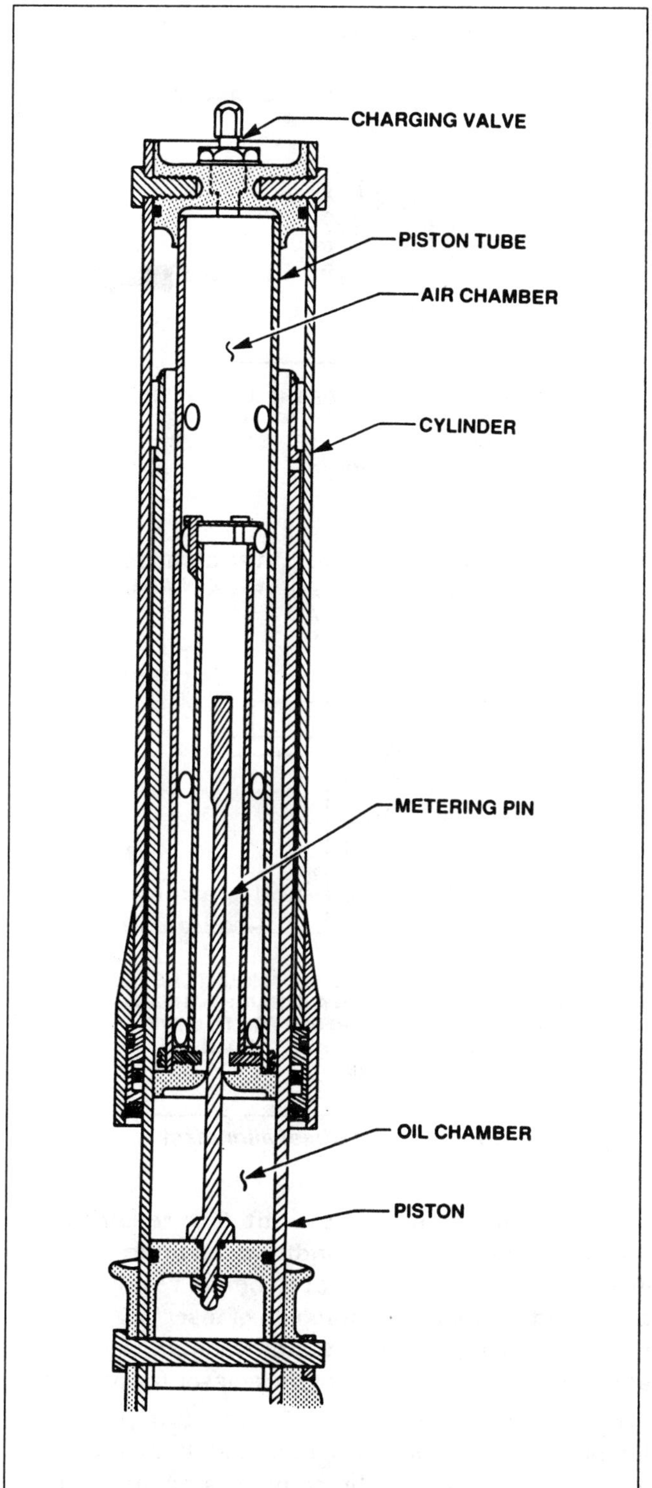

Figure 51.* *The oleo, or air-oil, strut absorbs the landing impact as oil transfers from the oil chamber into the air chamber around the tapered metering pin. Taxi shocks are taken up by the compressed air in the upper chamber.

Retractable landing gear may be operated manually, hydraulically, or electrically. An electric motor used to raise and lower a landing gear would most likely be a split field series-wound motor.

In addition to the retracting mechanism, additional provisions may be made to open and close landing gear doors at the proper time. The purpose of a sequence valve in a hydraulic retractable landing gear system is to ensure the operation at the proper time of the landing gear doors.

A landing gear position and warning system is required if the aircraft is equipped with retractable landing gear. This is to avoid gear up landings. A landing gear position and warning system will provide a warning in the cockpit when the throttle is retarded and gear is not down and locked.

On all aircraft equipped with retractable landing gear, some means must be provided to extend the landing gear if the normal operating mechanism fails.

After performing maintenance to an aircraft's landing gear retraction system which may have affected its ability to retract or extend properly, it is advisable to make an operational check with the aircraft placed on jacks.

Shock Absorbers

Some provision is generally made to absorb the impact at landing, and the shocks encountered during ground operations. Probably the simplest system utilizes landing gear mounted on flat steel springs, or steel tubes. These are effective and maintenance free.

Some aircraft use rubber bungee cords, or rubber "donuts" to absorb these forces. These will require regular inspection and periodic replacement.

The most widely used device is the air-oil shock strut, commonly referred to as an oleo strut. Shock struts are self-contained hydraulic units that support the aircraft on the ground and protect the aircraft structure by absorbing and dissipating the shock loads of landing. Shock struts must be inspected and serviced regularly to function efficiently.

The cylinder of this strut is attached to the aircraft structure, and a close fitting piston is free to move up and down inside the cylinder. It is kept in alignment and prevented from coming out of the cylinder by torque links. The upper link is hinged to the cylinder and the lower link to the piston. The wheel is mounted on the axle, which is attached to the piston. In retractable gear installations, a cam may be incorporated in the nose gear shock strut to center the nosewheel as it retracts.

When an air/oil type of landing gear shock strut is used, the initial shock of landing is cushioned by the fluid being forced through a metered opening. The metering pin in the strut serves to retard the flow of oil as the strut is compressed.

To prevent a very rapid extension of an oleo strut after immediate compression resulting from landing impact, various types of valves or orifices are used, which restrict the reverse fluid flow. A sleeve, spacer, or bumper ring is incorporated in an oleo strut to limit the extension stroke.

If an aircraft shock strut bottoms upon initial landing contact but functions correctly during taxi, the most probable cause is low fluid. On most aircraft, the oil level of an air and oil shock strut is checked by releasing the air and seeing that the oil is to the level of the filler plug. Use extreme caution when releasing the air from the strut. A raised "H" on the stem of an air valve denotes that the valve core is a high-pressure type.

The type of hydraulic fluid to be used in a shock strut depends upon the type of seals used in the shock strut. Shock struts utilizing natural rubber seals should not be serviced with mineral base fluid. Instructions concerning the type of fluid and amount of air pressure to be put in a shock strut are found in the aircraft manufacturer's service manual. Extension of an oleo shock strut is measured to determine the proper operating position of the strut. If a shock strut bottoms after it has been properly serviced, the strut should be removed, disassembled, and inspected.

When an empty shock strut is filled, care should be taken to extend and compress the strut completely at least two times to ensure proper packing ring seating and removal of air bubbles.

In shock struts, chevron seals are used to prevent oil from escaping. When tightening the packing nut on an air/oil shock strut installed on an aircraft, the aircraft should be jacked.

Wheel Alignment

Wheel alignment on an aircraft is an important consideration to minimize tire wear. As with an automobile, the wear pattern on the tires is a good indicator of an out-of-alignment situation.

Torque links keep the landing gear pointed in a straight-ahead direction. These torque links may also be referred to as "scissors" or "nut-crackers". These links may be used to adjust for a toe-in or toe-out condition.

Toe-in or toe-out is the amount the wheels deviate from a straight-ahead condition. If the extended longitudinal axis of the main landing gear wheel assemblies intersects aft of the aircraft, the wheels can be termed as having toe-out. The repair for an out-of-tolerance toe-in condition of main landing gear wheels (determined not to be the result of bent or twisted components) consists of inserting, removing, or changing the location of washers or spacers at the center pivotal point of the scissor torque links.

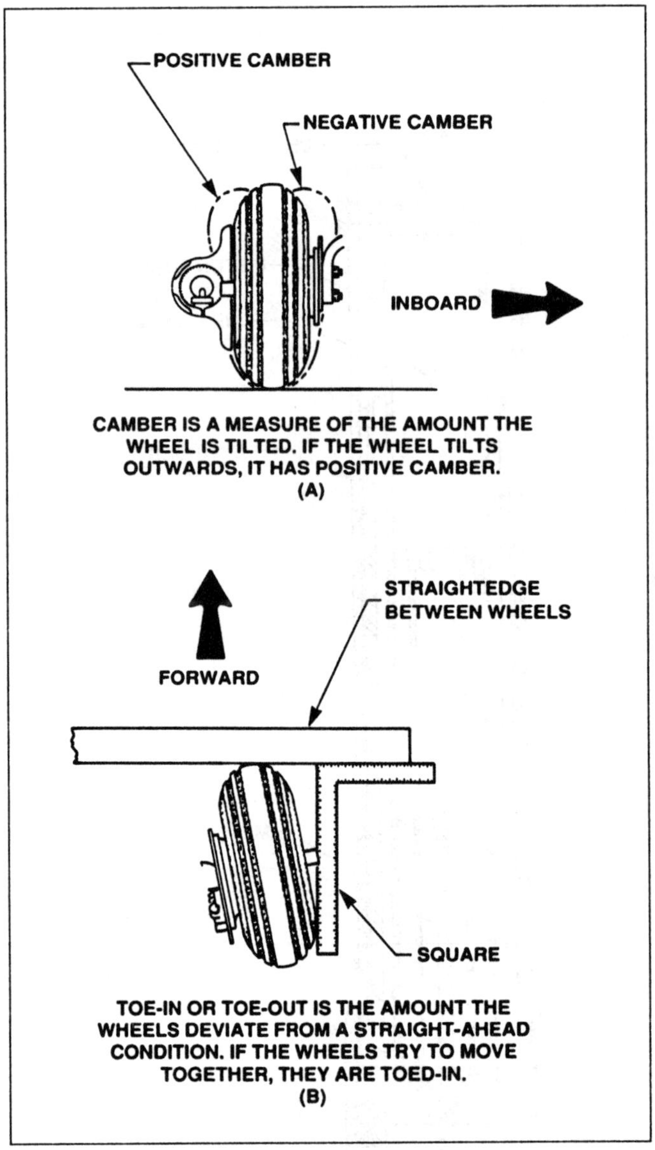

Figure 52. Conditions of wheel alignment.

Camber is a measure of the amount the wheel leans, as viewed from straight ahead. If the top of the wheel leans outward, the camber is positive, but if it leans inward, the camber is negative.

Nose Wheel Steering and Shimmy Dampers

Light aircraft are commonly provided nose wheel steering capabilities through a simple system of mechanical linkage hooked to the rudder pedals.

Large aircraft may utilize a power source for nose wheel steering.

The geometry of the nose wheel makes it possible for it to shimmy, or oscillate back and forth, at certain speeds. To prevent this, almost all nose wheels are equipped with a shimmy damper. A shimmy damper controls vibration, or shimmy, through hydraulic damping. The damper is either attached to or built in to the nose gear. There are three types of shimmy dampers commonly used on aircraft: (1) the piston type, (2) vane type, and (3) the steer damper, which accomplishes the two separate functions of steering and/or eliminating shimmy.

The piston type damper is common on small aircraft. It consists of a hydraulic cylinder with a controlled bleed of fluid between the two sides of the piston. The restricted flow prevents rapid movement of the piston, but has no effect on normal steering.

The vane type damper utilizes two movable vanes in a fluid filled chamber. Turning the nose wheel in either direction causes the rotating vanes to move in the housing. The rotating vanes can only be moved as fast as fluid can be displaced from one chamber to the other. This means that the shimmy damper offers little resistance to slow motion, but offers high resistance to shimmy on takeoff or landing.

A steer damper consists of a closed cylinder containing rotary vane-type working chambers (similar to the vane-type shimmy damper) and a valving system. An automatic damping action occurs at the steer dampener if for any reason the flow of high-pressure fluid is removed from the inlet of the steer damper.

Brake Systems

Aircraft brakes are used for slowing, stopping, holding, and steering the aircraft. Their proper function is essential to safe operations.

Brakes are installed in each main landing wheel and they may be operated independently of each other. The right-hand landing wheel is controlled by applying toe pressure to the right rudder pedal and the left-hand wheel is controlled by the left rudder pedal.

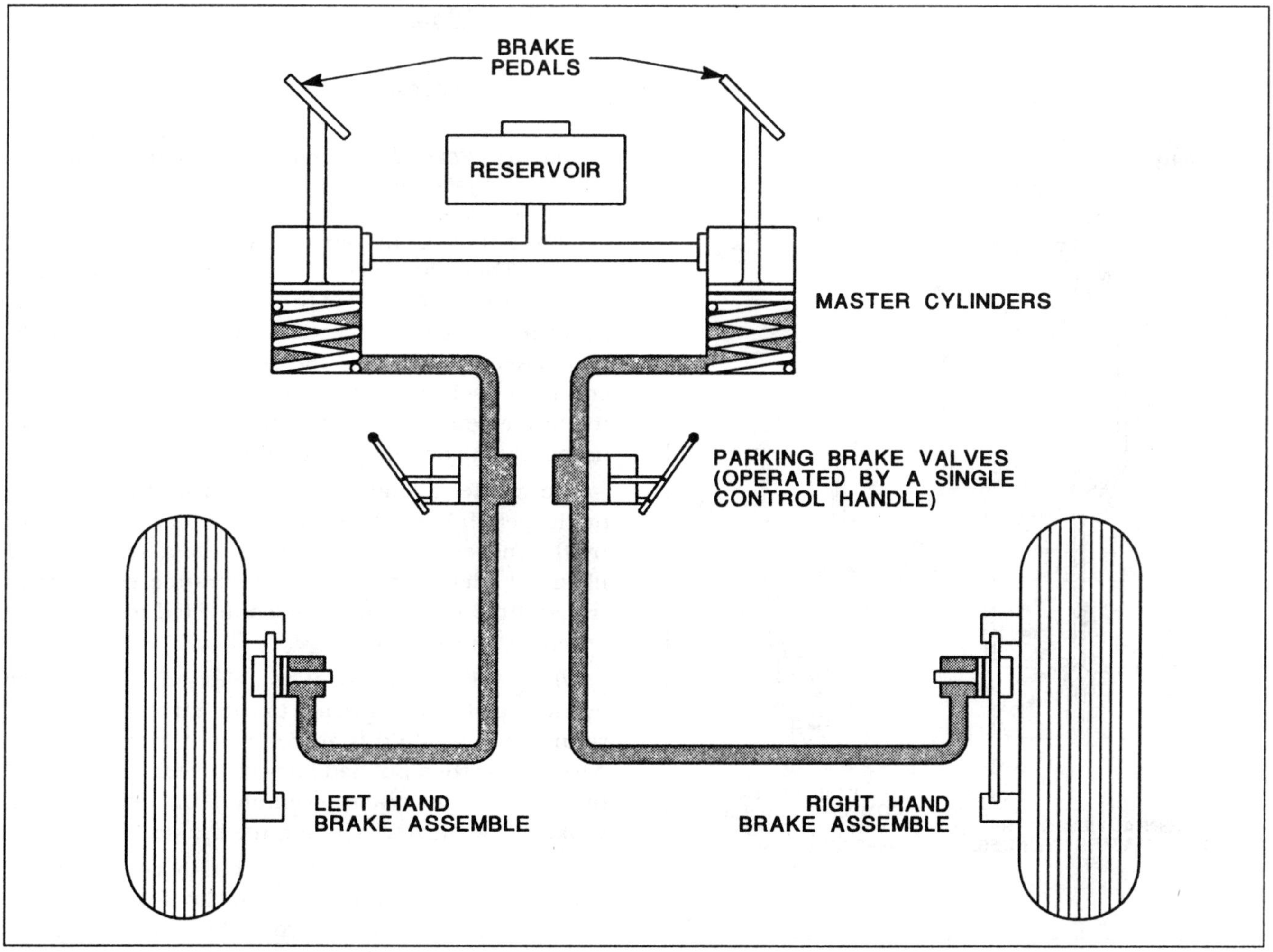

Figure 53. Typical independent brake system.

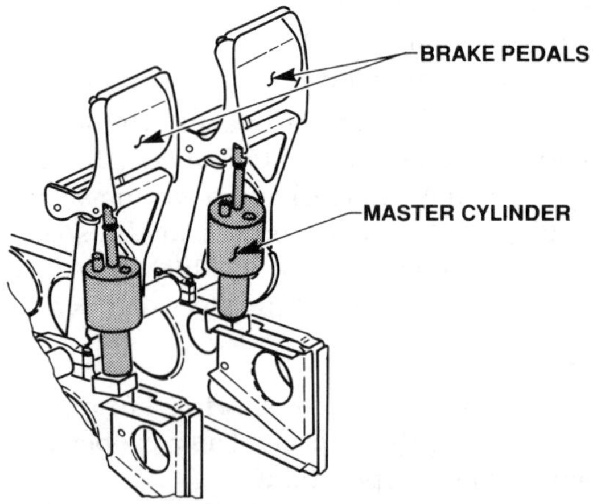

THE MASTER CYLINDERS MOUNT ON THE RUDDER PEDALS, AND PRESSURE ON THE TOP OF THE PEDALS DEPRESSES THE PISTONS IN THE MASTER CYLINDERS.
(A)

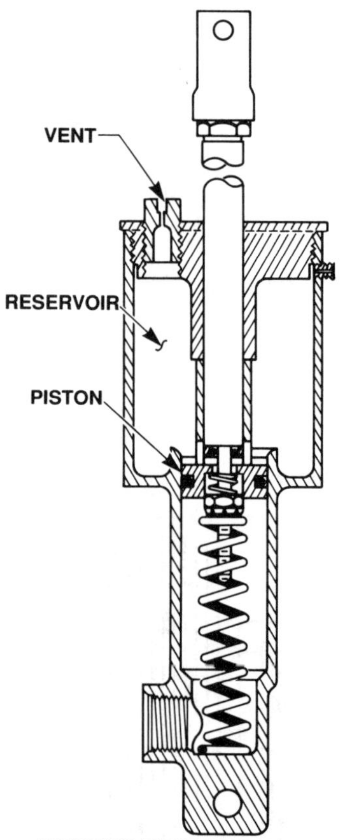

INTERNAL VIEW OF THE BRAKE MASTER CYLINDER.
(B)

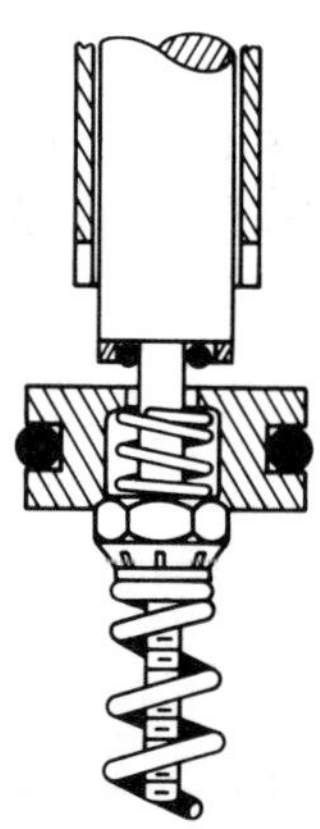
DETAIL OF THE COMPENSATOR PORT IN THE BRAKE MASTER CYLINDER.
(C)

Figure 54. Master cylinder for individual hydraulic brakes.

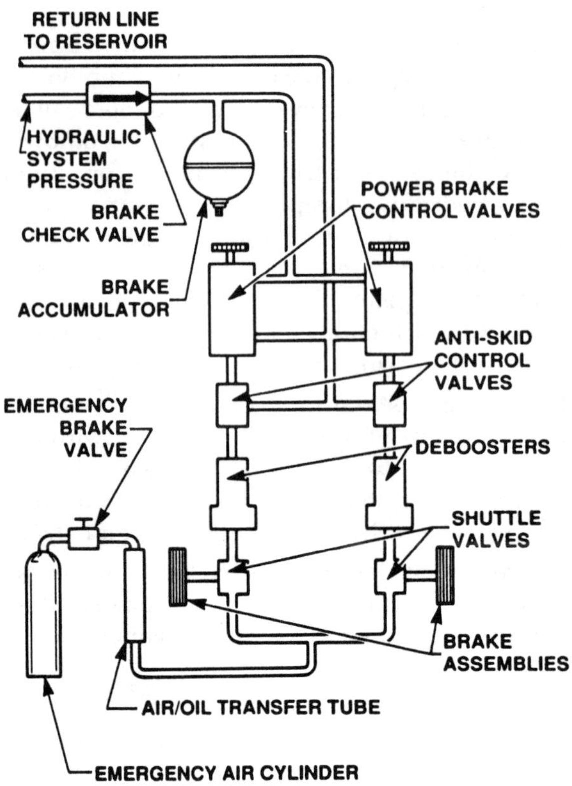

Figure 55. Power brake control system for a large jet aircraft.

Light aircraft generally use an independent brake system. This type of system has its own reservoir and is entirely independent of the aircraft's main hydraulic system. The system consists of a reservoir, one or two master cylinders, mechanical linkage which connects each master cylinder with its corresponding brake pedal, connecting fluid lines, and a brake assembly at each main landing gear wheel.

The master cylinder is actuated by toe pressure on the pedal. The master cylinder builds up pressure by the movement of a piston inside a sealed, fluid filled cylinder. The resulting hydraulic pressure is transmitted to the fluid line connected to the brake assembly in the wheel.

When the brake pedal is released, the master cylinder piston is returned to the "off" position by a return spring. Fluid that was moved into the brake assembly is then pushed back to the master cylinder by a piston in the brake assembly. If the return spring broke in the master cylinder, the brakes would drag.

The typical master cylinder has a compensating port, or valve, that permits fluid to flow from the brake chamber back to the reservoir when excessive pressure is developed in the brake line due to temperature change. The compensating port, or valve, in a brake

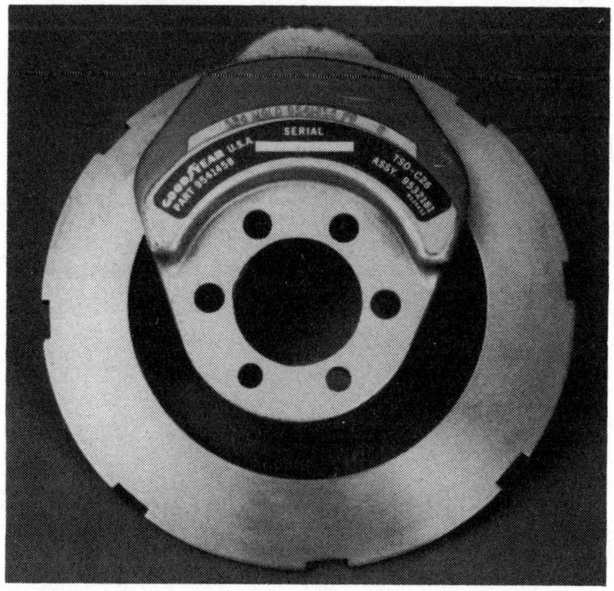

Figure 56. Single-disk brake used on a small general aviation aircraft.

Figure 57. Multiple-disk brake used in the wheel of a large jet-transport aircraft.

master cylinder of an independent brake system also permits reservoir fluid to fill the master cylinder.

It is very important that the compensating port remain clear and functional. A restricted compensating port of a master cylinder will cause slow release of brakes, and foreign particles stuck in the compensating port may cause brake dragging.

The seals on the piston in the master cylinder must be maintained in good condition. Internal leakage in a brake master cylinder unit can cause fading brakes. Aside from an external leak in the line, an internal leak in the master cylinder is the most likely cause for parking brakes creeping continually to the OFF position.

Aircraft brakes requiring a large volume of fluid to operate the brakes generally use power brake control valves. The pressure source for power brakes is the main hydraulic system.

Debooster valves are used in brake systems primarily to reduce the pressure and release the brakes rapidly. Deboosters also increase the volume of fluid flow. If a brake debooster is used in a hydraulic brake system, its position in the system will be between the brake control valve and the brake actuating cylinder.

Lockout deboosters generally allow full debooster piston travel without fluid from the high pressure side entering the low pressure chamber.

A shuttle valve is used in the brake actuating line to isolate the emergency brake system from the normal power brake control valve system. A relief valve is incorporated in the brake system to compensate for thermal expansion.

Brake Assemblies

Brake assemblies commonly used on aircraft are the single-disk, dual-disk, multiple-disk, semented rotor, or expander tube types. The single- and dual-disk types are most common on small aircraft; the multiple-disk type is normally used on medium-sized aircraft; and the segmented rotor and expander tube types are commonly found on heavier aircraft.

The braking action of single-disk brakes is accomplished by compressing a rotating brake disk between two opposite brake linings. Equal pressure on both sides of the disk is assured by keying the disk loosely in the wheel so it can move from side to side.

The lining material used in these brakes are made of an organic material which has a high coefficient of friction and good thermal qualities. If an airplane equipped with master cylinders and single-disk brakes has excessive brake pedal travel, but the brakes are hard and effective, the probable cause is worn brake linings.

Multiple-disk brakes are heavy-duty brakes, and may be used with power brake control valves or power boosted master cylinders. A series of small circular pistons force the pressure plate against the disk stack to apply the clamping force.

Automatic adjusters compensate for the disk wear each time the brake is applied, eliminating the need for manual adjustment. If an aircraft equipped with

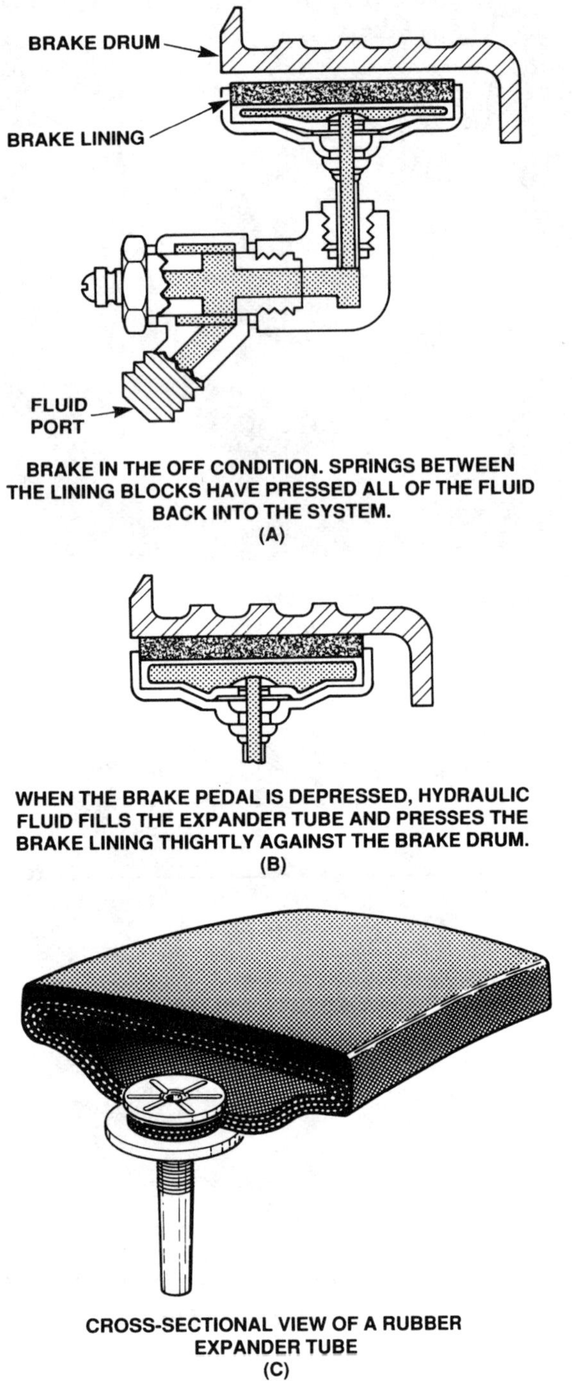

Figure 58. An expander tube brake.

multiple-disk brakes is reported to have sluggish and jerky brake operation, you should disassemble and clean the automatic adjuster.

With respect to an aircraft equipped with hydraulically operated multiple-disk-type brakes, do not set the parking brake when the brakes are hot.

The expander tube brake is a low-pressure brake with 360° of braking surface. It is light in weight, has few moving parts, and may be found on large aircraft as well as small. Many brake types can be adapted to operate mechanically or hydraulically. Expander-tube-type brakes are not adaptable to mechanical operation.

Expander tube brakes will probably not be encountered often in the maintenance shop today. There are, however, a few things to keep in mind when maintaining these units.

After replacing the blocks in an expander-tube type brake assembly, it will be necessary to ensure that the radial clearance is not less than the allowable minimum.

Pressure should not be applied to an expander tube brake without the wheel being installed to prevent damage to the brake assembly.

A pilot reports that the brakes are dragging on an aircraft equipped with expander-tube type brakes. On investigation, it is found that the aircraft has been in storage for the past 6 months. The expander tubes may have grown during the inoperative period.

Cracks in the surface of a brake drum may not be cause for rejection if the cracks are less than 1″ in length and do not extend to the edge of the drum.

Inspection and Maintenance of Brake Systems

Proper function of the brake system is essential to safe operation of the aircraft. Brake inspection must be conducted at frequent intervals, and needed maintenance must be performed promptly and carefully.

Look carefully to see that no fluid is leaking from the wheel cylinder onto the brake linings. This is a frequent cause of brake "grabbing".

Maintain the proper fluid level at all times to prevent brake failure or the introduction of air into the system. Spongy brakes are usually the result of air in the system. If air is present in the system, remove it by bleeding the system.

In brake service work, the term bleeding the brakes is the process of withdrawing fluid from the system for the purpose of removing air that has entered the system. There are two general methods of bleeding brake systems: bleeding from the top downward (gravity method) and bleeding from the bottom upward (pressure method). The method used generally depends on the type and design of the brake system.

In the gravity method, the air is expelled from the brake system through one of the bleeder valves provided on the brake assembly.

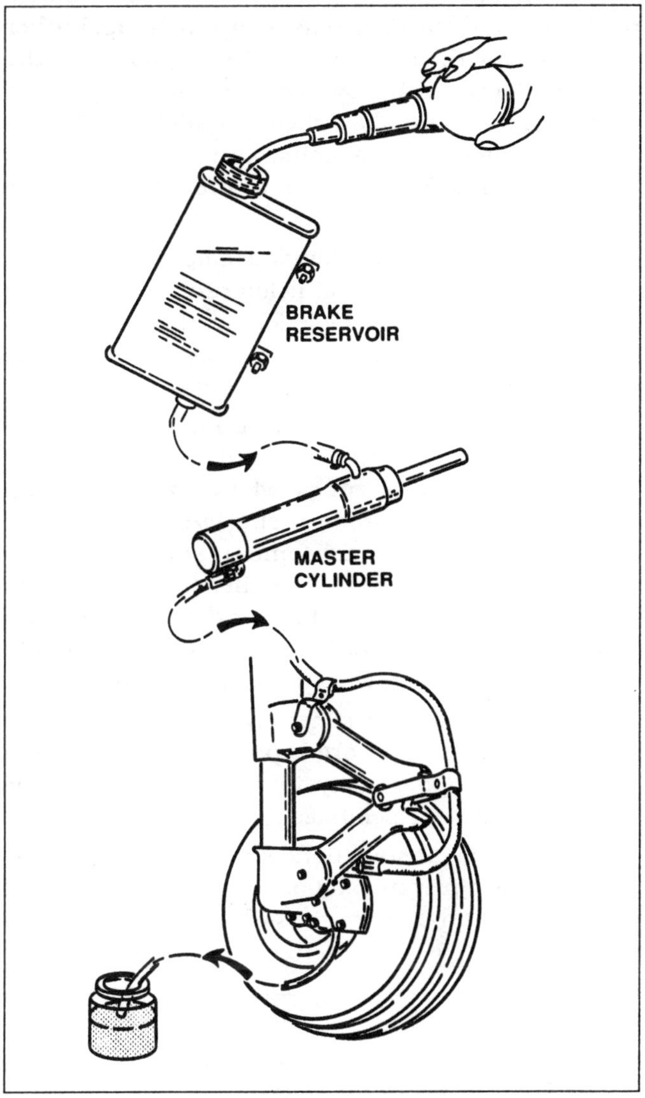

***Figure 59.** Master cylinder brakes may be bled from the top down by keeping the reservoir full of fluid and working the master cylinder to expel any fluid having air trapped in it.*

A bleeder hose is attached to the bleeder valve, and the free end of the hose is placed in a receptacle containing enough hydraulic fluid to cover the end of the hose. The air-laden fluid is then forced from the system by operating the brake. It can be determined that all air has been purged from the brake system by submerging the bleeder hose in hydraulic fluid and watching until all air bubbles have escaped. If all of the air is gone, the brakes will be firm, not spongy.

In the pressure method, the air is expelled through the brake system reservoir or other specially provided location. Perform this method of bleeding strictly in accordance with the specific manufacturer's instructions for the aircraft concerned.

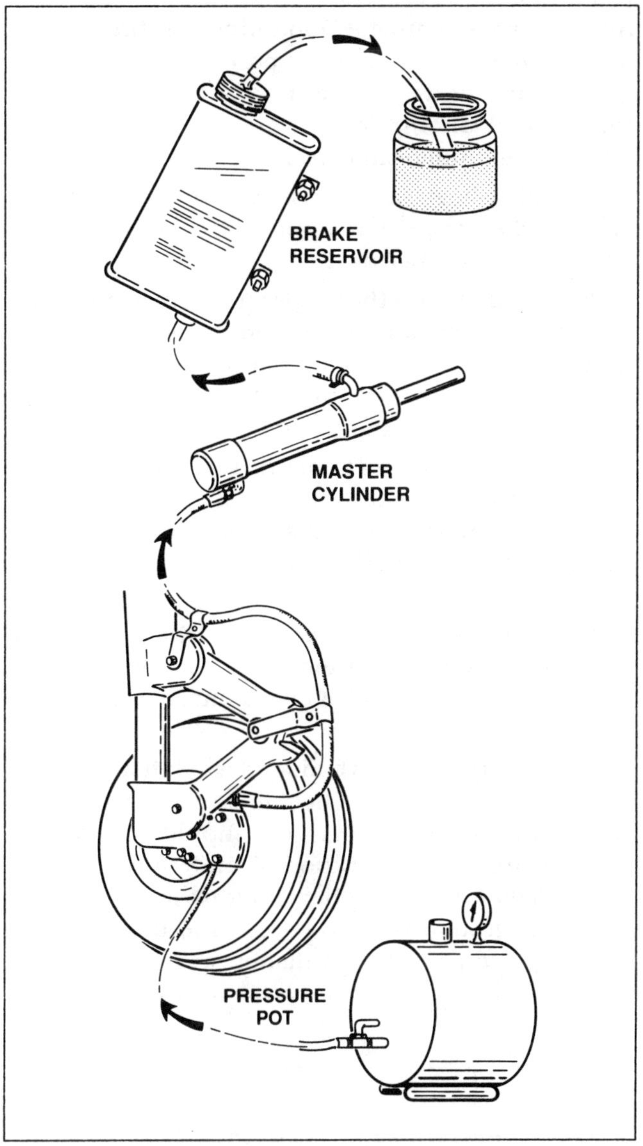

***Figure 60.** Master cylinder brakes may be bled of any air in the fluid by attaching a pressure pot to the bleeder valve on the brake and flowing fluid through the master cylinder and the resservoir until there are no bubbles in the fluid.*

One method of bleeding brakes which utilizes a power-brake control valve and a debooster is to charge the aircraft's hydraulic system and depress the brake pedal, with the bleeder valve at the brake open and the bleeder hose submerged in a fluid container, until all air has been purged from the system. An important consideration when bleeding by this method is to discontinue bleeding before the debooster reaches the end of its travel.

If it is determined that spongy brake action is not caused by air in the system, the next most likely cause would be deteriorated flexible hoses.

Aircraft Wheels and Tires

Aircraft wheels provide mounting for tires which absorb shock on landing, support the aircraft on the ground and assist in ground control during taxi, takeoff, and landing. Wheels are usually made from either aluminum or magnesium.

The split wheel is used on most aircraft today. This type of wheel may be manufactured to use either a tube-type or a tubeless tire.

Proper inflation is the single most important factor in obtaining maximum life from tires. The correct inflation pressure for an aircraft tire can be obtained from the operators manual. Ideally the pressure should be checked daily, and before each flight.

Wait at least 2 hours (3 hours in hot weather) after a flight before checking tire pressure. If there is a difference of more than 5 lbs. of air pressure in tires mounted as duals, correct the discrepancy and enter in the logbook.

Excessive wear in the center of the tread of an aircraft tire is an indication of overinflation. Overinflated aircraft tires may cause damage to the wheel flange.

While overinflation is bad, underinflation is worse, as it causes an excess of heat to be generated within the tire. Excessive wear in the shoulder area of an aircraft tire is an indication of underinflation. The best safeguards against heat buildup in aircraft tires are short ground rolls, slow taxi speeds, minimum braking, and proper tire inflation.

Some aircraft wheels are equipped with fusible plugs to deflate the tire in case of overheating. Rather than allowing the heat to build to the point that the tire will blow out, the low-melting point alloy in the plugs will melt and allow the tire to deflate. If a tire has been subjected to a temperature high enough to melt one of the fusible plugs in the wheel, the tire should be scrapped.

Nylon aircraft tires will develop temporary flat spots under static load. To avoid nylon tire spotting on an aircraft that is to remain idle for a period longer than 3 days, the aircraft should be blocked up so that no weight is on the tires, or moved every 48 hours.

To prevent rapid deterioration, aircraft tires should be stored in an area that is dry and cool. Never store tires or tubes where any petroleum product can contaminate them. Ozone and ozone producing electrical equipment should also be avoided in storage areas. Whether in storage or in service, if hydraulic fluid, fuel, or other solvents are spilled on tires they should be wiped with a dry cloth, followed by a washdown with soap and water.

Tire replacement is generally on an as-required basis. Whenever tires are replaced, the technician also has the opportunity to carefully inspect the wheel and brake assemblies.

The removal, installation, and repair of landing gear tires by a private pilot on an aircraft owned and operated by him/her are considered to be preventive maintenance.

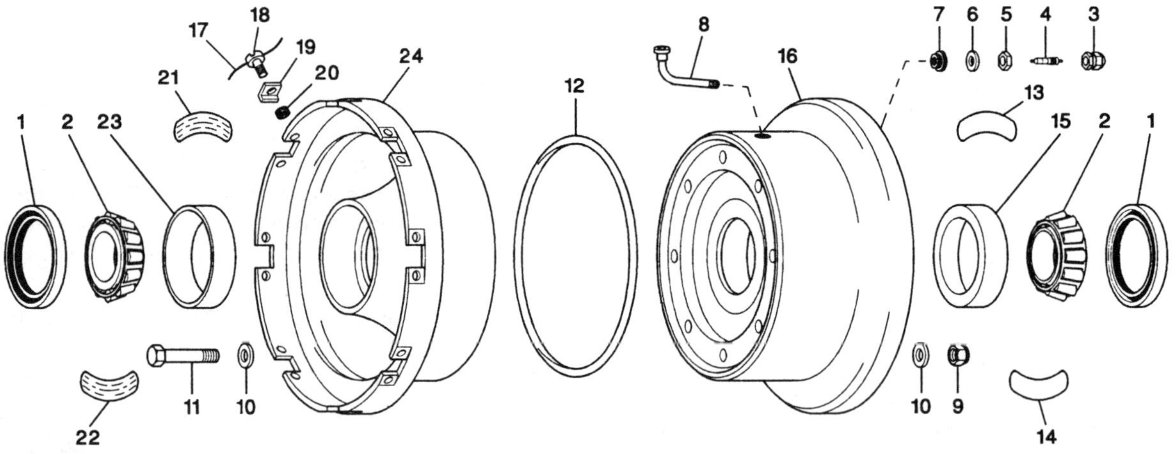

1–SEAL ASSEMBLY
2–CONE, BEARING
3–CAP, VALVE
4–CORE, VALVE
5–NUT
6–SPACER
7–GROMMET
8–STEM, VALVE
9–NUT
10–WASHER
11–BOLT
12–PACKING
13–PLATE, IDENTIFICATION
14–PLATE, INSTRUCTION
15–CUP, BEARING
16–WHEEL-HALF, OUTER
17–WIRE, LOCK
18–SCREW
19–KEY, TORQUE
20–INSERT, HELI-COIL
21–PLATE, IDENTIFICATION
22–PLATE, INSTRUCTION
23–CUP, BEARING
24–WHEEL-HALF, INNER

Figure 61. Split wheel-light aircraft.

Figure 62. Typical wheel balancing stand for aircraft wheels.

Safety in handing wheels which have tires mounted on them is very important. Tire and wheel manufacturers often recommend that the tires be deflated before removing the wheel from the axle. This is a safety precaution in case the bolts that hold the wheel halves together have been damaged or weakened.

Use caution removing the tire, a bead breaker is recommended to prevent damage to the bead seat area. Inspect and repair the wheels and brake assemblies as necessary. Install the new tire (and tube, if required) on the wheel following instructions outlined by the manufacturer.

Even on aircraft, the option of new or recapped tires is available. A high-speed, type VII or type VIII aircraft tire, with a sound cord and bead may be recapped an indefinite number of times.

Most aircraft tire manufacturers recommend that the tubes in newly installed tires be first inflated, fully deflated, and then reinflated to the correct pressure. This will allow the tube to position itself correctly inside the tire.

The wheel and tire assembly should be balanced prior to reinstallation on the aircraft. Even light aircraft will benefit from properly balanced wheel assemblies. The primary purpose for balancing aircraft wheel assemblies is to prevent heavy spots and reduce vibration.

When tire installation is complete, a stripe or mark extending from the rim of a wheel onto the tire may be painted on to indicate slippage. A newly mounted tire and/or tube should be checked for proper inflation for several days.

REVIEW QUESTIONS — Landing Gear Systems

81. *What should be done if an excessive quantity of hydraulic fluid is spilled on a main wheel tire?*
82. *Spongy brakes are usually caused by what?*
83. *Why do we bleed brakes?*
84. *What is the likely result of a wheel cylinder leaking fluid on the brake lining?*
85. *What type of brake system cannot be adapted to mechanical operation?*
86. *For what purpose is a stripe placed extending from the rim of a wheel onto the tire?*
87. *What type of damage may be caused by overinflated tires?*
88. *What is the primary purpose of balancing aircraft wheel assemblies?*
89. *What determines the type of hydraulic fluid used in a shock strut?*
90. *What may be caused by internal leakage in a brake master cylinder?*
91. *What is the pressure source for power brakes?*
92. *What does a shuttle valve do?*
93. *What is the purpose of a relief valve in a brake system?*
94. *If an oleo strut bottoms upon initial landing contact, but functions correctly during taxi, what is the likely cause?*
95. *Why are debooster cylinders used in power brake systems?*
96. *How many times may a high-speed aircraft tire with a sound cord, body, and bead be recapped?*
97. *What should be done with a tire removed after being subject to temperatures high enough to melt the fusible plugs?*

98. *How long should you wait after a flight before checking tire pressure?*

99. *Excessive wear in the center of the tread of an aircraft tire is an indication of what?*

100. *What type of seals are used in shock struts to prevent oil from escaping?*

Answer questions 8305 through 8373 in the FAA Test section of this book.

HYDRAULIC AND PNEUMATIC SYSTEMS

Hydraulic systems have many advantages as a power source for operating various aircraft units. They combine the advantages of light weight, ease of installation, simplification of inspections, and minimum maintenance requirements. Hydraulic operations are almost 100% efficient, with only a negligible loss due to fluid friction.

All hydraulic systems are essentially the same, regardless of their function. Each hydraulic system has a minimum number of components, and some type of hydraulic fluid. To assure proper system operation and to avoid damage to nonmetallic components of the hydraulic system, the correct fluid must be used.

All fluids transmitt pressure, but certain properties are required for selection and use in a hydraulic system. Desirable properties of a good hydraulic fluid include low viscosity, chemical stability, high flash point, and high fire point.

One of the most important properties of any hydraulic fluid is its viscosity. Viscosity is internal resistance to flow, and will increase as temperatures decrease. A satisfactory liquid for a given hydraulic system must have enough body to give a good seal; but it must not be so thick that it offers resistance to flow, leading to power loss and higher operating temperatures. Some systems with high pressures and/or high flow rates require a heat exchanger, or hydraulic fluid cooler, to maintain a safe working temperature.

Chemical stability is another property considered when selecting a hydraulic fluid. It is the liquid's ability to resist oxidation and deterioration for long periods of time.

Flash point is the temperature at which a liquid gives off vapor in sufficient quantity to ignite momentarily, or "flash" when a flame is applied. A high flash point is desirable for hydraulic fluids.

Fire point is the temperature at which vapor is given off in sufficient quantity to ignite and continue to burn when exposed to a flame. Like flash point, a high fire point is required of desirable hydraulic fluids.

When servicing aircraft hydraulic systems, use the type fluid specified in the aircraft manufacturer's maintenance manual or on the instruction plate affixed to the reservoir or unit. Hydraulic fluids are dyed a specific color for each type of fluid to aid in identification.

Types of Hydraulic Fluids

Three types of hydraulic fluids currently being used in civil aircraft are vegetable-base, mineral-base, and phosphate ester-base.

Vegetable-base hydraulic fluid (MIL-H-7644) is composed essentially of castor oil and alcohol. It has a pungent alcoholic odor and is generally dyed blue. Natural rubber seals are used with vegetable base hydraulic fluid. This type fluid is flammable. Alcohol may be used to flush systems normally serviced with vegetable base hydraulic fluid.

Mineral-base hydraulic fluid (MIL-H-5606) is processed from petroleum. It has an odor similar to penetrating oil and is dyed red. Synthetic rubber seals are used with petroleum-base fluids. An O-ring intended for use in a system using MIL-H-5606 fluid will be marked with a blue stripe or dot. Additives have improved the properties and characteristics of petroleum-base hydraulic fluid. Do not mix with vegetable-base or phosphate ester-base hydraulic fluids. This type of fluid is flammable under normal conditions. Naphtha or varsol may be used to flush a system normally serviced with mineral base hydraulic fluid.

Phosphate ester-base hydraulic fluids were developed to provide a fire-resistant hydraulic fluid for use in high performance aircraft. This type of synthetic hydraulic fluid has a high flash point. MIL-H-8466 (Skydrol® 500 A & B) hydraulic fluid is a light purple color and requires the use of butyl rubber seals.

Skydrol is not without its problems for the maintenance technician. It is quite susceptible to contamination by water from the atmosphere and must be kept tightly sealed. It will attack polyvinyl chloride, and must not be allowed to drip onto electrical wiring as it will break down the insulation.

When servicing a system using phosphate ester fluid, you must be careful to use only seals and hoses compatible with this fluid. Information about the compatibility of fire-resistant hydraulic fluid with aircraft materials can be obtained from the manufacturer's technical bulletins. Materials which are Skydrol compatible or resistant include most common aircraft metals and polyurethane and epoxy paints. Skydrol hydraulic fluid is compatible with natural fibers and also with nylon and polyester. A system normally serviced with Skydrol can be flushed with Trichlorethylene.

Generally speaking, if an aircraft hydraulic system is serviced with the wrong type of fluid, the remedial procedure should be to drain, flush, and usually change all affected seals. The system can then be serviced with the proper fluid.

A filter is a screening or straining device used to clean the hydraulic fluid, thus preventing foreign particles and contaminating substances from remaining in the system.

A filter incorporating specially treated cellulose paper is identified as a micronic filter. These filters are designed to prevent the passage of solids greater than 10 microns (0.000394") in size.

Hydraulic fluid filtering elements constructed of porous paper are normally discarded at regular intervals and replaced with new filtering elements, rather than being cleaned.

Basic Hydraulic System

Regardless of its function and design, every hydraulic system has a minimum number of basic components. A hydraulic system referred to as a "power pack" system will have all hydraulic power components located in one unit.

The first of the basic components is the reservoir. This unit stores the supply of hydraulic fluid for operation of the system. It replenishes the system fluid when needed, and provides room for thermal expansion.

A pump is necessary to create a flow of fluid. Most aircraft systems utilize an engine-driven or electric motor-driven pumps. A hand-pump may be provided for emergency operation.

The selector valve is used to direct the flow of fluid. The valves may be operated manually or by solenoid. This valve may direct pressurized fluid to one end of an actuating cylinder and simultaneously direct return flow from the other end.

An actuating cylinder converts fluid pressure into linear or reciprocating motion. The primary purpose of a hydraulic actuator is to transform fluid pressure into mechanical force. An hydraulic motor may be used when rotary mechanical motion is required.

Reservoirs

Aircraft that fly at the lower altitudes normally have hydraulic systems supplied with fluid from an unpressurized reservoir. The reservoir must be large enough to hold all of the fluid in any condition of the actuating cylinders. A means of determining the fluid level in the reservoir is usually provided.

Reservoirs are sometimes designed with a standpipe in one of the outlet ports in order to assure an emergency supply of fluid. The outlet port with the standpipe furnishes fluid to the normal system power pump. The quantity of fluid which is not available to the main system pump is retained to supply fluid to the auxiliary pump.

If fluid is added to a nonpressurized reservoir in a constant-pressure hydraulic system while the system is pressurized, the fluid level will increase when the system pressure is reduced. When the system pressure is reduced, the fluid which it takes to pressurize the system will be returned to the reservoir, possibly overfilling it; therefore reservoirs should not be overfilled.

Aircraft that operate at altitudes where there is not enough air pressure to assure a positive feed of fluid to the hydraulic pump will use a pressurized reservoir. The main purpose of a pressurized reservoir in a hydraulic system is to prevent pump cavitation.

In a hydraulic system that has a reservoir pressurized with turbine-engine compressor bleed air, an air pressure regulator will be used to reduce the air pressure between the engine and the reservoir.

Before removing the filler cap of a pressurized hydraulic reservoir, relieve the air pressure.

Hydraulic Pumps

Hydraulic power is transmitted by the movement of fluid by a pump. The pump does not create the pressure, but the pressure is produced when the flow of fluid is restricted. A variety of types of pumps are used in aircraft hydraulic systems.

There are two basic types of powered pumps: those having a constant displacement, and those whose displacement is variable. A hydraulic pump is a constant-displacement type if it is able to deliver a uniform rate of fluid flow.

Constant-displacement pumps move a specific volume of fluid each time its shaft turns. This type of pump must have some form of regulator or relief valve to relieve the pressure that will build up when the pump moves more fluid than the system can use.

The gear pump is the simplest and most common type of constant-displacement pumps. In a gear-type hydraulic pump, a mechanical safety device incorporated to protect the pump from overload is the shear pin in the pump drive.

Other types of constant-displacement pumps include the gerotor, the fixed-angle piston type, and the vane type.

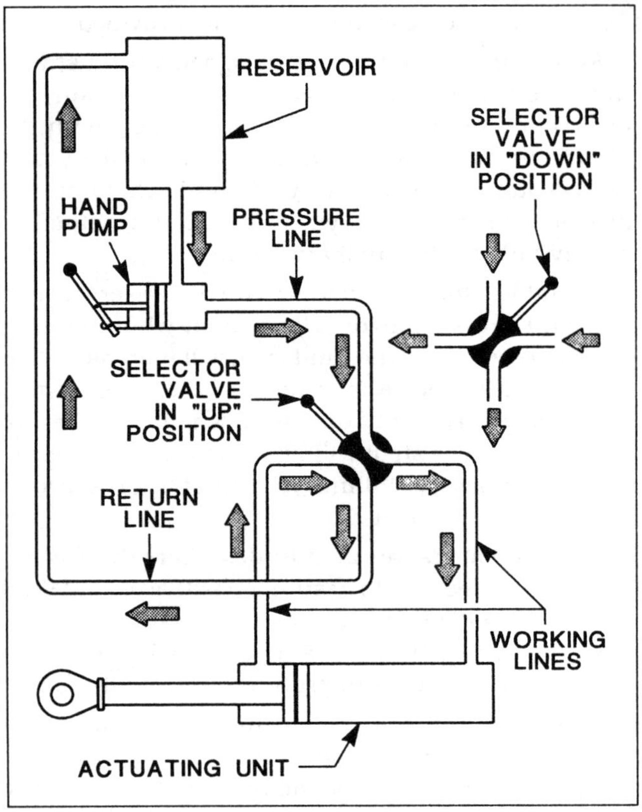

Figure 63. Basic hydraulic system with hand pump.

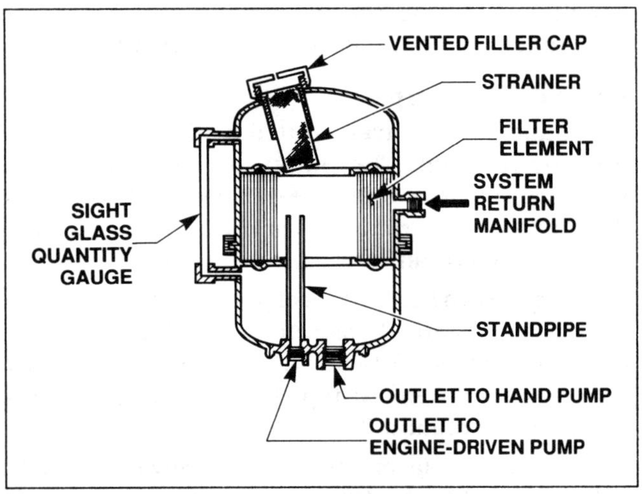

Figure 64. Nonpressurized hydraulic reservoir.

Most variable displacement hydraulic pumps of current design contain a built-in means of system pressure regulation. The same force that is used to control the pressure regulator in a system equipped with a constant-displacement pump may be used to regulate pump output. A piston-type pump may be equipped so that the regulating force will vary the effective length of the piston stroke and vary the output of the pump.

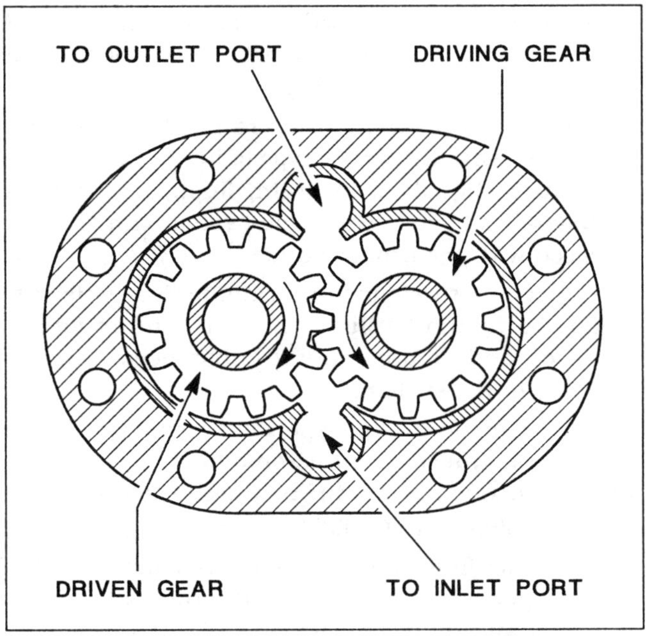

Figure 65. Gear-type power pump.

Hydraulic pumps may be provided with a shear section on the drive shaft which will allow the shaft to break if the pump should seize. This condition could occur if the hydraulic system pressure control and relief units fail to function properly. The over-pressurization of the system would cause the pump to be difficult to turn, and the shaft would shear.

Pressure Regulation

Hydraulic pressure must be regulated in order to use it to perform the desired tasks. Pressure regulating systems will always use three elemental devices; a pressure relief valve, a pressure regulator, and a pressure gage.

A pressure relief valve is used to limit the amount of pressure being exerted on a confined liquid. The pressure relief valve is, in effect, a system safety valve. The thermal relief valve installed in a hydraulic system will have the highest pressure setting.

The term "pressure regulator" is applied to a device used in hydraulic systems that are pressurized by constant-delivery type pumps. The spool-type hydraulic system pressure regulator is automatic in operation and normally requires no attention by pilot or flightcrew. The purpose of the pressure regulator is to maintain system operating pressure within a predetermined range and to unload the pump. The pressure regulator will allow circulation of the fluid when there are no demands on the system.

If the pressure regulator in a system equipped with a constant-displacement hydraulic pump diverts fluid from the system to the reservoir, the output pressure reduces but the volume remains the same.

The pressure gage is used to measure the pressure in the hydraulic system. This is normally a bourdon-tube-type gage and is calibrated to match the system it accommodates.

Excessive fluctuation of the pressure gage when the hydraulic pump is operating would most likely be caused by an inadequate supply of fluid.

Accumulators

We have learned before that liquids are incompressible, the accumulator uses an air charge to gain the effect of compressibility and provide an auxiliary source of hydraulic pressure. A fully charged accumulator provides a source for additional hydraulic power when heavy demands are placed on the system. The accumulator also provides a cushion against "hammering" during the system operation.

All accumulators consist of a strong container divided by some sort of partition. One side is open to the hydraulic system, and the other contains a charge of compressed air. Air is prevented from entering the fluid system in an accumulator by physically separating the air chamber from the oil chamber with a flexible or movable separator. There are basically three types of accumulators used: the piston type, the bladder type, and the diaphragm type.

After a hydraulic accumulator has been installed and the air chamber charged, the main system hydraulic pressure gage will not show a hydraulic pressure reading until the fluid side of the accumulator has been charged.

A hydraulic accumulator is charged with an air preload of 1,000 PSI. When a hydraulic system pressure of 3,000 PSI is developed, the pressure on the air side of the accumulator will also be 3,000 PSI.

To check the air charge in an accumulator, reduce all hydraulic pressure, then observe the reading on the accumulator air gage. The air charge may also be checked when the engine is not operating, but the system still has hydraulic pressure, by operating some hydraulic unit slowly and noting the pressure at which a rapid pressure drop begins as it goes toward zero. That value will reflect the air charge in the accumulator.

Accumulator maintenance is not difficult, but because of the high pressure air charge, caution must be observed. The first step in removing an accumulator from an aircraft is to relieve system pressure. Before disassembling any accumulator, make sure that all pre-load air (or nitrogen) pressure has been discharged. When discharging this pressure, make certain that you know what type of high pressure air valve is used, and that you use the correct tool for releasing the pressure. Do not release the pressure with your fingernail as you might do to a bicycle tire.

Fluid Control Valves

For hydraulic systems to operate as intended, the flow of fluid must be rigidly controlled. Many kinds of valve units are used in exercising this control. One of the simplest and most commonly used is the check valve. This valve allows free flow of fluid in one direction, but no flow or a restricted flow in the opposite direction.

The simplest type is known as the in-line check valve (often called check valve). A check valve allows free fluid flow in one direction and no fluid flow in the other direction.

The orifice-type check valve is used to allow normal operating speed of a mechanism in one direction, and limiting the operating speed in the opposite direction. An orifice check valve permits fluid to flow freely in one direction, but restricts the rate at which fluid is allowed to flow in the other direction.

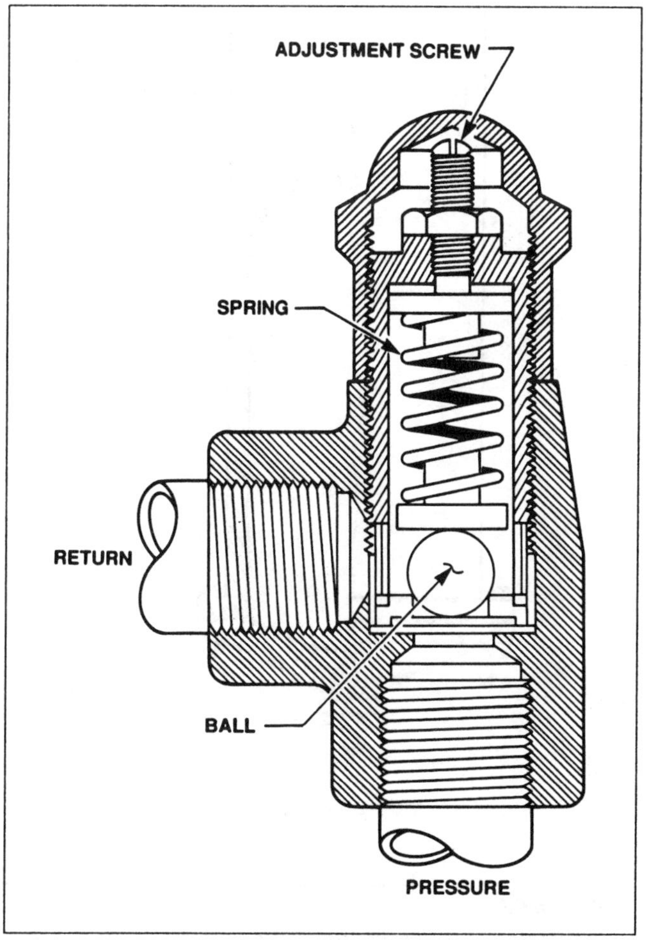

Figure 66. Simple ball-type pressure relief valve.

Line-disconnect or quick-disconnect valves are installed in hydraulic lines to prevent the loss of fluid when units are removed. These valves may be used to connect a hydraulic test stand to the system. Such valves are installed in the pressure and suction lines of the system just in front of and immediately behind the power pump.

A flap overload valve may be installed to prevent the flaps from being lowered at airspeeds which would impose excessive structural loads.

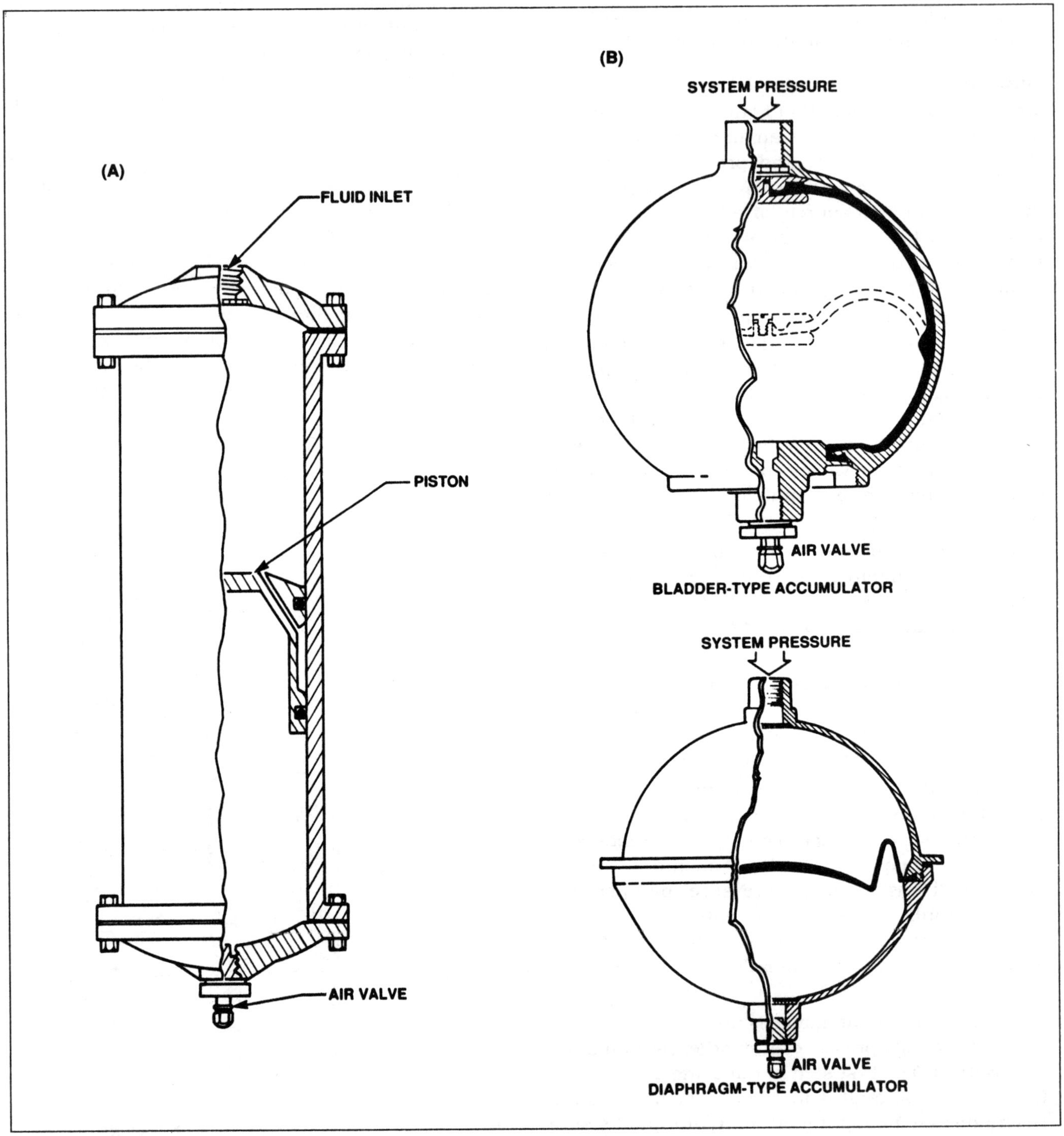

Figure 67.
(A) Piston-type accumulator.
(B) Spherical accumulators.

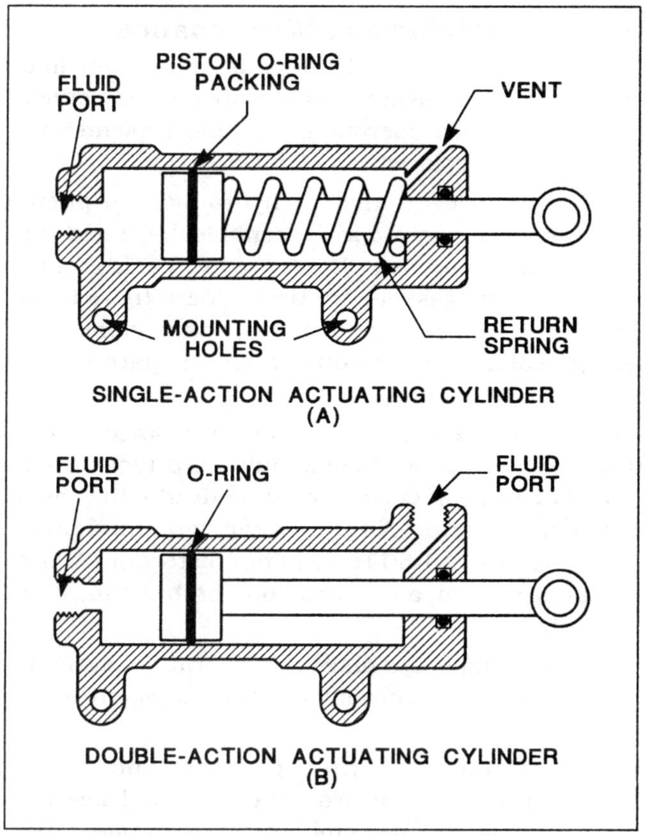

Figure 68. Double-action actuating cylinder.

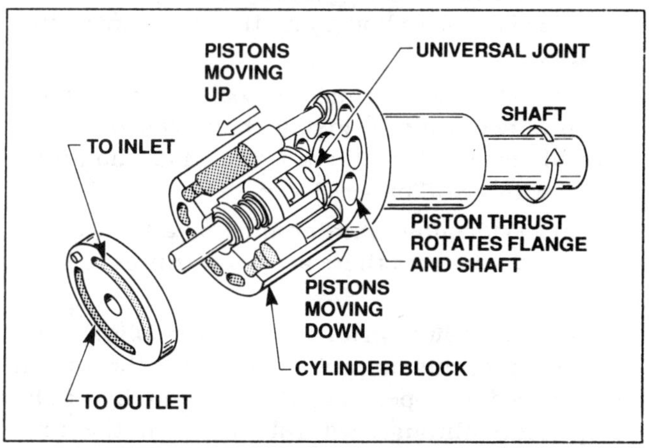

Figure 69. Piston-type rotary hydraulic motor.

A sequence valve will cause one hydraulic operation to follow another in a definite order. These valves are common in landing gear systems to sequence the opening and closing of the landing gear doors.

Some hydraulic systems incorporate a device which is designed to remain open to allow a normal fluid flow in the line, but close if the fluid flow increases above and established rate. This device is generally referred to as a hydraulic fuse. This unit will serve to block a line if a serious leak should occur.

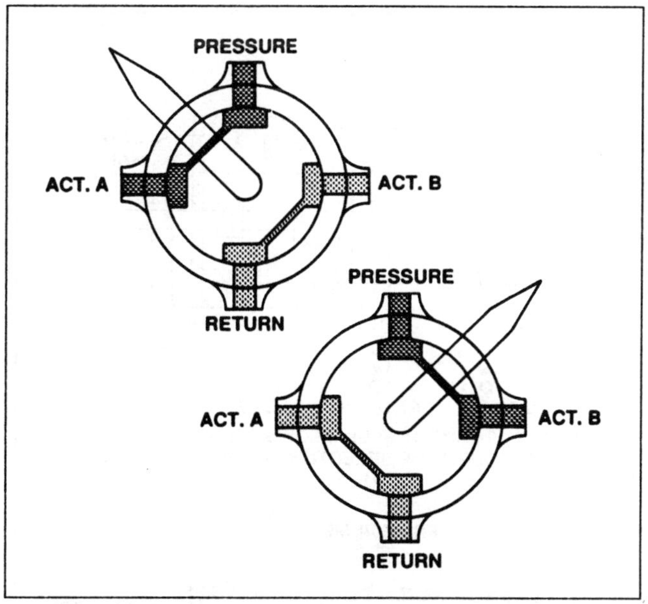

Figure 70. Closed center, plug-type, 4-part selector valve.

Hydraulic Actuators

An actuating cylinder is used to impart linear motion to some movable object or mechanism. Actuating cylinders are of two major types: (1) single-action, and (2) double-action. The single-action (single port) actuating cylinder is capable of producing powered movement in one direction only. The double-action (two port) actuating cylinder is capable of producing powered movement in two directions.

Hydraulic motors may be used to impart rotary motion, and may be substituted for electrical motors in some systems. One advantage of piston-type hydraulic motors over electric motors is that there is no fire hazard if the motor is stalled.

Selector Valves

Selector valves are used to control the direction of movement of an actuating unit. A selector valve provides a pathway for the simultaneous flow of hydraulic fluid into and out of a connected actuating unit.

Selector valves have various numbers of ports. The number of ports is determined by the particular requirement of the system. The four-port, closed center valve is one most commonly used to provide for simultaneous flow of fluid into and out of a connected actuating unit. The term four-way is often used instead of four-port in referring to selector valves.

The open center valve directs fluid through the center of the valve back to the reservoir when a unit is not being actuated. One of the distinguishing characteristics of an open-center selector valve used

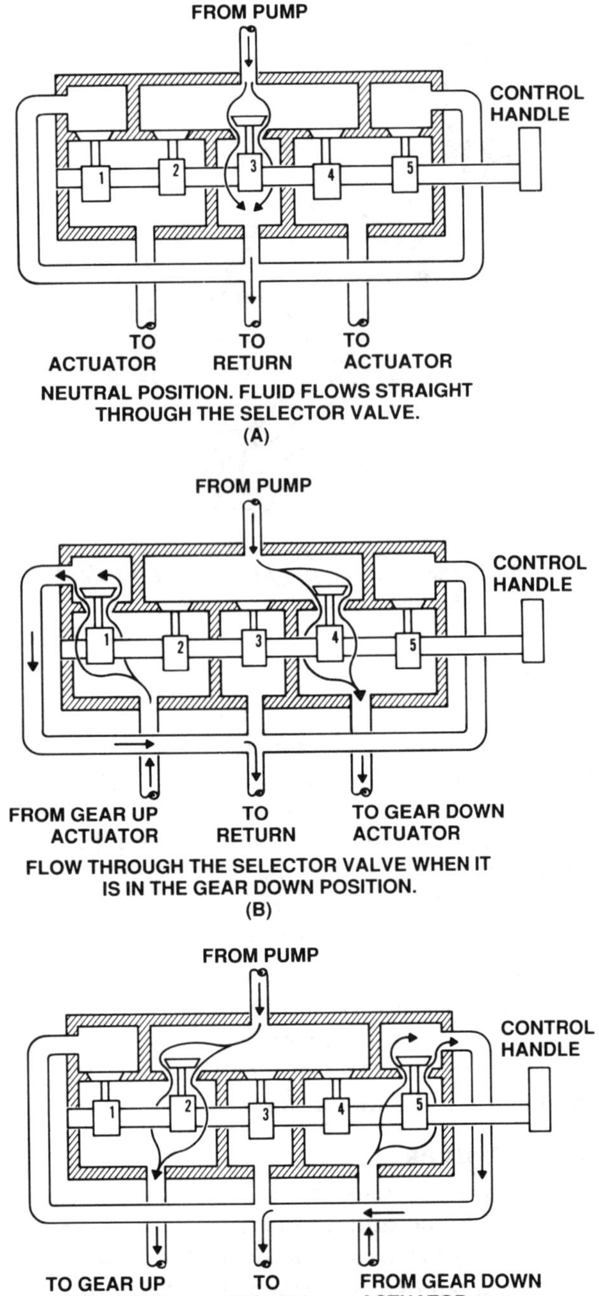

Figure 71. Open center poppet-type selector valve.

in a hydraulic system is that the fluid flows through the valve in the OFF position.

A crossflow valve which is designed to bypass fluid from one side of an actuating cylinder to the other side, under certain conditions, may be found in some aircraft installed in the landing gear system.

System Servicing and Maintenance

Hydraulic systems require regular inspection and maintenance to assure reliable operation. Service and maintenance operations include replenishing the fluid, and replacement of seals.

Seals are used throughout hydraulic and pneumatic systems to minimize internal leakage and the loss of system pressure. There are two basic groups of seals in use: gaskets are used where there is no relative motion between the parts, and packings are flexible sealing elements used when the part is subject to relative motion.

To prevent external and internal leakage in hydraulic units, the most commonly used type of seal is the O-ring seal. O-rings may be used either as a gasket or a packing. An O-ring can withstand pressures up to about 1,500 PSI without distortion. Above this pressure an anti-extrusion, or backup, ring should be used.

Back-up rings are used in hydraulic systems above 1,500 PSI to prevent the high pressure from extruding the seal.

Both chevron, or V-ring, packings and D-ring packings get their name from their shape. These are one-way seals, and will stop fluid flow in one direction only. When installing a chevron-type seal, the open side of the seal should face the direction of fluid pressure.

To protect packing rings or seals from damage when it is necessary to install them over or inside threaded sections, the threaded section should be covered with a stiff paper sleeve.

When performing maintenance or adjustments to hydraulic systems there are a few things to keep in mind:

- The removal of air from an aircraft hydraulic system generally presents no problem because the air will be removed by operating the various hydraulic components through several cycles. If the time required to operate the mechanism decreases with successive operations, it indicates the air is being worked out of the system.
- The open ends of hydraulic lines may be protected from contamination by sealing the lines with caps or plugs.
- Before removing a spool-type or a balanced-type pressure regulator, actuate the flap selector valve until no pressure remains in the system.
- If it is necessary to adjust several pressure regulating valves, the units with the highest pressure settings are adjusted first.

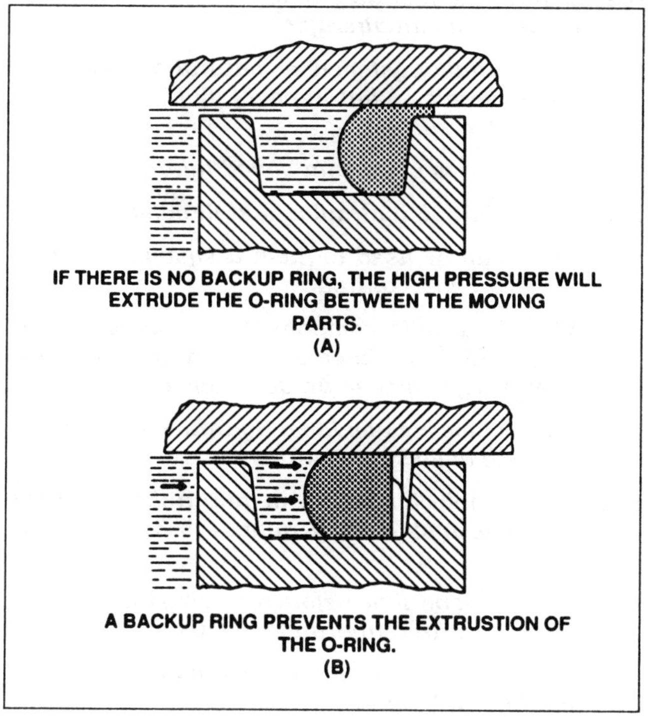

Figure 72. An O-ring must be backed up in a high-pressure system.

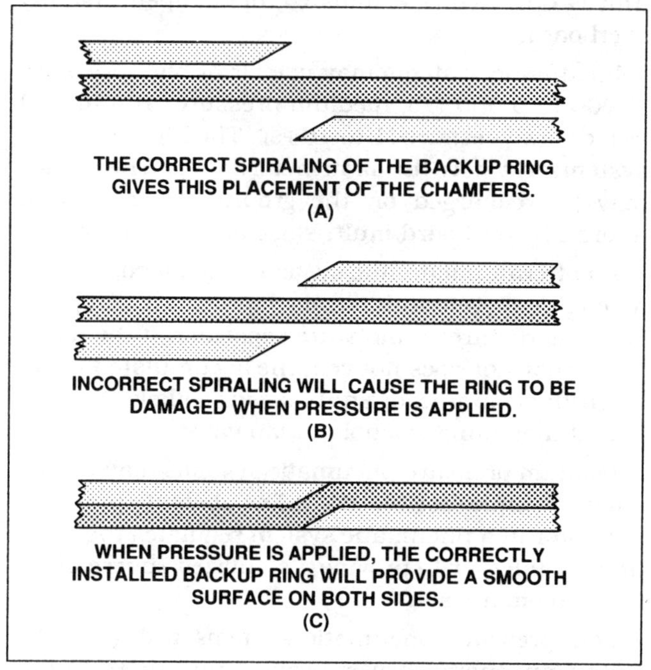

Figure 73. Proper spiraling of a Teflon® backup.

- The main system pressure relief valve in a system equipped with a power control valve should be adjusted with the power control valve held in the CLOSED position.
- Before adjusting the relief valve of a main hydraulic system incorporating a pressure regulator, eliminate the action of the unloading valve.
- If an aircraft's constant-pressure hydraulic system cycles more frequently than normal and no fluid leakage can be detected, the most probable cause is a low accumulator air preload.
- After installation of a rebuilt hydraulic hand pump, if the handle cannot be moved in the pumping direction (pressure stroke), the most likely cause is an incorrectly installed outport check valve. Severe kickback of the handle during the normal intake stroke will indicate the outport check valve is sticking open.
- If an operational check during ground runup indicates that the flaps cannot be lowered using the main hydraulic system, but can be lowered using the emergency hand pump, the probable cause is a low fluid level in the reservoir. The fluid level is below the standpipe.
- A worn hydraulic pump shaft seal can normally be detected by hydraulic fluid flowing from the pump drain line.
- If an engine-driven hydraulic pump fails to maintain normal system pressure during the operation of a low-demand system such as cowl flaps, the probable cause is a restriction in the pump outlet.
- Chattering of the hydraulic pump during engine runup is an indication of an air leak in the pump inlet line.
- If while the hydraulic pump is running, the pressure is normal, but when the pump stops no hydraulic pressure is available, it indicates a leaking accumulator air valve.
- If hydraulic fluid is released when the air valve core of the accumulator is depressed, it is evidence of a ruptured diaphragm.
- A common cause of slow actuation of hydraulic components is internal leakage in the actuating unit.
- Hydraulic hoses and tubing must be installed as outlined in AC 43.13-1A. Never pull a hose or tube into alignment by tightening the flare nut. The flare will be damaged and excess pressure will be placed on the whole system.

Pneumatic Systems

Some aircraft manufacturers equip their aircraft with a pneumatic system. These systems operate a great deal like a hydraulic system, except they employ air instead of a liquid for transmitting power. One advantage claimed for pneumatic systems is that no return lines are required. The air that is expended and no longer needed when an actuating unit is operated is exhausted, or dumped, usually overboard.

Pneumatic systems may use air at high pressure (1,000 to 3,000 PSI), medium pressure (100 to 150 PSI), or low pressure (1 to 10 PSI). The high pressure systems will store the air in metal bottles. The bottles may be recharged on the ground, or in certain aircraft by on-board multi-stage air compressors.

An aircraft pneumatic system which incorporates an engine-driven multistage reciprocating compressor, also requires a moisture separator. If the moisture separator does not vent the accumulated water when the compressor shuts down, a likely cause is a malfunctioning solenoid dump valve.

Medium pressure pneumatic systems may use jet engine compressor bleed air. The absolute pressure regulator in a pneumatic system regulates the compressor inlet air to provide a stabilized source of air for the compressor.

Low pressure pneumatic systems will generally utilize an engine driven, vane-type pump. These systems are used to power gyroscopic instruments.

As with hydraulic systems, a variety of valve and flow control devices are used on a pneumatic system. They include:

- Relief valves: used in pneumatic systems as damage-preventing units.
- Control valve: used to operate the various units.
- Check valve: permits flow in one direction only.
- Restrictor: a type of control valve that reduces the speed of operation of a unit.

REVIEW QUESTIONS — Hydraulic and Pneumatic Systems

101. *Which side of a chevron seal should face the direction of fluid pressure?*
102. *A filter incorporating specially treated cellulose paper is identified as what?*
103. *What is the purpose of an orifice check valve?*
104. *What is the most commonly used type of seal?*
105. *What unit will allow fluid flow in one direction, but no flow in the other direction?*
106. *What type of selector valve allows fluid to flow through the valve while it is in the OFF position?*
107. *For what purpose are relief valves installed in pneumatic systems?*
108. *How may extrusion of an O-ring seal be prevented?*
109. *What is the first step when removing an accumulator from an aircraft?*
110. *If a hydraulic system uses natural rubber packing materials, what type of fluid is used to service the system?*
111. *What color is petroleum based hydraulic fluid?*
112. *Is MIL-H-5606 hydraulic fluid flammable?*
113. *What may be used to flush a system normally serviced with Skydrol?*
114. *What hydraulic component automatically directs fluid from either the normal source or an emergency source to an actuating cylinder?*
115. *Which valve installed in a hydraulic system will have the highest pressure setting?*
116. *The unit which causes one hydraulic operation to follow another in a definite order is called what?*
117. *What must be done before removing the filler cap of a pressurized hydraulic reservoir?*
118. *At what point in the system would you expect to find hydraulic quick-disconnect fittings?*
119. *What is the main purpose of a pressurized hydraulic reservoir?*
120. *Chatter is a hydraulic system is caused by what?*

Answer questions 8374 through 8496 in the FAA Test section of this book

CABIN ATMOSPHERE CONTROL SYSTEMS

Modern aircraft take advantage of flight at high altitudes. At these altitudes much of the surface weather can be avoided, high speed "jet stream" winds can be taken advantage of, and jet aircraft can operate more efficiently. There are, however, problems when humans are subjected to the cold temperatures and oxygen deficient air at these altitudes.

The aircraft must provide, in the cabin, an atmosphere that is comfortable and supplies sufficient oxygen for the crew and passengers. The effects of cold temperatures are known to us all, but the effects of a reduced oxygen level are not commonly encountered. A reduction in normal oxygen supplies to the tissue of the body can produce important changes in body functions, thought processes, and degree of consciousness. This sluggish condition of mind and body caused by a deficiency or lack of oxygen is called hypoxia. The greatest danger from hypoxia during long exposure at 10,000 feet MSL altitude is headache and fatigue.

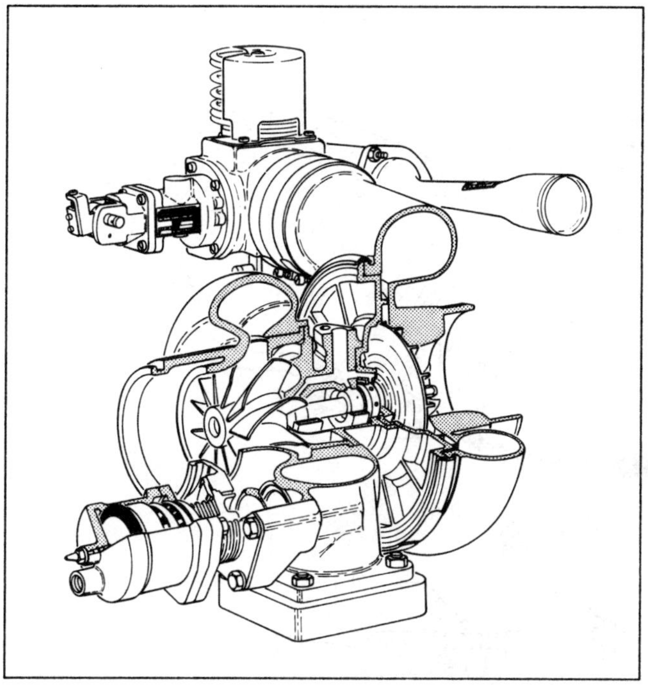

Figure 74. Turbochargers supply pressurization air through the venturi which serves as a flow limiter.

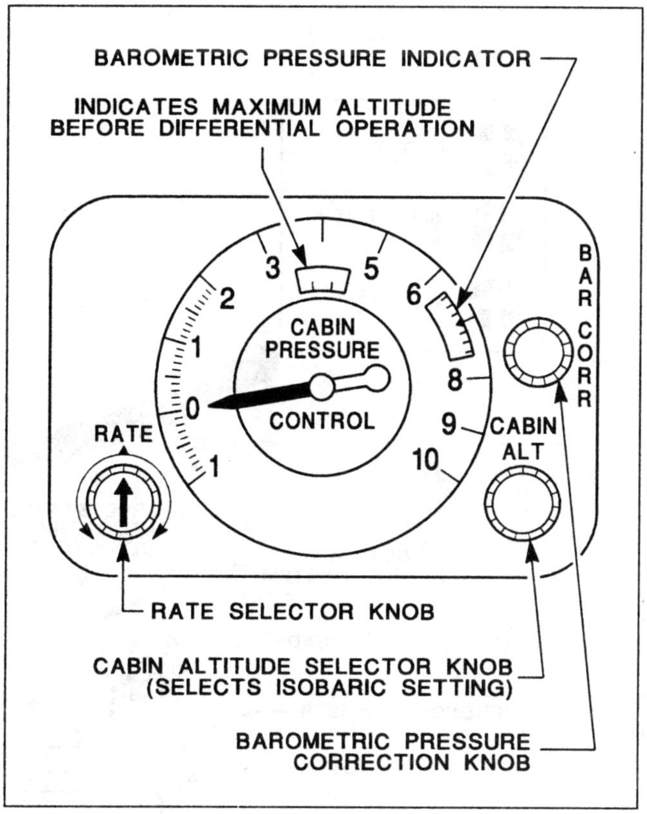

Figure 75. Pressurization controller.

Cabin Pressurization Systems

A cabin pressurization system may be designed which will provide the effects of a lower altitude in the aircraft's cabin. In the typical pressurization system the cabin, flight deck, and baggage compartments are incorporated into a sealed unit which is capable of containing air under a pressure higher than outside atmospheric pressure.

Pressurized air is pumped into this sealed pressure vessel at a relatively constant volume. Air is released by a device called an outflow valve. The primary function of the outflow valve is to maintain the desired cabin pressure. Should the pressurization system fail, the crew will inform passengers to don oxygen masks. The announcement is frequently a pre-recorded announcement.

The degree of pressurization and, therefore, the operating altitude of the aircraft are limited by several critical design factors. Primarily the fuselage is designed to withstand a particular maximum cabin differential pressure. Cabin differential pressure is the ratio between the internal and external air pressures acting on an aircraft. This is also a measure of the stress on the fuselage skin. The stress placed on the fuselage due to pressurization is tension.

The basic requirements for a cabin pressurization system are (1) a source of compressed air, (2) a means of controlling cabin pressure by regulating the outflow of air from the cabin, and (3) a method of limiting the maximum pressure differential to which the cabin will be subjected.

Cabin pressurization sources can be either engine-driven compressors, independent cabin superchargers, or air bled from the compressor section of a turbine engine.

The positive-displacement type compressor that is most suitable for use in large piston-engined aircraft for cabin pressurization is the roots blower. Oil free delivery of the air is achieved by using separate bearing chambers, rubber seals, and labyrinth seals in the compressor.

Independent cabin compressors can be engine-driven through accessory drive gearing or can be powered by bleed air from a turbine engine compressor.

The compressor of a turbojet engine may provide air for the pressurization and air conditioning systems of a jet aircraft. Usually, the air bled from a gas turbine engine compressor is free from contamination and can be used safely for cabin pressurization.

The outflow valve, by regulating the air exit, is the major controlling element in the pressurization system. This valve is placed in a pressurized portion of the fuselage, usually underneath the lower compart-

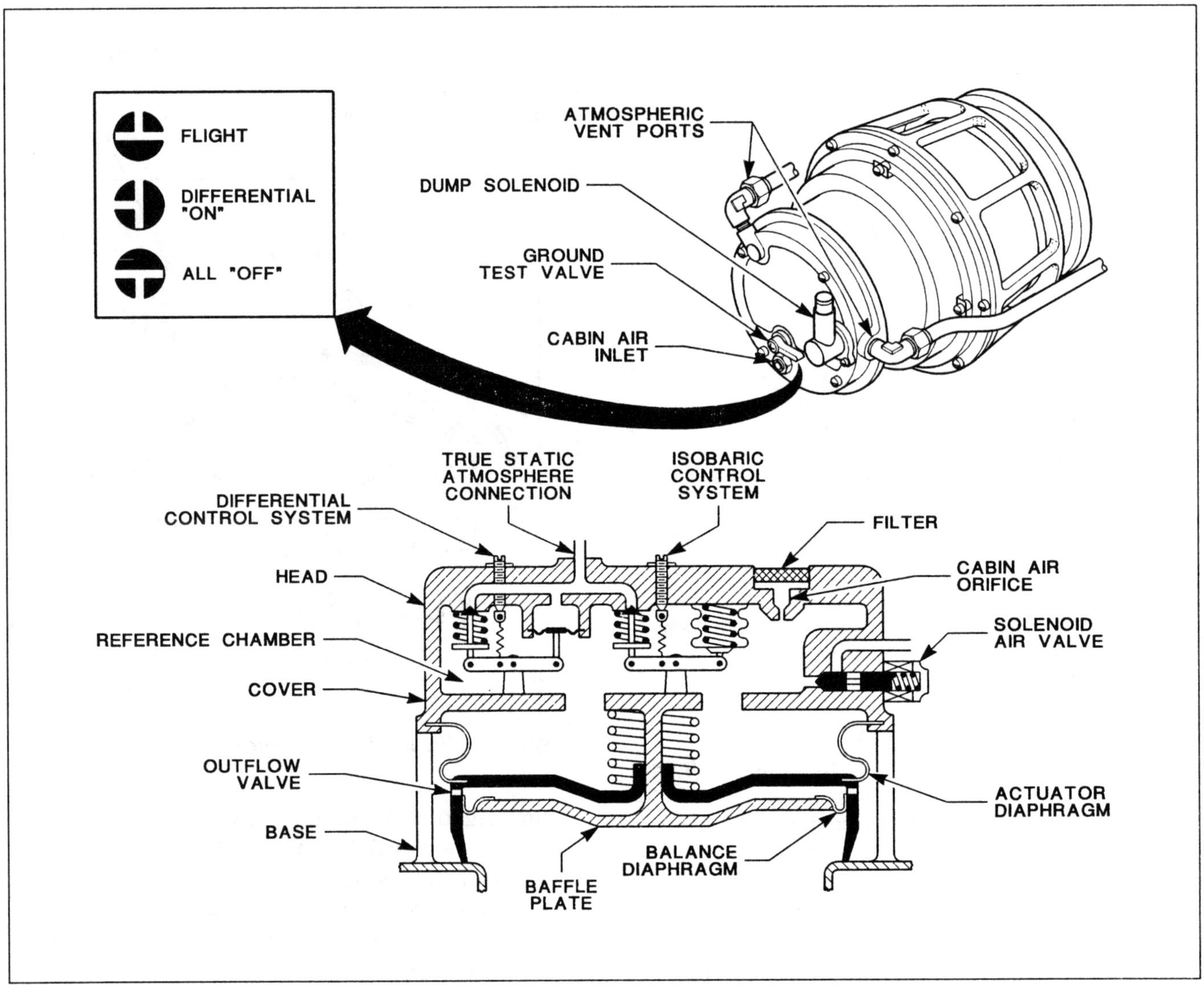

Figure 76. Cabin air pressure regulator.

ments. The purpose of this valve is to vent cabin air overboard.

In many aircraft, the outflow valve will be held fully open on the ground by a landing gear operated switch. During flight, as altitude is gained, the valve closes gradually to make a greater restriction to the outflow of cabin air. The cabin rate of climb or descent is determined by the rate of closing or opening of the outflow valve. During cruising flight, the cabin altitude is directly related to the degree of outflow valve opening.

The cabin air pressure safety valve is a combination pressure relief valve, vacuum relief, and dump valve. The pressure relief valve opens when the cabin differential pressure reaches a preset value, the negative pressure relief valve prevents the cabin altitude from becoming higher than the airplane altitude, and the dump valve is to relieve all positive pressure from the cabin.

The delivery air duct check valve prevents a loss of pressurization through a disengaged cabin air compressor.

The pressurization controller is the source of control signals for the pressurization system. The controller looks very much like an altimeter which has several added adjustment knobs. The controller uses barometric pressure, cabin altitude, and cabin rate of change. The cabin pressure control setting has a direct influence upon the outflow valve opening.

The operating cabin pressurization ranges of a pressurized aircraft are unpressurized, isobaric, and differential. The cabin pressure regulator controls cabin pressure to a selected value in the isobaric range and limits cabin pressure to a preset differential value in the differential range.

When the cabin pressure regulator is operating in the isobaric range, cabin pressure is maintained constant by the movement of the regulator bellows, limiting the flow of air from the reference chamber. If the isobaric metering valve in a pressurization control system moves toward the OPEN position, cabin pressure will decrease.

When the cabin pressure regulator is operating in the differential mode, reference pressure is vented to the atmosphere by the differential metering valve.

The cabin rate of change control establishes the rate of climb or descent of the cabin pressure. This allows pressure changes in the cabin to be made at a rate that is more comfortable to the occupants. If the cabin rate of climb is too great, the controls should be adjusted to cause the outflow valve to close faster.

A jet pump may be used to provide for augmentation of airflow in some areas of the aircraft.

If auxiliary ventilation is selected during pressurized flight while at cruising altitude, an increase in cabin altitude will occur.

Oxygen Systems

In case of failure of a cabin pressurization system, and in certain small and medium aircraft that have been designed without cabin pressurization, oxygen equipment may be installed for use by passengers and crew at high altitudes.

Oxygen used in aircraft systems is at least 99.5% pure and is practically water free. High pressure cylinders containing oxygen for aviation use can be identified by their green color and the words "AVIATOR'S BREATHING OXYGEN" stenciled in 1″ white letters.

The primary difference between aviation breathing oxygen and most other types of commercially available compressed oxygen is that the aviation breathing oxygen has had all water and water vapor removed.

Today, all gaseous oxygen is stored in steel cylinders under a pressure of 1,800 to 2,400 PSI. Some oxygen systems may use a pressure reducer valve to change high cylinder pressure to low system pressure. If the pressure reducer fails, the pressure relief valve will prevent high-pressure oxygen from entering the system downstream.

The oxygen system is identified by the type of regulator used. There are two basic types of regulators in use, and in each type there are variations. For low-demand systems we normally use continuous flow regulators that allow oxygen to flow from the storage cylinder, regardless of whether the user is inhaling or exhaling.

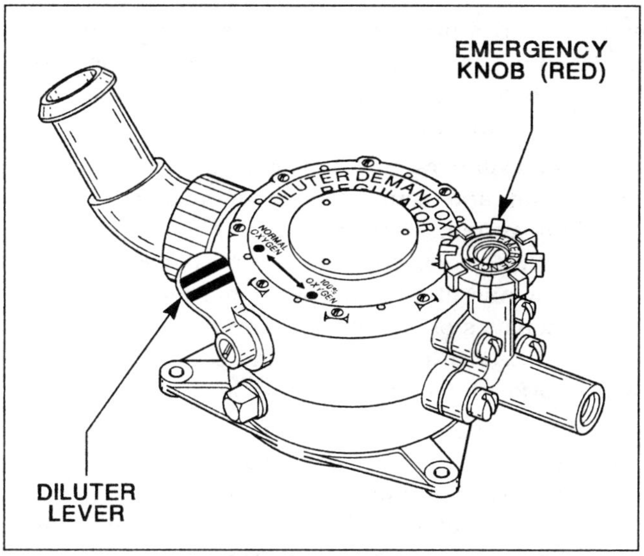

Figure 77. Diluter-demand regulator control.

The amount of oxygen delivered to the mask in a continuous flow oxygen system is controlled by a calibrated orifice. A rebreather bag is used with a continuous flow oxygen system. The emergency oxygen system that drops the mask to the passengers of a jet airliner is of the continuous flow type.

Demand type regulators allow a flow of oxygen only when the user is inhaling and shuts it off during exhale. Oxygen systems in unpressurized aircraft are generally of the continuous-flow and pressure-demand types.

The oxygen regulators used by the flight crews of most commercial jet aircraft are of the diluter-demand type. When the supply lever is turned on, oxygen can flow from the supply into the regulator. There is a pressure reducer at the inlet of the regulator. The demand valve shuts off all flow of oxygen until the user inhales, and operates when the user breathes.

A diluter-demand regulator dilutes the oxygen supplied to the mask with air from the cabin. The amount of cabin air mixed with the oxygen is reduced as altitude increases. At about 34,000 feet, pure oxygen is metered to the mask. The purpose of the airflow metering aneroid assembly found in oxygen diluter demand regulators is to regulate airflow in relation to cabin altitude when in the diluter demand position.

If the cabin atmosphere contains smoke or other dangerous fumes, the user may select the 100% OXYGEN position. This feeds only pure oxygen to the mask. In the event of a regulator failure, an emergency toggle is provided. When the emergency lever is placed in the EMERGENCY position, the demand valve is held open and oxygen flows continuously from the supply system to the mask.

Oxygen cylinders may be mounted permanently in the aircraft and connected to an installed oxygen plumbing system. Oxygen plumbing lines should be identified with color-coded tape applied to each end of the tubing and at intervals along its length. The tape coding consists of a green band overprinted with the words BREATHING OXYGEN and a black rectangle symbol overprinted on a white background.

Sealing compounds and lubricants used on oxygen equipment must be of an approved type. Never use any product containing oil, grease, or any other hydrocarbon. Teflon tape may be used as a lubricant on oxygen system tapered pipe thread connections.

A pressure relief valve is incorporated in the main supply line of a high-pressure system. The relief valve prevents high-pressure oxygen from entering the system downstream of the pressure reducer, if the pressure reducer fails. Pressure relief valves are vented to blow out plugs in the fuselage skin.

The main cause of contamination in a gaseous oxygen system is moisture. A contaminated oxygen system is normally purged with oxygen.

If a high-pressure oxygen cylinder is to be installed in an airplane, it must meet the specifications of the Interstate Commerce Commission or the Department of Transportation. Before a high-pressure oxygen cylinder is serviced, it must be the correct type and have been hydrostatically tested within the proper time.

Serviceability of an oxygen cylinder may be determined by a pressure test with water. Steel oxygen cylinders should be hydrostatically tested each 5 years.

The amount of oxygen in a portable, high-pressure cylinder may be determined by reading the pressure gage mounted on the cylinder. If an oxygen bottle pressure is allowed to drop below a specified minimum, it may cause moisture to collect in the bottle, resulting in corrosion.

Any leak, no matter how small, must be found and repaired. When it is suspected that the aircraft's oxygen system has developed a leak, the lines and fittings should be checked by bubble testing with a testing soap solution manufactured specifically for this purpose.

System leakage tests must be performed according to the manufacturers specifications, and charts are provided to calculate the allowable pressure drop. The system is allowed to cool, usually one hour after

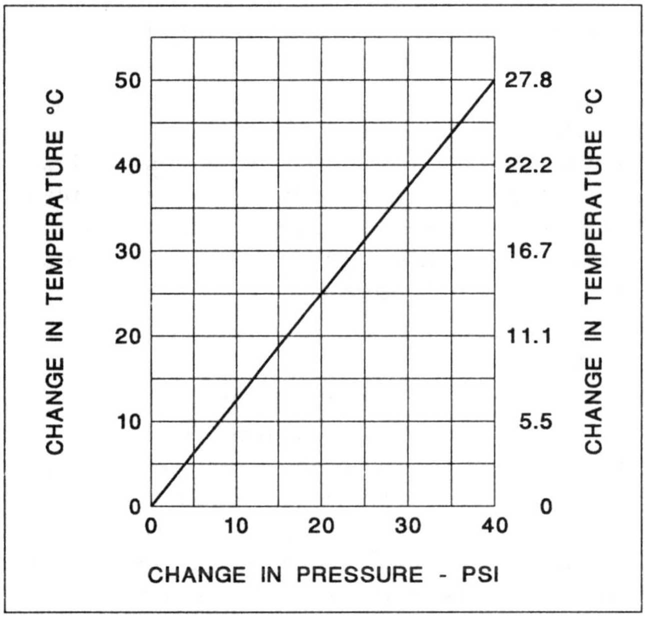

Figure 78. Correction of pressure during leakage test for change in temperature rises. Subtract pressure change if temperature fails.

filling before the pressures and temperatures are recorded. After several hours they are recorded again.

For emergency or backup use in pressurized aircraft, the chemical oxygen candle system is the least complicated and requires the least maintenance. On a volume basis, the storage capacity of oxygen in candles is about three times that of compressed gas. The units are safe to handle, and the generators are inert below 400°F, even under severe impact.

A new procedure for producing oxygen is its extraction from the air by a mechanical separation process Air is drawn through a patented material called a molecular sieve. As it passes through, the nitrogen and other gases are trapped in the sieve and only the oxygen passes through. Only part of the oxygen is breathed and the rest is used to purge the nitrogen from the sieve and prepare it for another cycle of filtering.

Aircraft Heaters

A large part of the heating requirements for pressurized aircraft may be accomplished automatically when the air is compressed for use in cabin pressurization. In many cases, additional heat need not be added. Compression of the air often provides more than the necessary heating. Consequently, cooling, to some degree, is required even when the outside air temperature is not so high.

When additional heat is necessary, one of the following types of systems may be employed: (1)

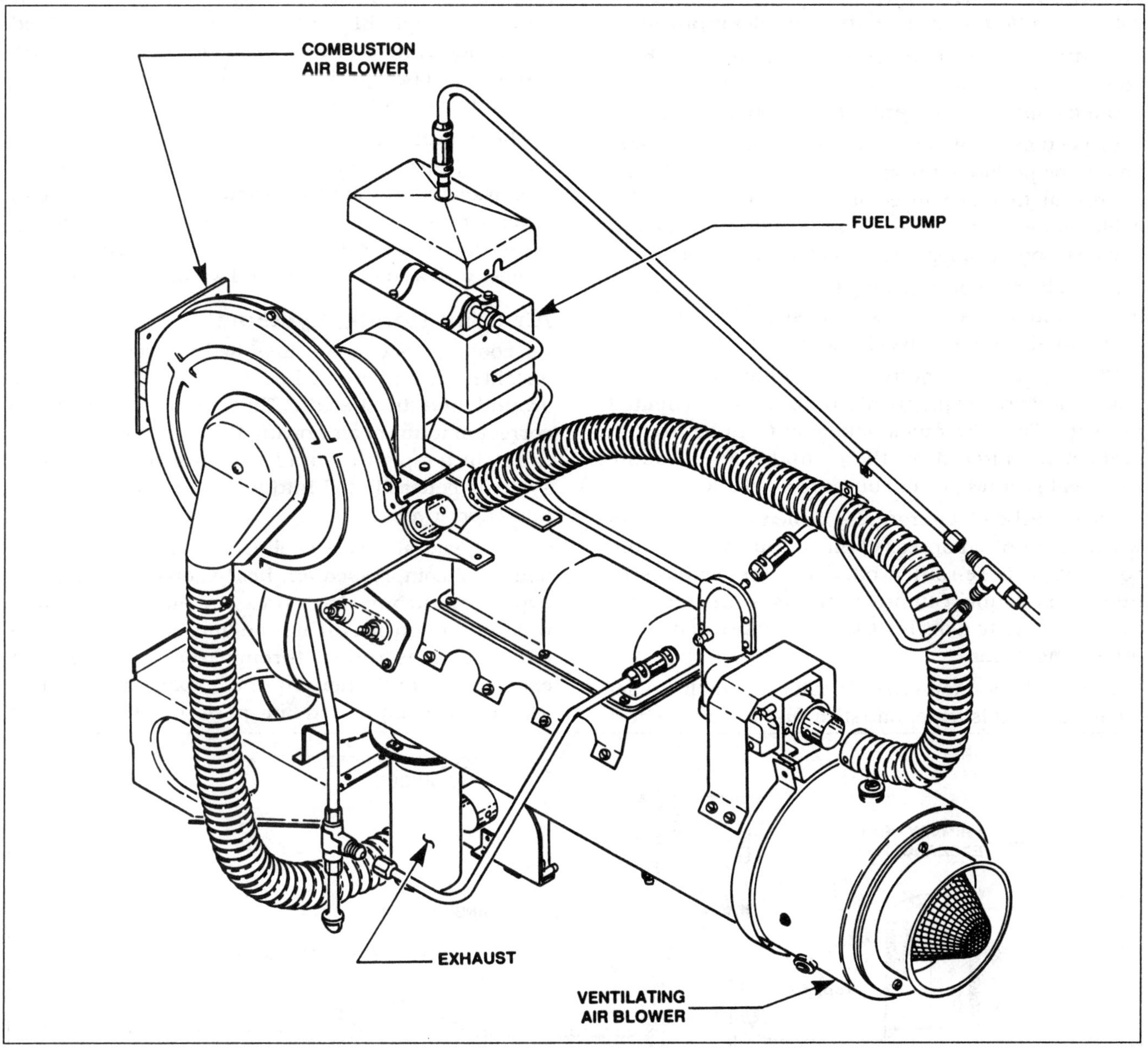

Figure 79. Combustion heater.

gasoline combustion heaters, (2) electric heaters, (3) recycling of compressed air, and (4) exhaust gas air-to-air heat exchanger.

Gasoline combustion heaters are for all practical purposes, a forced air furnace that burns aviation fuel. Regardless of size, every combustion heater needs four things to operate: (1) fuel to burn, (2) a source of ignition, (3) combustion air to provide the oxygen required to support the flame and, (4) ventilating air to carry the heat to the places where it is needed.

The fuel used in the heater is generally supplied from the same fuel tanks that supply the engines. After the fuel is filtered, it flows through a fuel solenoid valve and metering nozzle. The operation of an aircraft combustion heater is usually controlled by a thermostat circuit which alternately turns the fuel off and on, a process known as cycling.

Ignition may be provided either by a spark plug or a glow coil. The spark plug ignition usually operates continuously to prevent fouling of the plug. A thermal cutout switch breaks the circuit to the glow plug when the heater reaches a temperature sufficient to maintain the flame. This prolongs the life of the ignitor.

Combustion air is received through either the main air intake or through a separate scoop. This air is provided by ram pressure during flight and by ground blowers during ground operations. To prevent too much air from entering the heater as air pressure increases, either a combustion air relief valve or a differential pressure controller is provided.

Ventilating air may come from one of three sources: (1) a blower, (2) a ram-air inlet, or (3) the cabin compressors on pressurized aircraft.

Inspection and maintenance of combustion heaters should be performed regularly and with special care. Failure in this system could lead to uncomfortable cabin temperatures, or the possibility of carbon monoxide poisoning of flight crew and/or passengers.

After cleaning or replacing the filtering element in a combustion heater, the system should be pressurized and all connections checked for leaks.

Electric heaters require large amounts of electrical power, and are not practical if the aircraft has limited capacity. They are quick acting and can be used to preheat the aircraft on the ground if an adequate electrical ground power source is available.

Some turbojet aircraft use a heating system in which the hot compressed air output of the cabin compressor is re-routed back into the compressor inlet. This double compression raises the temperature of the air to a degree that other types of heating are not necessary.

The most common type of heater for small single-engine aircraft is the exhaust shroud heater. A sheet metal shroud is installed around the muffler and air moving through this muff picks up heat. This heated air is then routed to the cabin. This type of heater requires no energy other than that which would have normally been wasted.

These systems must be carefully inspected for cracks. Even the smallest crack could allow carbon monoxide fumes from the exhaust to enter the cabin. Some aircraft require inspection of these heaters by pressurizing the exhaust system and covering the muffler with a soap solution and watching for bubbles.

Air Conditioning Systems

Air cooling systems are installed to provide a comfortable atmosphere within the aircraft both on the ground and in the air. These systems keep the correct amount of air flowing through the aircraft at the right temperature and moisture content. These systems generally fall into two categories, air cycle and vapor cycle.

The basic air-cycle cooling system consists of a source of compressed air, heat-exchangers, and an expansion turbine. High pressure air from the compressor section of a turbine engine, or from the cabin compressor, is routed through the air-to-air heat exchangers and then to the expansion turbine. When the compressed air performs the work of

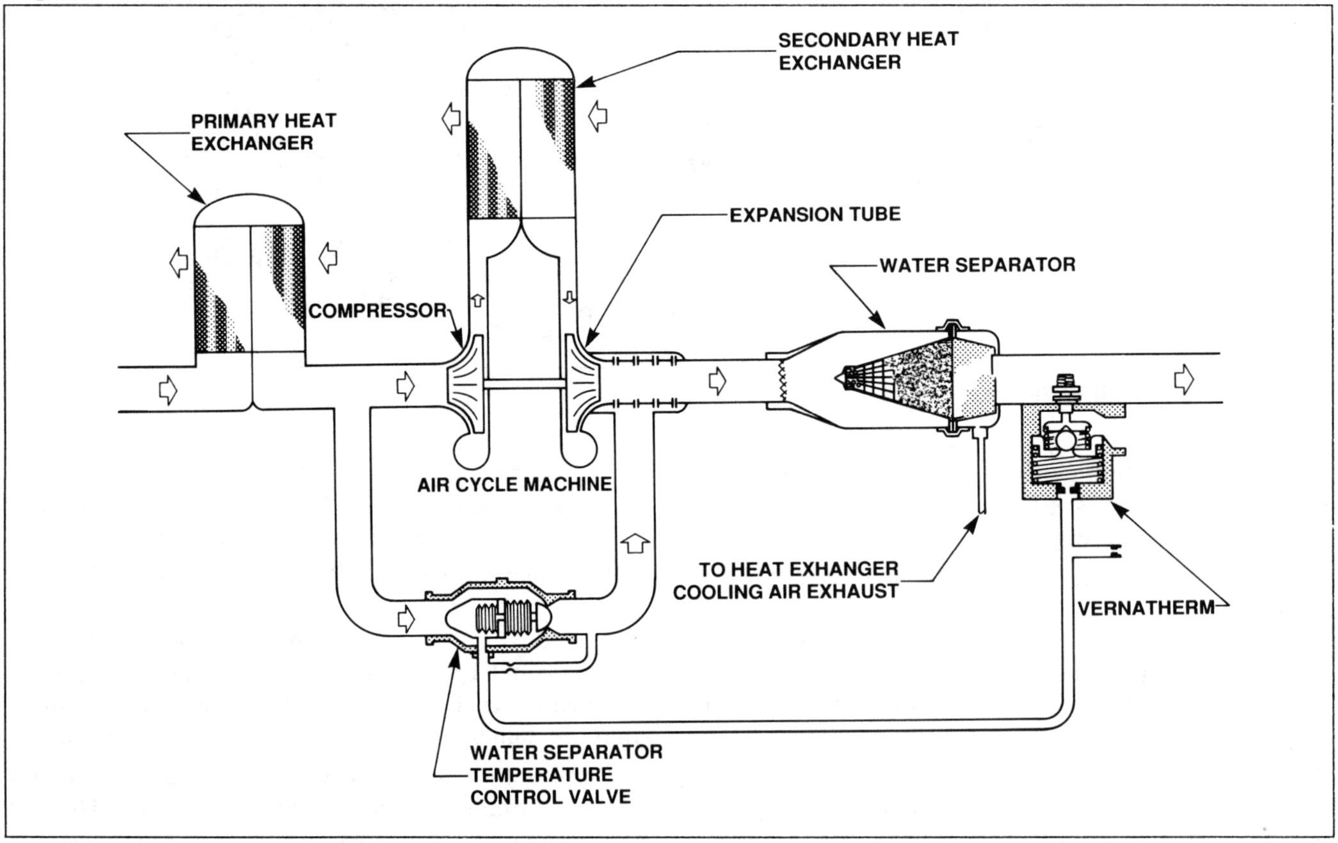

Figure 80. Air cycle conditioner.

turning the turbine, it undergoes a pressure and temperature drop. It is this temperature drop which produces the cold air used for air conditioning.

The rapid cooling of the air in the turbine causes moisture to condense in the form of fog. This air passes through a water separator where the water is removed and carried overboard.

A mixing valve is incorporated to control the supply of hot, cool, and cold air. The gasper vents over the seats of most airliners are connected only to the cold air supply.

The vapor-cycle air conditioning system operates on the same principle as the refrigerator in our kitchens. A refrigerant changes its state from a liquid to a vapor, and in doing so it absorbs heat from the cabin. This heat is taken outside of the aircraft and given off to the outside air as the refrigerant returns to its liquid state.

To understand this system, it is divided into two sides, one that accepts the heat and the other that disposes of it. The side that accepts the heat is called the low side, because here both the temperature and pressure of the refrigerant is low. The side where the heat is given up is known as the high side, and here the temperature and pressure of the refrigerant is high. The system is divided at the compressor (where the refrigerant is compressed, raising both its pressure and temperature), and the expansion valve where the pressure and temperature of the refrigerant are dropped.

We can trace the path of the refrigerant beginning with the receiver-dryer, which acts as a reservoir. The liquid refrigerant confined in the receiver at a high pressure is allowed to flow through the expansion valve into the evaporator. The expansion valve acts as a metering device and reduces the pressure of the liquid refrigerant. The refrigerant moves into the coils of the evaporator. The refrigerant is a low-pressure liquid as it enters the evaporator.

The point at which Freon flowing through a vapor-cycle cooling system absorbs heat and changes from a liquid to a gas is the evaporator. The pressure in the evaporator is low enough so that the boiling point of the liquid refrigerant is below the temperature of the air to be cooled. Heat flows from the space to be cooled to the liquid refrigerant, causing it to boil. The refrigerant is a low-pressure vapor as it leaves the evaporator. Frost or ice buildup on the evaporator is likely to be caused by inadequate airflow through the evaporator.

Cold vapor from the evaporator enters the compressor, where its pressure and temperature is raised. Ambient air is used for cooling the compressor. It leaves the compressor as a hot, high-pressure vapor.

This high-pressure vapor flows into the condenser. The point at which Freon flowing through a vapor-cycle cooling system gives up heat and changes from a gas to a liquid is the condenser. The function of the condenser is to transfer heat from the Freon gas to ambient air. The refrigerant leaves the condenser as a high-pressure liquid, and returns to the receiver to complete the cycle.

The basic system described above may be made more efficient with the addition of more sophisticated components. Some vapor cycle systems use a subcooler to reduce the temperature of the liquid refrigerant after it leaves the receiver. By cooling the refrigerant, premature vaporization can be prevented.

A practical system encounters varying heat loads and requires a refrigerant throttling device to prevent starvation or flooding of the evaporator. This variable-orifice effect is accomplished by the use of a thermostatic expansion valve. The position of the thermostatic expansion valve is determined by the temperature and pressure of the Freon in the outlet of the evaporator.

In an operating vapor-cycle system, if the two lines connected to the expansion valve are essentially the same temperature it indicates that the expansion valve is not metering Freon properly.

Compressors used in aircraft vapor-cycle systems are usually of the reciprocating type. On small aircraft, these are usually engine driven, through a belt drive. The compressors used on larger aircraft may be driven by electric or hydraulic motors, or by compressor bleed air.

Freon-12 is the most commonly used refrigerant. It is stable at low or high temperatures, does not react with any of the materials or seals used in air conditioning systems, and is not flammable. Freon 12 will boil at sea level pressure at –21.6°F. Any Freon-12 dropped on the skin will result in frostbite. Even a trace in the eye can cause damage. Goggles should be worn when working with Freon 12.

When Freon-12 passes over an open flame it becomes deadly phosgene gas. The effects of phosgene gas are cumulative, and may cause severe respiratory system damage.

Because this system is a sealed unit, the oil used to lubricate the expansion valve and compressor seals must be sealed in with the system. Only a special refrigeration oil is suitable for use in vapor-cycle systems. When a vapor-cycle system is not in operation, oil seepage is an indication that the system is leaking Freon.

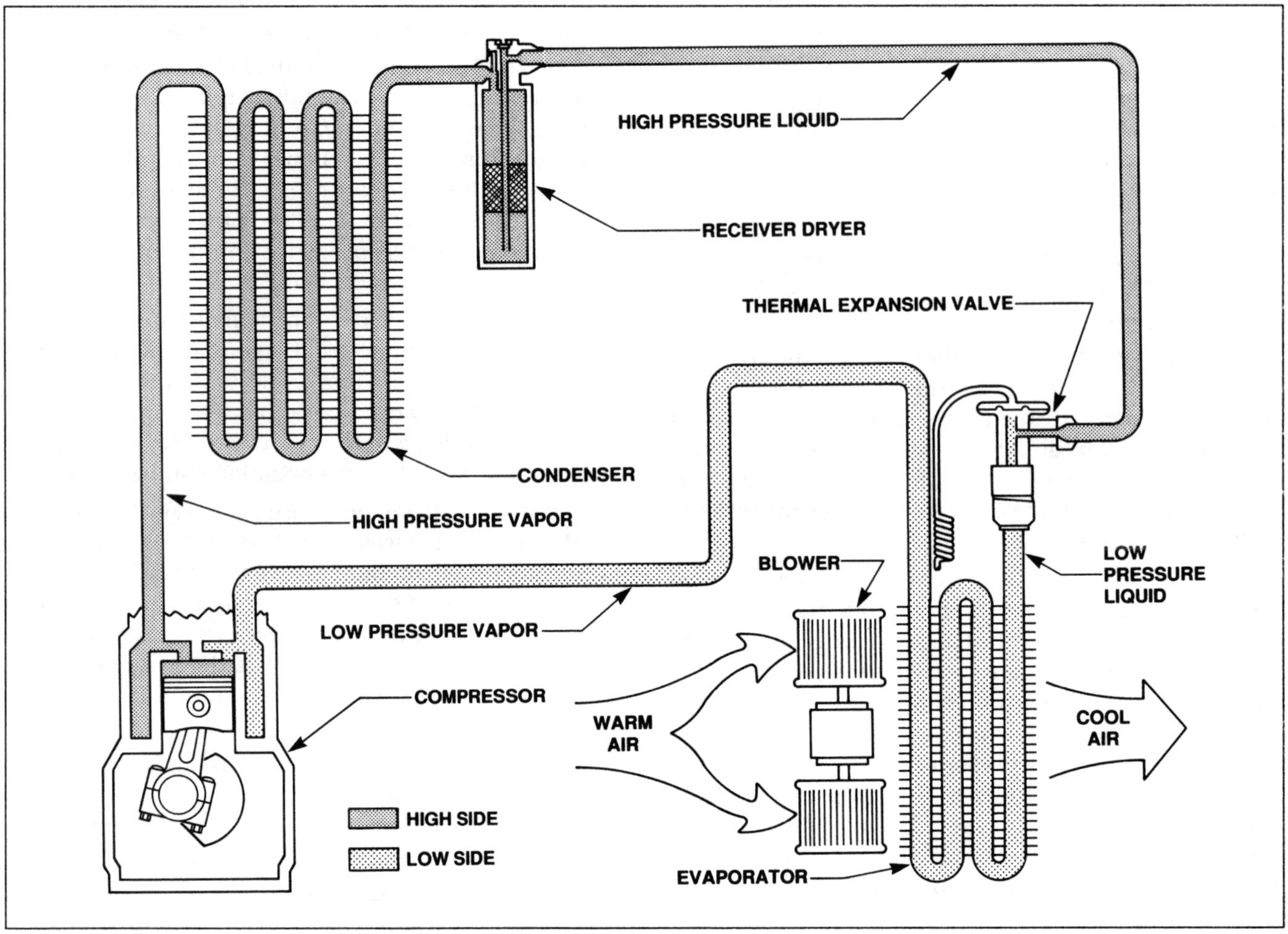

Figure 81. Vapor cycle of refrigeration.

With a small investment in special equipment, maintenance shops can perform most required service on these systems. One essential tool for troubleshooting and service is the manifold set.

The manifold set consists of a manifold with three fittings to which refrigerant service hoses are attached: two hand valves with O-ring type seals, two gages, one for the low pressure side of the system, and one for the high pressure side. The normal operating pressures of a Freon air conditioning system with an evaporator temperature of 40 to 50°F are: Low pressure of 20 to 30 PSI and high pressure of 225 to 300 PSI.

The low pressure gage is a compound gage, meaning it will read pressures either side of atmospheric. The high pressure gage usually has a range from zero up to 600 PSI. Special hoses are attached to the fittings of the manifold set for servicing the system.

To aid in determining whether servicing of the refrigerating unit is required, a liquid line sight gage is installed between the receiver/dryer and the expansion valve. The sight gage consists of a fitting with windows on both sides, permitting a view of fluid passage through the line.

When checking a Freon system, a steady stream of bubbles in the sight gage indicates the charge is low. If the gage indicates a low Freon charge, the system should be operated for a period of time to reach a stable condition, and then the Freon level rechecked.

Freon may be added to the system and the system operated for a period of time. A cooling system has been charged with the proper amount of Freon when the bubbles in the sight glass disappear.

The addition of Freon should be accomplished according to the instructions published by the aircraft manufacturer. If liquid refrigerant is introduced into the low side of a vapor-cycle cooling system when the pressure is too high or the outside air temperature is too low damage to the compressor may occur.

Whenever a Freon system is to be opened for maintenance, it is necessary to purge the system. When purging a Freon system it is important to release the charge at a slow rate to prevent excessive loss of refrigerant oil.

Anytime the system has been opened, it must be evacuated before recharging. Only a few drops of moisture will contaminate and completely block an air conditioning system. It this moisture freezes in the expansion valve, the action stops. Water will react with the refrigerant to form hydrochloric acid, which is highly corrosive to the metals in the system. Water is removed from the system by evacuation. The evacuation of the system removes any water that may be present by lowering the boiling point of the water and drawing out the vapor.

The system is evacuated using a vacuum pump, and then must be recharged with Freon. With the system under vacuum from evacuation, close all valves, connect the center hose to a refrigerant supply. The bottle should be placed vertical with the outlet at the bottom. Open the high pressure valve to permit Freon to flow into the system.

When charging the system after evacuation, if the low-pressure gage fails to come out of a vacuum, a blockage in the system is indicated. If the system will not take a Freon charge, the expansion valve is most likely the component at fault.

REVIEW QUESTIONS — Cabin Atmosphere Control Systems

121. *What section of a turbojet engine provides air for the pressurization system?*

122. *What is accomplished by the ventilating air on a combustion heater?*

123. *What is the purpose of the dump valve in a pressurized aircraft?*

124. *How can it be determined that a vapor-cycle cooling systems is charged with the proper amount of Freon?*

125. *What type of oxygen system uses a rebreather bag?*

126. *The main cause of contamination in gaseous oxygen systems is what?*

127. *The point at which Freon flowing through a vapor-cycle cooling system gives up heat and changes from a gas to a liquid is where?*

128. *What component of a pressurization system prevents the cabin altitude from becoming higher than the airplane altitude?*

129. *What is the purpose of evacuating a Freon air-conditioning system?*

130. *What is the primary control in a pressurization system?*

131. *What unit in a vapor-cycle cooling system serves as a reservoir for the refrigerant?*

132. *What drives the air-conditioning compressor on most single-engine aircraft?*

133. *In what position should the bottle be in when adding liquid Freon to a vapor-cycle cooling system?*

134. *What happens to Refrigerant-12 when it passes over an open flame?*

135. *In the diluter-demand oxygen regulator, when does the demand valve operate?*

Answer questions 8497 through 8585 in the FAA Test section of this book.

AIRCRAFT INSTRUMENTS

One way of grouping aircraft instruments is according to the job they perform. Within this grouping they can be classified as flight instruments, engine instruments, and navigation instruments.

The instruments that aid in controlling the in-flight attitude of the aircraft are known as the flight instruments. They are located on the main instrument panel within clear sight of the pilot. On larger aircraft they may be duplicated on a second instrument panel, in front of the copilot.

Basic flight instruments include the airspeed indicator, altimeter, and the magnetic compass. In addition, most aircraft have a rate-of-turn indicator, a bank indicator, a rate-of-climb indicator, and an artificial horizon indicator.

Engine instruments measure the quantity and pressure of fuel and oil, engine RPM, and other factors depending upon the type of engine used. The engine instruments usually include a tachometer, fuel and oil pressure gages, oil temperature, and a fuel quantity gage.

Many of the temperature measuring instruments are based on a wheatstone bridge or ratio meter. examples are free air, carburetor air, coolant and oil temperatures.

Navigational instruments provide information that enables the pilot to direct the aircraft accurately along the desired course. This group of instruments includes a clock, compasses (magnetic and directional gyro), radios, and other instruments for presenting navigational information to the pilot.

Instrument mechanisms are housed in some sort of case. These cases may be airtight and their integrity essential to the operation of the unit. Bakelite is the most commonly used plastic for instrument cases. Cases for electrically operated instruments are generally constructed of iron or steel to prevent interference from outside magnetic fields.

Instrument range markings indicate, at a glance, whether a particular system or component is operating in the desirable and safe range. Instruments should be marked and graduated in accordance with the Aircraft Specifications or Type Certificate Data Sheets, and the specific maintenance or flight manual. The person installing the instrument is responsible for making sure that it is properly marked.

A red radial line may be used to indicate maximum and minimum ranges. A blue arc indicates that operation is permitted under certain conditions. A green arc indicates normal operating range. Yellow is used to indicate caution. A blue radial line on an airspeed indicator shows the best rate-of-climb speed with one engine inoperative.

A white index marker is placed near the bottom of all instruments that have range markings on the cover glass. The index marker is a line extending from the cover glass onto the instrument case. The marker shows if glass slippage has occurred.

With a few exceptions, instruments are mounted on a panel in the cockpit so that the dials are in plain view of the pilot. Instrument panels are usually made of sheet aluminum alloy.

The method of mounting instruments on the panel depends upon the design of the instrument case. In one design, the bezel is flanged in such a manner that the instrument can be flush mounted in its cutout from the rear of the panel. The flanged type case can also be mounted from the front side of the panel. A flangeless instrument case can be mounted by an expanding-type clamp secured by a screw from the front of the instrument panel.

Instrument panels are usually shock mounted to absorb low-frequency, low-amplitude shocks. The type and number of shock mounts to be used are determined by the weight of the complete panel unit. Instrument panels will be electrically bonded to the aircraft structure to provide current return paths.

Three of the most important flight instruments are connected into a pitot-static system. These instruments are the altimeter, the airspeed indicator, and the rate-of-climb (vertical speed) indicator. The airspeed indicator is the only one connected to both pitot and static sources.

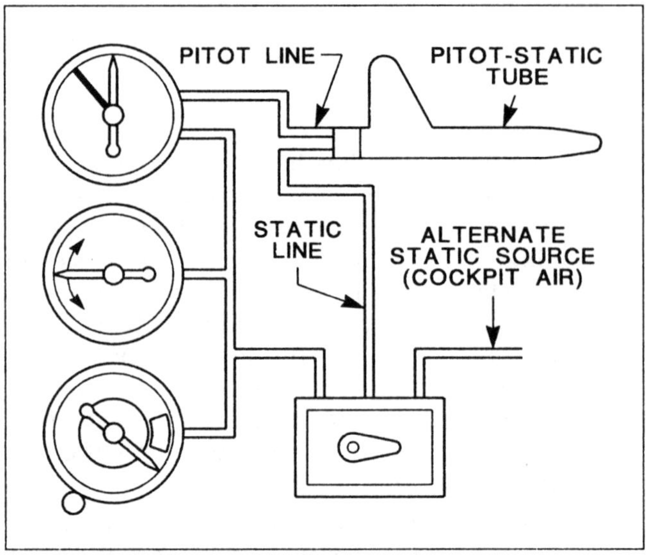

Figure 82. Pitot-static system with alternate source of static pressure.

The altimeter is constructed on the same principle as an aneroid barometer. The heart of an altimeter is the aneroid mechanism. The expansion or contraction of the aneroid with pressure changes actuates the linkage and the indicating hands show the altitude. Around the aneroid mechanism is a device called the bimetallic yoke. This compensates for the effect of temperature on the metals of the aneroid mechanism.

A barometric altimeter indicates pressure altitude when the barometric scale is set at 29.92 inches of mercury. This is one method of testing the accuracy of an altimeter. Another aneroid type instrument is the Manifold Air Pressure gage used with reciprocating engined aircraft. When the engine is not operating, the MAP gage should read the existing atmospheric pressure.

Airspeed indicators are sensitive pressure gages which measure the difference between the pitot and static pressures, and present this difference in terms of indicated airspeed.

A radio altimeter is a device which displays the aircraft's altitude as measured from a radio signal instead of atmospheric pressure. The signal is transmitted toward the ground and then returned. The time is then converted into feet of altitude above the ground.

The rate of climb or vertical speed indicator is a sensitive differential pressure gage that indicates the rate at which an aircraft is climbing or descending.

Aircraft pitot-static systems must be tested for leaks any time the system has been opened, when malfunction is suspected, and at the periods speci-

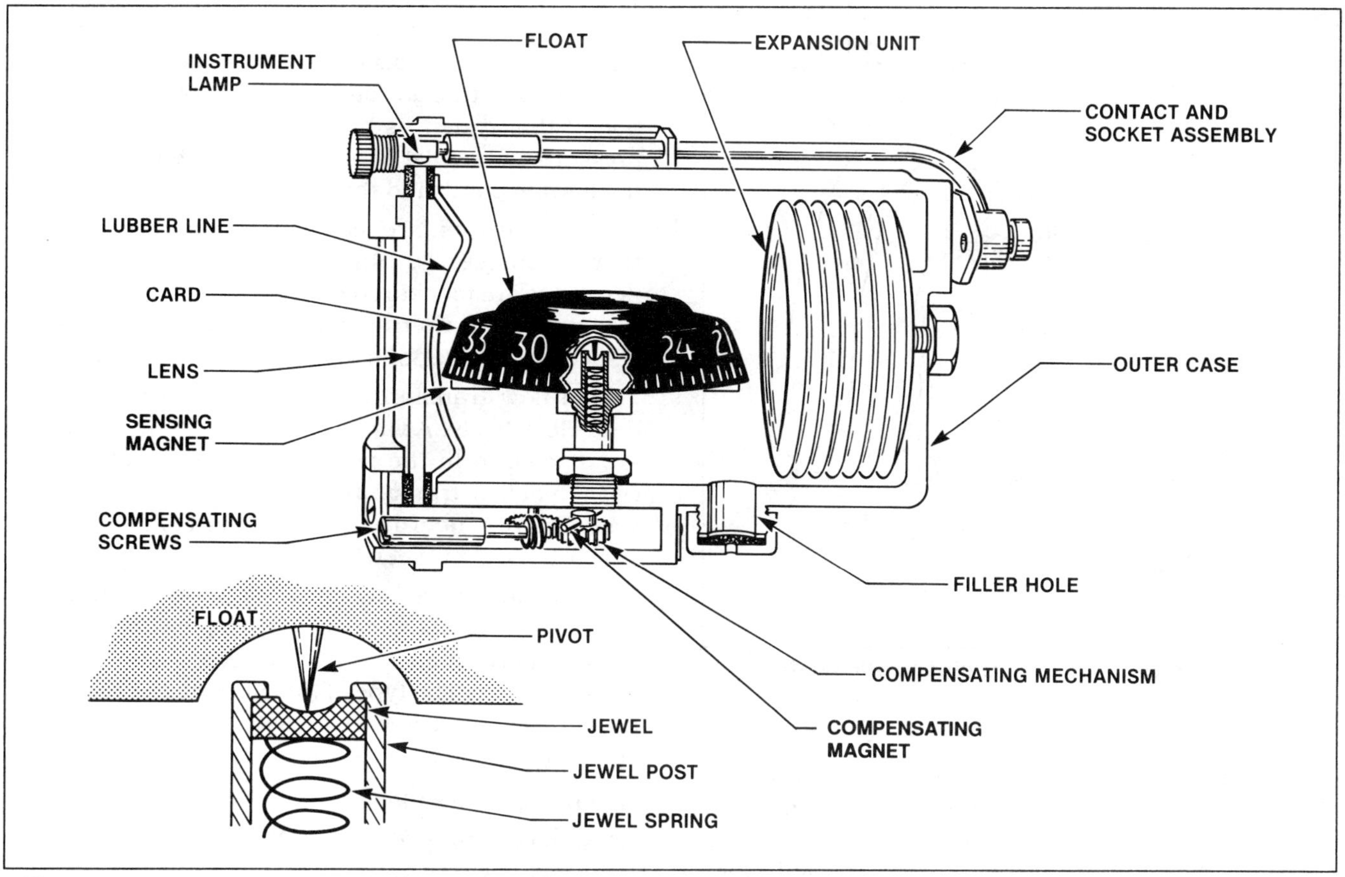

Figure 83. Internal mechanism of a floating magnet-type magnetic compass.

fied in the FARs. The minimum requirements for testing and inspection of instrument static pressure systems required by FAR Section 91.411 are contained in FAR Part 43, Appendix E.

The method of testing consists basically of applying pressure and suction to the pitot tube and static port respectively, using a leak tester and coupling adapters, or by observing the altimeter. The rate of leakage should be within tolerances prescribed for the system. The maximum altitude loss permitted during an unpressurized aircraft instrument static pressure system integrity test is 100 feet in one minute. Leaks can be located by isolating the instruments one at a time. Leak tests do not serve as calibration tests.

If the instrument static pressure line becomes disconnected inside a pressurized cabin during cruise flight, the altimeter and airspeed will both read low.

Three of the most common flight instruments (the attitude indicator, heading indicator, and the turn needle of the turn-and-bank indicator) are controlled by gyroscopes. The turn-and-bank instrument indicates trim and serves as an emergency source of bank information in case the attitude gyro fails.

The attitude indicator is also known as the artificial horizon. This basic flight instrument shows the relationship between the pitch and roll axes of the aircraft and a vertical line through the center of the earth. It gives us a reference so that we can keep the wings level, but tells us nothing about the direction in which the nose of the aircraft is pointing.

The directional gyro provides a stable heading indicator. The lubber line on a directional gyro is used to represent the nose of the aircraft, and the numbers on the card indicate the magnetic direction. The directional gyro offers many advantages over the float type magnetic compass.

Gyro instruments may be driven electrically, by air pressure, or by suction. Most instrument vacuum systems today use a dry-type vacuum pump utilizing carbon vanes. These are very reliable units, but are susceptible to damage from solid airborne particles and must take in filtered air. Excessive vacuum may result if the vacuum relief valve is improperly adjusted.

The magnetic compass is a simple, self-contained instrument which operates on the principle of mag-

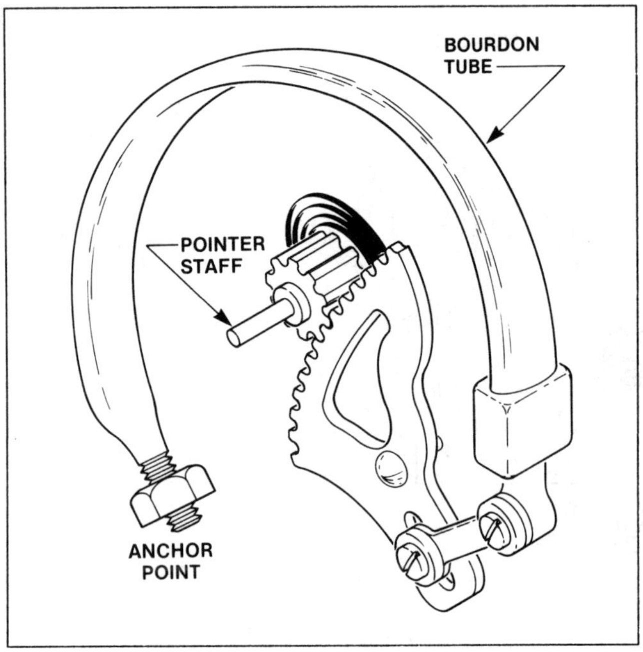

Figure 84. Bourdon tube-type pressure indicator mechanism.

netic attraction. The compass consists of a liquid-filled bowl containing a pivoted float element to which one or more magnets, called needles are fastened. A reference marker, known as the lubber line, is attached to the compass bowl. The liquid in the bowl dampens the oscillations of the float and decreases the friction of the pivot. A diaphragm and vent provide for expansion and contraction of the liquid as altitude and temperature change.

The compass needle is affected not only by the earth's magnetic field, but also by radio and other electrical disturbances, and by metal components in the aircraft. To reduce this deviation, each compass is checked and compensated periodically by adjustment of the N-S and E-W compensator magnets. This procedure is known as "swinging" the compass. The maximum deviation (during level flight) permitted in a magnetic direction indicator installed in an aircraft is 10°. Once compensated and swung, the deviations should be entered on a correction card and mounted on the compass.

Other instruments may utilize a variety of operating mechanisms. A bourdon tube-type instrument is generally thought of as measuring only pressure, but certain oil temperatures systems utilize this concept also. The operating mechanism of most hydraulic pressure gages is a bourdon tube. A typical deicing pressure gage uses a bourdon tube and the case is vented to keep the interior at atmospheric pressure.

The angle-of-attack indicating system measures the angle between the local airflow around the direction indicator and the fuselage reference plane. The operation of this system is based on the measurement of pressure differential at a point where the airstream is not flowing parallel to the true angle of attack.

Temperatures are commonly measured using thermocouples. Turbine engine exhaust gas temperatures are measured using chromel/alumel thermocouples The thermocouple leads are designed for a specific installations and may not be altered.

Fuel flow transmitters are designed to transmit data electrically. A variety of operating mechanisms are discussed in the powerplant section of this series. Synchro transmitters may be used in these systems. A synchro transmitter is connected to a synchro receiver electrically with wires.

A certificated technician may perform 100-hour inspections of instruments. Any repair or alteration must be performed by a certified repair station rated for such work. The type of instrument discrepancies that may be corrected by an aviation technician are:

· Red line missing
· Mounting screws loose
· Case paint chipped
· Leaking at line B-nut

The type of items that would require replacement of the instrument are:

· Case leaking
· Glass cracked
· Will not zero out
· Fogged

REVIEW QUESTIONS — Aircraft Instruments

136. *The number of shock mounts required for an instrument panel is determined by what?*

137. *Which instruments are connected to the pitot-static system?*

138. *What is indicated by the green arc on an instrument?*

139. *What will indicate if the cover glass of an instrument has slipped?*

140. *Why are most electrical instruments mounted in iron or steel cases?*

141. *Which instruments are usually controlled by gyroscopes?*

142. *What is the fixed reference marker attached to the compass bowl called?*

143. *What is the maximum deviation permitted in a compass?*

144. *What is the maximum altitude loss permitted in an unpressurized aircraft during a static systems test?*

145. *When an altimeter is set to 29.92 in. hg. on the ground, what will the altimeter indicate?*

146. *What does a reciprocating engine manifold pressure gage read when the engine is not operating?*

Answer questions 8586 through 8613 in the FAA Test section of this book.

AUTOPILOT SYSTEMS

The primary purpose of an autopilot is to relieve the pilot of control of the aircraft during long periods of flight. A fully integrated autopilot controls the aircraft about all three axes.

All autopilot systems contain the same basic components: (1) gyros, to sense what the airplane is doing; (2) servos, to move the control surfaces; and (3) an amplifier, to increase the strength of the gyro signals enough to operate the servos. On modern large aircraft the signal inputs are frequently derived from, and monitored by, the flight management computer. These inputs can also be received from the HSI (horizontal situation indicator). This gyroscopic instrument is really a combination of several; the VOR, directional gyro, electric compass and DME. When any function is turned off, the appropriate flag, such as NAV, HDG or GS, will be displayed. A controller is also provided to allow manual control of the aircraft through the autopilot system.

Modern autopilots can be described in terms of their three major channels: (1) the rudder, (2) aileron, and (3) the elevator channels.

The rudder channel receives two signals that determine when and how much the rudder will move. The first is a course signal derived from the compass system. As long as the aircraft remains on the heading selected, no signal will develop. Any deviation causes the compass system to send a signal to the rudder channel that is proportional to the angular displacement of the aircraft from the preset heading.

The second signal received by the rudder channel is the rate signal, which provides information anytime the aircraft is turning about the vertical axis. This information is provided by the turn-and-bank indicator gyro. The relative motion between the gyro and its supporting system is used to develop the signal. When the aircraft attempts to turn off course, the rate gyro develops a signal proportional to the rate of turn, and the course gyro develops a signal proportional to the amount of displacement. The two signals are sent to the rudder channel of the amplifier where they are combined and their strength increased. The amplified signal is then sent to the rudder servo. The servo will turn the rudder in the proper direction to return the aircraft to the preset course.

As the rudder surface moves, a follow-up signal is developed which opposes the input signal. When the two signals are equal in magnitude, the servo stops moving. As the aircraft arrives on course, the course signal will reach a zero value, and the rudder will be returned to the streamlined position.

If the attitude or direction were changed by the autopilot in order to correct a deviation and the involved control surfaces were returned to neutral immediately after the aircraft reached its correct position overshoot and oscillation would likely result.

The aileron channel receives its input signal from a transmitter located in the gyro horizon indicator. Any movement of the aircraft about its longitudinal axis will cause a signal to be developed.

The elevator channel of an autopilot controls the aircraft about the lateral axis, and detects changes in pitch of the aircraft.

Dutch roll, a combination yawing and rolling oscillation that affects many sweptwing aircraft, is counteracted with a yaw damper system. This system may be required to operate during all phases of flight.

Autopilot systems may be operationally checked as part of an inspection program, or in response to a pilot report. When operationally checking an autopilot system on the ground, after the aircraft's main power has been switched on, the autopilot should be engaged only after the gyros come up to speed and the amplifier warms up. Failure to follow this procedure could result in damage to the system.

REVIEW QUESTIONS — Autopilot Systems

147. *What is the primary purpose of an autopilot?*

148. *What unit allows manual controlling of the aircraft through the autopilot system?*

149. *What unit in an autopilot system senses heading changes?*

150. *What component of an autopilot system applies torque to the control surfaces of an aircraft?*

151. *What type of system may be installed to counteract the effects of dutch roll?*

152. *A fully integrated autopilot system controls the aircraft around how many axes?*

Answer questions 8647 through 8660 in the FAA Test section of this book.

NAVIGATION AND COMMUNICATION SYSTEMS

Communication and navigation are the two major functions of airborne radio equipment. Safe aircraft operation is dependent to a large degree on the satisfactory performance of this equipment.

Aircraft maintenance technicians' responsibilities include the installation and inspection of radios, antennas, navigational equipment, and associated wiring. In addition, the FAA has certified radio repair facilities to perform maintenance on radios, antennas, navigation, and radar equipment. Transmitting equipment is calibrated by persons licensed by the FCC.

The Federal Communications Commission issues the license required for the operation of aircraft radio transmitters. The aircraft's radio station license must be displayed in the aircraft. Traditionally all personnel operating aircraft transmitters must be holders of a Restricted Radiotelephone license issued by the FCC.

The most common communication system is use today is the VHF system. VHF provides VOR navigation and ATC communication. VHF airborne communications sets operate in the frequency range from 108.0 MHz to 139.5 MHz. VHF signals follow approximately straight lines, and are sometimes referred to as, "line of sight". This frequency range may be used for both communications and navigation.

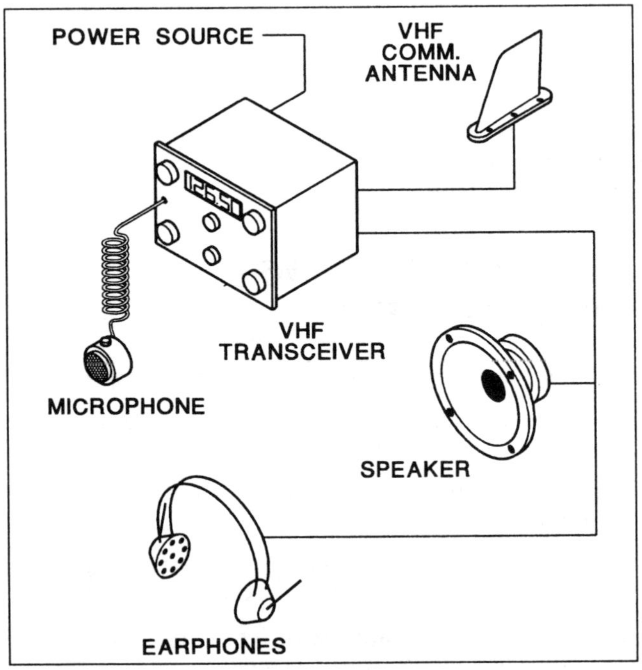

Figure 85. VHF system dagram.

VHF Omnirange System

The VHF Omnidirectional Range (VOR) is an electronic navigation systems. It produces 360 usable radials, or courses, any one of which is a radio path connected to the station. The typical airborne VOR receiving system consists of a receiver, visual indicator, antenna, and a power supply. The preferred location of a VOR antenna on light aircraft is on top of the cabin with the apex of the V pointing forward.

The Instrument Landing System

The instrument landing system (ILS), is one of the facilities of the Federal airways, and operates on the VHF band of frequencies. The ILS can be used to direct an aircraft safely to the runway. The entire systems consists of a runway localizer, a glide slope signal, and marker beacons for position location.

The localizer equipment produces a radio signal aligned with the center of an airport runway. The airborne VOR receiver functions as the localizer receiver during ILS operation.

The glide slope is a radio beam which provides vertical guidance to the pilot, establishing the correct angle of descent to the runway. The glideslope and VOR/Localizer receivers may be separate receivers or combined in a single case. The glide slope frequency is paired to the localizer and one frequency selector is used to tune both receivers.

The information from both localizer and glide slope receivers is presented on the Course Deviation Indi-

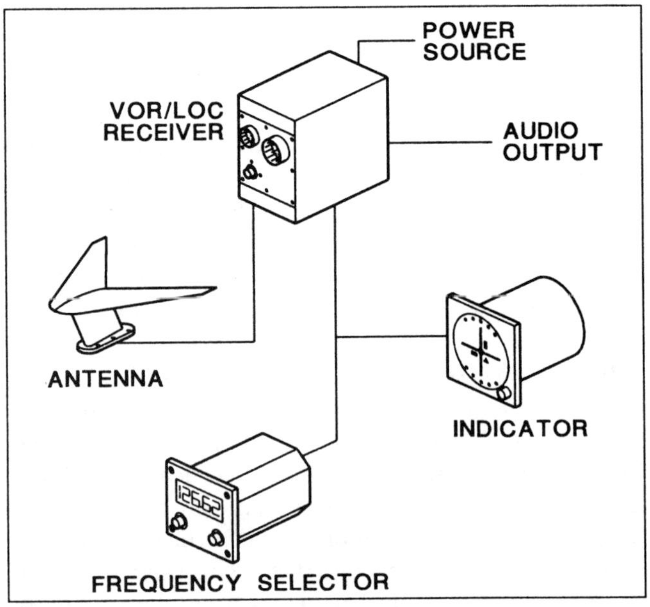

Figure 86. VOR system diagram.

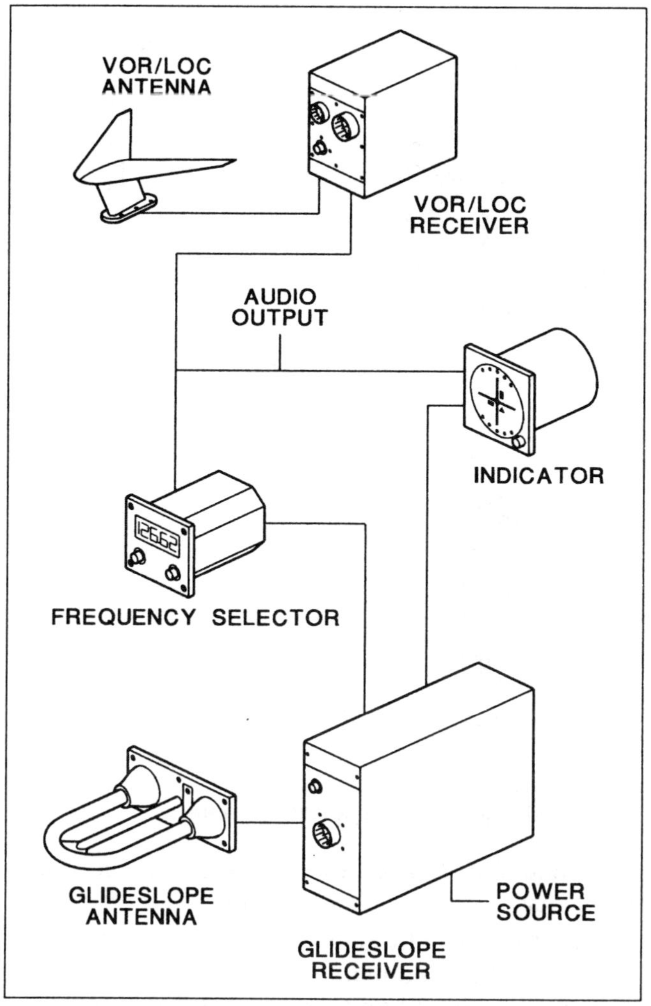

Figure 87. Component diagram of an ILS.

cator (CDI); the vertical needle displays localizer information and the horizontal needle displays glideslope information. When both needles are centered, the aircraft is on course and descending at the proper rate.

Marker beacons are used in connection with the ILS. The markers are signals which indicate the position of the aircraft along the approach to the runway. The outer marker marks the beginning of the approach path. The outer marker signal is a tone keyed in long dashes. In addition, a purple lamp in the cockpit is illuminated. The middle marker is usually about 3,500 feet from the end of the runway and produces a higher pitched tone, keyed with alternate dots and dashes. An amber lamp flashes to indicate passage over the middle marker.

Long Range Navigation (LORAN)

LORAN stations transmit accurately sequenced pulse signals at 100 KHz, producing ground waves that follow the curvature of the earth. By using one master and two or more secondary stations, the system can translate the signals into latitude and longitude.

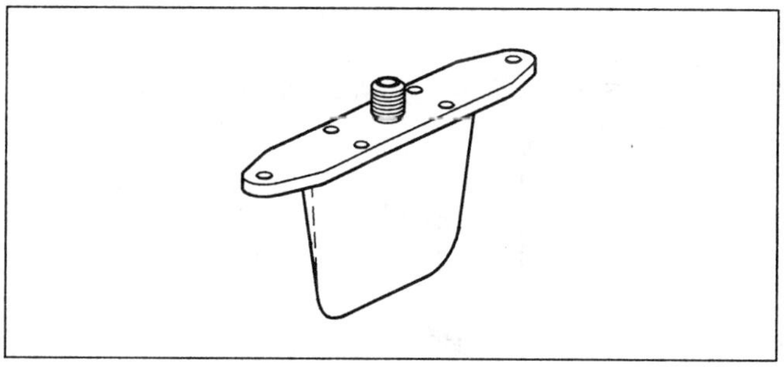

Figure 88. Typical DME antenna.

Ground Proximity Warning Systems (GPWS)

The main component in a GPWS system is a computer that monitors burometric and radar altitude, rate of climb or descent, flap and landing gear positions. When the computer parameters calculate an impending ground impact, it issues a voice warning of "Pull up"!

Distance Measuring Equipment

The purpose of distance-measuring equipment (DME) is to provide a constant visual indication of the distance the aircraft is from a ground station. DME operates in the UHF range. Transmitting and receiving frequencies are paired with a VOR channel. In some aircraft installations the DME channel selector is ganged with the VOR channel selector to simplify operation. When installing a DME antenna, it should be aligned with the centerline of the aircraft. A DME antenna should be located in a position on the aircraft that will not be blanked by the wing when the aircraft is banked.

Automatic Direction Finders

Automatic direction finders (ADF) are radio receivers equipped with directional antennas which are used to determine the direction from which signals are received. The ADF operates in the low and medium frequency spectrum. The direction to the station is displayed on an indicator located in the cockpit, as a relative bearing to the station.

The airborne equipment consists of a receiver, loop antenna, sense or nondirectional antenna, indicator, and control unit. After an ADF antenna has been installed, the loop must be calibrated. One antenna can be used for the radio range and standard broadcast bands in light aircraft because the two ranges are close together.

EFIS (Electronic Flight Instrument System)

Newer instrument systems send information to an electronic bus. The display is produced on a CRT,

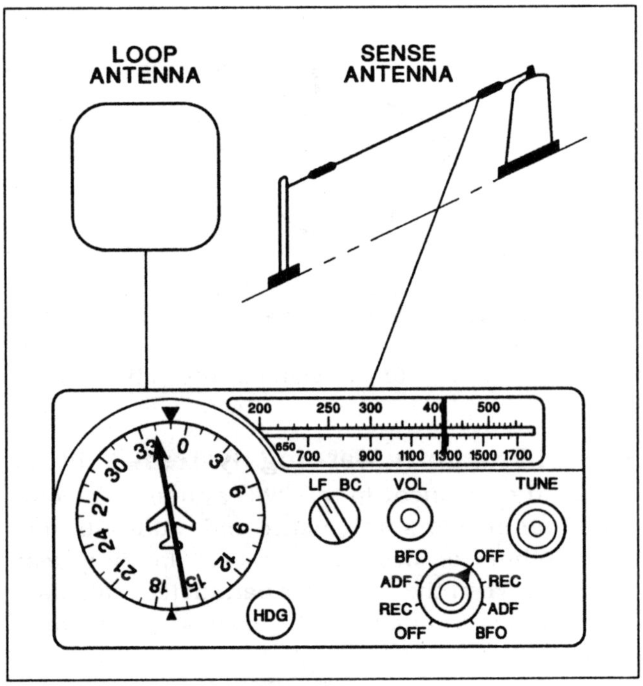

Figure 89. Typical ADF installation.

similar to a TV, that is connected to the bus. The EFIS processes digital data and displays it on the CRT as a representation of an aircraft instrument. This internal processing of digital data is a function of an internal symbol generator. An internal display controller can be changed by the pilot to display various systems; engine, flight, fuel management, etc. Most of the aircraft's instruments can be displayed on one CRT.

With the electronic instrumentation comes electronic testing. BITE (built in test equipment) can indicate individual component operation or failure using LED coded readouts.

Radar Transponders

The radar beacon transponder system is used in conjunction with ground based radar to provide positive aircraft identification directly on the controller's radar scope. Civil transponders operate in two modes labeled "Mode A" and "Mode AC", which are switch controlled. The selector switch is place in "Mode AC" when it is necessary to transmit altitude information. This is now required when operating in the vicinity of most major airports.

Emergency Locator Transmitters

One radio transmitter that operates automatically upon impact in a crash is the Emergency Locator Transmitter (ELT). FAR 91.52 requires most aircraft in the general aviation fleet to be equipped with one of these units. If activated, it will transmit a downsweeping tone on both

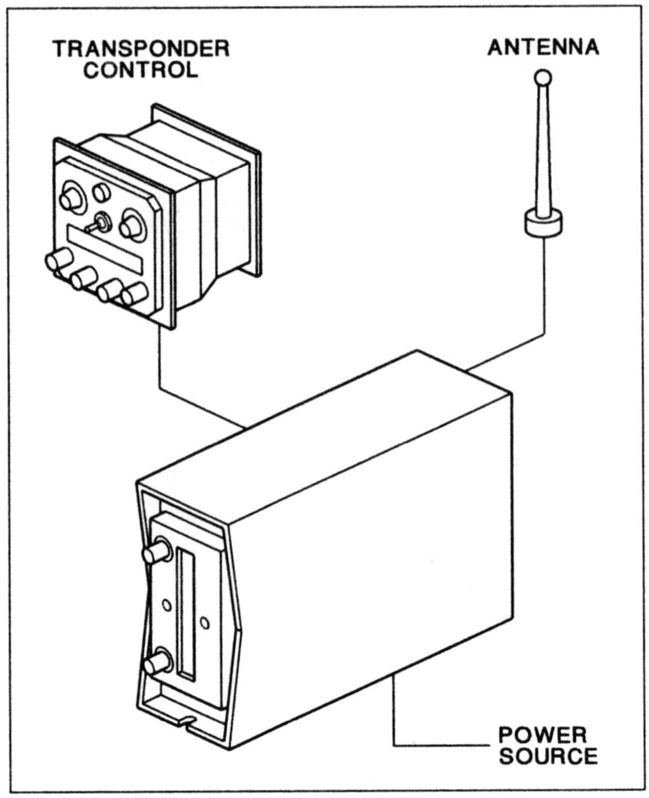

Figure 90. Typical transponder system.

121.5 and 243.0 MHz, which are universally used emergency frequencies for aviation.

An ELT is normally activated by an inertial switch or equivalent mechanism if subjected to a force of a prescribed intensity and duration. It must activate when the force is applied parallel to the longitudinal axis of the aircraft. The preferred location for mounting the ELT is as far aft as possible, but forward of the vertical fin.

The most common power source for an ELT is a self-contained battery The ELT battery must be capable of furnishing power for signal transmission for at least 48 hours. These batteries must be replaced when they reach 50% of their useful (shelf) life, as established by the manufacturer. This information is conveyed in the form of a replacement date marked on the outside of the transmitter.

The operation of an installed ELT may be verified during aircraft inspection by tuning a communications receiver to the civil emergency frequency and activating the ELT momentarily.

Radio Installations

There are many factors which the technician must consider prior to altering an aircraft by the addition of radio equipment. Each installation should be planned to allow easy access for inspection, maintenance, and exchange of units. Weight and balance changes should be considered, as adding radio equipment will affect the empty weight and useful load.

The radio equipment must be securely mounted to the aircraft. All mounting bolts must be secured by locking devices to prevent loosening from vibration. Shock mounts are used to protect radio equipment from vibration damage.

Adequate clearance between the radio equipment and the adjacent structure must be provided. When installing radio equipment under a seat, 1″ of clearance is required with the seat occupied and subjected to maximum downward seat spring deflection.

Suppression of radio interference is a task of first importance. Almost every component of the aircraft is a possible source of radio interference. Radio equipment is bonded to the aircraft in order to provide a low-impedance ground and to minimize radio interference from static electrical discharges.

Shielding is one of the most effective methods of suppressing radio noise. The primary object of shielding is to electrically contain the radio frequency noise energy. Ignition wiring and spark plugs are usually shielded to minimize radio interference.

A transmitting or receiving antenna is connected directly to its associated radio by wire(s) which are shielded. This interconnecting wire is called a coaxial cable.

When installing coaxial cable, it should be secured firmly along its entire length at 2 ft. intervals. To assure optimum operation, coaxial cables should not be routed or tied to other wire bundles. When bending coaxial cable, the bend radius should be at least 10 times the diameter of the cable.

Elimination of static electricity is desired in order to reduce radio receiver interference. Static dischargers are installed on aircraft to bleed off static electricity. Three major types of static dischargers are in use:

(1) Flexible vinyl-covered, silver or carbon-impregnated braid.
(2) Semiflexible metallic braid.
(3) Null-field.

Flexible and semi-flexible dischargers are attached to the aircraft by metal screws and should be periodically checked for tightness. At least 1″ of the inner braid of vinyl covered dischargers should extend beyond the vinyl covering. Null-field dischargers are riveted and epoxy bonded to the aircraft structure.

Antenna Installation

An antenna is a special type of electrical circuit designed to radiate and receive electromagnetic energy. For proper performance, it is important that the radio equipment manufacturer's instructions be carefully followed in matching and coupling the antenna to the radio equipment. Generic information on radio and antenna installation may be found in AC 43.13-2A.

When a rigid antenna is installed on the vertical stabilizer, evaluate the flutter and vibration characteristics of the installation.

The approximate drag load an antenna is required to withstand can be determined by the following formula:

$$D = .000327\ AV^2$$

Where:

D is the drag load on the antenna in lbs.

A is the frontal area of the antenna in sq. ft.

V is the V_{NE} of the aircraft in MPH.

When an antenna is installed, it should be fastened so that loads imposed are transmitted to the aircraft structure. The following steps are typical of surface mounted rigid antennas:

(1) Place a template on the fore and aft centerline at the desired location and drill the mounting holes and the correct diameter hole for the transmission line cable.
(2) Install a reinforcing doubler of sufficient strength to reinforce the aircraft skin. Doublers are used when antennas are installed to reinstate the structural strength of the aircraft skin.
(3) Install the antenna on the fuselage, making sure that the mounting bolts are tightened firmly against the reinforcing doubler, and that the mast is drawn tight against the gasket. A gasket is used to prevent the entry of moisture. When a gasket is not used, a sealer, such as zinc chromate paste, may be substituted.
(4) Route the transmission line cable to the receiver.

REVIEW QUESTIONS — Navigation and Communications Systems

153. *What is the most common communication system in use today?*

154. *What agency issues the Radio Station License for an aircraft?*

155. *At what intervals along its length must coaxial cable be secured?*

156. *Where is information about the expiration date of an ELT battery found?*

157. *What is the preferred location of an ELT?*

158. *For what purpose is a gasket or sealer used when installing an antenna?*

159. *What is the preferred location of a VOR antenna on a light aircraft?*

160. *What is the purpose of the localizer?*

161. *What is the minimum bend radius for coaxial cable?*

162. *How much clearance from the seat bottom is required when installing radio equipment under a seat?*

Answer questions 8661 through 8697 in the FAA Test Section of this book.

AIRCRAFT FUEL SYSTEMS

The aircraft fuel system stores fuel and delivers the proper amount of clean fuel at the right pressure to meed the demands of the engine. The basic components of a fuel system include tanks, lines, valves, pumps, filtering units, gages, warning signal and primer. Some systems will include central fueling provisions, fuel dump valves, and a means for transferring fuel.

Fuel Tanks

The location, size, shape, and construction of fuel tanks vary with the type and intended use of the aircraft. Fuel tanks are made of materials that will not react chemically with any aviation fuel. Aluminum alloy is widely used, and synthetic rubber bladder-type fuel cells are used in some installations.

Integral fuel cells are usually built into the wings of the aircraft structure, they are not removable. This type of construction is usually referred to as a "wet wing", and is used in many large aircraft to reduce weight.

All except the very smallest tanks are fitted with internal baffles to resist fuel surging caused by changes in attitude of the aircraft.

Usually, a sump and a drain are provided at the lowest point in the tank, as shown in figure 91. The primary purpose of a fuel tank sump is to provide a place where water and dirt accumulations in the tank can collect and be drained. When a sump or low point is provided in the tank, the main fuel supply is not drawn from the bottom of the tank.

If an aircraft is fueled from a truck or storage tank which is known to be uncontaminated with dirt or water, periodic checks of the aircraft's fuel tank sumps and system strainers are still necessary due to the possibility of contamination from other sources.

The top of each tank is vented to the outside air in order to maintain atmospheric pressure within the

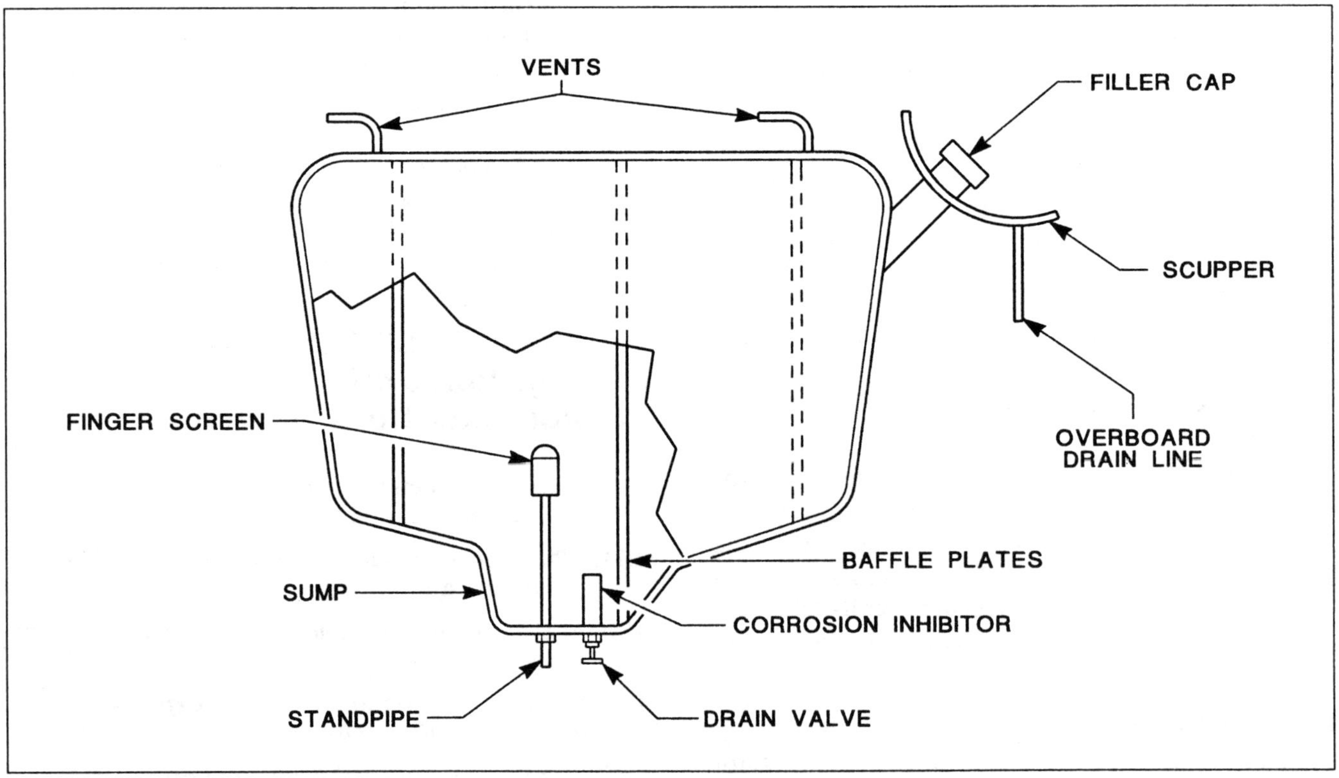

Figure 91. A typical metal fuel tank.

tank. If a gravity-feed fuel system is permitted to supply fuel to an engine from more than one tank at a time, the tank air spaces must be interconnected.

The filler neck and cap are usually located in a recessed well, equipped with a scupper and drain. The scupper is designed to prevent overflowing fuel from entering the wing or fuselage structure. Fuel filler openings are clearly marked with the word "fuel", the tank capacity, and the type of fuel to be used.

Fuel filler opening of transport category aircraft must be marked the word "fuel", the minimum fuel grade or permissible fuel designations, and the maximum permissible fueling and defueling pressure for pressure fueling systems.

A great deal of attention has been paid lately to aircraft accidents resulting from misfueling. Be certain that the filler openings are clearly marked. It is possible for a jet engine to operate on avgas, but a reciprocating engine will not operate on turbine fuel.

Servicing aircraft fuel tanks may be accomplished by filling each tank through the filler neck, or by pressure fueling. Pressure fueling is more practical for large aircraft than the over-the-wing method because it saves time by providing a single point to fuel the entire aircraft. Pressure fueling also reduces the chances for fuel contamination.

Pressure fueling systems vary with the aircraft manufacturers; therefore, the fueling operator should consult the aircraft manufacturer's instructions for detailed procedures. Most aircraft pressure fueling systems consist of a pressure fueling hose and a panel of controls and gages that permit one person to fuel or defuel any or all fuel tanks of an aircraft.

Aircraft pressure fueling instructions are normally placarded on the fuel control panel access door. Before fueling an aircraft using the pressure fueling method, the truck pump pressure must be correct for that refueling system. The aircraft should be bonded to the truck, and the truck grounded to drain off static charges.

Defueling may be accomplished using the pressure fueling connection, or through the over-the-wing filler openings. When defueling aircraft with sweptback wings, defuel the outboard wing tanks first. This is to maintain lateral stability. Aircraft defueling should be accomplished in the open air for good ventilation.

The use of turbine fuels in aircraft has resulted in some problems not associated with aviation gasolines. One of these problems is microbial contaminants. These growths produce a brown slime inside of the fuel tanks, and may require periodic cleaning.

If it is necessary to enter an aircraft's fuel tank: (1) defuel and purge the tank in an open and safe environment, (2) continue purging the tank during the entire work period, (3) station an assistant outside the fuel tank access to perform rescue operations if required and, (4) before entering the tank, touch a static discharge plate.

Fuel Pumps

Fuel boost pumps are operated to provide a positive flow of fuel to the engine. The electrically driven centrifugal booster pump supplies fuel under pressure to the inlet of the engine driven fuel pump. This type of pump is an essential part of the fuel system, particularly at high altitudes, to keep the pressure on the suction side of the engine-driven pump from becoming low enough for the fuel to boil.

The booster pump is mounted at the tank outlet with a detachable sump, or is submerged in fuel at the bottom of the tank. As fuel enters the pump from the tank, a high-speed impeller throws the fuel outward at high velocity. This high rotational velocity swirls the fuel and produces an action that separates the air and vapor from the fuel. This results in practically vapor free fuel delivery. Since a centrifugal-type pump is not a positive-displacement pump, no relief valve is necessary. The outlet fuel pressure on a submerged, single-speed, centrifugal-type fuel pump is regulated by the pump's design and internal clearances.

Sliding-vane type boost pumps may still be found in service. Their construction and operation are identical to engine driven fuel pumps. They are a positive-displacement pump and will require a pressure-relief valve.

Flapper-type check valves may be used in integral fuel tanks to prevent the fuel supply from flowing away from the boost pumps.

The purpose of the engine-driven fuel pump is to deliver a continuous supply of fuel at the proper pressure at all times during engine operation. The pump widely used at the present time is the positive-displacement, rotary-vane-type.

A fuel pressure relief valve is required on an positive-displacement fuel pump. This valve is designed to open at the set pressure regardless of the pressure of the fuel entering the pump.

To maintain the proper relation between fuel pressure and carburetor inlet air pressure, the chamber above the fuel pump relief valve is vented either to the atmosphere or through a balanced line to carburetor air inlet pressure. The diaphragm will compensate fuel pressures to altitude changes.

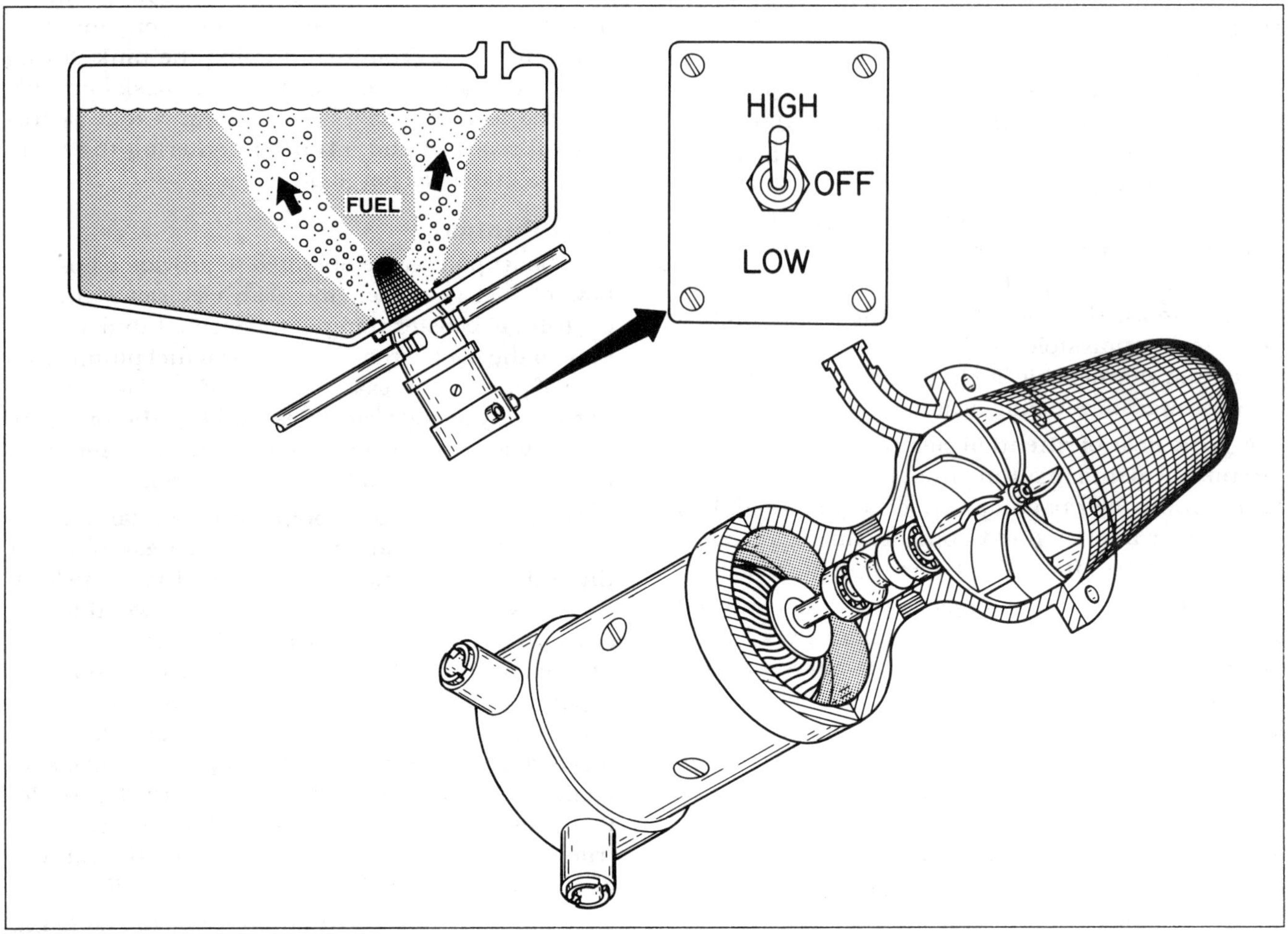

Figure 92. Centrifugal fuel booster pump.

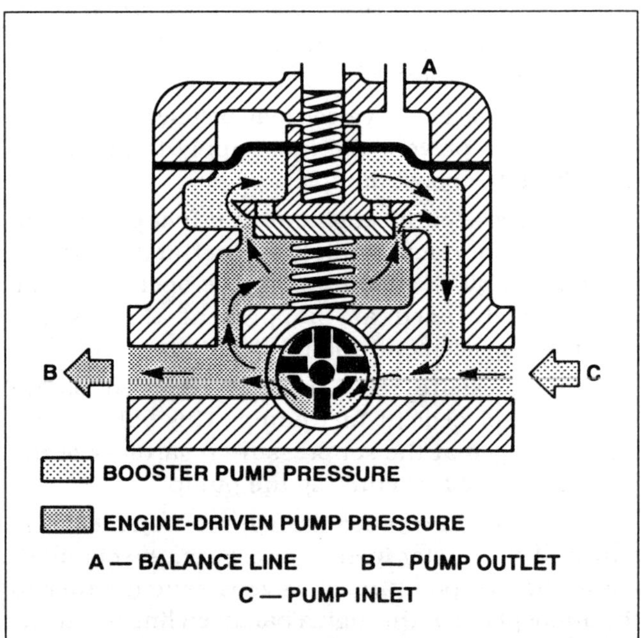

Figure 93. Engine-driven fuel pump (pressure delivery).

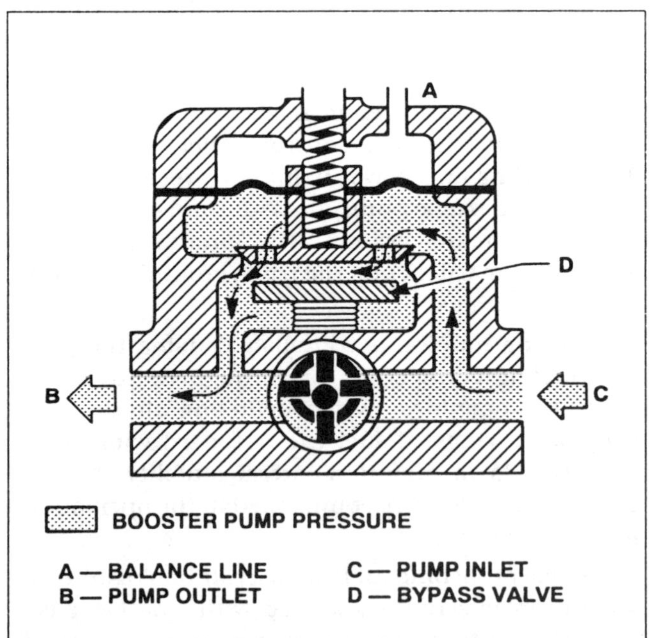

Figure 94. Engine-driven fuel pump (bypass flow).

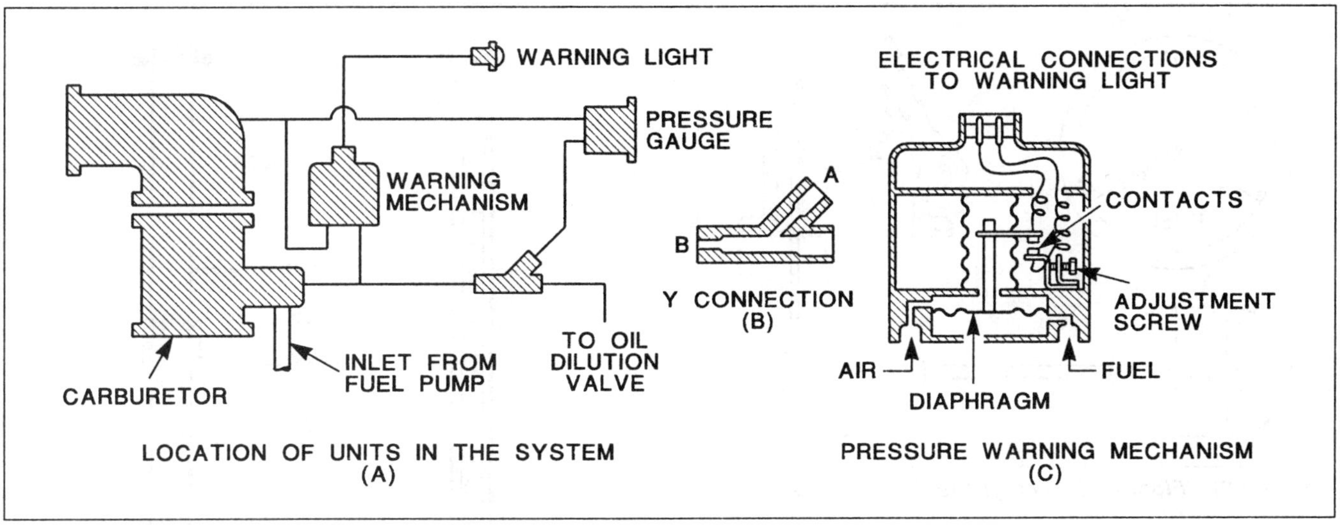

Figure 95. Fuel pressure indicating system.

In addition to the relief valve, the fuel pump has a bypass valve that permits fuel to flow around the pump rotor whenever the pump is inoperative. When fuel is needed for starting the engine, or in the event of an engine-driven pump failure, fuel supplied by the booster pump is allowed to enter the carburetor.

The fuel pressure gage is a differential pressure indicator with two connections on the back of the indicator housing. The air connection is vented to the carburetor air inlet, and the fuel connection is attached to the fuel inlet chamber of the carburetor.

In some aircraft with several fuel tanks, the possible danger of allowing the fuel supply in one tank to become exhausted before the selector valve is switched to another tank is prevented by the installation of a fuel pressure warning system. The first positive indication that a change-over from one fuel tank to another would come from the fuel pressure warning system.

A pressure-sensitive mechanism is used to actuate the fuel pressure warning system. A fuel pressure warning switch contacts close and a warning light is turned on when the fuel pressure drops below specified limits.

Fuel pressure for the pressure warning signal is taken at the fuel pressure line of the carburetor. A transmitter in the fuel pressure warning system converts fluid pressure to an electrical signal. This mechanism can be adjusted to change the fuel pressure warning limits. The manufacturer's maintenance manuals are necessary to effectively troubleshoot the fuel pressure warning system.

Fuel System Components

In an aircraft fuel system, the various tanks and other components are usually joined together by fuel lines made of metal tubing connected by lengths of flexible hose. The diameter of the tubing is governed by the fuel flow requirements of the engine.

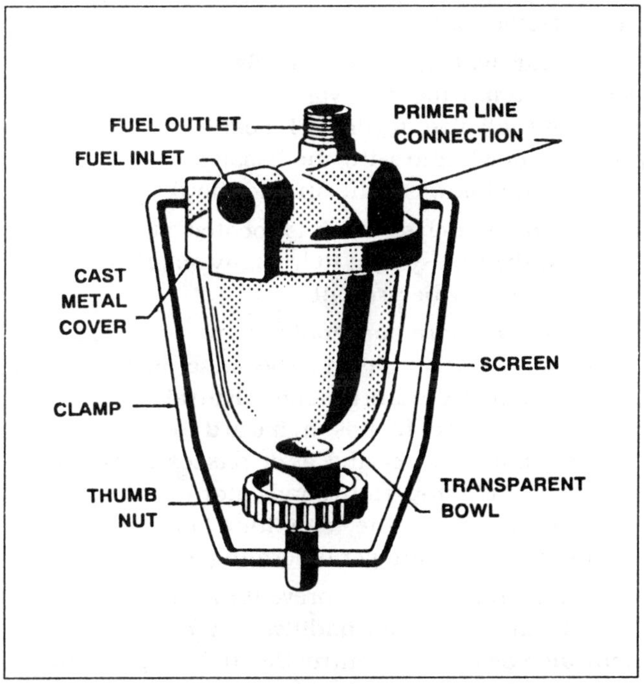

Figure 96. Main fuel strainer for light aircraft.

Selector valves are installed in the fuel system to provide a means for shutting off the fuel flow, for tank and engine selection, for crossfeed, and for fuel transfer. The valve must accommodate the full flow capacity of the fuel line, must not leak, and must operate freely with a definite "feel" or "click" when it is in the correct position.

The location of leaks and defects within the internal portions of the fuel system can usually be deter-

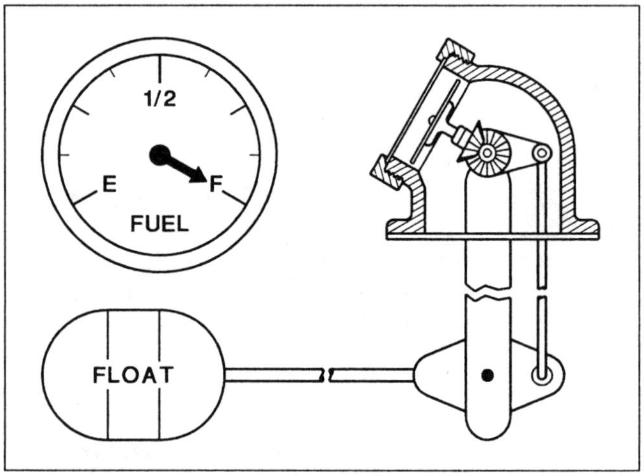

Figure 97. Float-and-level-type fuel level gage.

mined by observing the pressure gage and operating the selector valves.

Internal leakage of a fuel valve may be checked without removing the valve from the aircraft by placing the valve in the OFF position, draining the strainer bowl, and with the boost pump on, watch to see if fuel flows to the strainer bowl.

The main fuel strainer is located at the lowest point in the fuel system to trap any small amount of water that may be present.

Jet fuel is more susceptible to water contamination than aviation gasoline because jet fuel has a higher viscosity than gasoline. Entrained water in aviation turbine fuel is a hazard because of its susceptibility to freezing as it passes through the filters. A fuel temperature indicator may be used to determine when the condition of the fuel is approaching the danger of forming ice crystals.

A common method of preventing this icing is the use of anti-icing fuel additives. A heat exchanger may also be used to control the fuel temperature on some turbine-powered aircraft. Heat exchangers may use engine bleed air, or engine lubricating oil, to warm the fuel.

Fuel Quantity Indicators

Fuel quantity gages are necessary so that the operator may know the amount of fuel remaining in the tanks. Four general types of fuel quantity gages may be found on aircraft. They include: (1) sight glass, (2) mechanical, (3) electrical and, (4) electronic. Each fuel quantity indicator must be calibrated to read zero when the quantity of fuel remaining is equal to the unusable fuel supply.

The sight glass is the simplest form of fuel quantity gage. The indicator is a glass or plastic tube placed

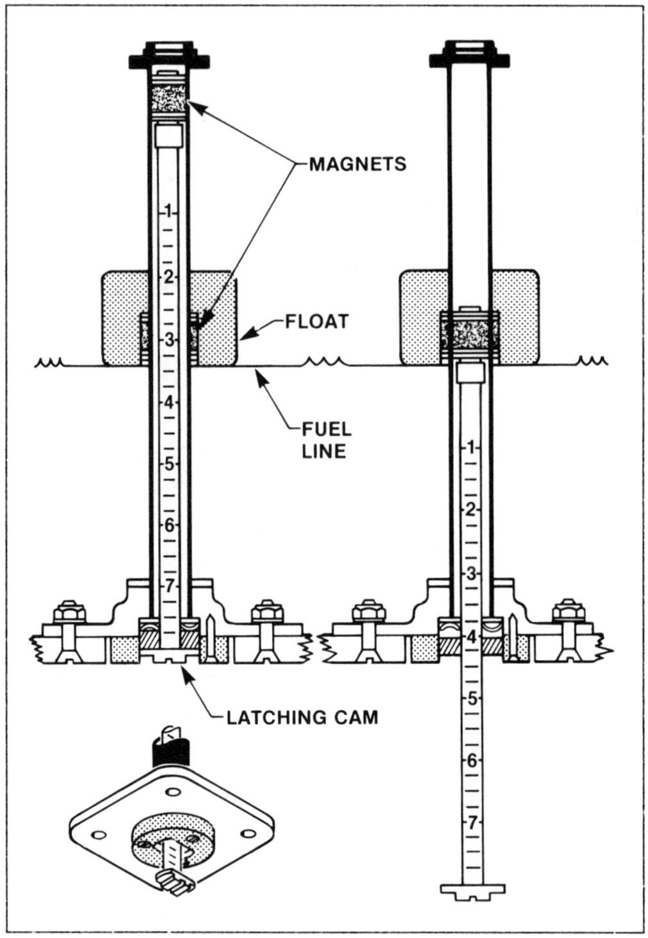

***Figure 98.** Under-wing fuel quanity indicating stick used to manually determine the quanity of fuel in a tank.*

on the same level as the tank. It operates on the principle that a liquid seeks its own level.

The mechanical-type fuel quantity gage is usually located in the tank and is known as a direct reading gage. One type of mechanical gage is illustrated in figure 97.

An electrical-type fuel quantity indicating system consists of an indicator in the cockpit and a float-operated transmitter installed in the tank. This transmitter sends an electric signal to the fuel quantity indicator.

One advantage of electrical and electronic fuel quantity indicating systems is that several fuel tank levels can be read on one indicator. Another advantage is that the indicator can be located any distance from the tank.

The electronic-type (capacitance) fuel quantity gage differs from the other types in that it has no moving parts in the fuel tank. The probe of a capacitance-type fuel level gage is essentially a condenser with fuel and air acting as a dielectric. Several probes may be installed in each fuel tank.

The electronic-type fuel quantity indicating system consists of a bridge circuit, an amplifier, an indicator, and a tank unit. This system is more accurate in measuring fuel level because it measures weight instead of gallons.

When fuel quantity is measured in pounds instead of gallons, the measurement will be more accurate because fuel volume varies with temperature change. Cold fuel is heavier per gallon than warm fuel.

Most large transport category aircraft have provisions for ground personnel to gage the amount of fuel in a tank, manually. The drip gage is one means of accomplishing this task.

When the cap and hollow drip tube are drawn out from the lower wing surface, the fuel enters the open top of the tube when it reaches the level of the fuel. A steady drip signifies that the tube is properly positioned, and the fuel quantity can be read directly.

A fuel totalizer is a component which indicates the amount of fuel in all the fuel tanks.

Multi-engine Fuel Systems

The design of the fuel system for an aircraft having two or more engines presents problems not normally encountered in single-engine fuel systems. A large number of fuel tanks are often required to carry the necessary fuel. These tanks may be located in widely separated parts of the aircraft. The individual engine fuel systems must be interconnected so that fuel can be fed from any tank to any engine.

Crossfeed systems are used in multiengine aircraft fuel systems to maintain a balanced fuel condition. Their primary purpose is allow the feeding of any engine from any tank.

Certain large aircraft may use a fuel manifold system. This is actually a variation on the crossfeed system. Advantages to this system include: (1) any engine can be fed from any tank, (2) all engines can be fed from all tanks simultaneously, (3) a damaged tank can be isolated from the rest of the fuel system and, (4) all tanks can be serviced through a single connection.

Fuel Jettison Systems

A fuel jettison system is required for transport category and general aviation aircraft if the maximum takeoff weight is 105% of the maximum landing weight. It must be able to jettison enough fuel within 10 minutes for general aviation or 15 minutes for transport category aircraft, to meet the requirements of the specification.

Fuel jettisoning past the limits prescribed by the FARs is prevented by dump valve limit valves or a low level circuit. The fuel jettison valve must be designed to allow flight personnel to close the valve during any part of the jettisoning operation.

The system consists of lines, valves, dump chutes, and chute-operating mechanisms. Each wing contains either a fixed or extendable dump chute. In either case the fuel must discharge clear of the airplane. The system and its operation must be free of fire hazards.

The fuel jettison system is usually divided into two separate, independent systems, one for each wing. This is employed to maintain lateral stability when jettisoning fuel. Fuel is moved overboard in most fuel jettisoning systems by boost pumps, and is accomplished through a common manifold and outlet in each wing.

Fuel System Maintenance

Welded steel tanks are most common in the smaller single and twin engine aircraft. When leaks are found, the tank must be drained and inerted. Inerting (purging) the tank may be accomplished by slowly discharging a carbon dioxide fire extinguisher into the tank. If the tank is to be welded, removal is necessary.

Before welding a fuel tank, the interior should be steam cleaned for a minimum of 8 hours. This is to remove any traces of fuel. If aluminum tanks are welded using oxyacetylene equipment, a flux must be used. After welding, a 5% solution of nitric or sulfuric acid should be used to remove any remaining flux. Federal Aviation Regulations define the test procedures to be followed when testing a repaired fuel tank of a small aircraft.

Fuel cell leaks will usually appear on the lower skin of the aircraft. A fuel stain in any area should be investigated immediately. Fuel cells suspected of leaking should be drained and removed. All fuel cell maintenance should be accomplished in accordance with the manufacturer's specifications. If it is necessary to leave a bladder-type fuel tank empty for an extended period of time, the inside of the tank should be coated with engine oil.

The integral tank is a nonremovable part of the aircraft. Because of the nature of an integral tank, some leaks allow fuel to escape directly to the atmosphere. This makes is feasible to disregard certain minute leaks that do not represent a fire hazard or too great a loss of fuel. Fuel leaks located in unvented areas of the aircraft are considered flight hazards. Leaks in an integral tank must be repaired in accordance with the aircraft manufacturer's specifications. Riveting may be used to repair most integral fuel tanks.

A fuel system schematic would be most useful to locate and troubleshoot an internal fuel leak in a large aircraft fuel system. When inspecting a fuel system, you should check all valves located downstream of boost pumps with the boost pumps operating.

When moving the mixture control on a normally operating engine into the idle cutoff position, engine RPM should slightly increase before the engine starts to die. This will indicate a proper idle mixture control setting.

One of the first items to be checked when troubleshooting a rough running fuel-injected engine is excessive fuel pressure. The presence of fuel stains around a fuel nozzle would indicated a clogged fuel nozzle.

When installing a rigid fuel line, ½" in diameter, the lines should be supported each 16".

There are three types of aviation gasoline in use today. They are identified as 80, 100, and 100LL. These identifiers represent performance ratings and are related to the anti-knock characteristics of the gasoline. The antiknock characteristics of a fuel may be improved by adding a knock inhibitor. This is usually in the form of tetraethyl lead (TEL). In the past aromatic fuels were used with ratings exceeding these listed. One disadvantage of using aromatic fuels is that they deteriorate rubber parts.

One very important characteristic of aviation gasoline is its ability to vaporize. If aviation gasoline does not vaporize readily enough it can cause hard starting, poor acceleration, uneven fuel distribution and excessive crankcase dilution. If it vaporizes too readily, it can cause vapor lock.

A measure of a gasoline's tendency to vapor lock is obtained from the Reid vapor pressure test. The maximum allowable vapor pressure for aviation gasoline is 7 PSI. The vapor pressure of aviation gasoline is lower than the vapor pressure of automotive gasoline.

REVIEW QUESTIONS — Aircraft Fuel Systems

163. *When is a fuel jettison system required?*

164. *How is fuel moved overboard in most fuel jettison systems?*

165. *What procedure should be followed when defueling an aircraft with swept wings?*

166. *What is the primary purpose of the crossfeed system?*

167. *What are the advantages of pressure fueling?*

168. *What repair method may be used to repair most integral fuel tanks?*

169. *What gas is used for purging an aircraft fuel tank?*

170. *What is the purpose of the diaphragm in a vane-type fuel pump?*

171. *What should be used to remove the flux from an aluminum fuel tank after a welded repair?*

172. *What type of remote-reading fuel quantity indicating system has no moving parts in the tank?*

173. *What kind of probe is used in an electronic fuel quantity system?*

174. *Which weighs more per gallon, warm fuel or cold fuel?*

175. *What is the purpose of flapper type check valves in integral fuel tanks?*

176. *What sources of heat may be used for turbine engine fuel heaters?*

177. *Where should aircraft defueling be accomplished?*

178. *What are the purpose of baffle plates in fuel tanks?*

179. *What markings should be next to the fuel filler opening?*

180. *What type of boost pump requires a pressure-relief valve?*

Answer questions 8698 through 8802 in the FAA Test section of this book.

ELECTRICAL SYSTEM COMPONENTS

The satisfactory performance of any modern aircraft depends to a very great deal on the continuing reliability of the electrical systems. The continued proper performance of electrical systems depends on the knowledge and techniques of the technician who installs and maintains the equipment.

Electrical Equipment Installation

The systems are designed to maintain high reliability in the wire termination to prevent circuit failures. Wire is manufactured in sizes according to a standard known as the AWG (American Wire Gage). The AWG system of numbers is used to designate electrical wire sizes, the number assigned to a size is related to its cross-sectional area.

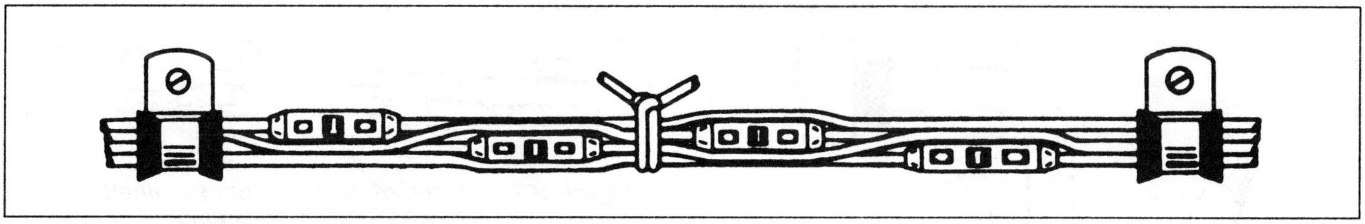

Figure 99. Staggered splices in a wire bundle.

Wire size must be determined by considering a number of factors, including:

(1) Permissible voltage drop

(2) Current carrying capacity of the conductor

(3) Type of load

(4) Allowable power loss

Allowable voltage drop may be calculated with the following formula:

$$VD = RLA$$

Where:

VD = Voltage drop

R = Resistance per foot

L = Length of wire

A = Amperes

The two most generally used conductors are copper and aluminum. When replacing copper electrical wire with aluminum, the aluminum wire must be two wire sizes larger than the copper. A No. 10 copper electrical cable would be replaced with a No. 6 aluminum cable.

Aluminum wire must be stripped very carefully because individual strands will break easily after being nicked.

Wiring may be installed in a variety of manners, depending on the size and function of the wire. The following definitions must be kept in mind:

(1) Open wiring — any wire, wire group, or wire bundle not enclosed in conduit.

(2) Wire group — two or more wires going to the same location tied together to retain identity of the group.

(3) Wire bundle — two or more wire groups tied together because they are going the same direction at the point where the tie is located.

Electric wiring installed in aircraft without special enclosing means (open wiring) offers the advantages of ease of installation, simple maintenance, and reduced weight.

When bundling open wiring, the bundles should be limited as to the number of cables to minimize damage from a single electrical fault. The minimum bend radii for electrical wire in groups or bundles is 10 times the outside diameter of the group or bundle.

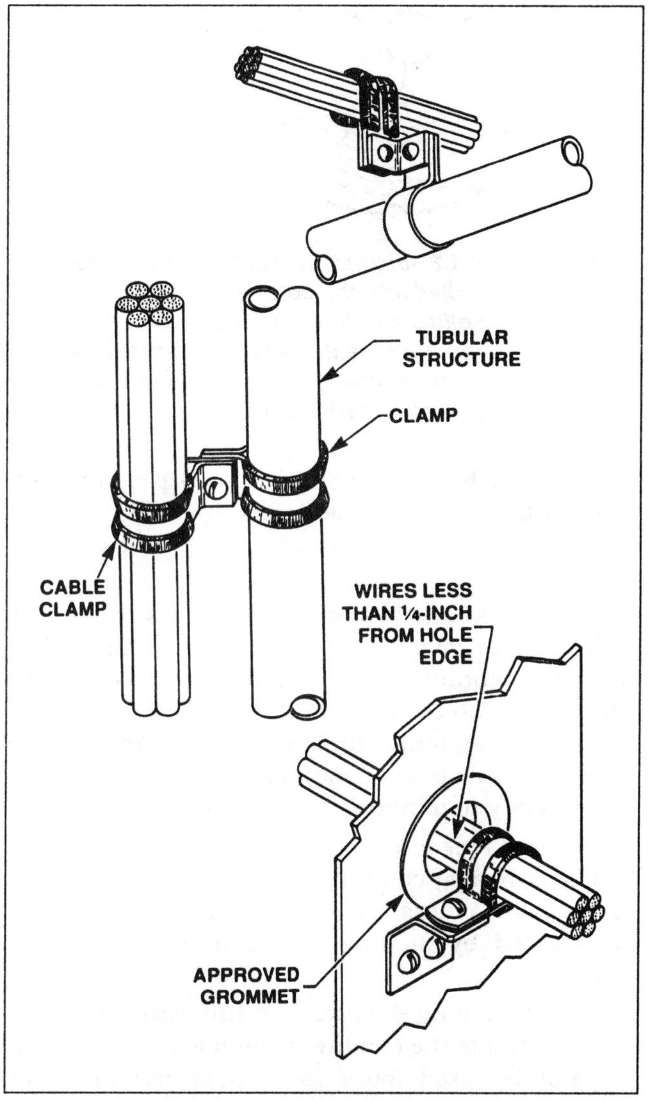

Figure 100. Method of attaching wire bundles to an aircraft structure, using cushion clamps.

Spliced connections in wire groups or bundles should be located so that they can be easily inspected. When splicing electrical wire, stagger the splices along the length of the bundle with no more than one splice between any two connectors.

Wires and wire groups should be protected against chafing or abrasion in those locations where contact with sharp surfaces or other wires would damage the

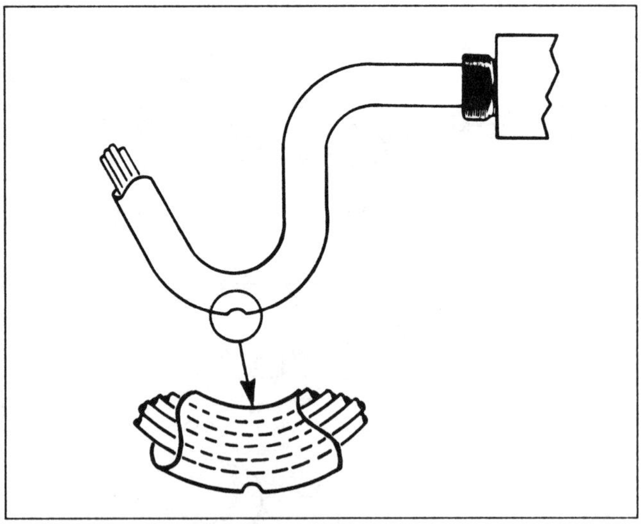

Figure 101. A 1/8"-diameter drain hole should be drilled into the bottom of all runs of solid conduit at the lowest point in the run. Be sure that all burrs are removed from the inside of the conduit before putting the wire in it.

insulation. Where electric cables must pass through holes in bulkheads, formers, ribs, firewalls, etc., the wires should be protected by using a rubber grommet.

If a wire is installed so that it comes in contact with some moving parts, it should pass through a conduit. Conduit offers mechanical protection to wires and cables. It is available in metallic and nonmetallic materials and in both rigid and flexible forms. Flexible brass conduit may be used where it is necessary to minimize radio interference.

If several long lengths of electrical cable are to be installed in rigid conduit, the possibility of damage to the cable as it is pulled through the conduit will be reduced by dusting the cable with powdered soapstone.

When installing electrical conduit, care should be taken to locate the conduit in such a manner that it will not be used for a footstep or handhold for personnel.

Groups of wires may be terminated with quick-disconnect plugs, and individual wires will have terminals crimped onto the wire. These terminals are then attached to a screw-type lug on a terminal strip, or on the electrical component itself.

When attaching a terminal to the end of an electric cable, it should be determined that the strength of the cable-to-terminal joint is at least equal to the tensile strength of the cable itself. If it is necessary to use an electrical connector where it may be exposed to moisture, the technician should use a

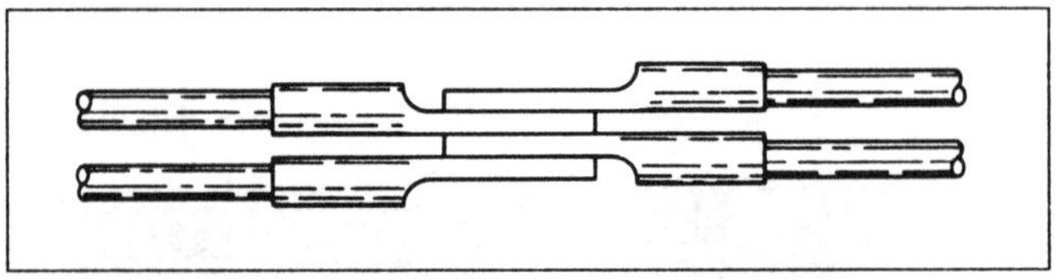

Figure 102. Method of stacking four terminals for attachment to a single lug on a terminal strip.

special moisture-proof type. Many wires are terminated in AN or MS electrical connectors. These are pin and socket connectors. Older AN and MS connectors were soldered. Newer connectors use special crimping tools and insertable pins and sockets.

Most of the wires that are not terminated in quick-disconnect plugs are connected to terminal strips housed in junction boxes.These wires use crimped ring terminals. Terminal strips normally have barriers between the adjacent studs to keep the wires separated. Ideally, there should only be two wires attached to any stud, but in no case can there be more than four terminals stacked on any stud. During inspection of the terminal strips it should be determined that the terminal studs are anchored against rotation.

The terminal strips are normally mounted in junction boxes made of either aluminum alloy or fiberglass. Aircraft electrical junction boxes located in a fire zone are usually constructed of stainless steel. These boxes are mounted where there is a minimum chance of their being damaged. All boxes must have drain holes to prevent any accumulation of water. They should be mounted either vertically or facing downward, to minimize the possibility of washers and nuts dropping between the terminals. It is important to prevent oil canning in electrical junction boxes to prevent internal short circuits.

Bonding is the electrical connecting of two or more conducting objects not otherwise adequately connected. Aircraft components are bonded to allow electrical charges to move through the aircraft structure without causing sparks. Grounding is the electrical connecting of a conducting object to the primary structure for a return path for current.

Electrolytic action can rapidly corrode a bonding connection if suitable precautions are not observed. Aluminum alloy jumpers are recommended for most cases; however copper jumpers can be used to bond together parts made of stainless steel, cadmium plated steel, copper, brass, or bronze.

When selecting hardware for attaching bonding connections to an aircraft structure, the amount of current to be carried, mechanical strength, and ease of installation should be considered.

The resistance of all bond and ground connections should be tested after connections are made. The resistance of each connection should normally not exceed 0.003 ohm.

Static dischargers are installed to dissipate static electricity collected by bonded structure. The static is discharged to the atmosphere as a low level current.

Conductors should be protected with circuit breakers or fuses located as close as possible to the electrical power source. The circuit breaker or fuse should open before the conductor emits smoke.

Aircraft fuse capacity is rated in amperes. The two kinds of fuse holders used most commonly in aircraft circuits are plug-in, and clip types. Fuse selection should be so that the capacity matches the needs of the circuit.

Most fuses have a strip of metal which is designed to melt at a set temperature. Current limiters are a type of fuse that will stand a considerable overload for a short period of time before they open.

Circuit breakers have an advantage over fuses in that they are resettable and reusable. All resettable circuit breakers should open the circuit in which they are installed, regardless of the position of the operating control, when an overload or circuit fault exists. This type of circuit breaker is referred to as "trip-free". Automatic reset circuit breakers should not be used as circuit protective devices in aircraft.

Switches control the current flow in most aircraft electrical circuits. A switch is used to start, stop, or change the direction of the current flow in the circuit. The switch in each circuit must be insulated heavily enough for the voltage of the circuit.

The nominal current rating on an aircraft switch is normally stamped on the switch housing. Nominal rating of switches refers to continuous current rating with the contacts closed. For known continuous load current applications, derated switches may be selected to obtain reasonable switch efficiency and service life. Inadvertent operation of a switch may be prevented by the installation of a suitable guard over the switch.

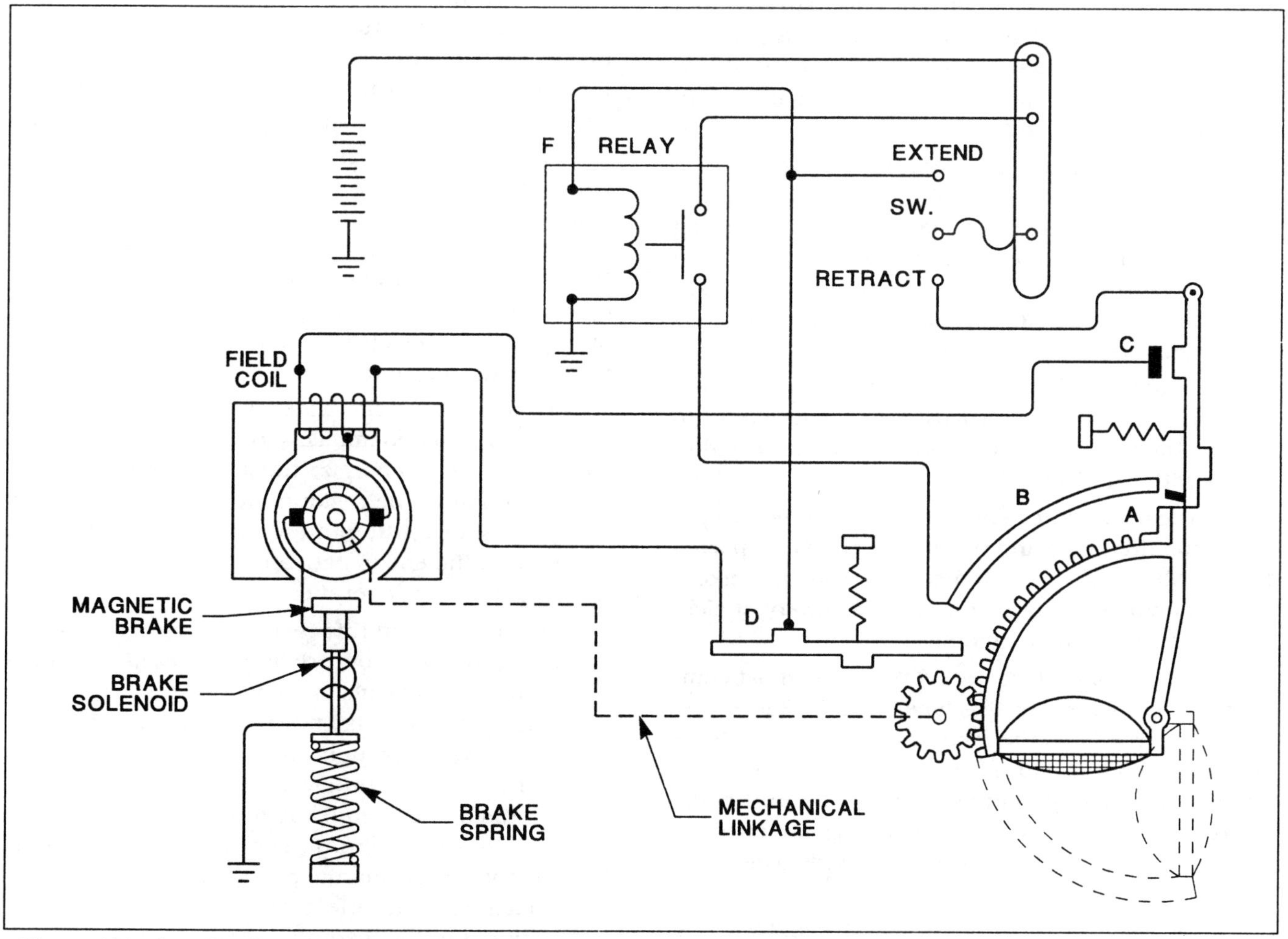

Figure 103. Landing light mechanism and circuit.

Aircraft lighting systems provide illumination for both exterior and interior use. Aircraft operating at night must be equipped with position lights. The Federal Aviation Administration establishes the minimum standards concerning the number, color, location, and intensity of position lights for navigation on civil aircraft. The position lights are located forward (L & R) and aft.

The green light unit is always mounted at the extreme tip of the right wing. The red unit is mounted in a similar position on the left wing. The white unit is usually located on the vertical stabilizer in a position where it is clearly visible through a wide angle from the rear of the aircraft. If one switch is used to control all navigation lights, the lights are most likely connected parallel to each other and in series with the switch.

Anticollision lighting is also used on aircraft for night operations. When installing an anticollision light, install a switch independent of the position light switch.

Landing lights are installed in aircraft to illuminate runways during night landings. These lights may be streamlined into the leading edge of the aircraft's wing, or may be a retracting type. When the lamps are not in use, a motor retracts them into receptacles in the wing.

As shown in figure 103, one type of retractable landing light motor has a split field winding so the motor can be operated in either direction. A landing light retraction motor may employ an armature brake to stop the motor. When a motor is equipped with this type of brake, a friction brake is applied by a spring and released by a magnet.

Rheostats may be used in some cockpit lighting circuits. When adding a rheostat to a light circuit to control the light intensity, it should be connected in series with the light.

Electrical equipment installations will require periodic maintenance and troubleshooting of reported faults. Much of this maintenance will be accomplished as a result of testing with a variety of electrical measuring instruments.

The electrical multimeter (VOM) is the most common tool employed for this type of work. Remember for voltage readings the meter is installed parallel to the unit being measured, and for current readings the meter is installed in series with the unit being measured. When very high resistances are encountered, a Megohmmeter may be necessary to measure these value.

An electrolytic capacitor can be checked with an ohmmeter to determine its condition. Caution should be observed when handling a high-voltage capacitor in an electrical circuit because it may retain its charge after power is removed.

When a diode is checked for an open circuit or a short circuit, it should be disconnected from the circuit.

A thermocouple type ammeter may be used to measure radio frequency alternating current. The most accurate type of frequency measuring instrument utilizes an integrated circuit chip having a clock circuit.

Do not overlook simple logic when troubleshooting. If a transformer has some of its turns shorted together, it will get hot.

Generators

All rotating generators produce alternating current, but in order to use it to charge batteries and to operate direct current machinery, we must convert it into DC. We do this before it leaves the generator, with a mechanical switch called the commutator. The commutator is mounted on the end of the shaft that supports the rotating coils. This assembly is known as the armature. Carbon brushes, mounted in spring holders, ride on the commutator as it rotates to pick up the current as it is produced.

The magnetic field which is cut by the rotating coils is produced by an electromagnet. These electromagnets are known as the field coils and are wound on soft iron cores known as the pole shoes. These iron cores are constructed of many laminations, to reduce the effects of eddy currents. The field coils are mounted to the generator frame. It is the strength of this electromagnetic field that will determine the output of the generator.

The soft iron pole shoes will retain a certain amount of magnetism. This residual magnetism is the force used to energize the generating system on engine startup. The residual magnetism present in the pole shoes will produce a "residual voltage" which is sufficient to activate the voltage regulator and begin system operation.

If for some reason this residual magnetism is lost, it may be restored by briefly connecting a voltage source to the field coils to create a surge of magnetic force. This operation is known as "flashing the field". Current flowing through the armature sets up electromagnetic fields in the windings. These new fields tend to distort, or bend, the magnetic flux between the poles of the generator from a straight line path. This distortion of the magnetic field is called armature reaction. The effect of armature reaction is proportional to the load on the generator.

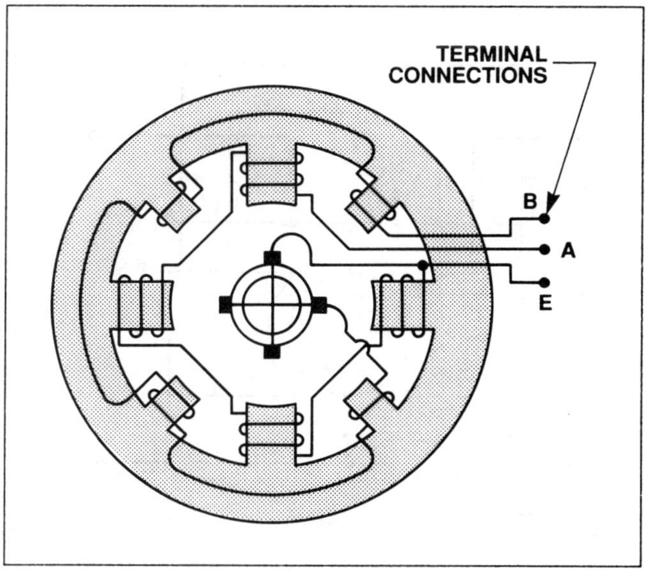

Figure 104. Generator with interpoles.

The method most often used in overcoming the effect of armature reaction is through the use of interpoles. An interpole is a pole placed between the main poles of a generator. Interpoles counteract field distortion and reduce arcing at the brushes.

There are three types of DC generators: series-wound, shunt-wound, and compound wound. The difference in type depends on the relationship of the field winding to the external circuit. Generators are rated in amperes at rated voltage, and this information is usually stamped on the generator.

The field winding of a series generator is connected in series with the external circuit, called the load. Series generators have very poor voltage regulation and are never employed as an airplane generator.

A generator having a field winding connected in parallel with the external circuit is called a shunt generator. Shunt-type generators may be found on light to medium aircraft.

A compound-wound generator combines a series winding and a shunt winding in such a way that the characteristics of each are used to advantage. This type of generator may be used for heavy duty applications.

Generator overhaul may be accomplished in many maintenance facilities. Certain pieces of test equipment are required. For testing armatures, a device known as a growler is used. This unit is basically an electromagnet forming a V into which the armature can be placed.

The growler may be used to determine the presence of a shorted armature. To test for a shorted circuit in the winding, a thin strip of steel (a hacksaw blade works well) is placed on the armature segments and the armature slowly rotated on the growler. One or more shorted coils will cause a strong vibration of the metal strip.

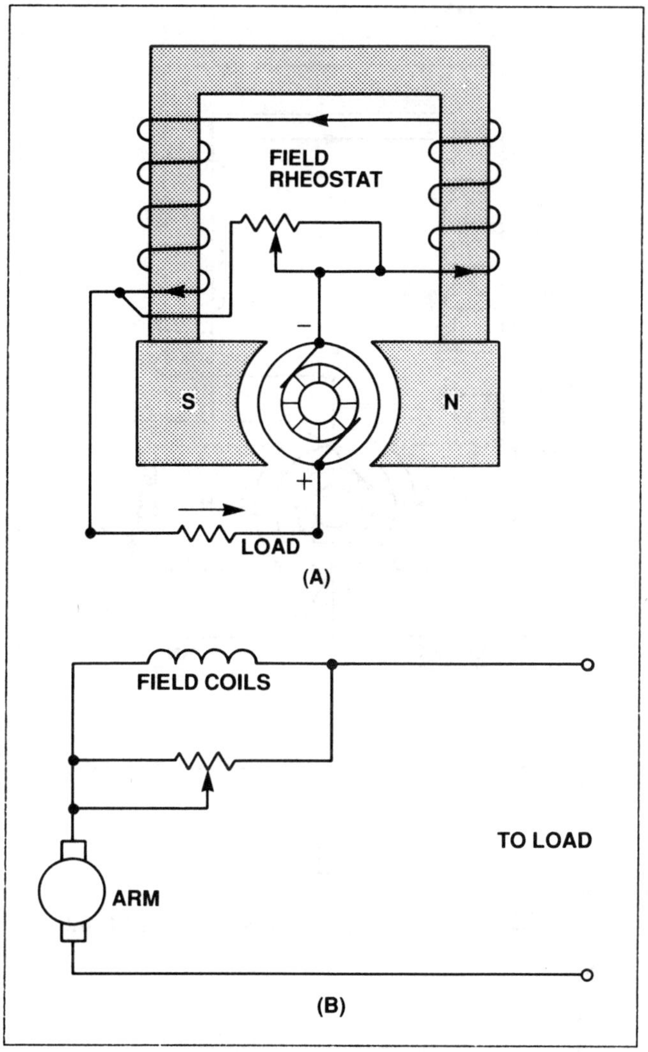

Figure 105. Diagram and schematic of a series-wound generator.

To test generator or motor armature winding for opens, place the armature in a growler and connect a 110 VAC test light on adjacent segments; the light should light.

When using an ohmmeter to check the continuity of a generator field coil, the coil should show very low resistance if it is a series field coil.

Commutator surfaces may be restored by turning the armature on a lathe and smoothing the surface of the commutator. After turning the armature, the mica insulation between the commutator bars of may be undercut to a depth equal to the width of the mica.

A spring scale may be used to measure brush spring tension. Excessive pressure will cause the

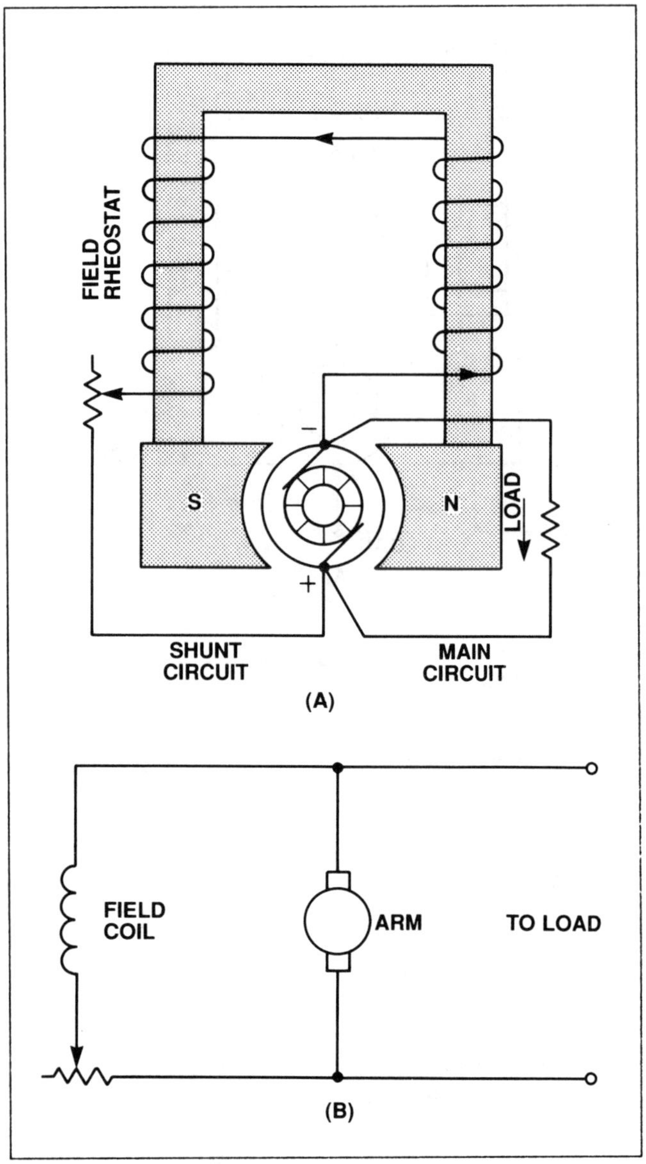

Figure 106. Shunt-wound generator.

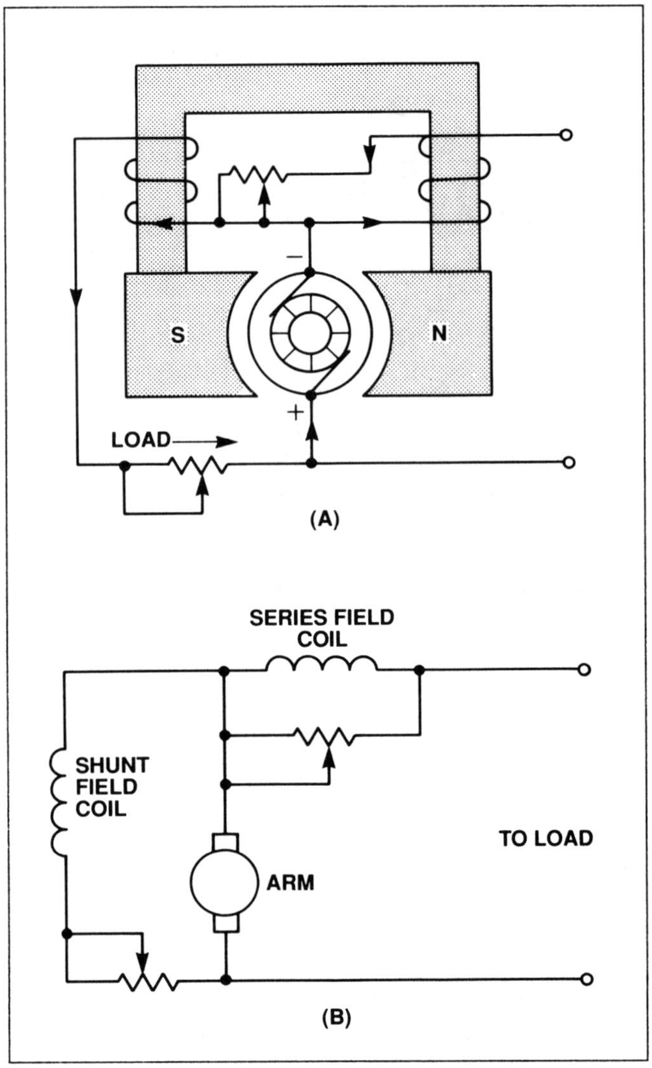

Figure 107. Compound-wound generator.

brushes to wear quickly, and too little pressure will allow "bouncing" of the brushes, resulting in burned or pitted surfaces on the commutator.

Flexible low-resistance pigtails are provided on most brushes and their connections should be secure. The brush pigtail eliminates any possible sparking to the brush guides caused by the movement of the brushes within the holder.

The only practical method of maintaining a constant voltage output from an aircraft generator under varying conditions of speed and load is to vary the strength of the magnetic field. A voltage regulator controls generator output by varying current flow to the generator field coil.

While a variety of electromechanical and electronic voltage regulators are in use today, it is the three unit vibrator type voltage regulator that is often used to understand regulator operating principles.

The three units of this type of voltage regulator are the voltage regulator, the current limiter, and the reverse current cut-out relay.

The voltage regulator section controls voltage output by regulating the strength of the field. When the regulator points are closed current is directed to the field coils and output is increased. When the points are open only a small amount of current is available to the coils. If a generator is equipped with a vibrator-type voltage regulator, the actual time the voltage regulator points remain open depends on the load carried by the generator.

The current limiter decreases the voltage output of the generator by opening its field circuit and inserting a resistor anytime the current output of the generator exceeds the manufacturer's rating.

The reverse current cutout relay will open the main generator circuit whenever the generator voltage drops below the battery voltage. This is to prevent the battery from "motorizing" the generator.

If the reverse current cutout relay contact points fail to open after the generator output has dropped below battery potential, current will flow through the generator armature opposite the normal direction and through the shunt field in the normal direction.

On multi-engine aircraft, generators may be connected in parallel. When DC generators are operated in parallel to supply power for a single load, their controls include an equalizer circuit to assure that all generators share the load equally. The equalizer circuit operates by increasing the output of the low generator and decreasing the output of the high generator until they are equal.

If any one generator in a 24-volt DC system shows low voltage, the most likely cause is an out-of-adjustment voltage regulator.

Certain turbine powered aircraft may use a single unit which will serve as both the starter and the generator. During starts, current is directed to the unit from the battery or a ground power source. After engine start, the unit will supply current to operate the aircraft systems, and charge the battery.

The high current flow in the starter-generator creates a great deal of heat. During ground operation, a starter-generator is normally cooled by an integral fan. During flight operations it is cooled by the integral fan and ram air.

Troubleshooting generator systems will depend heavily on the technicians understanding of how the system operates, and what voltages should appear at various points in the circuit. The following are some troubleshooting hints:

(1) If a short circuit occurs between the positive armature lead and the field lead of a shunt generator which has the voltage regulator located in the positive side of the field circuit, the generator voltage will increase.
(2) If the positive field lead between a generator and generator control panel breaks and is shorted while the engine is running, a voltmeter connected to generator output would indicate residual voltage.

A battery-generator system provides direct current. On installations requiring alternating current from the battery-generator systems, it is necessary to have an inverter to convert a portion of the aircraft's DC power to AC. This AC is used mainly for instruments, radio, radar, lighting, and other accessories.

There are two basic types of inverters; the rotary and the static. Either type can be single-phase or multi-phase. One type of rotary inverter is the inductor-type.

The frequency of the inductor-type inverter is determined by the number of poles and the speed of the motor. The voltage is controlled by the DC stator field current. A cutaway of this type of inverter is shown in figure 108.

Since static inverters use solid-state components, they are considerably smaller and lighter in weight than rotary inverters. Static inverters are commonly used to provide power for such frequency sensitive instruments as the attitude gyro and directional gyro.

The use of alternating current (AC) in aircraft offers several advantages over DC. Alternating current systems operate at higher voltages than DC systems and therefore use less current and may use smaller and lighter weight wiring. In an AC electrical power system, voltages may be easily stepped up and down using transformers. Whenever DC is required, the alternating current can be rectified to produce the needed DC. Diodes are a common device used as a rectifier.

Alternators are rated in volt-amps, which is a measure of the apparent power being produced by the generator. In an AC circuit with no phase lead or lag, real power equals apparent power.

Most aircraft alternating current systems operate at a frequency of 400 Hertz. Maintaining a constant frequency is a necessary task when generating AC. This constant speed is maintained by a CSD (constant speed drive) driven from the engine gearcase and drives the generator using a variable speed hydraulic drive. CSDs are cooled by internal oil spray. CSDs are fully controllable from the cockpit. Once shut down in flight, a CSD can only be put back on-line on the ground. The frequency of the output of an AC generator (alternator) depends upon the speed of rotation and the number of field poles. Three cycles of AC voltage are produced in a 6-pole alternator of the revolving field type during each revolution of the rotor.

AC generators used on aircraft may be of the brush-type or the brushless type. In the brush-type alternator, the rotor has an exciter winding as well as the field winding. The exciter field is controlled by the voltage regulator to vary the current produced by the exciter armature. The AC produced in this armature is rectified by a brush-and-commutator arrangement, and the resulting DC output is fed

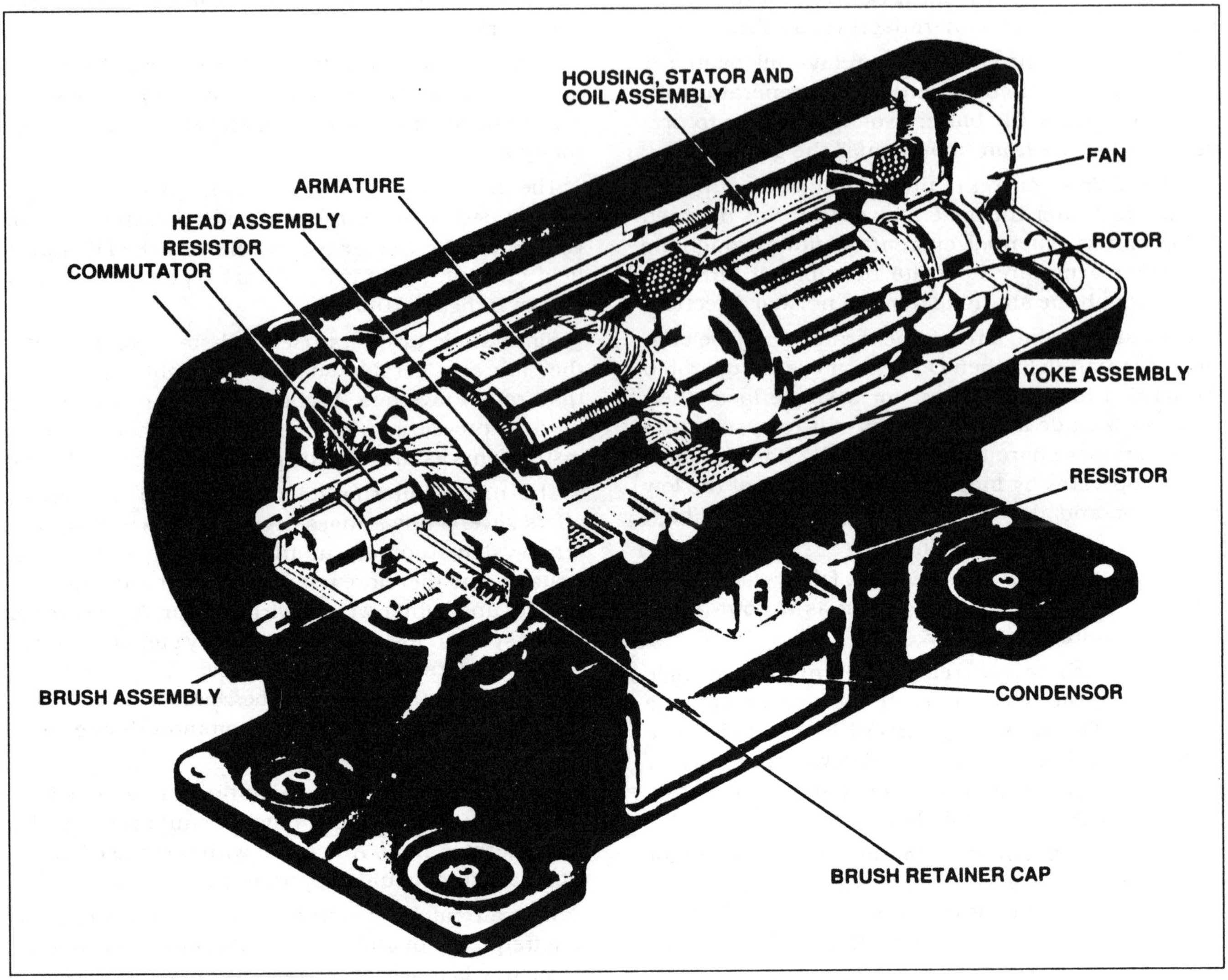

Figure 108. Cutaway view of inductor-type rotary inverter.

through slip rings into the field windings of the rotor. By using this arrangement, the needed amount of alternator field current can be controlled by a much smaller current through the voltage regulator.

Most of the AC generators used in the large jet-powered aircraft are of the Integrated Drive (IDG) brushless type. Since brushless alternators have no current flow between brushes and slip rings, they are quite efficient at high altitudes, where brush arcing could be a problem.

On multi-engine aircraft using AC generators, a paralleling system must be installed. When AC generators are operated in parallel, the frequency and voltage must both be equal.

Transformers may be used to step-up and step-down alternating current. A transformer uses two coils of wire wound on a common core. The relationship between the number of turns of wire in each coil will determine the relationship of the voltage and current in and out of the transformer. The voltage in an AC transformer secondary that contains twice as many loops as the primary will be greater and the amperage less than in the primary.

An ammeter is installed in a battery charging system to indicate the rate of current being used to charge the battery.

In installations where the ammeter is in the generator or alternator lead, and the regulator system does not limit the maximum current that the generator or alternator can deliver, the ammeter can be redlined at 100% of the generator or alternator rating.

Major adjustments on equipment such as regulators, contactors, and inverters are best accomplished outside the airplane on test benches with necessary instruments and equipment. Adjustment procedures should be as outlined by the equipment manufacturer.

Motors

DC motors are constructed in much the same manner as DC generators, their components and maintenance practices are identical. The obvious difference being that generators convert rotary motion to electrical energy, and motors convert electrical energy into rotary motion.

There are three basic types of DC motors: series, shunt, and compound. They differ largely in the method in which their field and armature are connected.

In the series motor, the field windings consisting of relatively few turns of heavy wire are connected in series with the armature winding. The same current flowing through the field winding also flows through the armature winding. Any increase in current, therefore, strengthens the magnetism of both the field and the armature.

The principal advantage of the series-wound DC motor is its high starting torque. A series-wound DC motor will normally require more current at low RPM than at high RPM. Series motors are often used in aircraft as engine starters and for raising and lowering flaps and landing gear.

By reversing the direction of current flow in either the armature or the field windings, the direction of a motor's rotation may be reversed. This will reverse the magnetism of either the armature or the magnetic field in which the armature rotates.

Because of their advantages, many types of aircraft motors are designed to operate on alternating current. In general AC motors are less expensive than comparable DC motors. In many instances, AC motors do not use brushes and commutators avoiding sparking common to DC motors. AC motors are very reliable and very little maintenance is needed.

The two general types of AC motors used in aircraft systems are (1) induction, and, (2) synchronous. Either type may be single-phase or three-phase.

REVIEW QUESTIONS — Electrical System Components

181. *Why do some DC motors employ two sets of field windings, wound in opposite directions?*
182. *What is the principal advantage of the series-wound DC motor?*
183. *What is the cause of generator brush arcing?*
184. *Where is the nominal current rating for a switch normally found?*
185. *What are the locations and colors of the navigation lights on an aircraft?*
186. *What type of DC generator is not used as an aircraft generator?*
187. *How is a starter-generator cooled on the ground?*
188. *What type of instrument is used for measuring very high values of resistance?*
189. *What type of capacitor can be checked with an ohmmeter?*
190. *What are the three basic types of DC motors?*
191. *What are the two general types of AC motors use in aircraft systems?*
192. *What is the minimum bend radii for an electrical wire bundle?*
193. *In what units is aircraft fuse capacity rated?*
194. *What is the maximum number of terminals that may be installed on any single stud of a terminal strip?*
195. *What type of conduit may be used to minimize radio interference?*
196. *How should a voltmeter be connected?*
197. *What is the advantage of a current limiter?*
198. *What are the iron cores of most induction coils laminated?*
199. *What method is used to restore residual magnetism in a DC generator?*
200. *How can the direction of a DC motor be changed?*
201. *How are generators rated?*

Answer questions 8803 through 8930 in the FAA Test section of this book.

POSITION AND WARNING SYSTEMS

A synchro-type indicating system is an electrical system used for transmitting information from one point to another. Most position-indicating instruments are designed around a synchro system. The three most common types of synchro systems are (1) autosyn, (2) selsyn, and (3) magnesyn. These three systems are all similar in construction and all operate on identical electrical and mechanical principles.

The DC selsyn system is a widely used electrical method of indicating a remote mechanical movement or position. It may be used to indicate the position of retractable landing gear, wing flaps, cowl

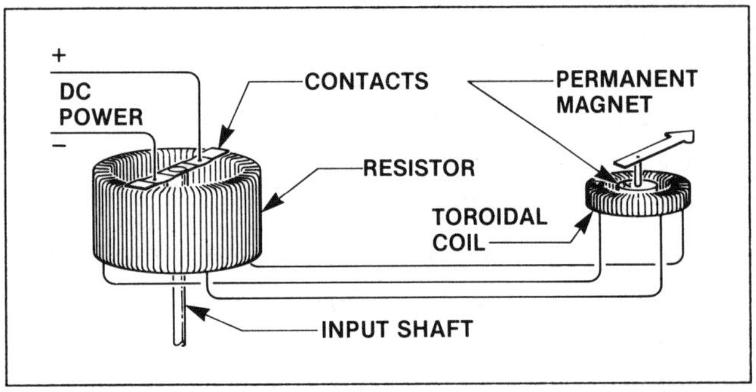

Figure 109. Direct current remote indicating system.

flap, or oil cooler doors. Fuel quantity indicating systems may also use this principal.

A selsyn system consists of a transmitter, an indicator, and connecting wires. The voltage required is supplied from the aircraft's electrical system.

The magnesyn system is an electrical self-synchronous device used to transmit the direction of a magnetic field from one coil to another. The basic difference between an autosyn and a magnesyn indicating system is the rotor. The rotor in a magnesyn system uses a permanent magnet and the rotor in an autosyn uses an electromagnet.

To provide a visual indication of landing gear position, indicators are installed in the cockpit. Gear warning devices are incorporated on all retractable gear aircraft and usually consist of a horn or some other aural device and a red warning light. The horn blows and the light comes on when one or more throttles are retarded and the landing gear is in any position other than down and locked.

Several types of gear position indicators are available. A common systems uses a red light for unsafe gear, a green light for gear down, and no light for gear up.

Accidental retraction of a landing gear may be prevented by such safety devices as mechanical downlocks, safety switches, and ground locks. Mechanical downlocks are built-in parts of a gear retraction system and are operated automatically by the retraction mechanism.

A landing gear safety switch (squat switch) is usually located on one of the main landing gear shock struts. This switch is actuated by linkage through the landing gear torque links. When the strut is compressed (aircraft on the ground), the torque links are close together, causing the adjusting link to open the safety switch. During takeoff, as the weight of the aircraft leaves the struts, the struts and torque links extend, causing the switch to close, allowing the gear to be raised.

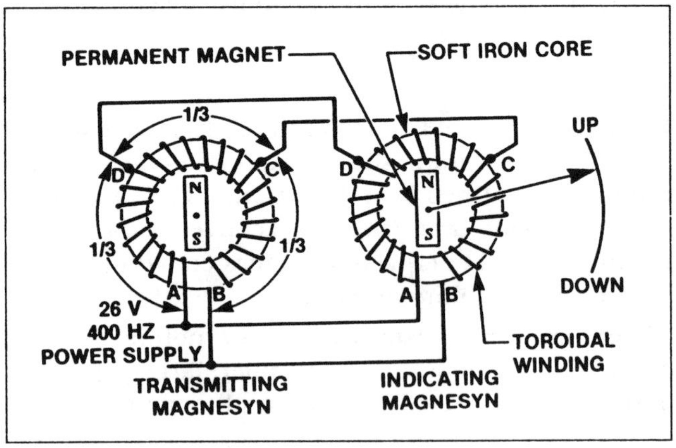

Figure 110. Magnesyn-type AC remote indicating system.

An antiskid system is an electrohydraulic system designed to bring the aircraft to a stop without skidding. A wheel going into a skid will greatly reduce the value of the aircraft's brakes. The system consists basically of three components; wheel speed sensors, control box, and control valves.

The skid control system performs four functions: (1) normal skid control, (2) locked wheel control, (3) touchdown protection, and (4) fail-safe protection.

The antiskid brake system in most installations is activated or armed by a switch in the cockpit. A warning lamp lights when the system is turned off or if there is a system failure. The skid control generator is the unit which measures the wheel rotational speed. It also senses any change in the speed. There is one of these small generators mounted in the axle of each wheel.

The skid control box reads the signal from the generator and senses any changes in signal strength. It can interpret them as developing skids, locked wheels, brake applications, and brake releases. It analyses this information and then sends appropriate signals to the skid control valves.

The skid control valves mounted on the brake control valve are solenoid operated. If a skid develops, a signal is sent to the skid control valve solenoid and brake application is relaxed.

When the airplane is slowed below approximately 20 MPH, the antiskid system automatically deactivates to give the pilot full control of the brakes for maneuvering and parking.

When an airplane takes off and the ground safety switch closes, it completes the circuit for the retraction of the gear. Depending upon the airplane, it may also deactivate the antiskid system. What this actually does is deactivate the brakes so they cannot be locked up upon landing.

The primary purpose of a takeoff warning system is to alert the crew that a flight control is not properly set prior to takeoff. It is actuated by the closing of the thrust lever switch. The control indicating system indicates the controls trimmed position. An aural warning is provided whenever the setting of the speed brake, flaps, or stabilizer is unsafe for takeoff.

Microswitches are used primarily as limit switches to control electrical units automatically. Repairs to the landing gear microswitches would require a landing gear retraction test to verify proper operation.

Certain aircraft may be equipped with angle-of-attack indicators to provide the flight crew with important information used to optimize performance and prevent stalls. The angle-of-attack detector operates from differential pressure when the airstream is not parallel to the true angle of attack of the aircraft and warns that a stall is imminent. Some light aircraft use a pneumatic, or reed, type of stall warning indicator that is activated by negative pressure.

REVIEW QUESTIONS — Position and Warning Systems

202. *What flight controls or systems are connected to the takeoff warning system?*

203. *What activates a takeoff warning system?*

204. *What are the three basic components of the antiskid system?*

205. *What conditions will activate the antiskid warning light in the cockpit?*

206. *Where is the landing gear safety switch usually located?*

207. *The basic difference between an autosyn and a magnesyn indicating system is what?*

Answer questions 9834 through 8968 in the FAA Test section of this book.

ICE AND RAIN CONTROL SYSTEMS

Ice may be expected to form whenever there is visible moisture in the air and the temperature is near or below freezing. An exception is carburetor icing, which can occur during warm weather with no visible moisture present. If ice is allowed to accumulate on the wings and empennage leading edges, it destroys the lift characteristics of the airfoil. Ice or rain accumulations on the windshield interfere with vision.

There are two types of ice control systems used on aircraft: anti-icing and deicing systems. Anti-icing systems prevent the formation of ice, and deicing systems remove the ice after it has formed.

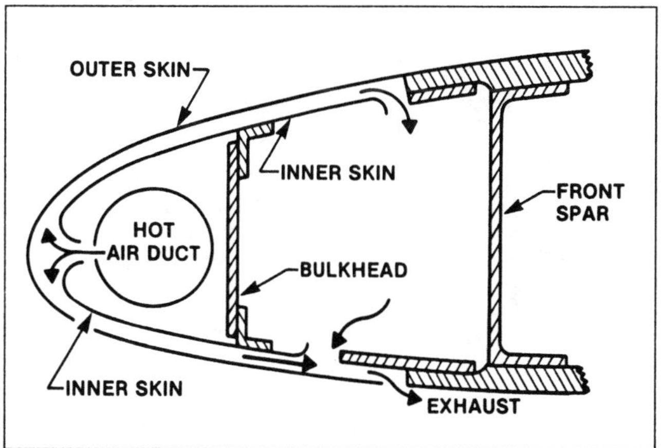

***Figure 111.** Thermal deicing is accomplished by flowing hot compressor bleed air from the engine through a duct in the leading edge of the wing.*

Some aircraft are protected against airframe icing by heating the leading edges of the airfoils and intake ducts. This type of anti-ice system is operated whenever icing conditions are first encountered, or expected to occur.

Three possible sources of hot air for the operation of a thermal anti-icing system are engine bleed air, combustion heaters, and augmentor tubes.

Thermocycling switches are used to control the temperature of anti-icing systems using surface combustion heaters.

Most aircraft that use thermal anti-icing systems today are turbine powered, and it is a simple matter to use compressor bleed air for ice control.

Anti-icing systems using combustion heaters usually have a separate system for each wing and the empennage. The anti-icing system is automatically controlled by overheat switches, thermal cycling switches, a balance control, and a duct pressure safety switch.

Pneumatic deicing systems use rubber deicers, called boots or shoes, attached to the leading edge of the wing and stabilizers. The deicers are composed of a series of inflatable tubes. During operation, the tubes are inflated with pressurized air, and deflated in an alternating cycle as shown in figure 112. This inflation and deflation causes the ice to crack and break off. The ice is then carried away by the slipstream.

Deicing boots may be inflated by pressure supplied by a vane-type pump or by air bled from turbine engine compressors. If the older "wet-type" pump is used, an oil separator will be installed to protect the deicer boots from oil deterioration.

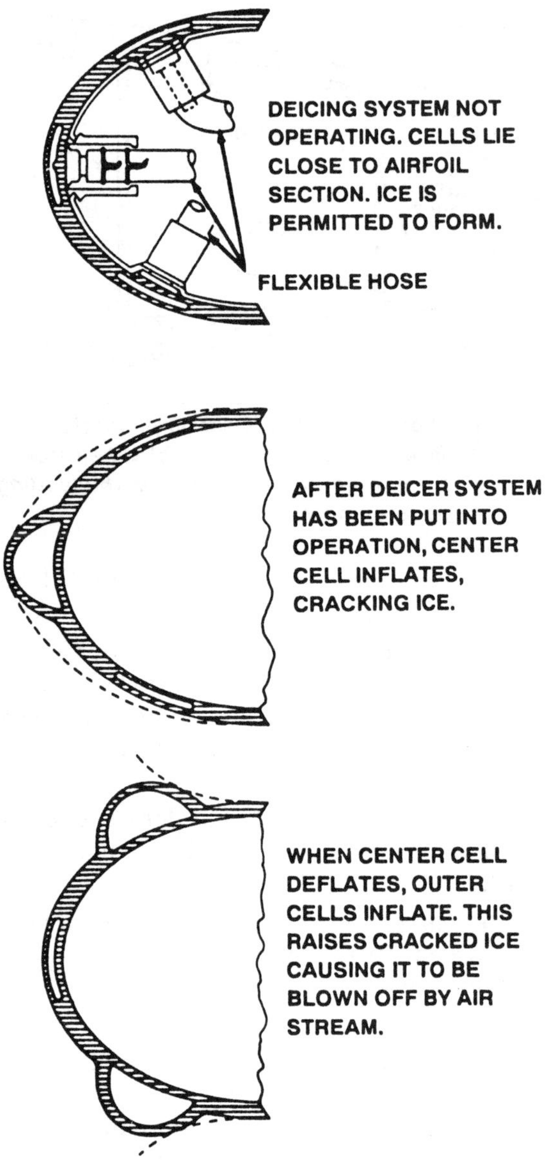

Figure 112. Deicer boot inflation cycle.

The tubes in deicer boots are inflated alternately to keep disturbance of the airflow to a minimum. The inflation sequence is controlled by a centrally located distributor valve. The distributor valve also directs vacuum to the deicer boots for holddown in flight. A suction relief valve will regulate the vacuum of the air pump to hold the deicing boots deflated when the pneumatic deicing system is off.

During an operational check of a pneumatic deicer system, fluctuating readings on the pressure gage and relatively steady readings on the vacuum gage will indicate normal operation.

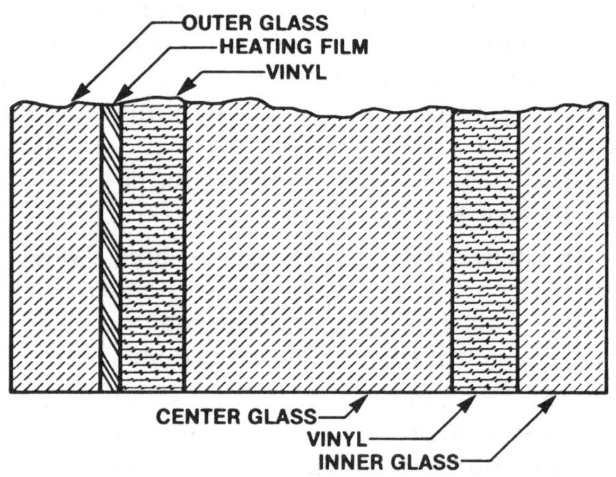

Figure 113. The windshield of a modern jet aircraft is made up of laminations of glass and vinyl. It is heated by electric current flowing through a conductive film on the inside of the outer layer of glass.

Deicer boots are made of soft, pliable rubber, or rubberized fabric, and contain tubular air cells. The outer ply of the deicer is of conductive neoprene to provide resistance to deterioration by the elements. The neoprene also provides a conductive surface to dissipate static electricity charges.

The deicer boots are attached to the leading edge of wing and tail surfaces with cement fairing strips and screws, or a combination of both. The new type deicer boots are completely bonded to the surface with cement. When installing pneumatic surface-bonded type deicer boots, remove all paint from the area to be covered by the boot.

Cleaning of deicing boots should be done at the same time the aircraft is washed, using a mild soap and water solution. Grease and oil can be removed with a cleaning agent, such as naphtha, followed by a soap and water scrubbing.

The presence of ice on an aircraft sitting on the ground may be the result of direct precipitation, formation of frost on integral fuel tanks after prolonged flight at high altitude, or accumulations on the landing gear following taxiing through snow or slush.

Any deposits of ice, snow, or frost on the external surfaces of an aircraft may drastically affect its performance. Any measures taken to remove frozen deposits while the aircraft is on the ground must also prevent the possible re-freezing of the liquid.

Ethylene glycol and isopropyl alcohol may be used as a deicing fluid to remove frost from an aircraft surface. Wet snow should be removed using a brush or a squeegee.

Snow or ice can enter turbine engine intakes and freeze in the compressor. If the compressor is immobile because of ice, hot air should be blown through the engine until the rotating parts are free.

In order to keep window areas free of ice, frost, etc., window anti-icing, deicing, defogging, and demisting systems are used.

Three methods of anti-icing may be used on aircraft windshields: (1) windshield wipers and anti-icing fluid, (2) heated air circulating systems, and (3) an electric heating element in the windshield.

One of the most common methods for controlling ice formation and fog on modern aircraft windows is the use of an electrical heating element built into the window. An electrically heated windshield system includes:

(1) Windshield auto-transformers and heat control relays.
(2) Heat control toggle switches.
(3) Indicating lights.
(4) Windshield control units.
(5) Temperature sensing elements (thermistors) laminated in the panel.

The systems receives power from the 115 VAC buses through the windshield heat control circuit breaker.

There are several problems associated with electrically heated windshields. These include delamination, scratches, arcing, and discoloration.

Delamination (separation of the plies) although undesirable, is not harmful structurally provided it is within the limits established by the aircraft manufacturer, and is not in an area where it affects the optical qualities of the panel.

Arcing in an electrically heated windshield panel usually indicates a breakdown in the conductive coating. Where arcing exists, there is invariably a certain amount of local overheating which can cause further damage to the panel.

The window defrost system directs heated air from the cabin heating system (or from an auxiliary heater) to the pilot's and copilot's windshield and side windows by means of a series of ducts and outlet. In warm weather when heated air is not needed for defrosting, the system can be used to defog the windows. This is usually done by blowing ambient air on the windows using the blowers.

An alcohol deicing system is provided on some aircraft to remove ice from the windshield and the carburetor. Fluid from the alcohol supply tank is controlled by a solenoid valve which is energized when any of the alcohol pumps are on. Alcohol flow from the solenoid valve is filtered and directed through the alcohol pumps to the carburetors and windshield.

Carburetor heat systems may also be used to provide warm air to the intake of carburetors as an ice control measure. These system utilize air that has been warmed in a heat exchanger using the heat energy that would have normally been vented overboard with the exhaust gases.

An electric heating element built into the pitot head may be used to prevent ice formation on the pitot tube. A switch located in the cockpit controls the power to the heater.

Heating elements should be checked for functioning by ensuring that the pitot head begins to warm up when power is applied. If an ammeter is installed in the circuit, the heater operation can be verified by noting the current consumption when the heater is turned on.

When rain forms on a windshield during flight, it becomes a hazard and must be eliminated. To provide a clear windshield, rain is eliminated by wiping

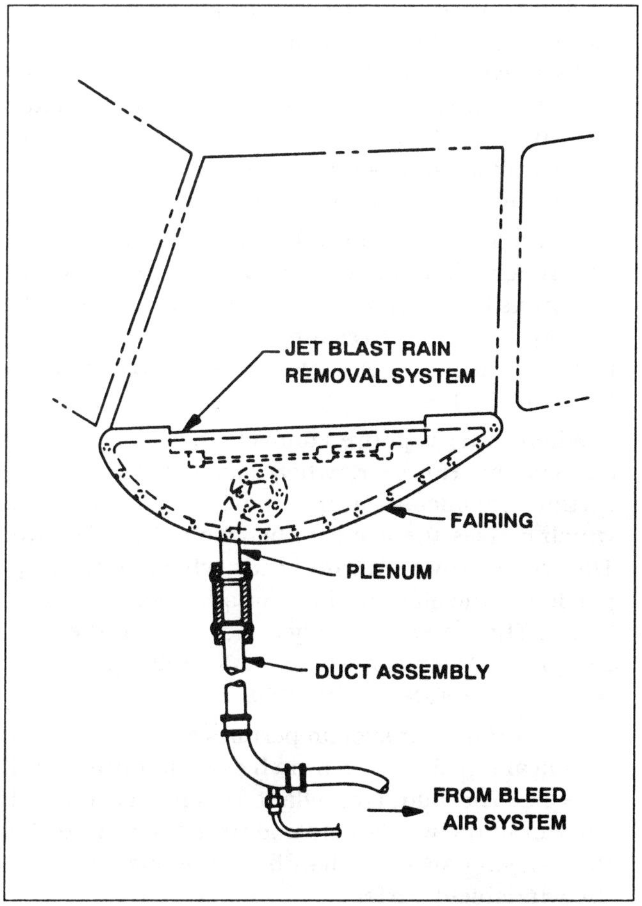

Figure 114. Typical pneumatic rain removal system.

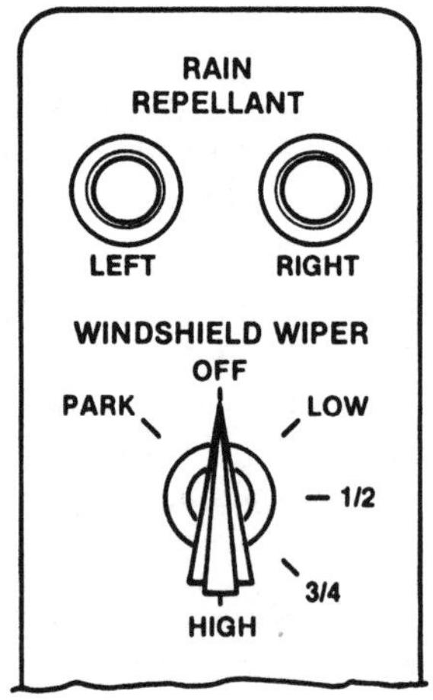

Figure 115. Rain control panel on a jet transport airplane.

if off or blowing it off. A third method of rain removal involves chemical rain repellents.

Windshield wiper systems may be powered electrically or hydraulically. These systems characteristically have two basic problem areas. One is the tendency of the slipstream aerodynamic forces to reduce the wiper blade loading pressure on the window, causing inefficient wiping or streaking. The other is achieving fast enough wiper oscillation to keep up with high rain impingement rates during heavy rain falls. As a result, most wiper systems fail to provide satisfactory vision in heavy rains.

Pneumatic rain removal systems for windshields us an air blast to form a barrier that prevents raindrops from striking the windshield surface. With this system high pressure, high temperature air from the turbine engine compressor is blown across the windshields.

When water is poured onto clean glass, it spreads out evenly. However, when glass is treated with certain chemicals, a transparent film is formed which causes the water to behave quite differently. The water draws up into beads which cover only a portion of the glass and the area between the beads is dry. The water is readily removed from the glass. The principle lends itself quite naturally to removing rain from aircraft windshields.

A rain repellant system permits application of the chemical repellent by a switch or push button in the cockpit. The liquid repellant is squirted onto the exterior of the windshield and uses the rain itself as the carrying agent to distribute the chemicals over the windshield surface.

Chemical rain repellent should not be used on a dry windshield. It will restrict visibility. Should the system be operated inadvertently, do not operate the windshield wipers or the rain clearing system, as this tends to increase smearing.

REVIEW QUESTIONS — Ice and Rain Control Systems

208. *How may a pitot/static tube heater be checked for operation?*
209. *What controls the inflation sequence in a pneumatic deicer boot system?*
210. *What is the source of pressure for inflating deicer boots on reciprocating engine aircraft?*
211. *Where are the heat sensors located on most aircraft with electrically heated windshields?*
212. *What are three possible sourced of hot air for the operation of thermal wing anti-icing systems?*
213. *What are two methods carburetor ice may be eliminated?*
214. *Why should a chemical rain repellant not be used on a dry windshield?*
215. *What may be used to remove wet snow deposits from an airplane?*

Answer questions 8969 through 8996 in the FAA Test section of this book.

FIRE DETECTION AND EXTINGUISHING SYSTEMS

Fire is a major hazard to both the occupants and the aircraft. The prevention, detection, and quick extinguishing of aircraft fires should be a prime concern to aircraft designers, operators, maintenance personnel, and flight crews.

Fire Detection Systems

The function of a fire detection system is to activate a warning device in the event of a fire. There are a number of types of fire detection systems available to detect the presence of a fire before it is visible to the flight crew. We will discuss here the most common methods, old and new.

The thermal switch fire detection system is a spot-type system that uses a number of thermally activated switches to warn of fire. The spot detector sensors operate using a bimetallic thermoswitch that closes when heated to a high temperature. This is often referred to as the Fenwall fire detection system after its manufacturer.

The Fenwall fire detection system uses spot detectors wired parallel between two separate circuits.

With this arrangement, a short may exist in either circuit without causing a false fire warning.

The spot detector system operates without a control unit. When an overheat condition, or a fire, causes the switch in a detector to close, the alarm bell sounds and a warning light for the affected area is lighted. When used in fire detection systems having a single indicator light, thermal switches are wired in parallel with each other and in series with the light.

The thermocouple fire detection system measures temperature rise compared to a reference temperature. This system may also be known as the Edison fire detection system, after its manufacturer.

If a fire breaks out in the compartment protected by this system, the temperature of one or more of the active thermocouples will suddenly rise much higher than the reference. A thermocouple in a fire detection system causes the warning system to operate because it generates a small current when heated. This current will energize the sensitive relay, and when its contacts close the slave relay is energized and the warning light illuminates.

The continuous-loop system works on the same basic principal as the spot-type fire detection system, except that instead of using individual thermal switches the continuous-loop system has as its sensor a long inconel tube. The continuous-loop fire detector is an overheat-type detector.

The operating principal of the continuous loop fire detection system sensor is based on the core resistance material which prevents current flow at normal temperatures. The inconel tube is filled with this material which surrounds the conductor or conductors. At normal temperatures, this material acts as an insulator, but as temperatures increase the resistance will decrease and current will flow, causing the alarm to sound.

Maintenance of fire detection systems includes the replacement of damaged sensing elements. The most common cause of false fire warnings in a continuous-loop fire detection systems are dents, kinks, or crushed sensor sections.

Fire Extinguishing Systems

The National Fire Protection Association has classified fires according to their fuel, and the fire extinguishing agents must be matched to the class of fire that they are used to extinguish.

- *Class-A* fires are those that have ordinary combustible materials such as paper, cloth, or wood as their fuel.
- *Class-B* fires have flammable liquids as their fuel. A gasoline or oil fire is defined as a class B fire.
- *Class-C* fires involve energized electrical equipment.
- *Class-D* fires involve flammable metals such as magnesium. The use of water on class-D fires has no effect.

The three types of fire extinguishing agents for aircraft interior fires are water, carbon dioxide, and dry-chemical. A carbon dioxide hand-held fire extinguisher may be used on an electrical fire if the horn is nonmetallic. A carbon dioxide fire extinguisher can be weighed to determine its charge.

Aircraft wheel and brake assemblies may contain components made of magnesium. This is an extremely dangerous material when burning, and requires special handling. The proper fire extinguishing agent to use on an aircraft brake fire is dry powder chemical.

Carbon dioxide may be used in an installed system. An installed system will have one or more carbon dioxide cylinders in the fuselage, connected by aluminum tubing to the discharge nozzles. Installed systems have two blowout disks on the side of the fuselage near the storage bottles. Evidence that the fire extinguishing system has been intentionally discharged is indicated by the absence of a yellow disk on the side of the fuselage. Thermal discharge of an installed carbon dioxide fire extinguishing system will be indicated by the rupture of the red plastic disk in the thermal discharge line.

Many systems today use Bromotrifluoromethane (Halon 1301). This is the safest fire extinguishing agent to use from a standpoint of toxicity and corrosion hazards. This is commonly used in high-rate discharge (HRD) systems.

In a HRD system, the extinguishing agent is stored in a pressurized container. The nitrogen charge and fire-extinguisher agent in a HRD system are retained by a breakable disk and fusible plug. The fusible plug is set to rupture is certain temperature conditions are met. The breakable disk is ruptured when the system is activated.

When the emergency shutoff valves are closed during an engine fire, the fuel flow to the engine will be blocked.

System maintenance and inspection should be performed at regular intervals and any time that the system has been discharged. Color coding for fire extinguisher system components and lines is brown.

The pressurized containers holding the extinguishing agent in an HRD system will have a pres-

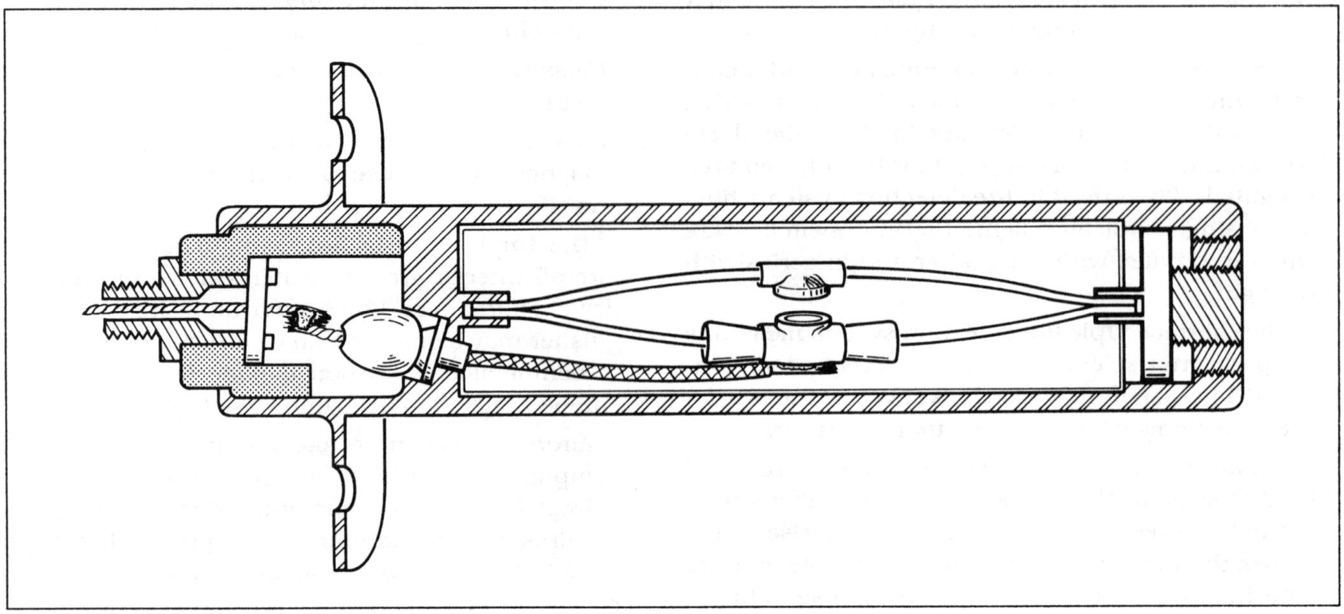

Figure 116. Fenwal spot dectector.

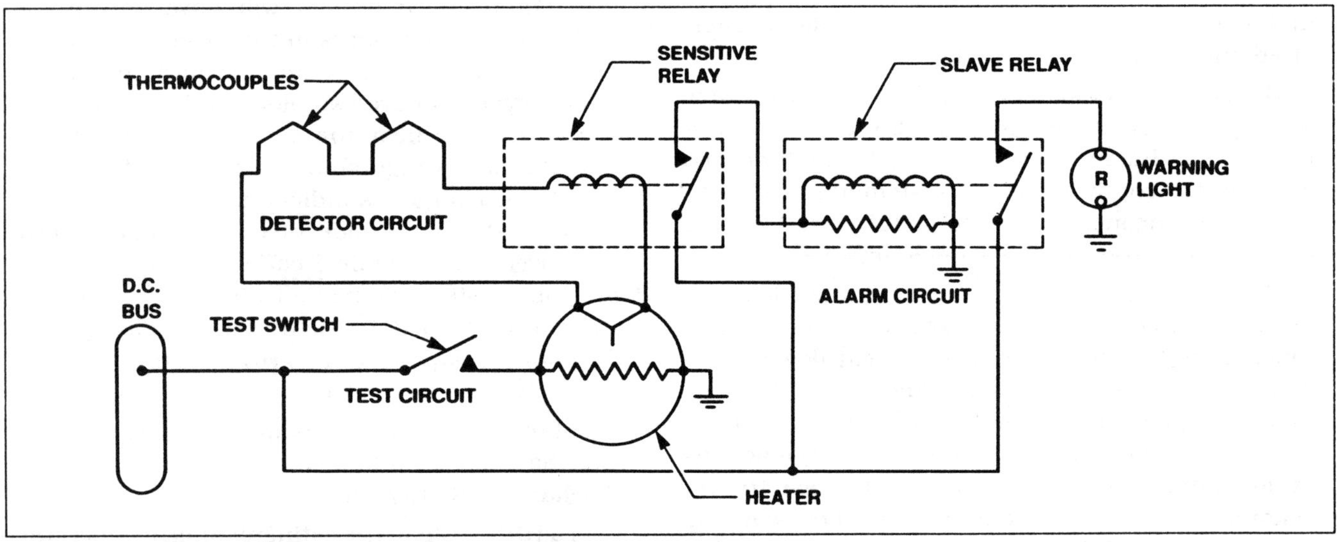

Figure 117. Thermocouple fire warning circuit.

sure gage to indicate the contents. If the indicator is not between the minimum and maximum limits, you should replace the extinguisher container.

When a fire extinguisher system using a Freon agent has been discharged, you must purge the system with air and replace the container.

A squib, as used in a fire protection system, is a device for causing the fire extinguishing agent to be released. If a fire extinguisher cartridge is removed from a discharge valve for any reason, it is recommended that the cartridge be used only on the original discharge valve assembly.

During removal or installation of discharge cartridges, the terminals should be grounded or shorted to prevent accidental discharge. Before connecting the cartridge terminals to the electrical system, the system should be checked with a voltmeter to see that no voltage exists at the terminal connections.

Smoke Detection Systems

A smoke detection system monitors the cargo and baggage compartments for the presence of smoke, which is indicative of a fire condition. Smoke detection instruments are classified by their method of detection.

Smoke detectors which use a measurement of light transmissibility in the air are called photoelectric devices. A photoelectric smoke detector will warn only when smoke is present. Photoelectric cell smoke

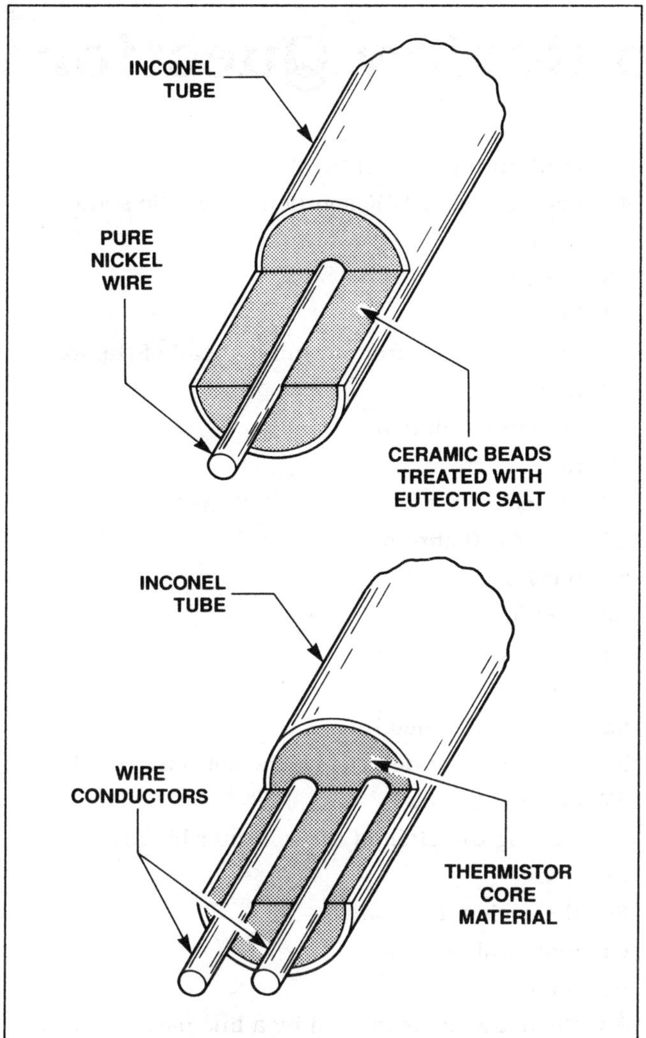

Figure 118. Typical elements for a continuous-loop-type fire detection system.

detectors are used in the cargo and/or baggage compartment of an aircraft.

On a few aircraft, visual smoke detectors provide the only means of smoke detection. Indication is provided by drawing smoke through a line into the indicator, using either a suitable suction device or cabin pressurization.

When a visual smoke detector is activated, a lamp within the indicator senses light reflected from the smoke and illuminates automatically. The light is scattered so that the smoke is rendered visible in the appropriate window of the display.

The CO detectors, which detect concentrations of carbon monoxide gas, are rarely used to monitor cargo and baggage compartments. However, they have gained widespread use in conducting tests for the presence of carbon monoxide gas in aircraft cabins and cockpits.

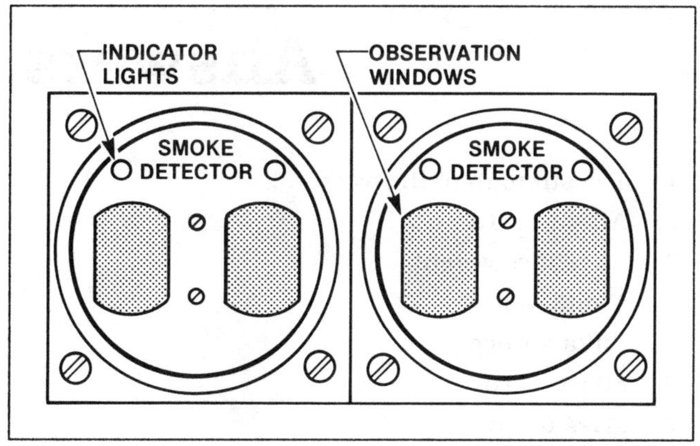

Figure 119. Smoke detectors located on the flight engineer's panel of a jet transport airplane.

There are several types of portable testers (sniffers) in use. One type has a replaceable indicator tube which contains a yellow silica gel, impregnated with a complex silicon-molybdate compound and is catalyzed using palladium sulfate.

In use, a sample of air is drawn through the detector tube. When the air samples containing carbon monoxide, the yellow silica gel turns to a shade of green. A contaminated CO portable test unit may be returned to service by installing a new indicating element.

REVIEW QUESTIONS — Fire Detection and Extinguishing Systems

216. *In what area of an aircraft would you find a carbon monoxide detector?*

217. *What fire detection system measures temperature rise compared to a reference temperature?*

218. *What is the proper fire extinguishing agent to use on a brake fire?*

219. *What is the color code for fire-extinguisher lines?*

220. *What evidence will indicate that an installed fire-extinguishing system has been intentionally discharged?*

221. *Which fire extinguishing agent is considered to be the least toxic?*

Answer questions 8997 through 9028 in the FAA Test section of this book.

When you have completed this text, review questions and FAA test questions, you should prepare for the sample FAA Airframe Mechanic Exam.

Answers To Review Questions

1. The strength of the wood
2. AC 43.13-1A
3. To reduce weight
4. Cantilever
5. Sitka spruce
6. 80 PSI warp
7. glass fabric
8. baseball
9. blush
10. acetone
11. too much material applied in one coat
12. exclude sunlight from the fabric
13. shear
14. three time the thickness of the thickest sheet.
15. one and on-half D by one-half D
16. bolts of equal diameter
17. 1⁄16″
18. the metallic ring test
19. perpendicular
20. zinc chloride will have no effect
21. 1″ in diameter
22. delaminations
23. methyl ethyl ketone (MEK)
24. crazing
25. TSO C22
26. The soak method
27. 90°
28. 140°
29. because they are less subject to corrosion than uncoated aluminum alloys
30. cantilever
31. relieve stress concentration and prevent cracking
32. 1⁄8″ in diameter and 1⁄2″ long
33. dimples
34. a drill one size smaller than the rivet shank diameter
35. 5056 aluminum alloy
36. Monel
37. Start by tapping the edges and gradually work toward the center
38. perpendicular
39. adding approximately one-half of the stock thickness to the bend radius
40. minimum radius of bend
41. testing with a 10% solution of caustic soda
42. neutral line
43. relief holes
44. heat treatment
45. prevent oxides from forming ahead of the weld
46. neutral
47. welded patch plate
48. no
49. tinned
50. SAE 4130 chrome/moly steel
51. 5 PSI
52. acetone
53. slightly carburizing
54. 15 PSI
55. corrode the metal
56. increasing or decreasing the collective pitch
57. angle of attack
58. varying the pitch of the tail rotor blades
59. track
60. the cycle pitch control
61. longitudinal axis
62. pitch
63. the acute angle formed by a line parallel to the wing chord and a line parallel to the longitudinal axis of the aircraft.
64. the rudder
65. lateral axis
66. from leading idge to trailing edge
67. span and chord
68. reduce stalling speed
69. stall the inboard portions of the wing first
70. provide directional stability
71. to reduce the possibility of control surface flutter
72. ailerons
73. assist the pilot in moving the control surfaces
74. extra-flexible
75. retain a set cable tension
76. yes
77. if necessary to reach a place where the inspection can be done
78. 10 hours
79. if a special permit has been issued for the aircraft

80. a certificated technician with Airframe and Powerplant ratings
81. wipe with a dry cloth and wash with soap and water
82. air in the system
83. to remove air from the system
84. brake grabbing
85. expander tube
86. to indicate slippage
87. damage to the wheel flange
88. prevent heavy spots and reduce vibration
89. the type of seals used
90. fading brakes
91. the main hydraulic system
92. isolate the emergency brake system from the normal system
93. compensate for thermal expansion
94. low fluid
95. reduce pressure to the brake and increase the volume of fluid flow
96. an indefinite number of times
97. the tire should be scrapped
98. at least 2 hours (3 in hot water)
99. underinflation
100. chevron seals
101. the open side
102. micronic filter
103. restrict flow in one direction
104. O-ring
105. check valve
106. open-center
107. as damage-preventing units
108. use a backup ring on the side of the O-ring away from the pressure
109. relieve system pressure
110. vegetable base oil
111. red
112. yes
113. trichlorethylene
114. shuttle valve
115. thermal relief valve
116. sequence valve
117. relieve the air pressure
118. at the hydraulic pumps
119. prevent hydraulic pump cavitation
120. air in the system
121. compressor
122. carries the heat to the places where it is needed
123. relieve all positive pressure from the cabin
124. air bubbles in the sight glass disappear
125. continuous flow
126. moisture
127. condenser
128. negative pressure relief valve
129. remove moisture
130. outflow valve
131. receiver-dryer
132. the aircraft engine, through a belt
133. vertical with the outlet at the bottom
134. changes to phosgene gas
135. when the user breathes
136. the weight of the complete panel unit
137. altimeter, airspeed, vertical speed indicator
138. the normal operating range
139. the white slippage marks
140. to prevent interference from outside magnetic fields
141. attitude indicator, heazding indicator, turn needle of turn and bank
142. lubber line
143. 10°
144. 100 feet in 1 minute
145. pressure altitude
146. ambient pressure
147. to relieve the pilot of control of the aircraft during long periods of flight
148. flight controller
149. directional gyro
150. servo
151. a yaw damper system
152. three
153. VHF
154. Federal Communications Commission
155. 2 feet
156. on the outside of the transmitter
157. as far aft as possible, but forward of the vertical fin
158. to prevent the entry of moisture
159. on the top of the cabin with the apex of the V pointing forward
160. align the aircraft with the center of the runway
161. 10 times the diameter of the cable
162. 1″
163. when the max takeoff weight exceeds the max landing weight
164. boost pumps

165. defuel the outboard wing tanks first
166. to allow the feeding of any engine from any tank
167. reduce the chances for fuel contamination, faster
168. riveting
169. carbon dioxide
170. compensate fuel pressures to altitude changes
171. 5% solution of nitric or sulfuric acid
172. electronic (capacitance)
173. a capacitor
174. cold fuel is heavier per gallon
175. to prevent fuel from flowing away from the boost pumps
176. engine oil, or bleed air
177. in the open air
178. resist fuel surging within the fuel tank
179. the word "FUEL" and the minimum fuel grade or designation for the engine
180. sliding-vane
181. so the motor can be operated in either direction
182. high starting torque
183. low spring tension
184. stamped on the switch housing
185. green on the right wing tip, red on the left wing tip, and white on the tail
186. series
187. an integral fan
188. megohmmeter
189. an electrolytic capacitor
190. series, shunt, compound
191. induction and synchronous
192. ten times the outside diameter of the bundle
193. amperes
194. four
195. flexible brass
196. in parallel with the load
197. it will take an overload for a short period
198. reduce the effects of eddy currents
199. flashing the field
200. reverse the electrical connections to either the field or armature windings
201. amperes at rated voltage
202. speed brades, flaps, stabilizer
203. thrust lever switch
204. wheel speed sensored, control box, and control valves
205. if the system is turned off or if there is a system failure
206. on the main landing gear shock strut
207. the rotor in a magnesyn system uses a permanent magnet
208. activate the system while watching the ammeter or loadmeter
209. distributor valve
210. vane-type pump
211. imbedded in the glass
212. engine bleed air, combustion heaters, and augmentor tubes
213. alcohol spray and heated induction air
214. it will restrict visibility
215. a brush or squeegee
216. cockpit and/or cabin
217. thermocouple
218. dry powder chemical
219. brown
220. yellow disk blown out
221. Bromotrifluoromethane (Halon 1301)

A&P Technician Airframe Study Guide Practice Exam #1

Answer the following questions in the FAA Exam section of this book. Mark answers by filling in the appropriate circle next to the question number. When you have completed the exam, you may check your answers with the key provided. Remember a score of 70 percent is passing.

		A	B	C			A	B	C			A	B	C
1.	8002.	O	O	O	34.	8434.	O	O	O	67.	8653.	O	O	O
2.	8005.	O	O	O	35.	8454.	O	O	O	68.	8656.	O	O	O
3.	8019.	O	O	O	36.	8472.	O	O	O	69.	8661.	O	O	O
4.	8025.	O	O	O	37.	8496.	O	O	O	70.	8662.	O	O	O
5.	8046.	O	O	O	38.	8507.	O	O	O	71.	8666.	O	O	O
6.	8049.	O	O	O	39.	8519.	O	O	O	72.	8670.	O	O	O
7.	8054.	O	O	O	40.	8522.	O	O	O	73.	8674.	O	O	O
8.	8061.	O	O	O	41.	8527.	O	O	O	74.	8677.	O	O	O
9.	8074.	O	O	O	42.	8534.	O	O	O	75.	8680.	O	O	O
10.	8092.	O	O	O	43.	8541.	O	O	O	76.	8684.	O	O	O
11.	8103.	O	O	O	44.	8545.	O	O	O	77.	8686.	O	O	O
12.	8115.	O	O	O	45.	8551.	O	O	O	78.	8692.	O	O	O
13.	8124.	O	O	O	46.	8555.	O	O	O	79.	8694.	O	O	O
14.	8130.	O	O	O	47.	8563.	O	O	O	80.	8706.	O	O	O
15.	8142.	O	O	O	48.	8567.	O	O	O	81.	8713.	O	O	O
16.	8151.	O	O	O	49.	8571.	O	O	O	82.	8716.	O	O	O
17.	8168.	O	O	O	50.	8575.	O	O	O	83.	8718.	O	O	O
18.	8178.	O	O	O	51.	8585.	O	O	O	84.	8726.	O	O	O
19.	8185.	O	O	O	52.	8592.	O	O	O	85.	8732.	O	O	O
20.	8189.	O	O	O	53.	8596.	O	O	O	86.	8741.	O	O	O
21.	8219.	O	O	O	54.	8605.	O	O	O	87.	8758.	O	O	O
22.	8235.	O	O	O	55.	8612.	O	O	O	88.	8773.	O	O	O
23.	8242.	O	O	O	56.	8614.	O	O	O	89.	8794.	O	O	O
24.	8305.	O	O	O	57.	8617.	O	O	O	90.	8806.	O	O	O
25.	8326.	O	O	O	58.	8631.	O	O	O	91.	8823.	O	O	O
26.	8318.	O	O	O	59.	8633.	O	O	O	92.	8831.	O	O	O
27.	8321.	O	O	O	60.	8636.	O	O	O	93.	8851.	O	O	O
28.	8332.	O	O	O	61.	8641.	O	O	O	94.	8869.	O	O	O
29.	8370.	O	O	O	62.	8645.	O	O	O	95.	8890.	O	O	O
30.	8381.	O	O	O	63.	8646.	O	O	O	96.	8903.	O	O	O
31.	8400.	O	O	O	64.	8647.	O	O	O	97.	8951.	O	O	O
32.	8413.	O	O	O	65.	8648.	O	O	O	98.	8969.	O	O	O
33.	8426.	O	O	O	66.	8651.	O	O	O	99.	8997.	O	O	O
										100.	9041.	O	O	O

Practice Exam #1 Answer Key

No.	Question	A	B	C
1.	8002.	○	●	○
2.	8005.	●	○	○
3.	8019.	●	○	○
4.	8025.	○	●	○
5.	8046.	○	○	●
6.	8049.	○	○	●
7.	8054.	○	○	●
8.	8061.	●	○	○
9.	8074.	○	●	○
10.	8092.	○	○	●
11.	8103.	●	○	○
12.	8115.	○	●	○
13.	8124.	○	○	●
14.	8130.	○	○	●
15.	8142.	○	○	●
16.	8151.	○	●	○
17.	8168.	○	●	○
18.	8178.	○	○	●
19.	8185.	○	○	●
20.	8189.	○	○	●
21.	8219.	○	○	●
22.	8235.	●	○	○
23.	8242.	○	●	○
24.	8305.	○	○	●
25.	8326.	●	○	○
26.	8318.	●	○	○
27.	8321.	○	○	●
28.	8332.	○	○	●
29.	8370.	○	○	●
30.	8381.	○	●	○
31.	8400.	○	○	●
32.	8413.	○	○	●
33.	8426.	○	●	○
34.	8434.	●	○	○
35.	8454.	○	○	●
36.	8472.	○	●	○
37.	8496.	○	○	●
38.	8507.	●	○	○
39.	8519.	●	○	○
40.	8522.	○	●	○
41.	8527.	○	●	○
42.	8534.	○	○	●
43.	8541.	○	○	●
44.	8545.	○	●	○
45.	8551.	○	○	●
46.	8555.	○	●	○
47.	8563.	○	●	○
48.	8567.	○	○	●
49.	8571.	○	○	●
50.	8575.	○	○	●
51.	8585.	●	○	○
52.	8592.	○	○	●
53.	8596.	●	○	○
54.	8605.	○	○	●
55.	8612.	○	○	●
56.	8614.	●	○	○
57.	8617.	○	●	○
58.	8631.	○	●	○
59.	8633.	○	●	○
60.	8636.	●	○	○
61.	8641.	●	○	○
62.	8645.	○	●	○
63.	8646.	○	○	●
64.	8647.	●	○	○
65.	8648.	○	○	●
66.	8651.	○	●	○
67.	8653.	●	○	○
68.	8656.	○	○	●
69.	8661.	●	○	○
70.	8662.	○	○	●
71.	8666.	○	○	●
72.	8670.	●	○	○
73.	8674.	○	●	○
74.	8677.	●	○	○
75.	8680.	○	○	●
76.	8684.	○	○	●
77.	8686.	○	●	○
78.	8692.	○	○	●
79.	8694.	●	○	○
80.	8706.	○	●	○
81.	8713.	○	●	○
82.	8716.	●	○	○
83.	8718.	○	○	●
84.	8726.	●	○	○
85.	8732.	●	○	○
86.	8741.	●	○	○
87.	8758.	○	○	●
88.	8773.	●	○	○
89.	8794.	●	○	○
90.	8806.	●	○	○
91.	8823.	○	○	●
92.	8831.	○	●	○
93.	8851.	○	○	●
94.	8869.	○	○	●
95.	8890.	○	●	○
96.	8903.	○	●	○
97.	8951.	●	○	○
98.	8969.	●	○	○
99.	8997.	○	●	○
100.	9026.	○	●	○

A&P Technician Airframe Study Guide Practice Exam #2

Answer the following questions in the FAA Exam section of this book. Mark answers by filling in the appropriate circle next to the question number. When you have completed the exam, you may check your answers with the key provided. Remember a score of 70 percent is passing.

		A	B	C			A	B	C			A	B	C
1.	8003.	❍	❍	❍	34.	8393.	❍	❍	❍	67.	8627.	❍	❍	❍
2.	8008.	❍	❍	❍	35.	8404.	❍	❍	❍	68.	8632.	❍	❍	❍
3.	8022.	❍	❍	❍	36.	8405.	❍	❍	❍	69.	8639.	❍	❍	❍
4.	8030.	❍	❍	❍	37.	8420.	❍	❍	❍	70.	8643.	❍	❍	❍
5.	8035.	❍	❍	❍	38.	8424.	❍	❍	❍	71.	8664.	❍	❍	❍
6.	8060.	❍	❍	❍	39.	8429.	❍	❍	❍	72.	8668.	❍	❍	❍
7.	8073.	❍	❍	❍	40.	8436.	❍	❍	❍	73.	8688.	❍	❍	❍
8.	8094.	❍	❍	❍	41.	8441.	❍	❍	❍	74.	8695.	❍	❍	❍
9.	8109.	❍	❍	❍	42.	8445.	❍	❍	❍	75.	8699.	❍	❍	❍
10.	8114.	❍	❍	❍	43.	8452.	❍	❍	❍	76.	8710.	❍	❍	❍
11.	8127.	❍	❍	❍	44.	8456.	❍	❍	❍	77.	8719.	❍	❍	❍
12.	8132.	❍	❍	❍	45.	8462.	❍	❍	❍	78.	8721.	❍	❍	❍
13.	8148.	❍	❍	❍	46.	8469.	❍	❍	❍	79.	8725.	❍	❍	❍
14.	8164.	❍	❍	❍	47.	8477.	❍	❍	❍	80.	8729.	❍	❍	❍
15.	8176.	❍	❍	❍	48.	8481.	❍	❍	❍	81.	8733.	❍	❍	❍
16.	8190.	❍	❍	❍	49.	8494.	❍	❍	❍	82.	8740.	❍	❍	❍
17.	8212.	❍	❍	❍	50.	8500.	❍	❍	❍	83.	8754.	❍	❍	❍
18.	8219.	❍	❍	❍	51.	8515.	❍	❍	❍	84.	8766.	❍	❍	❍
19.	8222.	❍	❍	❍	52.	8523.	❍	❍	❍	85.	8780.	❍	❍	❍
20.	8234.	❍	❍	❍	53.	8528.	❍	❍	❍	86.	8792.	❍	❍	❍
21.	8241.	❍	❍	❍	54.	8533.	❍	❍	❍	87.	8798.	❍	❍	❍
22.	8293.	❍	❍	❍	55.	8538.	❍	❍	❍	88.	8828.	❍	❍	❍
23.	8303.	❍	❍	❍	56.	8543.	❍	❍	❍	89.	8838.	❍	❍	❍
24.	8309.	❍	❍	❍	57.	8546.	❍	❍	❍	90.	8849.	❍	❍	❍
25.	8313.	❍	❍	❍	58.	8556.	❍	❍	❍	91.	8857.	❍	❍	❍
26.	8319.	❍	❍	❍	59.	8562.	❍	❍	❍	92.	8871.	❍	❍	❍
27.	8324.	❍	❍	❍	60.	8568.	❍	❍	❍	93.	8893.	❍	❍	❍
28.	8331.	❍	❍	❍	61.	8583.	❍	❍	❍	94.	8915.	❍	❍	❍
29.	8337.	❍	❍	❍	62.	8595.	❍	❍	❍	95.	8921.	❍	❍	❍
30.	8351.	❍	❍	❍	63.	8606.	❍	❍	❍	96.	8929.	❍	❍	❍
31.	8369.	❍	❍	❍	64.	8613.	❍	❍	❍	97.	8937.	❍	❍	❍
32.	8376.	❍	❍	❍	65.	8616.	❍	❍	❍	98.	8970.	❍	❍	❍
33.	8385.	❍	❍	❍	66.	8622.	❍	❍	❍	99.	9004.	❍	❍	❍
										100.	9043.	❍	❍	❍

Practice Exam #2 Answer Key

No.	Question	A	B	C
1.	8003.	○	○	●
2.	8008.	○	○	●
3.	8022.	●	○	○
4.	8030.	○	○	●
5.	8035.	○	●	○
6.	8060.	●	○	○
7.	8073.	○	○	●
8.	8094.	●	○	○
9.	8109.	○	○	●
10.	8114.	●	○	○
11.	8127.	○	●	○
12.	8132.	○	●	○
13.	8148.	○	○	●
14.	8164.	●	○	○
15.	8176.	○	●	○
16.	8190.	●	○	○
17.	8212.	○	○	●
18.	8219.	○	○	●
19.	8222.	●	○	○
20.	8234.	○	●	○
21.	8241.	○	●	○
22.	8293.	○	○	●
23.	8303.	●	○	○
24.	8309.	○	●	○
25.	8313.	○	○	●
26.	8319.	●	○	○
27.	8324.	○	○	●
28.	8331.	○	○	●
29.	8337.	●	○	○
30.	8351.	○	○	●
31.	8369.	○	●	○
32.	8376.	○	○	●
33.	8385.	○	○	●
34.	8393.	○	●	○
35.	8404.	○	●	○
36.	8405.	○	○	●
37.	8420.	●	○	○
38.	8424.	○	○	●
39.	8429.	○	○	●
40.	8436.	○	●	○
41.	8441.	○	●	○
42.	8445.	●	○	○
43.	8452.	○	●	○
44.	8456.	●	○	○
45.	8462.	●	○	○
46.	8469.	○	○	●
47.	8477.	○	○	●
48.	8481.	●	○	○
49.	8494.	○	○	●
50.	8500.	○	●	○
51.	8515.	○	●	○
52.	8523.	○	●	○
53.	8528.	●	○	○
54.	8533.	●	○	○
55.	8538.	○	●	○
56.	8543.	○	○	●
57.	8546.	○	○	●
58.	8556.	○	○	●
59.	8562.	○	○	●
60.	8568.	○	●	○
61.	8583.	○	○	●
62.	8595.	○	○	●
63.	8606.	○	●	○
64.	8613.	○	○	●
65.	8616.	○	○	●
66.	8622.	○	●	○
67.	8627.	○	●	○
68.	8632.	○	●	○
69.	8639.	○	●	○
70.	8643.	●	○	○
71.	8664.	○	○	●
72.	8668.	○	●	○
73.	8688.	●	○	○
74.	8695.	●	○	○
75.	8699.	●	○	○
76.	8710.	○	○	●
77.	8719.	○	○	●
78.	8721.	●	○	○
79.	8725.	○	●	○
80.	8729.	○	○	●
81.	8733.	○	●	○
82.	8740.	●	○	○
83.	8754.	●	○	○
84.	8766.	●	○	○
85.	8780.	○	●	○
86.	8792.	○	○	●
87.	8798.	●	○	○
88.	8828.	○	○	●
89.	8838.	●	○	○
90.	8849.	○	●	○
91.	8857.	○	●	○
92.	8871.	○	●	○
93.	8893.	○	●	○
94.	8915.	●	○	○
95.	8921.	○	●	○
96.	8929.	○	○	●
97.	8937.	○	●	○
98.	8970.	○	○	●
99.	9004.	○	○	●
100.	9028.	○	●	○

SECTION 2

1994-96 Airframe Mechanics FAA Exam Book

This section contains a reprint of the FAA's *Aviation Mechanic Airframe Written Test Book* and IAP's own answers, explanations and references.

PREFACE

This written test book has been developed by the Federal Aviation Administration (FAA) to be use by FAA testing centers and FAA designated written test examiners for testing applicants in the following knowledge area:

Aviation Mechanic Airframe

Applicants may use this written test book as a study guide. It is issued as FAA-T-8080-12E, Aviation Mechanic Airframe Written Test Book, and is available to the public from:

Superintendent of Documents U.S. Government Printing Office Washington, DC 20402	or	IAP, Inc. P.O. Box 10000 Casper, WY 82602-1000

or from U.S. Government Printing Office bookstores located in major cities throughout the United States.

The questions included in this publication are predicated on regulations, principles, and practices that were valid at the time of publication. The question selection sheets prepared for use with this written test book are security items and are revised at frequent intervals.

The FAA does NOT publish, supply, or make available the correct answers to questions included in this written test book. Students should determine the correct answers through research and study of appropriate subject material, by working with instructors, or by attending appropriate schools. The FAA is NOT responsible for either the content of commercial reprints of this written test book, or the accuracy of the answers they may supply.

Comments regarding the answers, explanations and references should be directed to:

IAP, Inc.
P.O. Box 10000
Casper, WY 82602-1000
(800) 443-9250 ❖ (307) 266-3838 ❖ 307-472-5106

CONTENTS

APPENDIX 1

APPENDIX 2

APPENDIX 3

GENERAL INSTRUCTIONS

MAXIMUM TIME ALLOWED FOR TEST: 4 HOURS

Maximum time allowed for each test is based upon previous experience and educational statistics. This time is considered more than adequate for applicants with proper preparation and instruction.

MATERIALS

Materials to be used with this written test book when used for airman certification testing:

1. AC Form 8080-3, Airman Written Test Application, which includes the answer sheet.

2. Question selection sheet which identifies the questions to be answered.

3. Plastic overlay sheet which can be placed over electrical drawings, graphs, and charts for plotting purposes.

TEST INSTRUCTIONS

1. Read the instructions on page 1 of AC Form 8080-3, and complete page 4 of the form. Incomplete or erroneous personal information entered on this form delays the scoring process.

2. The questions in this written test book are numbered consecutively beginning with 8001. Refer to the question selection sheet to determine which questions to answer.

3. For each question number on the answer sheet, find the appropriate question in the written test book.

4. Mark your answer in the space provided for each question number on the answer sheet. Spaces 1, 2, or 3 left unmarked will be counted by the computer scanner as a miss.

5. The test questions are of the multiple-choice type. Until revised, answer sheets contain selections listed as 1, 2, 3, and 4 and should be interpreted as A, B, and C respectively. Selection 4 should never be used.

6. The supplementary material required to answer the questions will be found in appendix 2.

7. Read each question carefully and avoid hasty assumptions. Do not answer until you understand the question. Do not spend too much time on any one question. Answer all of the questions that you readily know and then reconsider those you find difficult. Be careful to make any necessary conversions.

8. If a regulation or procedure is changed after this written test book is printed, you will receive credit for the affected question.

DO NOT USE THIS BOOK UNLESS IT CORRESPONDS WITH THE BOOK NUMBER ON THE TEST.

THE MINIMUM PASSING GRADE IS 70.

WARNING

§65.18 Written tests: Cheating or other unauthorized conduct.

(a) Except as authorized by the Administrator, no person may—

(1) Copy, or intentionally remove, a written test under this part;

(2) Give to another, or receive from another, any part or copy of that test;

(3) Give help on that test to, or receive help on that test from, any person during the period that test is being given;

(4) Take any part of that test in behalf of another person;

(5) Use any material or aid during the period that test is being given; or

(6) Intentionally cause, assist, or participate in any act prohibited by this paragraph.

(b) No person who commits an act prohibited by paragraph (a) of this section is eligible for any airman or ground instructor certificate or rating under this chapter for a period of 1 year after the date of that act. In addition, the commission of that act is a basis for suspending or revoking any airman or ground instructor certificate or rating held by that person.

INTRODUCTION TO THE AVIATION MECHANIC AIRFRAME WRITTEN TEST BOOK

This written test book presents the FAA Aviation Mechanic Airframe written tests. The requirements for a mechanic certificate and ratings, and the privileges, limitations, and general operating rules for certificated mechanics are prescribed in Federal Aviation Regulation (FAR) Part 65, Certification: Airmen Other Than Flight Crewmembers. Any person who applies and meets the requirements is entitled to a mechanic certificate.

Question selection sheets are used in conjunction with this written test book to administer the proper written test to each applicant. Each test is constructed from the questions included in this written test book.

The written test book is scheduled for revision each 24 months. Associated question selection sheets will be revised periodically, as required.

Testing and Scoring

The written test may be taken at FAA testing centers, FAA written test examiner's facilities, or other designated places.

The applicant is issued a "clean copy" of this written test book, an appropriate question selection sheet indicating the specific questions to be answered, and AC Form 8080-3, Airman Written Test Application, which includes the answer sheet. The written test book contains all supplementary material required to answer the questions. Supplementary material is located in appendix 2.

Instructions for completing the test are contained on page vii of this written test book.

Upon completion of the test, the applicant must surrender the issued written test book, question selection sheet, answer sheet, and any papers used for computations or notations to the monitor before leaving the test room.

The answer sheet is sent to the Mike Monroney Aeronautical Center in Oklahoma City, Oklahoma, where it is scored by computer. The applicant is then issued AC Form 8080-2, Airman Written Test Report. This form will list the test score and subject matter knowledge codes referencing the subjects in which the applicant is deficient. Retain AC Form 8080-2 to be presented for oral and practical testing, or for retesting in the event of written test failure.

The written test subject matter knowledge codes are then matched to the corresponding subject matter knowledge areas published in appendix 1 of this written test book. The applicant should review those subject areas until proficient. The applicant must obtain a written statement from an appropriately certificated mechanic stating that he/she has satisfactory knowledge of the subject areas shown to be deficient on AC Form 8080-2. In addition, the applicant may be retested in those subject areas by the mechanic.

The applicant should be aware that a subject matter code on AC Form 8080-2 appears only once even though more than one question may have been missed in that subject area. Therefore, the number of subject matter codes on AC Form 8080-2 may not represent the number of questions missed on the test.

When taking the test, the applicant should keep the following points in mind:

1. Answer each question in accordance with the latest regulations and procedures.

2. Read each question carefully before looking at the possible answers. You should clearly understand the problem before attempting to solve it.

3. After formulating an answer, determine which of the alternatives most nearly corresponds with that answer. The answer chosen should completely resolve the problem.

4. From the answers given, it may appear that there is more than one possible answer; however, there is only one answer that is correct and complete. The other answers are either incomplete or are derived from popular misconceptions.

5. If a certain question is difficult for you, it is best to proceed to other questions. After you answer the less difficult questions, return to those which gave you difficulty. Be sure to indicate on your question selection sheet the questions to which you wish to return.

6. When solving a calculator problem, select the answer nearest your solution. If you have solved it correctly, your answer will be closer to the correct answer than to any of the other choices.

7. To aid in scoring, enter personal data in the appropriate spaces on the test answer sheet in a complete and legible manner. Be sure to enter the test number exactly as printed on the question selection sheet.

Retesting after Failure—FAR Section 65.19

An applicant for a written, oral, or practical test for a certificate and rating, or for an additional rating under this part, may apply for retesting–

(a) After 30 days after the date the applicant failed the test; or

(b) Before the 30 days have expired if the applicant presents a signed statement from an airman holding the certificate and rating sought by the applicant, certifying that the airman has given the applicant additional instruction in each of the subjects failed and that the airman considers the applicant ready for retesting.

QUESTIONS

8001. Laminated wood spars may be substituted for solid rectangular wood spars

A— only in certain instances where the primary load is shared by one or more other original structural member.
B— if the same quality wood is used in both.
C— only upon specific approval by the manufacturer or the FAA.

8002. The strength of a well-designed and properly prepared wood splice joint is provided by the

A— bearing surface of the wood fibers.
B— glue.
C— reinforcement plates.

8003. Where is information found concerning acceptable species substitutions for wood materials used in aircraft repair?

A— Aircraft Specifications or Type Certificate Data Sheets.
B— Technical Standard Orders.
C— AC 43.13-1A.

8004. In cases of elongated boltholes in a wood spar or cracks in the vicinity of boltholes,

A— it is permissible to ream the hole, plug with hardwood, and redrill.
B— the spar may be reinforced by using hardwood reinforcing plates.
C— a new section of spar should be spliced in or the spar replaced entirely.

8005. A faint line running across the grain of a wood spar generally indicates

A— compression failure.
B— shear failure.
C— decay.

8006. Which statement about wood decay is correct?

A— Decay that occurs before the wood is seasoned does not affect the strength of the finished piece.
B— A limited amount of certain kinds of decay is acceptable in aircraft woods since decay affects the binding between the fibers and not the fibers themselves.
C— Decay is not acceptable in any form or amount.

8007. Which of the following conditions will determine acceptance of wood with mineral streaks?

A— Careful inspection fails to reveal any decay.
B— They produce only a small effect on grain direction.
C— Local irregularities do not exceed limitations specified for spiral and diagonal grain.

8008. The I-beam wooden spar is routed to

A— increase strength.
B— obtain uniform strength.
C— reduce weight.

8009. Pin knot clusters are permitted in wood aircraft structure provided

A— they produce a small effect on grain direction.
B— they have no mineral streaks.
C— no pitch pockets are within 12 inches.

8010. The cantilever wing uses

A— external struts or wire bracing.
B— no external bracing.
C— the skin to carry most of the load to the wing butt.

8011. Laminated wood is sometimes used in the construction of highly stressed aircraft components. This wood can be identified by its

A— parallel grain construction.
B— similarity to standard plywood construction.
C— perpendicular grain construction.

8012. When patching a plywood skin, abrupt changes in cross-sectional areas which will develop dangerous stress concentration should be avoided by using

A— circular or elliptical patches.
B— square patches.
C— doublers with any desired shaped patches.

8013. Glue deterioration in wood aircraft structure is indicated

A— when a joint has separated and the glue surface shows only the imprint of the wood with no wood fibers clinging to the glue.
B— when a joint has separated and the glue surface shows pieces of wood and/or wood fibers clinging to the glue.
C— by any joint separation.

8014. Compression failures in wood aircraft structures are characterized by buckling of the fibers that appear as streaks on the surface

A— at right angles to the growth rings.
B— parallel to the grain.
C— at right angles to the grain.

8015. When and how is finishing tape applied on a fabric-covered aircraft?

A— Sewed or laced on before dope is applied.
B— Doped on immediately prior to the finish coat.
C— Doped on after the first or second coat of dope.

8016. The determining factor(s) for the selection of the correct weight of textile fabric to be used in covering any type of aircraft is the

A— maximum wing loading.
B— speed of the aircraft.
C— speed of the aircraft and the maximum wing loading.

8017. How many fabric thicknesses will be found in a French-fell seam?

A— Five.
B— Three.
C— Four.

8018. Prior to application of surface tape to the trailing edge of a wing or control surface, the tape should be notched at intervals to

A— make it easier to apply to severe curvatures.
B— increase the length of the tape's edges for better doping.
C— prevent the entire tape from loosening in the event the tape begins to separate from the surface.

8019. Moisture, mildew, chemicals, and acids have no effect on

A— glass fabric.
B— linen fabric.
C— dacron fabric.

8020. The best method of repair for a fabric-covered surface which has an L-shaped tear, each leg of which is approximately 14 inches long, is to

A— re-cover the entire bay in which the tear is located.
B— sew from the end of each leg to the center of the tear with a baseball stitch and then dope on a patch.
C— sew with a baseball stitch from the center of the tear out toward the extremity of each leg and then dope on a patch.

8021. The strength classification of fabrics used in aircraft covering is based on

A— bearing strength.
B— shear strength.
C— tensile strength.

8022. Fabric rejuvenator

A— restores the condition of dope coatings.
B— restores fabric strength and tautness to at least the minimum acceptable level.
C— penetrates the fabric and restores fungicidal resistance.

8023. (1) Machine-sewn seams in aircraft covering fabrics may be of the folded-fell or French-fell types.

(2) A plain lapped seam is never permissible.

Regarding the above statements,

A— both No. 1 and No. 2 are true.
B— only No. 1 is true.
C— only No. 2 is true.

8024. When testing the strength of Grade A cotton fabric covering an aircraft that requires only intermediate grade, the minimum acceptable strength the fabric must have is

A— 70 percent of its original strength.
B— 70 percent of the original strength for intermediate fabric.
C— 56 pounds per inch warp and fill.

8025. When dope proofing the parts of the aircraft structure that come in contact with doped fabric, which of the following provide an acceptable protective coating?

1. Aluminum foil.
2. Resin impregnated cloth tape.
3. Any organic-type protective coating.
4. Cellulose tape.

A— 1 and 2.
B— 1 and 4.
C— 3 and 4.

8026. If registration numbers are to be applied to an aircraft with a letter height of 12 inches, what is the minimum space required for the registration mark N1683C?

Note:
2/3 x height = character width.
1/6 x height = width for 1.
1/4 x 2/3 height = spacing.
1/6 x height = stroke or line width.

A— 52 inches.
B— 48 inches.
C— 57 inches.

8027. If masking tape is applied to an aircraft such as for trim spraying, and is left on for several days and/or exposed to heat, it is likely that the tape will

A— not seal out the finishing material if the delay or heating occurs before spraying.
B— be weakened in its ability to adhere to the surface.
C— cure to the finish and be very difficult to remove.

8028. What is used to slow the drying time of some finishes and to prevent blush?

A— Reducer.
B— Retarder.
C— Rejuvenator.

8029. Which type of coating typically includes phosphoric acid as one of its components at the time of application?

A— Wash primer.
B— Epoxy primer.
C— Zinc chromate primer.

8030. Which properly applied finish topcoat is the most durable and chemical resistant?

A— Synthetic enamel.
B— Acrylic lacquer.
C— Polyurethane.

8031. Aluminum-pigment in dope is used primarily to

A— provide a silver color.
B— aid in sealing out moisture from the fabric.
C— exclude sunlight from the fabric.

8032. A correct use for acetone is to

A— thin zinc chromate primer.
B— remove grease from fabric.
C— thin dope.

8033. Which of the following is a hazard associated with sanding on fabric covered surfaces during the finishing process?

A— Overheating of the fabric/finish, especially with the use of power tools.
B— Static electricity buildup.
C— Embedding of particles in the finish.

8034. What is likely to occur if unhydrated wash primer is applied to unpainted aluminum and then about 30 to 40 minutes later a finish topcoat, when the humidity is low?

A— Corrosion.
B— A glossy, blush-free finish.
C— A dull finish due to the topcoat ''sinking in'' to primer that is still too soft.

8035. Fungicidal dopes are used in aircraft finishing as the

A— first coat to prevent fabric rotting and are applied extremely thin to saturate the fabric.
B— first full-bodied brushed-on coat to prevent fungus damage.
C— final full-bodied brushed-on coat to reduce blushing.

8036. Before spraying any finishing materials on unpainted clean aluminum,

A— wipe the surface with avgas or kerosene.
B— remove any conversion coating film.
C— avoid touching the surface with bare hands.

8037. What is likely to occur if hydrated wash primer is applied to unpainted aluminum and then about 30 to 40 minutes later a finish topcoat, when the humidity is low?

A— Corrosion.
B— A glossy, blush-free finish.
C— A dull finish due to the topcoat ''sinking in'' to primer that is still too soft.

8038. What is the usual cause of runs and sags in aircraft finishes?

A— Too much material applied in one coat.
B— Material is being applied too fast.
C— Low atmospheric humidity.

8039. Which defect in aircraft finishes may be caused by adverse humidity, drafts, or sudden changes in temperature?

A— Orange peel.
B— Blushing.
C— Pinholes.

8040. Which statement is true regarding paint system compatibility?

A— Old-type zinc chromate primer may not be used directly for touchup of bare metal surfaces.
B— Acrylic nitrocellulose lacquers may be used over old nitrocellulose finishes.
C— Old wash primer coats may be overcoated directly with epoxy finishes.

8041. A well-designed rivet joint will subject the rivets to

A— compressive loads.
B— shear loads.
C— tension loads.

8042. A main difference between Lockbolt/Huckbolt tension and shear fasteners (other than their application) is in the

A— number of locking collar grooves.
B— shape of the head.
C— method of installation.

8043. Alloy 2117 rivets are heat treated

A— by the manufacturer and do not require heat treatment before being driven.
B— by the manufacturer but require reheat treatment before being driven.
C— to a temperature of 910 to 930 °F and quenched in cold water.

8044. The general rule for finding the proper rivet diameter is

A— three times the thickness of the materials to be joined.
B— two times the rivet length.
C— three times the thickness of the thickest sheet.

8045. The bucked head of a rivet should be

A— one and one-half D by one-half D.
B— one-half D by one D.
C— one and one-half D by two D.

8046. One of the main advantages of Hi-Lok type fasteners over earlier generations is that

A— they can be removed and reused again.
B— the squeezed on collar installation provides a more secure, tighter fit.
C— they can be installed with ordinary hand tools.

8047. The markings on the head of a Dzus fastener identify the

A— body diameter, type of head, and length of the fastener.
B— body type, head diameter, and type of material.
C— manufacturer and type of material.

8048. The Dzus turnlock fastener consists of a stud, grommet, and receptacle. The stud length is measured in

A— hundredths of an inch.
B— tenths of an inch.
C— sixteenths of an inch.

8049. The Dzus turnlock fastener consists of a stud, grommet, and receptacle. The stud diameter is measured in

A— tenths of an inch.
B— hundredths of an inch.
C— sixteenths of an inch.

8050. Threaded rivets (Rivnuts) are commonly used to

A— join two or more pieces of sheet metal where shear strength is desired.
B— attach parts or components with screws to sheet metal.
C— join two or more pieces of sheet metal where bearing strength is desired.

8051. Cherrymax and Olympic-Lok rivets

A— utilize a rivet gun, special rivet set, and bucking bar for installation.
B— utilize a pulling tool for installation.
C— may be installed with ordinary hand tools.

8052. Hole filling fasteners (for example, MS20470 rivets) should not be used in composite structures primarily because of the

A— possibility of causing delamination.
B— increased possibility of fretting corrosion in the fastener.
C— difficulty in forming a proper shop head.

8053. Metal fasteners used with carbon/graphite composite structures

A— may be constructed of any of the metals commonly used in aircraft fasteners.
B— must be constructed of material such as titanium or corrosion resistant steel.
C— must be constructed of high strength aluminum-lithium alloy.

8054. Sandwich panels made of metal honeycomb construction are used in high-speed aircraft because this type construction

A— is lighter than single sheet skin of the same strength and is more corrosion resistant.
B— may be repaired by gluing replacement skin to the inner core material with thermoplastic resin.
C— has a high strength-to-weight ratio and greater stiffness than a single sheet.

8055. (1) When performing a ring (coin tap) test on composite structures, a change in sound may be due to damage or to transition to a different internal structure.

(2) The extent of separation damage in composite structures is most accurately measured by a ring (coin tap) test.

Regarding the above statements,

A— both No. 1 and No. 2 are true.
B— only No. 1 is true.
C— only No. 2 is true.

8056. Which of these methods may be used to inspect fiberglass/honeycomb structures for entrapped water?

1. Acoustic emission monitoring.
2. X-ray.
3. Backlighting.

A— 1 and 2.
B— 1 and 3.
C— 2 and 3.

8057. When balsa wood is used to replace a damaged honeycomb core, the plug should be cut so that

A— the grain is parallel to the skin.
B— it is about 1/8 inch undersize to allow sufficient bonding material to be applied.
C— the grain is perpendicular to the skin.

8058. When repairing puncture-type damage of a metal faced laminated honeycomb panel, the edges of the doubler should be tapered to

A— two times the thickness of the metal.
B— 100 times the thickness of the metal.
C— whatever is desired for a neat, clean appearance.

8059. One of the best ways to assure that a properly prepared batch of matrix resin has been achieved is to

A— perform a chemical composition analysis.
B— have mixed enough for a test sample.
C— test the viscosity of the resin immediately after mixing.

8060. Composite inspections conducted by means of acoustic emission monitoring

A— pick up the "noise" of corrosion or other deterioration occurring.
B— analyze ultrasonic signals transmitted into the parts being inspected.
C— create sonogram pictures of the areas being inspected.

8061. What precaution, if any, should be taken to prevent corrosion inside a repaired metal honeycomb structure?

A— Prime the repair with a corrosion inhibitor and seal from the atmosphere.
B— Paint the outside area with several coats of exterior paint.
C— None. Honeycomb is usually made from a man-made or fibrous material which is not susceptible to corrosion.

8062. One method of inspecting a laminated fiberglass structure that has been subjected to damage is to

A— strip the damaged area of all paint and shine a strong light through the structure.
B— use dye-penetrant inspection procedures, exposing the entire damaged area to the penetrant solution.
C— use an eddy current probe on both sides of the damaged area.

8063. When inspecting a composite panel using the ring test/tapping method, a dull thud may indicate

A— less than full strength curing of the matrix.
B— separation of the laminates.
C— an area of too much matrix between fiber layers.

8064. How many of the following are benefits of using microballoons when making repairs to laminated honeycomb panels?

1. Less filler shrinkage.
2. Greater flexibility.
3. Less density.
4. Lower stress concentrations.

A— Two.
B— Three.
C— Four.

8065. The length of time that a catalyzed resin will remain in a workable state is called the

A— pot life.
B— shelf life.
C— service life.

8066. A category of plastic material that is capable of softening or flowing when reheated is described as a

A— thermoplastic.
B— thermocure.
C— thermoset.

8067. The classification for high tensile strength fiberglass used in aircraft structures is

A— E.
B— S.
C— G.

8068. Which is an identifying characteristic of acrylic plastics?

A— Zinc chloride will have no effect.
B— Has a yellowish tint when viewed from the edge.
C— Acetone will soften plastic, but will not change its color.

8069. Superficial scars, scratches, surface abrasion, or rain erosion on fiberglass laminates can generally be repaired by applying

A— a piece of resin-impregnated glass fabric facing.
B— one or more coats of suitable resin (room-temperature catalyzed) to the surface.
C— a sheet of polyethylene over the abraded surface and one or more coats of resin cured with infrared heat lamps.

8070. The classification for fiberglass reinforcement material that has high resistivity and is the most common is

A— E.
B— S.
C— G.

8071. A potted compound repair on honeycomb can usually be made on damages less than

A— 4 inches in diameter.
B— 2 inches in diameter.
C— 1 inch in diameter.

8072. Composite fabric material is considered to be the strongest in what direction?

A— Fill.
B— Warp.
C— Bias.

8073. What reference tool is used to determine how the fiber is to be oriented for a particular ply of fabric?

A— Fill clock (or compass).
B— Bias clock (or compass).
C— Warp clock (or compass).

8074. The strength and stiffness of a properly constructed composite buildup depends primarily on

A— a 60 percent matrix to 40 percent fiber ratio.
B— the orientation of the plies to the load direction.
C— the ability of the fibers to transfer stress to the matrix.

8075. Which fiber to resin (percent) ratio for advanced composite wet lay-ups is generally considered the best for strength?

A— 40:60.
B— 50:50.
C— 60:40.

8076. What is the material layer used within the vacuum bag pressure system to absorb excess resin during curing called?

A— Bleeder.
B— Breather.
C— Release.

8077. Proper pre-preg composite lay-up curing is generally accomplished by

1. applying external heat.
2. room temperature exposure.
3. adding a catalyst or curing agent to the resin.
4. applying pressure.

A— 2 and 3.
B— 1 and 4.
C— 1, 3, and 4.

8078. When repairing large, flat surfaces with polyester resins, warping of the surface is likely to occur. One method of reducing the amount of warpage is to

A— add an extra amount of catalyst to the resin.
B— use short strips of fiberglass in the bonded repair.
C— use less catalyst than normal so the repair will be more flexible.

8079. When making repairs to fiberglass, cleaning of the area to be repaired is essential for a good bond. The final cleaning should be made using

A— MEK (methyl ethyl ketone).
B— soap, water, and a scrub brush.
C— a thixotropic agent.

8080. When necessary, what type of cutting fluid is usually acceptable for machining composite laminates?

A— Water soluble oil.
B— Water displacing oil.
C— Water only.

8081. Fiberglass laminate damage not exceeding the first layer or ply can be repaired by

A— filling with a putty consisting of a compatible resin and clean, short glass fibers.
B— sanding the damaged area until aerodynamic smoothness is obtained.
C— trimming the rough edges and sealing with paint.

8082. Fiberglass damage that extends completely through a laminated sandwich structure

A— may be repaired.
B— must be filled with resin to eliminate dangerous stress concentrations.
C— may be filled with putty which is compatible with resin.

8083. Fiberglass laminate damage that extends completely through one facing and into the core

A— cannot be repaired.
B— requires the replacement of the damaged core and facing.
C— can be repaired by using a typical metal facing patch.

8084. Repairing advanced composites using materials and techniques traditionally used for fiberglass repairs is likely to result in

A— restored strength and flexibility.
B— improved wear resistance to the structure.
C— an unairworthy repair.

8085. The preferred way to make permanent repairs on composites is by

A— bonding on metal or cured composite patches.
B— riveting on metal or cured composite patches.
C— laminating on new repair plies.

8086. Which of the following, when added to wet resins, provide strength for the repair of damaged fastener holes in composite panels?

1. Microballoons.
2. Flox.
3. Chopped fibers.

A— 2 and 3.
B— 1 and 3.
C— 1, 2, and 3.

8087. The part of a replacement honeycomb core that must line up with the adjacent original is the

A— cell side.
B— ribbon direction.
C— cell edge.

8088. Which of the following are generally characteristic of aramid fiber (Kevlar) composites?

1. High tensile strength.
2. Flexibility.
3. Stiffness.
4. Corrosive effect in contact with aluminum.
5. Ability to conduct electricity.

A— 1 and 2.
B— 2, 3 and 4.
C— 1, 3, and 5.

8089. Which of the following are generally characteristic of carbon/graphite fiber composites?

1. Flexibility.
2. Stiffness.
3. High compressive strength.
4. Corrosive effect in contact with aluminum.
5. Ability to conduct electricity.

A— 1 and 3.
B— 2, 3, and 4.
C— 1, 3, and 5.

8090. If an aircraft's transparent plastic enclosures exhibit fine cracks which may extend in a network over or under the surface or through the plastic, the plastic is said to be

A— hazing.
B— brinelling.
C— crazing.

8091. When installing transparent plastic enclosures which are retained by bolts extending through the plastic material and elastic stop nuts, the stop nuts should be

A— tightened to a firm fit plus one full turn.
B— tightened to a firm fit then backed off one full turn.
C— tightened to a firm fit.

8092. If a new safety belt is to be installed in an aircraft, the belt must conform to the strength requirements in which document?

A— STC 1282.
B— FAR Part 39.
C— TSO C22.

8093. Which is considered good practice concerning the installation of acrylic plastics?

A— When nuts and bolts are used, the plastic should be installed hot and tightened to a firm fit before the plastic cools.
B— When rivets are used, adequate spacer or other satisfactory means to prevent excessive tightening of the frame to the plastic should be provided.
C— When rivets or nuts and bolts are used, slotted holes are not recommended.

8094. The coefficient of expansion of most plastic enclosure materials is

A— greater than both steel and aluminum.
B— greater than steel but less than aluminum.
C— less than either steel or aluminum.

8095. If no scratches are visible after transparent plastic enclosure materials have been cleaned, their surfaces should be

A— polished with rubbing compound applied with a damp cloth.
B— buffed with a clean, soft, dry cloth.
C— covered with a thin coat of wax.

8096. Cabin upholstery materials installed in current standard category airplanes must

A— be fireproof.
B— be at least flame resistant.
C— meet the requirements prescribed in FAR Part 43.

8097. What is the most common method of cementing transparent plastics?

A— Heat method.
B— Soak method.
C— Bevel method.

8098. When holes are drilled completely through Plexiglas, a

A— standard twist drill should be used.
B— specially modified twist drill should be used.
C— wood drill should be used.

8099. What is the purpose of a gusset or gusset plate used in the construction and repair of aircraft structures?

A— To hold structural members in position temporarily until the permanent attachment has been completed.
B— To provide access for inspection of structural attachments.
C— To join and reinforce intersecting structural members.

8100. Select the alternative which best describes the function of the flute section of a twist drill.

A— Prevents overheating of the drill point.
B— Forms the area where the drill bit attaches to the drill motor.
C— Forms the cutting edges of the drill point.

8101. How many MS20470 AD-4-6 rivets will be required to attach a 10" x 5" plate, using a single row of rivets, minimum edge distance, and 4D spacing?

A— 56.
B— 54.
C— 52.

8102. Longitudinal (fore and aft) structural members of a semi-monocoque fuselage are called

A— spars and ribs.
B— longerons and stringers.
C— spars and stringers.

8103. Shallow scratches in sheet metal may be repaired by

A— burnishing.
B— buffing.
C— stop drilling.

8104. What should be the included angle of a twist drill for soft metals?

A— 118°.
B— 90°.
C— 65°.

8105. When comparing the machining techniques for stainless steel sheet material to those for aluminum alloy sheet, it is normally considered good practice to drill the stainless steel at a

A— higher speed with less pressure applied to the drill.
B— lower speed with more pressure applied to the drill.
C— lower speed with less pressure applied to the drill.

8106. A single-lap sheet splice is to be used to repair a section of damaged aluminum skin. If a double row of 1/8-inch rivets is used, the minimum allowable overlap will be

A— 1/2 inch.
B— 3/4 inch.
C— 1 inch.

8107. Which statement is true regarding the inspection of a stressed skin metal wing assembly known to have been critically loaded?

A— If rivets show no visible distortion, further investigation is unnecessary.
B— If bearing failure has occurred, the rivet shanks will be joggled.
C— If genuine rivet tipping has occurred, groups of consecutive rivet heads will be tipped in the same direction.

8108. What is the minimum edge distance for aircraft rivets?

A— Two times the diameter of the rivet shank.
B— Two times the diameter of the rivet head.
C— Three times the diameter of the rivet shank.

8109. When drilling stainless steel, the drill used should have an included angle of

A— 90° and turn at a low speed.
B— 118° and turn at a high speed.
C— 140° and turn at a low speed.

8110. What is the minimum spacing for a single row of aircraft rivets?

A— Two times the diameter of the rivet shank.
B— Three times the length of the rivet shank.
C— Three times the diameter of the rivet shank.

8111. (Refer to figure 1.) Which of the rivets shown will accurately fit the conical depression made by a 100° countersink?

A— 1.
B— 2.
C— 3.

8112. Which is correct concerning the use of a file?

A— Apply pressure on the forward stroke, only, except when filing very soft metals such as lead or aluminum.
B— A smoother finish can be obtained by using a double-cut file than by using a single-cut file.
C— The terms "double-cut" and "second-cut" have the same meaning in reference to files.

8113. What is one of the determining factors which permits machine countersinking when flush riveting?

A— Thickness of the material and rivet diameter are the same.
B— Thickness of the material is less than the thickness of the rivet head.
C— Thickness of the material is greater than the thickness of the rivet head.

8114. When repairing a small hole on a metal stressed skin, the major consideration in the design of the patch should be

A— the shear strength of the riveted joint.
B— to use rivet spacing similar to a seam in the skin.
C— that the bond between the patch and the skin is sufficient to prevent dissimilar metal corrosion.

8115. Which procedure is correct when using a reamer to finish a drilled hole to the correct size?

A— Turn the reamer in the cutting direction when enlarging the hole and in the opposite direction to remove from the hole.
B— Turn the reamer only in the cutting direction.
C— Apply considerable pressure on the reamer when starting the cut and reduce the pressure when finishing the cut.

8116. Repairs or splices involving stringers on the lower surface of stressed skin metal wings are usually

A— not permitted.
B— permitted only if the damage does not exceed 6 inches in any direction.
C— permitted but are normally more critical in reference to strength in tension than similar repairs to the upper surface.

8117. When straightening members made of 2024-T4, you should

A— straighten cold and reinforce.
B— straighten cold and anneal to remove stress.
C— apply heat to the inside of the bend.

8118. Clad aluminum alloys are used in aircraft because they

A— can be heat treated much easier than the other forms of aluminum.
B— are less subject to corrosion than uncoated aluminum alloys.
C— are stronger than unclad aluminum alloys.

8119. Which statement is true regarding a cantilever wing?

A— It has nonadjustable lift struts.
B— No external bracing is needed.
C— It requires only one lift strut on each side.

8120. Aircraft structural units, such as spars, engine supports, etc., which have been built up from sheet metal, are normally

A— repairable, using approved methods.
B— repairable, except when subjected to compressive loads.
C— not repairable, but must be replaced when damaged or deteriorated.

8121. A factor which determines the minimum space between rivets is the

A— length of the rivets being used.
B— diameter of the rivets being used.
C— thickness of the material being riveted.

8122. What should be the included angle of a twist drill for hard metal?

A— 118°.
B— 100°.
C— 90°.

8123. When fabricating parts from Alclad 2024-T3 aluminum sheet stock,

A— bends should be made with a small radius to develop maximum strength.
B— all bends must be 90° to the grain.
C— all scratches, kinks, tool marks, nicks, etc., must be held to a minimum.

8124. Monocoque fuselages derive their principal strength from

A— bulkheads and longerons.
B— longerons and formers.
C— the actual covering, metal or plywood.

8125. Which part(s) of a semi-monocoque fuselage prevent(s) tension and compression from bending the fuselage?

A— The fuselage covering.
B— Longerons and stringers.
C— Bulkheads and skin.

8126. Rivet gauge, or transverse pitch is the distance between the

A— centers of rivets in adjacent rows.
B— centers of adjacent rivets in the same row.
C— heads of rivets in the same row.

8127. Rivet pitch is the distance between the

A— centers of rivets in adjacent rows.
B— centers of adjacent rivets in the same row.
C— heads of rivets in the same row.

8128. (Refer to figure 2.) Select the preferred drawing for proper countersinking.

A— All are acceptable.
B— 2.
C— 1.

8129. What is indicated by a black ''smoky'' residue streaming back from some of the rivets on an aircraft?

A— The rivets were excessively work hardened during installation.
B— Exfoliation corrosion is occurring inside the structure.
C— Fretting corrosion is occurring between the rivets and the skin.

8130. The identifying marks on the heads of aluminum alloy rivets indicate the

A— degree of dimensional and process control observed during manufacture.
B— head shape, shank size, material used, and specifications adhered to during manufacture.
C— specific alloy used in the manufacture of the rivets.

8131. When an MS20470D rivet is installed, its full shear strength is obtained

A— only after a period of age hardening.
B— by the cold working of the rivet metal in forming a shop head.
C— by heat treating just prior to being driven.

8132. Which of the following need not be considered when determining minimum rivet spacing?

A— Rivet diameter.
B— Rivet length.
C— Type of material being riveted.

8133. What is the purpose of refrigerating 2017 and 2024 aluminum alloy rivets after heat treatment?

A— To accelerate age hardening.
B— To relieve internal stresses.
C— To retard age hardening.

8134. Under certain conditions, type A rivets are not used because of their

A— low strength characteristics.
B— high alloy content.
C— tendency toward embrittlement when subjected to vibration.

8135. A rivet set used to drive MS20470 rivets should

A— have the same radius as the rivet head.
B— have a slightly greater radius than the rivet head.
C— be nearly flat on the end, with a slight radius on the edge to prevent damage to the sheet being riveted.

8136. Heat-treated rivets in the D and DD series that are not driven within the prescribed time after heat treatment or removal from refrigeration

A— must be reheat treated before use.
B— must be discarded.
C— may be returned to refrigeration and used later without reheat treatment.

8137. The dimensions of an MS20430AD-4-8 rivet are

A— 1/8 inch in diameter and 1/4 inch long.
B— 1/8 inch in diameter and 1/2 inch long.
C— 4/16 inch in diameter and 8/32 inch long.

8138. The primary alloying agent of 2024-T36 is indicated by the number

A— 2.
B— 20.
C— 24.

8139. Which part of the 2017-T36 aluminum alloy designation indicates the primary alloying agent used in its manufacture?

A— 2.
B— 17.
C— 20.

8140. A sheet metal repair is to be made using two pieces of 0.040-inch aluminum riveted together. All rivet holes are drilled for 3/32-inch rivets. The length of the rivets to be used will be

A— 1/8 inch.
B— 1/4 inch.
C— 5/16 inch.

8141. Most rivets used in aircraft construction have

A— dimples.
B— smooth heads without markings.
C— a raised dot.

8142. MS20426AD-6-5 indicates a countersunk rivet which has

A— a shank length of 5/16 inch (excluding head).
B— a shank length of 5/32 inch (excluding head).
C— an overall length of 5/16 inch.

8143. Which rivet may be used as received without further treatment?

A— 2024-T4.
B— 2117-T3.
C— 2017-T3.

8144. (Refer to figure 3.) Which is the grip length of the flush rivet?

A— 1.
B— 2.
C— 3.

8145. Which rivets should be selected to join two sheets of .032-inch aluminum?

A— MS20425D-4-3.
B— MS20470AD-4-4.
C— MS20455DD-5-3.

8146. A sheet metal repair is to be made using two pieces of 0.0625-inch aluminum riveted together. All rivet holes are drilled for 1/8-inch rivets. The length of the rivets to be used will be

A— 5/32 inch.
B— 3/16 inch.
C— 5/16 inch.

8147. Mild steel rivets are used for riveting

A— nickel-steel parts.
B— magnesium parts.
C— steel parts.

8148. A DD rivet is heat treated before use to

A— harden and increase strength.
B— relieve internal stresses.
C— soften to facilitate riveting.

8149. When riveting dissimilar metals together, what precautions must be taken to prevent an electrolytic action?

A— Treat the surfaces to be riveted together with a process called anodic treatment.
B— Place a protective separator between areas of potential electrical difference.
C— Avoid the use of dissimilar metals by redesigning the unit according to the recommendations outlined in AC 43.13 1A.

8150. The length of a rivet to be used to join a sheet of .032-inch and .064-inch aluminum alloy should be equal to

A— two times the rivet diameter plus .064 inch.
B— one and one-half times the rivet diameter plus .096 inch.
C— three times the rivet diameter plus .096 inch.

8151. What is generally the best procedure to use when removing a solid shank rivet?

A— Drill through the manufactured head and shank with a shank size drill and remove the rivet with a punch.
B— Drill to the base of the manufactured rivet head with a drill one size smaller than the rivet shank and remove the rivet with a punch.
C— Drill through the manufactured head and shank with a drill one size smaller than the rivet and remove the rivet with a punch.

8152. Joggles in removed rivet shanks would indicate partial

A— bearing failure.
B— torsion failure.
C— shear failure.

8153. What type loads cause the most rivet failures?

A— Shear.
B— Bearing.
C— Head.

8154. Which rivet is used for riveting magnesium alloy structures?

A— Mild steel.
B— 5056 aluminum.
C— Monel.

8155. Which rivet is used for riveting nickel-steel alloys?

A— 2024 aluminum.
B— Mild steel.
C— Monel.

8156. The length of rivet to be chosen when making a structural repair that involves the joining of 0.032-inch and 0.064-inch aluminum sheet, drilled with a No. 30 drill, is

A— 7/16 inch.
B— 5/16 inch.
C— 1/4 inch.

8157. (Refer to figure 4.) The length of flat A is

A— 3.750 inches.
B— 3.875 inches.
C— 3.937 inches.

8158. (Refer to figure 4.) The amount of material required to make the 90° bend is

A— 0.3436 inch.
B— 0.3717 inch.
C— 0.3925 inch.

8159. (Refer to figure 5.) What is the length of flat A?

A— 3.7 inches.
B— 3.8 inches.
C— 3.9 inches.

8160. (Refer to figure 5.) What is the flat layout dimension?

A— 7.0 inches.
B— 6.8 inches.
C— 6.6 inches.

8161. If a streamline cover plate is to be hand formed using a form block, a piece of dead soft aluminum should first be placed over the hollow portion of the mold and securely fastened in place. The bumping operation should be

A— distributed evenly over the face of the aluminum at all times rather than being started at the edges or center.
B— started by tapping the aluminum lightly around the edges and gradually working down into the center.
C— started by tapping the aluminum in the center until it touches the bottom of the mold and then working out in all directions.

8162. A piece of flat stock that is to be bent to a closed angle of 15° must be bent through an angle of

A— 165°.
B— 105°.
C— 90°.

8163. When a piece of aluminum alloy is to be bent using a minimum radius for the type and thickness of material,

A— the piece should be bent slowly to eliminate cracking.
B— the layout should be made so that the bend will be 90° to the grain of the sheet.
C— less pressure than usual should be applied with the movable (upper) clamping bar.

8164. The flat layout or blank length of a piece of metal from which a simple L-shaped bracket 3 inches by 1 inch is to be bent depends upon the radius of the desired bend. The bracket which will require the greatest amount of material is one which has a bend radius of

A— 1/8 inch.
B— 1/2 inch.
C— 1/4 inch.

8165. If it is necessary to compute a bend allowance problem and bend allowance tables are not available, the neutral axis of the bend can be

A— represented by the actual length of the required material for the bend.
B— found by adding approximately one-half of the stock thickness to the bend radius.
C— found by subtracting the stock thickness from the bend radius.

8166. Unless otherwise specified, the radius of a bend is the

A— inside radius of the metal being formed.
B— inside radius plus one-half the thickness of the metal being formed.
C— radius of the neutral axis plus one-half the thickness of the metal being formed.

8167. The sharpest bend that can be placed in a piece of metal without critically weakening the part is called the

A— bend allowance.
B— minimum radius of bend.
C— maximum radius of bend.

8168. The most important factors needed to make a flat pattern layout are

A— radius, thickness, and mold line.
B— radius, thickness, and degree of bend.
C— the lengths of the legs (flat sections).

8169. A piece of sheet metal is bent to a certain radius. The curvature of the bend is referred to as the

A— bend allowance.
B— neutral line.
C— bend radius.

8170. You can distinguish between aluminum and aluminum alloy by

A— filing the metal.
B— testing with an acetic acid solution.
C— testing with a 10 percent solution of caustic soda.

8171. The purpose of a joggle is to

A— allow clearance for a sheet or an extrusion.
B— increase obstruction for a sheet or an extrusion.
C— decrease the weight of the part and still retain the necessary strength.

8172. When bending metal, the material on the outside of the curve stretches while the material on the inside of the curve compresses. That part of the material which is not affected by either stress is the

A— mold line.
B— bend tangent line.
C— neutral line.

8173. (Refer to figure 6.) Determine the dimensions of A, B, and C in the flat layout.

Setback = .252
Bend allowance = .345

A— A = .748
B = 2.252
C = 2.004.
B— A = .748
B = 1.496
C = 1.248.
C— A = 1.252
B = 2.504
C = 1.752.

8174. (Refer to figure 6.) What is dimension D?

Setback = .252
Bend allowance = .345

A— 3.492.
B— 4.182.
C— 3.841.

8175. The sight line on a sheet metal flat layout to be bent in a cornice or box brake is measured and marked

A— one-half radius from either bend tangent line.
B— one radius from either bend tangent line.
C— one radius from the bend tangent line that is placed under the brake.

8176. (Refer to figure 7.) What is dimension F?

Setback at D = .095
Setback at E = .068
Bend allowance at D = .150
Bend allowance at E = .112

A— 4.836.
B— 5.936.
C— 5.738.

8177. On a sheet metal fitting layout with a single bend, allow for stretching by

A— adding the setback to each leg.
B— subtracting the setback from one leg.
C— subtracting the setback from both legs.

8178. The aluminum alloys used in aircraft construction are usually hardened by which method?

A— Cold-working.
B— Aging.
C— Heat treatment.

8179. In Gas Tungsten Arc (GTA) welding, a stream of inert gas is used to

A— prevent the formation of oxides in the puddle.
B— concentrate the heat of the arc and prevent its dissipation.
C— lower the temperature required to properly fuse the metal.

8180. Which statement best describes magnesium welding?

A— Magnesium can be welded to other metals.
B— Filler rod should be nickel-steel.
C— Filler rod should be the same composition as base metal.

8181. Which statement is true in regard to welding heat-treated magnesium?

A— The welded section does not have the strength of the original metal.
B— Flux should not be used because it is very difficult to remove and is likely to cause corrosion.
C— Magnesium cannot be repaired by fusion welding because of the high probability of igniting the metal.

8182. The oxyacetylene flame for silver soldering should be

A— oxidizing.
B— neutral.
C— carburizing.

8183. Why is it necessary to use flux in all silver soldering operations?

A— To chemically clean the base metal of oxide film.
B— To prevent overheating of the base metal.
C— To increase heat conductivity.

8184. Engine mount members should preferably be repaired by using a

A— larger diameter tube with fishmouth and no rosette welds.
B— larger diameter tube with fishmouth and rosette welds.
C— smaller diameter tube with fishmouth and rosette welds.

8185. What method of repair is recommended for a steel tube longeron dented at a cluster?

A— Welded split sleeve.
B— Welded outer sleeve.
C— Welded patch plate.

8186. Welding over brazed or soldered joints is

A— not permitted.
B— permissible for mild steel.
C— permissible for most metals or alloys that are not heat treated.

8187. Which statement concerning soldering is correct?

A— Joints in electric wire to be soldered should be mechanically secure prior to soldering.
B— Changeable shades of blue can be observed on the surface of a copper soldering tip when the proper temperature for soldering has been reached.
C— If the soldering temperature is too high, the solder will form in lumps and not produce a positive bond.

8188. A resurfaced soldering iron cannot be used effectively until after the working face has been

A— fluxed.
B— polished.
C— tinned.

8189. Which of the following can normally be welded without adversely affecting strength?

1. Aircraft bolts.
2. SAE 4130 chrome/molybdenum tubing.
3. Spring steel struts.
4. Most heat-treated steel/nickel alloy components.

A— 2 and 4.
B— 1 and 3.
C— 2.

8190. In selecting a torch tip size to use in welding, the size of the tip opening determines the

A— amount of heat applied to the work.
B— temperature of the flame.
C— melting point of the filler metal.

8191. Why should a carburizing flame be avoided when welding steel?

A— It removes the carbon content.
B— It hardens the surface.
C— A cold weld will result.

8192. The most important consideration(s) when selecting welding rod is/are

A— current setting or flame temperature.
B— material compatibility.
C— ambient conditions.

8193. The oxyacetylene flame used for aluminum welding should

A— be neutral and soft.
B— be slightly oxidizing.
C— contain an excess of acetylene and leave the tip at a relatively low speed.

8194. A very thin and pointed tip on a soldering copper is undesirable because it will

A— transfer too much heat to the work.
B— have a tendency to overheat and become brittle.
C— cool too rapidly.

8195. Filing or grinding a weld bead

A— may be performed to achieve a smoother surface.
B— reduces the strength of the joint.
C— may be necessary to avoid adding excess weight or to achieve uniform material thickness.

8196. Acetylene at a line pressure above 15 PSI

A— is dangerously unstable.
B— should be used when a reducing flame is necessary.
C— is usually necessary when welding metal over 3/8-inch thick.

8197. Cylinders used to transport and store acetylene

A— are pressure tested to 3,000 PSI.
B— are green in color.
C— contain acetone.

8198. A welding torch backfire may be caused by

A— a loose tip.
B— using too much acetylene.
C— a tip temperature that is too cool.

8199. Which statement concerning a welding process is true?

A— The inert-arc welding process uses an inert gas to protect the weld zone from the atmosphere.
B— In the metallic-arc welding process, filler material, if needed, is provided by a separate metal rod of the proper material held in the arc.
C— In the oxyacetylene welding process, the filler rod used for steel is covered with a thin coating of flux.

8200. Where should the flux be applied when oxyacetylene welding aluminum?

A— Painted only on the surface to be welded.
B— Painted on the surface to be welded and applied to the welding rod.
C— Applied only to the welding rod.

8201. What purpose does flux serve in welding aluminum?

A— Removes dirt, grease, and oil.
B— Minimizes or prevents oxidation.
C— Ensures proper distribution of the filler rod.

8202. Why are aluminum plates 1/4 inch or more thick usually preheated before welding?

A— Reduces internal stresses and assures more complete penetration.
B— Reduces welding time.
C— Prevents corrosion and ensures proper distribution of flux.

8203. How should a welding torch flame be adjusted to weld stainless steel?

A— Slightly carburizing.
B— Slightly oxidizing.
C— Neutral.

8204. Oxides form very rapidly when alloys or metals are hot. It is important, therefore, when welding aluminum to use a

A— solvent.
B— filler.
C— flux.

8205. In gas welding, the amount of heat applied to the material being welded is controlled by the

A— amount of gas pressure used.
B— size of the tip opening.
C— distance the tip is held from the work.

8206. Oxygen and acetylene cylinders are made of

A— seamless aluminum.
B— steel.
C— bronze.

8207. When inspecting a butt-welded joint by visual means,

A— the penetration should be 25 to 50 percent of the thickness of the base metal.
B— the penetration should be 100 percent of the thickness of the base metal.
C— look for evidence of excessive heat in the form of a very high bead.

8208. Annealing of aluminum

A— increases the tensile strength.
B— makes the material brittle.
C— removes stresses caused by forming.

8209. Edge notching is generally recommended in butt welding above a certain thickness of aluminum because it

A— helps hold the metal in alignment during welding.
B— aids in the removal or penetration of oxides on the metal surface.
C— aids in getting full penetration of the metal and prevents local distortion.

8210. If too much acetylene is used in the welding of stainless steel,

A— a porous weld will result.
B— the metal will absorb carbon and lose its resistance to corrosion.
C— oxide will be formed on the base metal close to the weld.

8211. The shielding gases generally used in the Gas Tungsten Arc (GTA) welding of aluminum consist of

A— a mixture of nitrogen and carbon dioxide.
B— nitrogen or hydrogen, or a mixture of nitrogen and hydrogen.
C— helium or argon, or a mixture of helium and argon.

8212. The auxiliary (tail) rotor of a helicopter permits the pilot to compensate for and/or accomplish which of the following?

A— Attitude and airspeed.
B— Lateral and yaw position.
C— Torque and directional control.

8213. The vertical flight of a helicopter is controlled by

A— collective pitch changes.
B— cyclic pitch changes.
C— increasing or decreasing the RPM of the main rotor.

8214. A decrease in pitch angle of the tail rotor blades on a helicopter

A— causes the tail to pivot in the opposite dircction of torque rotation around the main rotor axis.
B— causes the tail to pivot in the direction of torque rotation around the main rotor axis.
C— is required to counteract main rotor torque produced by takeoff RPM.

8215. In rotorcraft external-loading, the ideal location of the cargo release is where the line of action passes

A— aft of the center of gravity at all times.
B— forward of the center of gravity at all times.
C— through the center of gravity at all times.

8216. The acute angle formed by the chord line of a wing and the relative wind is known as the

A— longitudinal dihedral angle.
B— angle of incidence.
C— angle of attack.

8217. A helicopter in forward flight, cruise configuration, changes direction by

A— varying the pitch of the main rotor blades.
B— changing rotor RPM.
C— tilting the main rotor disk in the desired direction.

8218. The purpose in checking main rotor blade tracking is to determine the

A— relative position of the blades during rotation.
B— flight path of the blades during rotation.
C— extent of an out of balance condition during rotation.

8219. In a hovering helicopter equipped with a tail rotor, directional control is maintained by

A— changing the tail rotor RPM.
B— tilting the main rotor disk in the desired direction.
C— varying the pitch of the tail rotor blades.

8220. If a single-rotor helicopter is in forward horizontal flight, the angle of attack of the advancing blade is

A— more than the retreating blade.
B— equal to the retreating blade.
C— less than the retreating blade.

8221. Main rotor blades that do not cone by the same amount during rotation are said to be out of

A— balance.
B— collective pitch.
C— track.

8222. One purpose of the clutch provided between the engine and the helicopter transmission is to

A— disconnect the rotor from the engine to relieve the starter load.
B— automatically disengage the rotor from the engine in case of an engine failure.
C— permit practice of autorotation landings.

8223. Which statement is correct concerning torque effect on helicopters?

A— Torque direction is the same as rotor blade rotation.
B— As horsepower decreases, torque increases.
C— Torque direction is the opposite of rotor blade rotation.

8224. What is the purpose of the free-wheeling unit in a helicopter drive system?

A— It disconnects the rotor whenever the engine stops or slows below the equivalent of rotor RPM.
B— It releases the rotor brake for starting.
C— It relieves bending stress on the rotor blades during starting.

8225. Movement about the longitudinal axis (roll) in a helicopter is effected by movement of the

A— collective pitch control.
B— cyclic pitch control.
C— tail rotor pitch control.

8226. Movement about the lateral axis (pitch) in a helicopter is effected by movement of the

A— collective pitch control.
B— cyclic pitch control.
C— tail rotor pitch control.

8227. Lateral dihedral, a rigging consideration on most airplanes of conventional design, contributes most to stability of the airplane about its

A— longitudinal axis.
B— vertical axis.
C— lateral axis.

8228. As the result of an accident, it is necessary to replace the fuselage fittings for the wing attachment. Prior to welding the fittings in place, a great deal of care should be exercised in locating them in their proper position because the

A— angle of incidence is one of the factors to be re-established at this time.
B— main spar attachment fittings must line up with fuselage fittings; otherwise, the spar location in the wing will have to be changed.
C— aspect ratio of the wing will be controlled by the fuselage fittings.

8229. If a pilot reports that an airplane flies left wing heavy, this condition may be corrected by

A— increasing the angle of incidence of the left wing, or decreasing the angle of incidence of the right wing, or both.
B— increasing the dihedral angle of the left wing, or decreasing the dihedral angle of the right wing, or both.
C— adjusting the dihedral angle of the left wing so that differential pressure between the upper and lower wing surfaces is increased.

8230. If the vertical fin of a single-engine, propeller-driven airplane is rigged properly, it will generally be parallel to

A— the longitudinal axis but not the vertical axis.
B— the vertical axis but not the longitudinal axis.
C— both the longitudinal and vertical axes.

8231. An airplane which has good longitudinal stability should have a minimum tendency to

A— roll.
B— pitch.
C— yaw.

8232. As the angle of attack of an airfoil increases, the center of pressure will

A— move toward the trailing edge.
B— remain stationary because both lift and drag components increase proportionally to increased angle of attack.
C— move toward the leading edge.

8233. The angle of incidence is that acute angle formed by

A— the angular difference between the setting of the main airfoil and the auxiliary airfoil (horizontal stabilizer) in reference to the longitudinal axis of the aircraft.
B— a line parallel to the wing chord and a line parallel to the longitudinal axis of the aircraft.
C— a line parallel to the wing from root to tip and a line parallel to the lateral axis of the aircraft.

8234. An airplane's center of lift is usually located aft of its center of gravity

A— so that the airplane will have a tail-heavy tendency.
B— so that the airplane will have a nose-heavy tendency.
C— to improve stability about the longitudinal axis.

8235. An airplane is controlled directionally about its vertical axis by the

A— rudder.
B— elevator(s).
C— ailerons.

8236. The elevators of a conventional airplane are used to provide rotation about the

A— longitudinal axis.
B— lateral axis.
C— vertical axis.

8237. Washing-in the left wing of a monoplane, for purposes of rigging corrections after flight test, will have what effect on the lift and drag of that wing?

A— Both drag and lift will decrease due to decreased angle of attack.
B— Both drag and lift will increase due to increased angle of attack.
C— The drag will decrease due to the effect of the lift increase.

8238. Flaps increase the effective lift of an airfoil by

A— increasing the camber of the airfoil.
B— introducing drag aft of the center of pressure.
C— increasing the angle of attack of the airfoil.

8239. If the right wing of a monoplane is improperly rigged to a greater angle of incidence than designated in the manufacturer's specifications, it will cause the

A— airplane to be off balance both laterally and directionally.
B— airplane to pitch and roll about the lateral axis.
C— right wing to have both an increased lift and a decreased drag.

8240. The chord of a wing is measured from

A— wingtip to wingtip.
B— wing root to the wingtip.
C— leading edge to trailing edge.

8241. When the lift of an airfoil increases, the drag will

A— decrease.
B— also increase.
C— increase while the lift is changing but will return to its original value.

8242. What physical factors are involved in the aspect ratio of airplane wings?

A— Thickness and chord.
B— Span and chord.
C— Dihedral and angle of attack.

8243. Improper rigging of the elevator trim tab system will affect the balance of the airplane about its

A— lateral axis.
B— longitudinal axis.
C— vertical axis.

8244. An airplane that has a tendency to gradually increase a pitching moment that has been set into motion has

A— poor longitudinal stability.
B— good lateral stability.
C— poor lateral stability.

8245. The purpose of wing slats is to

A— reduce stalling speed.
B— decrease drag.
C— increase speed on takeoff.

8246. The angle of incidence of an airplane

A— affects the dihedral of the wings.
B— is that angle between the relative wind and the chord of the wing.
C— does not change in flight.

8247. Buffeting is the intermittent application of forces to a part of an airplane. It is caused by

A— incorrect rigging of flaps.
B— an unsteady flow from turbulence.
C— incorrect rigging of ailerons.

8248. Movement of an airplane along its lateral axis (roll) is also movement

A— around or about the longitudinal axis controlled by the elevator.
B— around or about the lateral axis controlled by the ailerons.
C— around or about the longitudinal axis controlled by the ailerons.

8249. The primary purpose of stall strips is to

A— provide added lift at slow speeds.
B— stall the inboard portion of the wings first.
C— provide added lift at high angles of attack.

8250. Rigging and alignment checks should not be undertaken in the open; however, if this cannot be avoided, the aircraft should be positioned

A— obliquely into the wind.
B— facing any direction since it makes no difference if the wind is steady (not gusting).
C— with the nose into the wind.

8251. The correct dihedral angle can be determined by

A— measuring the angular setting of each wing at the rear spar with a bubble protractor.
B— placing a straightedge and bubble protractor across the spars while the airplane is in flying position.
C— using a dihedral board and bubble level along the front spar of each wing.

8252. The dihedral angle of a wing may be measured by placing a straightedge and level protractor on the

A— front spar.
B— wing root.
C— wing chord.

8253. Where would you find precise information to perform a symmetry alignment check for a particular aircraft?

A— Aircraft Specification or Type Certificate Data Sheet.
B— Manufacturer's service bulletins.
C— Aircraft service or maintenance manual.

8254. Where is the buttock line or buttline of an aircraft?

A— A height measurement left or right of, and perpendicular to, the horizontal centerline.
B— A width measurement left of, and perpendicular to, the vertical centerline.
C— A width measurement left or right of, and parallel to, the vertical centerline.

8255. Where is fuselage station No. 137 located?

A— 137 centimeters aft of the nose or fixed reference line.
B— 137 inches aft of the zero or fixed reference line.
C— Aft of the engine.

8256. Proper wing twist in a sheet metal constructed wing can usually be checked by utilizing a

A— plum bob, string, and straightedge.
B— bubble level and special fixtures described by the manufacturer.
C— straightedge, tape measure, and carpenter's square.

8257. The control cable terminals on most late model aircraft are swaged and a painted band is placed around the cable adjacent to the terminal in order to

A— show that the terminal was checked with a go-no-go gauge.
B— disclose twisting of the cable in the fitting.
C— detect slippage of the cable in the fitting.

8258. What nondestructive checking method is normally used to ensure that the correct amount of swaging has taken place when installing swaged-type terminals on aircraft control cable?

A— Measure the finished length of the terminal barrel and compare with the beginning length.
B— Use an after gauge to check the diameter of the swaged portion of the terminal.
C— Check the surface of the swaged portion of the terminal for cracks which indicate incomplete swaging.

8259. When inspecting a control cable turnbuckle for proper installation, determine that

A— no more than four threads are exposed on either side of the turnbuckle barrel.
B— the terminal end threads are visible through the safety hole in the barrel.
C— the safety wire ends are wrapped a minimum of four turns around the terminal end shanks.

8260. If all instructions issued by the swaging tool manufacturer are followed when swaging a cable terminal, the resultant swaged terminal strength should be

A— the full rated strength of the cable.
B— 80 percent of the full rated strength of the cable.
C— 70 percent of the full rated strength of the cable.

8261. Which is an acceptable safety device for a castle nut when installed on secondary structures?

A— Star washer.
B— Lockwasher.
C— Cotter pin.

8262. When used in close proximity to magnetic compasses, cotter pins are made of what material?

A— Corrosion resisting steel.
B— Anodized aluminum alloy.
C— Cadmium-plated low carbon steel.

8263. When a fiber stop nut can be threaded on a bolt or stud through the fiber with only the fingers, it should be

A— re-torqued frequently.
B— rejected.
C— reused only in a different location.

8264. The purpose of the vertical fin is to provide

A— directional stability.
B— longitudinal stability.
C— lateral stability.

8265. How are changes in direction of a control cable accomplished?

A— Pulleys.
B— Bell cranks.
C— Fairleads.

8266. What is the smallest size cable that may be used in aircraft primary control systems?

A— 1/4 inch.
B— 5/16 inch.
C— 1/8 inch.

8267. After repairing or re-covering a rudder, the surface should be rebalanced

A— to its spanwise axis.
B— in its normal flight position.
C— to manufacturer's specifications.

8268. Placing a piece of cloth around a stainless steel control cable and running it back and forth over the length of the cable is generally a satisfactory method of

A— applying par-al-ketone.
B— inspecting for broken wires.
C— inspecting for wear or corrosion.

8269. The cable-operated control system of an all-metal aircraft, not incorporating a temperature compensating device, has been rigged to the correct tension in a heated hangar. If the aircraft is operated in very cold weather, the cable tension will

A— decrease when the aircraft structure and cables become cold.
B— increase when the aircraft structure and cables become cold.
C— be unaffected if stainless steel cable is installed.

8270. Very often, repairs to a control surface require static rebalancing of the control surface. Generally, flight control balance condition may be determined by

A— checking for equal distribution of weight throughout the control surface.
B— the behavior of the trailing edge when the surface is suspended from its hinge points.
C— suspending the control surface from its leading edge in the streamline position and checking weight distribution.

8271. Excessive wear on both of the sides of a control cable pulley groove is evidence of

A— pulley misalignment.
B— cable misalignment.
C— excessive cable tension.

8272. Fairleads should never deflect the alignment of a cable more than

A— 12°.
B— 8°.
C— 3°.

8273. When installing a control cable, it is found that one end of a splice comes within 1/2 inch of entering a pulley groove with the loaded cable at the end of its travel. This condition

A— can be corrected by installing a smaller diameter pulley.
B— must be corrected by adjustment or replacement.
C— does not require further adjustment or replacement.

8274. With which system is differential control associated?

A— Trim.
B— Aileron.
C— Elevator.

8275. Which statement concerning the 100-hour inspection of an airplane equipped with a push-pull tube-type control system is true?

A— The threaded rod ends should not be adjusted in length for rigging purposes because the rod ends have been properly positioned and staked during manufacture.
B— The terminal end threads of the turnbuckles should be visible through the safety hole in the barrel.
C— The threaded rod ends should be checked for the amount of thread engagement by means of the inspection hole provided.

8276. If control cables are adjusted properly and the control surfaces tend to vibrate, the probable cause is

A— worn attachment fittings.
B— oil can effects on the control surfaces.
C— excessive cable tension.

8277. Aircraft flight control trim systems must be designed and installed so that the

A— pilot can determine the relative position of the trim tab from the cockpit.
B— operating control and the trim tab will always move in the same direction.
C— trim system will disengage or become inoperative if the primary flight control system fails.

8278. Stability about the axis which runs parallel to the line of flight is referred to as

A— longitudinal stability.
B— lateral stability.
C— directional stability.

8279. The purpose of spring tabs or servo tabs is to

A— assist the pilot in moving the control surfaces.
B— contribute to the static balance of the control surface.
C— make in-flight trim adjustments possible.

8280. If the control stick of an aircraft with properly rigged flight controls is moved rearward and to the left, the right aileron will move

A— down and the elevator will move down.
B— up and the elevator will move down.
C— down and the elevator will move up.

8281. Movement of the cockpit control toward the nosedown position during a ground operational check of the elevator trim tab system will cause the trailing edge of the trim tab to move in which direction?

A— Downward regardless of elevator position.
B— Upward regardless of elevator position.
C— Downward if the elevator is in the UP position and upward if the elevator is in the DOWN position.

8282. If the control stick of an aircraft with properly rigged flight controls is moved forward and to the right, the left aileron will move

A— up and the elevator will move down.
B— down and the elevator will move up.
C— down and the elevator will move down.

8283. If the travel of an airplane's controls is correct but the cables are rigged exceptionally tight, what probable effect will this have when flying the airplane?

A— The airplane will tend to fall off on one wing.
B— The airplane will be heavy on the controls.
C— The pilot will be unable to fly the airplane hands-off.

8284. During inspection of the flight control system of an airplane equipped with differential-type aileron control, side-to-side movement of the control stick will cause

A— each aileron to have a greater up travel (from the streamlined position) than down travel.
B— each aileron to have greater down travel (from the streamlined position) than up travel.
C— the left aileron to move through a greater number of degrees (from full up to full down) than the right aileron.

8285. A universal propeller protractor used to measure the degrees of aileron travel should be zeroed

A— with the aileron in the NEUTRAL position.
B— with the aileron in the DOWN position.
C— when the aircraft is in a level flight attitude.

8286. The universal propeller protractor can be used to measure

A— propeller track.
B— aspect ratio of a wing.
C— degrees of flap travel.

8287. (Refer to figure 8.) Identify the cable that is used in primary control systems and in other places where operation over pulleys is frequent.

A— 2.
B— 1.
C— 3.

8288. A tension regulator in the flight control cable system of a large all-metal aircraft is used primarily to

A— increase the cable tension in cold weather.
B— provide a means of changing cable tension in flight.
C— retain a set tension.

8289. (Refer to figure 9.) When the outside air temperature is 80 °F, select the acceptable 3/16 cable tension range.

A— 130 pounds minimum, 140 pounds maximum.
B— 117 pounds minimum, 143 pounds maximum.
C— 120 pounds minimum, 140 pounds maximum.

8290. Differential control on an aileron system means that

A— the down travel is more than the up travel.
B— the up travel is more than the down travel.
C— one aileron on one wing travels further up than the aileron on the opposite wing to adjust for wash-in and wash-out.

8291. Why is it generally necessary to jack an aircraft indoors for weighing?

A— So aircraft may be placed in a level position.
B— To stabilize the weighing scales.
C— So weighing scales may be calibrated to 0 pounds.

8292. Which should be accomplished before jacking an aircraft?

A— Install critical stress panels or plates.
B— Determine that the fuel tanks are empty.
C— Make sure the aircraft is leveled laterally.

8293. Which statement about Airworthiness Directives (AD's) is true?

A— AD's are information alert bulletins issued by the airframe, powerplant, or component manufacturer.
B— Compliance with an AD is not mandatory unless the aircraft affected is for hire.
C— Compliance with an applicable AD is mandatory and must be recorded in the permanent maintenance records.

8294. When overhauling electrical equipment, all necessary information should be obtained from

A— the applicable sections of AC 43.13-1A.
B— maintenance instructions published by the aircraft and/or equipment manufacturer.
C— Aircraft Specifications and/or Type Certificate Data Sheets.

8295. Which statement is correct when an aircraft is not approved for return to service after an annual inspection because of an item requiring major repair?

A— An appropriately rated mechanic may repair the defect, and an IA may approve the aircraft for return to service.
B— An appropriately rated mechanic may repair the defect and approve the aircraft for return to service.
C— Only the person who performed the annual inspection may approve the aircraft for return to service.

8296. Radio equipment installations made in accordance with Supplemental Type Certificate data require approval

A— using minor alteration procedures.
B— by an airframe and powerplant mechanic.
C— using major alteration procedures.

8297. An aircraft that is due a 100-hour inspection may be flown

A— if necessary to reach a place at which the inspection can be done.
B— if necessary to reach a place at which the inspection can be done, and a special flight permit is issued.
C— when it is operated under a progressive inspection program, and a special flight permit is issued.

8298. Where would you find the recommended statement for recording the approval or disapproval for return to service of an aircraft after a 100-hour or annual inspection?

A— FAR Part 65.
B— FAR Part 43.
C— FAR Part 91.

8299. The maximum time a 100-hour inspection may be extended is

A— 10 hours.
B— 10 hours with a special flight permit.
C— 12 hours with a special flight permit.

8300. Which statement is correct when an aircraft has not been approved for return to service after an annual inspection because of several items requiring minor repair?

A— Only the person who performed the annual inspection may approve the aircraft for return to service.
B— An appropriately rated mechanic may repair the defects and approve the aircraft for return to service.
C— An appropriately rated mechanic may repair the defects, but an IA must approve the aircraft for return to service.

8301. An aircraft that is due an annual inspection may be flown

A— if a special permit has been issued for the aircraft.
B— for the purpose of performing maintenance.
C— for a period of time not to exceed 10 hours.

8302. For an individual (not a repair station) to conduct a complete 100-hour inspection on an aircraft and approve it for return to service requires that he or she hold a mechanic certificate with

A— an airframe or powerplant rating.
B— airframe and powerplant ratings.
C— airframe and powerplant ratings and an inspection authorization (IA).

8303. Where would you find the operating conditions that make a 100-hour inspection mandatory?

A— FAR Part 91.
B— FAR Part 43.
C— AC 43.13-2A.

8304. Large airplanes and turbine-powered multiengine airplanes operated under FAR Part 91, General Operating and Flight Rules, must be inspected

A— in accordance with an inspection program authorized under FAR Part 91, Subpart E.
B— annually in accordance with FAR Part 43.
C— in accordance with progressive inspection requirements of FAR Section 91.411 and FAR Part 43.

8305. Exposure to and/or storage near which of the following is considered harmful to aircraft tires?

1. Low humidity.
2. Fuel.
3. Oil.
4. Ozone.
5. Helium.
6. Electrical equipment.
7. Hydraulic fluid.
8. Solvents.

A— 2, 3, 4, 5, 6, 7, 8.
B— 1, 2, 3, 5, 7, 8.
C— 2, 3, 4, 6, 7, 8.

8306. What would be the effect if the piston return spring broke in a brake master cylinder?

A— The brakes would become spongy.
B— The brake travel would become excessive.
C— The brakes would drag.

8307. In brake service work, the term "bleeding brakes" is the process of

A— withdrawing air only from the system.
B— withdrawing fluid from the system for the purpose of removing air that has entered the system.
C— replacing small amounts of fluid in reservoir.

8308. To prevent a very rapid extension of an oleo shock strut after initial compression resulting from landing impact,

A— various types of valves or orifices are used which restrict the reverse fluid flow.
B— the metering pin gradually reduces the size of the orifice as the shock strut extends.
C— the air is forced through a restricted orifice in the reverse direction.

8309. A pilot reports the right brake on an aircraft grabs when the brake pedal is depressed in a normal manner. The probable cause is that the right

A— master cylinder piston rod is bent.
B— wheel actuating cylinder is leaking fluid on the brake lining.
C— master cylinder piston return spring is weak.

8310. Aside from an external leak in the line, what will cause parking brakes to creep continually to the OFF position?

A— An internal leak in the master cylinder.
B— Insufficient hydraulic fluid in the reservoir.
C— Glazed brake linings.

8311. Why do most aircraft tire manufacturers recommend that the tubes in newly installed tires be first inflated, fully deflated, and then reinflated to the correct pressure?

A— To allow the tube to position itself correctly inside the tire.
B— To eliminate all the air between the tube and the inside of the tire.
C— To test the entire assembly for leaks.

8312. The metering pins in oleo shock struts serve to

A— lock the struts in the DOWN position.
B— retard the flow of oil as the struts are compressed.
C— meter the proper amount of air in the struts.

8313. After performing maintenance on an aircraft's landing gear system which may have affected the system's operation, it is usually necessary to

A— conduct a flight test.
B— re-inspect the area after the first flight.
C— make an operational check with the aircraft on jacks.

8314. Why do tire and wheel manufacturers often recommend that the tires on split rim wheels be deflated before removing the wheel from the axle?

A— To relieve the strain on the wheel retaining nut and axle threads.
B— As a safety precaution in case the bolts that hold the wheel halves together have been damaged or weakened.
C— To remove the static load imposed upon the wheel bearings by the inflated tire.

8315. The braking action of single-disk brakes is accomplished by compressing a rotating brake disk between two opposite brake linings. How is equal pressure on both sides of the rotating disk assured?

A— By hydraulically interconnecting the brake linings located on opposite sides of the rotating disk.
B— By keying the disk loosely in the wheel so that it can move from side to side.
C— By letting the brake linings automatically equalize because of the greater rate of wear on the lining with the most pressure.

8316. If it is determined that spongy brake action is not caused by air in the brake system, what is the next most likely cause?

A— Worn brake lining.
B— Internal leakage in the master cylinder.
C— Deteriorated flexible hoses.

8317. Many brake types can be adapted to operate mechanically or hydraulically. Which type is not adaptable to mechanical operation?

A— Single-disk spot type.
B— Single-servo type.
C— Expander-tube type.

8318. After replacing the blocks in an expander-tube type brake assembly, it will be necessary to

A— ensure that the radial clearance is not less than the allowable minimum.
B— make sure the movement of the actuating piston presses the brake blocks against the disk.
C— replace the tube as well as the blocks.

8319. A stripe or mark applied to a wheel rim and extending onto the sidewall of a tire is a

A— slippage mark.
B— wheel-to-tire balance mark.
C— wheel alignment reference mark.

8320. When bleeding aircraft brakes, one of the indications that the air has been purged from the system is

A— partial brake pedal travel.
B— full brake pedal travel.
C— firm brake pedals.

8321. Overinflated aircraft tires may cause damage to the

A— brake linings.
B— wheel hub.
C— wheel flange.

8322. Debooster valves are used in brake systems primarily to

A— ensure rapid application and release of the brakes.
B— reduce brake pressure and maintain static pressure.
C— reduce the pressure and release the brakes rapidly.

8323. The repair for an out-of-tolerance toe-in condition of main landing gear wheels determined not to be the result of bent or twisted components consists of

A— shimming the axle in the oleo trunnion.
B— inserting, removing, or changing the location of washers or spacers at the center pivotal point of the scissor torque links.
C— placing shims or spacers behind the bearing of the out-of-tolerance wheel or wheels.

8324. An embossed letter ''H'' on an air valve core stem

A— is the manufacturer's trademark.
B— indicates hydraulic type.
C— indicates high-pressure type.

8325. The primary purpose for balancing aircraft wheel assemblies is to

A— prevent heavy spots and reduce vibration.
B— distribute the aircraft weight properly.
C— reduce excessive wear and turbulence.

8326. Pressure should not be applied to an expander tube brake without the wheel being installed to prevent

A— damage to the brake assembly.
B— excessive loss of hydraulic fluid.
C— damage to the brake relief valve.

8327. On all aircraft equipped with retractable landing gear, some means must be provided to

A— retract and extend the landing gear if the normal operating mechanism fails.
B— extend the landing gear if the normal operating mechanism fails.
C— prevent the throttle from being reduced below a safe power setting while the landing gear is retracted.

8328. An automatic damping action occurs at the steer damper if for any reason the flow of high-pressure fluid is removed from the

A— outlet of the steer damper.
B— inlet of the steer damper.
C— replenishing check valve.

8329. What is the purpose of the torque links attached to the cylinder and piston of a landing gear oleo strut?

A— Limit compression stroke.
B— Hold the strut in place.
C— Maintain correct wheel alignment.

8330. The removal, installation, and repair of landing gear tires by the holder of a private pilot certificate only, on an aircraft owned and operated by him/her is

A— a violation of the Federal Aviation Regulations.
B— a minor repair.
C— preventive maintenance.

8331. Aircraft brakes requiring a large volume of fluid to operate the brakes generally

A— use independent master cylinder systems.
B— do not use brake system accumulators.
C— use power brake control valves.

8332. What is one effect a restricted compensator port of a master cylinder will have on a brake system?

A— The brakes will operate normally.
B— The reservoir will be filled by reverse flow.
C— The restriction will cause slow release of the brakes.

8333. When an air/oil type of landing gear shock strut is used, the initial shock of landing is cushioned by

A— compression of the air charge.
B— the fluid being forced through a metered opening.
C— compression of the fluid.

8334. Internal leakage in a brake master cylinder unit can cause

A— slow release of brakes.
B— fading brakes.
C— low fluid in the master cylinder reservoir.

8335. A sleeve, spacer, or bumper ring is incorporated in a landing gear oleo shock strut to

A— limit the extension of the torque arm.
B— limit the extension stroke.
C— reduce the rebound effect.

8336. The purpose of a sequence valve in a hydraulic retractable landing gear system is to

A— prevent heavy landing gear from falling too rapidly upon extension.
B— provide a means of disconnecting the normal source of hydraulic power and connecting the emergency source of power.
C— ensure operation of the landing gear and gear doors in the proper order.

8337. The pressure source for power brakes is

A— the main hydraulic system.
B— the power brake reservoir.
C— a master cylinder.

8338. Which statement is true with respect to an aircraft equipped with hydraulically operated multiple-disk type brake assemblies?

A— There are no minimum or maximum disk clearance checks required due to the use of self-compensating cylinder assemblies.
B— Do not set parking brake when brakes are hot.
C— No parking brake provisions are possible for this type of brake assembly.

8339. What type of valve is used in the brake actuating line to isolate the emergency brake system from the normal power brake control valve system?

A— A bypass valve.
B— An orifice check valve.
C— A shuttle valve.

8340. When tightening the packing nut on an air/oil shock strut installed on an aircraft, the

A— aircraft should be jacked.
B— fluid should be removed from the strut.
C— aircraft should be jacked and leveled.

8341. Instructions concerning the type of fluid and amount of air pressure to be put in a shock strut are found

A— on the airplane data plate.
B— in the aircraft operations limitations.
C— in the aircraft manufacturer's service manual.

8342. The purpose of a relief valve in a brake system is to

A— reduce pressure for brake application.
B— prevent the tire from skidding.
C— compensate for thermal expansion.

8343. Aircraft tire pressure should be checked

A— using only a push on stick-type gauge having 1-pound increments.
B— at least once a week or more often.
C— as soon as possible after each flight.

8344. If the extended longitudinal axis of the main landing gear wheel assemblies intersects aft of the aircraft, the wheels can be termed as having

A— toe-out.
B— toe-in.
C— negative camber.

8345. What is the purpose of a compensating port or valve in a brake master cylinder of an independent brake system?

A— Assists in the master cylinder piston return.
B— Prevents fluid from flowing back to the reservoir.
C— Permits reservoir fluid to fill the master cylinder.

8346. If an aircraft shock strut (air/oil type) bottoms upon initial landing contact, but functions correctly during taxi, the most probable cause is

A— low fluid.
B— low air charge.
C— a restricted metering pin orifice.

8347. What is the function of a cam incorporated in a nose gear shock strut?

A— Provides an internal shimmy damper.
B— Straightens the nosewheel.
C— Provides steering of aircraft during ground operation.

8348. Extension of an oleo shock strut is measured to determine the

A— amount of oil in the strut.
B— physical condition of the strut itself.
C— proper operating position of the strut.

8349. Debooster cylinders are used in brake systems primarily to

A— reduce brake pressure and maintain static pressure.
B— relieve excessive fluid and ensure a positive release.
C— reduce the pressure to the brake and increase the volume of fluid flow.

8350. If a shock strut bottoms after it has been properly serviced, the

A— strut should be disassembled and the metering pin orifice plate replaced.
B— air pressure should be increased.
C— strut should be removed, disassembled, and inspected.

8351. A high-speed aircraft tire with a sound cord body and bead may be recapped

A— a maximum of three times.
B— only by the tire manufacturer.
C— an indefinite number of times.

8352. If an airplane equipped with master cylinders and single-disk brakes has excessive brake pedal travel, but the brakes are hard and effective, the probable cause is

A— the master cylinder one-way cup is leaking.
B— worn brake linings.
C— worn brake disk causing excessive clearance between the notches on the perimeter of the disk and the splines or keys on the wheel.

8353. The correct inflation pressure for an aircraft tire can be obtained from

A— tire manufacturer's specifications.
B— the operator's manual.
C— the information stamped on the wheel.

8354. What should be checked when a shock strut bottoms during a landing?

A— Air pressure.
B— Packing seals for correct installation.
C— Fluid level.

8355. How can it be determined that all air has been purged from a master cylinder brake system?

A— By operating a hydraulic unit and watching the system pressure gauge for smooth, full-scale deflection.
B— By noting whether the brake is firm or spongy.
C— By noting the amount of fluid return to the master cylinder upon brake release.

8356. The left brake is dragging excessively on an airplane on which no recent brake service work has been performed. The most probable cause is

A— foreign particles stuck in the master cylinder compensating port.
B— excessively worn brake linings.
C— low fluid supply in the brake system reservoir.

8357. If a brake debooster is used in a hydraulic brake system, its position in the system will be

A— between the pressure manifold of the main hydraulic system and the power brake control valve.
B— between the brake control valve and the brake actuating cylinder.
C— in the brake pressure line between the brake pedal and the brake accumulator.

8358. The rubber seals used in a landing gear shock strut

A— are generally designed to be compatible with more than one type of fluid.
B— are kept from direct contact with fluid by teflon or nylon backup rings.
C— may be used only with a specific type of fluid.

8359. Lockout deboosters generally

A— allow full debooster piston travel without fluid from the high pressure side entering the low pressure chamber.
B— cannot allow full debooster piston travel without fluid from the high pressure side entering the low pressure chamber.
C— must be bled separately after brake bleeding has been completed.

8360. When a properly operating fusible plug has allowed a tire to deflate, the tire should be

A— replaced.
B— externally inspected for damage.
C— removed from the wheel and inspected for carcass and tread damage.

8361. To avoid nylon tire spotting on an aircraft that is to remain idle for a period longer than 3 days,

A— the aircraft should be moved every 36 hours.
B— no action is necessary.
C— the aircraft should be blocked up so that no weight is on the tires or moved every 48 hours.

8362. The best safeguards against heat buildup in aircraft tires are

A— proper tire inflation, minimum braking, and ground rolls into the wind.
B— short ground rolls, slow taxi speeds, minimum braking, and proper tire inflation.
C— minimum braking, proper tire inflation, and long ground rolls.

8363. The fusible plugs installed in some aircraft wheels will

A— indicate tire tread separation.
B— prevent overinflation.
C— melt at a specified elevated temperature.

8364. What action, if any, should be taken when there is a difference of more than 5 pounds of air pressure in tires mounted as duals?

A— Replace both tires.
B— Correct the discrepancy and enter in logbook.
C— Replace the tire with the lowest pressure.

8365. How long should you wait after a flight before checking tire pressure?

A— At least 2 hours (3 hours in hot weather).
B— At least 3 hours (4 hours in hot weather).
C— At least 4 hours (5 hours in hot weather).

8366. Excessive wear in the shoulder area of an aircraft tire is an indication of

A— overinflation.
B— excessive toe-in.
C— underinflation.

8367. Excessive wear in the center of the tread of an aircraft tire is an indication of

A— incorrect camber.
B— excessive toe-out.
C— overinflation.

8368. When an empty shock strut is filled with fluid, care should be taken to extend and compress the strut completely at least two times to

A— thoroughly lubricate the piston rod.
B— force out any excess fluid.
C— ensure proper packing ring seating and removal of air bubbles.

8369. In shock struts, chevron seals are used to

A— absorb bottoming effect.
B— prevent oil from escaping.
C— serve as a bearing surface.

8370. On most aircraft, the oil level of an air and oil shock strut is checked by

A— removing the oil filler plug and inserting a gauge.
B— measuring the length of the strut extension with a certain air pressure in the strut.
C— releasing the air and seeing that the oil is to the level of the filler plug.

8371. A pilot reports that the brakes are dragging on an aircraft equipped with expander-tube type brakes. On investigation, it is found that the aircraft has been in storage for the past 6 months and the

A— brake linings have swelled due to age.
B— expander tubes have grown during the inoperative period.
C— brake lining return springs have weakened with age.

8372. A landing gear position and warning system will provide a warning in the cockpit when the throttle is

A— retarded and gear is not down and locked.
B— advanced and gear is down and locked.
C— retarded and gear is down and locked.

8373. An electric motor used to raise and lower a landing gear would most likely be a

A— shunt field series-wound motor.
B— split field shunt-wound motor.
C— split field series-wound motor.

8374. When installing a chevron-type seal in an aircraft hydraulic cylinder, the open side of the seal should face

A— opposite the direction of fluid pressure.
B— up or forward when the unit is installed in a horizontal position.
C— the direction of fluid pressure.

8375. Nose gear centering cams are used in many retractable landing gear systems. The primary purpose of the centering device is to

A— align the nosewheel prior to touchdown.
B— engage the nosewheel steering.
C— center the nosewheel before it enters the wheel well.

8376. What device in a hydraulic system with a constant-delivery pump allows circulation of the fluid when no demands are on the system?

A— Pressure relief valve.
B— Shuttle valve.
C— Pressure regulator.

8377. A fully-charged hydraulic accumulator provides

A— air pressure to the various hydraulic components.
B— a source for additional hydraulic power when heavy demands are placed on the system.
C— positive fluid flow to the pump inlet.

8378. A hydraulic system referred to as a "power pack" system will

A— have an engine-driven pump for greater pressure.
B— have all hydraulic power components located in one unit.
C— have a pressurized reservoir.

8379. A hydraulic hose identified as MIL-H-8794 will have a yellow stripe running the length of the hose. This stripe

A— is used to ensure that the hose is installed without excessive twisting.
B— identifies that the hose is for hydraulic fluid only.
C— identifies that the hose is constructed of synthetic rubber and may be suitable for a wide range of applications.

8380. An O-ring intended for use in a hydraulic system using MIL-H-5606 (mineral base) fluid will be marked with

A— a blue stripe or dot.
B— one or more white dots.
C— a white and yellow stripe.

8381. What condition would most likely cause excessive fluctuation of the pressure gauge when the hydraulic pump is operating?

A— Accumulator air pressure low.
B— Inadequate supply of fluid.
C— System relief valve sticking closed.

8382. A filter incorporating specially treated cellulose paper is identified as a

A— sediment trap.
B— cuno filter.
C— micronic filter.

8383. The purpose of an orifice check valve is to

A— relieve pressure to a sensitive component.
B— restrict flow in one direction and allow free flow in the other.
C— relieve pressure in one direction and prevent flow in the other direction.

8384. (Refer to figure 10.) The trunnion nut on an aircraft landing gear requires a torque of 320 inch-pounds. To reach the nut, a 2-inch straight adapter must be used on an 18-inch torque wrench. How many foot-pounds will be indicated on the torque wrench when the required torque of the nut is reached?

A— 24.
B— 28.8.
C— 22.

8385. A special bolt in a landing gear attachment requires a torque value of 440 inch-pounds. How many foot-pounds are required?

A— 36.8.
B— 38.
C— 36.6.

8386. To protect packing rings or seals from damage when it is necessary to install them over or inside threaded sections, the

A— threaded section should be coated with a heavy grease.
B— packings should be stretched during installation to avoid contact with the threads.
C— threaded section should be covered with a stiff paper sleeve.

8387. To prevent external and internal leakage in aircraft hydraulic units, the most commonly used type of seal is the

A— O-ring seal.
B— gasket seal.
C— chevron seal.

8388. Which allows free fluid flow in one direction and no fluid flow in the other direction?

A— Check valve.
B— Metering piston.
C— Shutoff valve.

8389. Select the valve used in a hydraulic system that directs pressurized fluid to one end of an actuating cylinder and simultaneously directs return fluid to the reservoir from the other end.

A— Sequence.
B— Shuttle.
C— Selector.

8390. What function does the absolute pressure regulator perform in the pneumatic power system?

A— Regulates the compressor outlet air pressure to stabilize the system pressure.
B— Regulates the pneumatic system pressure to protect the moisture separator from internal explosion.
C— Regulates the compressor inlet air to provide a stabilized source of air for the compressor.

8391. (1) Relief valves are used in pneumatic systems as damage-preventing units.

(2) Check valves are used in both hydraulic and pneumatic systems.

Regarding the above statements,

A— both No. 1 and No. 2 are true.
B— neither No. 1 nor No. 2 is true.
C— only No. 1 is true.

8392. One of the distinguishing characteristics of an open-center selector valve used in a hydraulic system is that

A— fluid flows through the valve in the OFF position.
B— fluid flows in three directions in the ON position.
C— a limited amount of fluid flows in one direction and no fluid flows in the opposite direction.

8393. What type of packings should be used in hydraulic components to be installed in a system containing Skydrol?

A— AN packings made of natural rubber.
B— Packing materials made for ester base fluids.
C— AN packings made of neoprene.

8394. Relief valves are used in pneumatic systems

A— for one direction flow control.
B— to reduce the rate of airflow.
C— as damage-preventing units.

8395. An aircraft pneumatic system, which incorporates an engine-driven multistage reciprocating compressor, also requires

A— an oil separator.
B— a surge chamber.
C— a moisture separator.

8396. The removal of air from an aircraft hydraulic system is generally accomplished

A— through automatic bleed valves on individual components during system operation.
B— by operating the various hydraulic components through several cycles.
C— by allowing the system to remain inoperative for several hours.

8397. Pneumatic systems utilize

A— return lines.
B— relief valves.
C— diluter valves.

8398. The component in the hydraulic system that is used to direct the flow of fluid is the

A— check valve.
B— orifice check valve.
C— selector valve.

8399. What type of selector valve is one of the most commonly used in hydraulic systems to provide for simultaneous flow of fluid into and out of a connected actuating unit?

A— Four-port, closed-center valve.
B— Three-port, four-way valve.
C— Two-port, open-center valve.

8400. What is the purpose of using backup rings with O-rings in hydraulic systems above 1,500 PSI?

A— Prevent internal and external leakage of all moving parts within a hydraulic system.
B— Provide a seal between two parts of a unit which move in relation to each other.
C— Prevent high pressure from extruding the seal between the moving and stationary part.

8401. The purpose of the pressure regulator in a hydraulic system is to

A— maintain system operating pressure within a predetermined range and to unload the pump.
B— regulate the amount of fluid flow to the actuating cylinders within the system.
C— prevent failure of components or rupture of hydraulic lines under excessive pressure.

8402. A flexible sealing element subject to motion is a

A— compound.
B— packing.
C— gasket.

8403. Which characteristics apply to aircraft hydraulic systems?

1. Minimum maintenance requirements.
2. Lightweight.
3. About 80 percent operating efficiency (20 percent loss due to fluid friction).
4. Simple to inspect.

A— 1, 2, 3, 4.
B— 1, 3, 4.
C— 1, 2, 4.

8404. If a rigid tubc is too short for the flare to reach its seat before tightening, pulling it into place by tightening

A— is acceptable.
B— may distort the flare.
C— may distort the cone.

8405. The installation of a new metal hydraulic line should be made with

A— a straight tube to withstand the shocks and vibration to which it will be subjected.
B— a straight tube to permit proper alignment of the fitting and thereby reduce fluid loss through leakage.
C— enough bends to allow the tube to expand and contract with temperature changes and to absorb vibration.

8406. Extrusion of an O-ring seal is prevented in a high-pressure system by the use of a

A— backup ring on the side of the O-ring next to the pressure.
B— U-ring on the side of the O-ring away from the pressure.
C— backup ring on the side of the O-ring away from the pressure.

8407. What is one advantage of piston-type hydraulic motors over electric motors?

A— They are considerably quieter in operation.
B— There is no fire hazard if the motor is stalled.
C— They work satisfactorily over a wider temperature range.

8408. Generally, the first step in removing an accumulator from an aircraft is to

A— relieve system pressure.
B— discharge the preload.
C— drain the reservoir.

8409. (Refer to figure 11.) Which fitting is an AN flared-tube fitting?

A— 1.
B— 2.
C— 3.

8410. (Refer to figure 12.) Which illustration(s) show(s) the correct spiral for teflon backup rings?

A— 1 and 2.
B— 3.
C— 1 and 3.

8411. If a hydraulic brake system uses neoprene rubber packing materials, the correct hydraulic fluid to service the system is

A— mineral base oil.
B— vegetable base oil.
C— phosphate ester base oil.

8412. The internal resistance of a fluid which tends to prevent it from flowing is called

A— volatility.
B— viscosity.
C— acidity.

8413. What is the viscosity of hydraulic fluid?

A— The increase in volume of a fluid due to temperature change.
B— The fluid's ability to resist oxidation and deterioration for long periods.
C— The internal resistance of a fluid which tends to prevent it from flowing.

8414. Which is a characteristic of petroleum base hydraulic fluid?

A— Flammable under normal conditions.
B— Compatible to natural rubber seals and packings.
C— Nonflammable under all conditions.

8415. (1) When servicing aircraft hydraulic systems, use the type fluid specified in the aircraft manufacturer's maintenance manual or on the instruction plate affixed to the reservoir or unit.

(2) Hydraulic fluids for aircraft are dyed a specific color for each type of fluid.

Regarding the above statements,

A— only No. 1 is true.
B— only No. 2 is true.
C— both No. 1 and No. 2 are true.

8416. Petroleum base hydraulic fluid is which color?

A— Purple.
B— Blue.
C— Red.

8417. Which of the following is adversely affected by atmospheric humidity if left unprotected?

1. MIL-H-5606 hydraulic fluid.
2. Skydrol hydraulic fluid.
3. None of the above.

A— 1 and 2.
B— 3.
C— 2.

8418. Which is a characteristic of synthetic base hydraulic fluid?

A— Low moisture retention.
B— High flash point.
C— Low flash point.

8419. Which statement about fluids is correct?

A— Any fluid will completely fill its container.
B— All fluids are considered to be highly compressible.
C— All fluids readily transmit pressure.

8420. Three types of hydraulic fluids currently being used in civil aircraft are

A— mineral base, vegetable base, and phosphate ester base.
B— mineral base, phosphate ester base, and mixed mineral and phosphate ester base.
C— mineral base, phosphate ester base, and mixed vegetable and alcohol base.

8421. Which of the following lists only desirable properties of a good hydraulic fluid?

A— High viscosity, low flash point, chemical stability, high fire point.
B— High flash point, low viscosity, chemical stability, low fire point.
C— Low viscosity, chemical stability, high flash point, high fire point.

8422. Characteristics of MIL-H-8446 (Skydrol 500 A & B) hydraulic fluid are

A— blue color, phosphate ester base, fire resistant, butyl rubber seals.
B— light purple color, phosphate ester base, fire resistant, butyl rubber seals.
C— light green color, phosphate ester base, fire resistant, butyl rubber seals.

8423. Where can information be obtained about the compatibility of fire-resistant hydraulic fluid with aircraft materials?

A— Manufacturer's technical bulletins.
B— Aircraft manufacturer's specifications.
C— AC 43.13-1A.

8424. Characteristics of MIL-H-5606 hydraulic fluid are

A— light purple color, phosphate ester base, fire resistant, butyl rubber seals.
B— blue color, vegetable base, will burn, natural rubber seals.
C— red color, petroleum base, will burn, synthetic rubber seals.

8425. Characteristics of MIL-H-7644 hydraulic fluid are

A— red color, petroleum base, will burn, synthetic rubber seals.
B— light purple color, phosphate ester base, fire resistant, butyl rubber seals.
C— blue color, vegetable base, will burn, natural rubber seals.

8426. If an aircraft hydraulic system requires mineral base hydraulic fluid, but phosphate ester base hydraulic fluid is used, what will be the effect on the system?

A— No effect.
B— System will be contaminated, fluids will not blend, and the seals will fail.
C— System will be contaminated, fluids will not blend, but there will be no seal problem.

8427. What is used to flush a system normally serviced with MIL-H-5606 hydraulic fluid?

A— Methyl ethyl ketone or kerosene.
B— Naphtha or varsol.
C— Lacquer thinner or trichlorethylene.

8428. What is used to flush a system normally serviced with Skydrol hydraulic fluid?

A— Trichlorethylene.
B— Naphtha.
C— Stoddard solvent.

8429. How can the proper hydraulic fluid to be used in an airplane be determined?

A— Refer to the aircraft parts manual.
B— Consult the aircraft Type Certificate Data Sheet.
C— Consult the aircraft manufacturer's service manual.

8430. Phosphate ester base hydraulic fluid is very susceptible to contamination from

A— teflon seal material.
B— water in the atmosphere.
C— ethylene-propylene elastomers.

8431. (1) Materials which are Skydrol compatible or resistant include most common aircraft metals and polyurethane and epoxy paints.

(2) Skydrol hydraulic fluid is compatible with natural fibers and also with nylon and polyester.

Regarding the above statements,

A— neither No. 1 nor No. 2 is true.
B— both No. 1 and No. 2 are true.
C— only No. 1 is true.

8432. The hydraulic component that automatically directs fluid from either the normal source or an emergency source to an actuating cylinder is called a

A— bypass valve.
B— shuttle valve.
C— crossflow valve.

8433. The primary purpose of a hydraulic actuating unit is to transform

A— fluid motion into mechanical pressure and back again.
B— fluid pressure into useful work.
C— energy from one form to another.

8434. The primary function of the flap overload valve is to

A— prevent the flaps from being lowered at airspeeds which would impose excessive structural loads.
B— cause the flap segments located on opposite sides of the aircraft centerline to extend and retract together so that the aircraft will not become aerodynamically unbalanced to the extent that it becomes uncontrollable.
C— boost normal system pressure to the flaps in order to overcome the air loads acting on the relatively large flap area.

8435. A unit which transforms hydraulic pressure into linear motion is called

A— an actuating cylinder.
B— an accumulator.
C— a hydraulic pump.

8436. If it is necessary to adjust several pressure regulating valves in a hydraulic system, what particular sequence, if any, should be followed?

A— Units most distant from the hydraulic pump should be adjusted first.
B— Units with the highest pressure settings are adjusted first.
C— Units are independent of each other, and therefore, no particular sequence is necessary.

8437. If an aircraft's constant-pressure hydraulic system cycles more frequently than usual and no fluid leakage can be detected, the most probable cause is

A— a too high relief valve setting.
B— pump volume output too high.
C— low accumulator air preload.

8438. Unloading valves are used with many engine-driven hydraulic pumps to

A— dampen out pressure surges.
B— relieve the pump pressure.
C— relieve system pressure.

8439. What safety device is usually located between the driving unit and hydraulic pump drive shaft?

A— Thermal relief valve.
B— Pump motor safety switch.
C— Pump drive coupling shear section.

8440. Which valve installed in a hydraulic system will have the highest pressure setting?

A— Pressure regulator valve.
B— Main relief valve.
C— Thermal relief valve.

8441. Excluding lines, which components are required to make up a simple hydraulic system?

A— Actuator, pressure reservoir, accumulator, and selector valve.
B— Pump, reservoir, selector valve, and actuator.
C— Pump, reservoir, relief valve, and shuttle valve.

8442. Most variable displacement hydraulic pumps of current design

A— must be driven at a nearly constant speed in order to be practical for use.
B— are not practical for use with a closed-center hydraulic system.
C— contain a built-in means of system pressure regulation.

8443. In a gear-type hydraulic pump, a mechanical safety device incorporated to protect the pump from overload is the

A— bypass valve.
B— check valve.
C— shear pin.

8444. After installation of a rebuilt hydraulic hand pump, it is found that the handle cannot be moved in the pumping direction (pressure stroke). The most likely cause is an incorrectly installed

A— hand pump inport check valve.
B— inport/outport orifice check valve.
C— hand pump outport check valve.

8445. Pressure is a term used to indicate the force per unit area. Pressure is usually expressed in

A— pounds per square inch.
B— pounds per inch.
C— pounds per cubic inch.

8446. If two actuating cylinders which have the same cross-sectional area but different lengths of stroke are connected to the same source of hydraulic pressure, they will exert

A— different amounts of force but will move at the same rate of speed.
B— equal amounts of force but will move at different rates of speed.
C— equal amounts of force and will move at the same rate of speed.

8447. Using a hand pump, pressure of 100 PSI has been built up in a hydraulic system. The hand pump piston is 1 inch in diameter. A 1/2-inch line connects the hand pump to an actuating cylinder 2 inches in diameter. What is the pressure in the line between the hand pump and the actuator?

A— 100 PSI.
B— 150 PSI.
C— 200 PSI.

8448. Heat exchanger cooling units are required in some aircraft hydraulic systems because of

A— fluid flammability.
B— high pressures and high rates of fluid flow.
C— the high heat generated from braking.

8449. Which is true regarding the ground check of a flap operating mechanism which has just been installed?

A— If the time required to operate the mechanism increases with successive operations, it indicates the air is being worked out of the system.
B— If the time required to operate the mechanism decreases with successive operations, it indicates the air is being worked out of the system.
C— All hydraulic lines and components should be checked for leaks by applying soapy water to all connections.

8450. A hydraulic system operational check during ground runup of an aircraft indicates that the wing flaps cannot be lowered using the main hydraulic system, but can be lowered by using the emergency hand pump. Which is the most likely cause?

A— The flap selector valve has a severe internal leak.
B— The pressure accumulator is not supplying pressure to the system.
C— The fluid level in the reservoir is low.

8451. Many hydraulic reservoirs contain a small quantity of fluid which is not available to the main system pump. This fluid is retained to

A— prime the main system.
B— supply fluid to the auxiliary pump.
C— supply fluid to the pressure accumulator.

8452. The unit which causes one hydraulic operation to follow another in a definite order is called a

A— selector valve.
B— sequence valve.
C— shuttle valve.

8453. The purpose of a hydraulic pressure regulator is to

A— prevent the system pressure from rising above a predetermined amount due to thermal expansion.
B— boost the pressure in portions of the system.
C— relieve the pump of its load when no actuating units are being operated.

8454. Severe kickback of the emergency hydraulic hand pump handle during the normal intake stroke will indicate which of the following?

A— The hand pump inport check valve is sticking open.
B— The main system relief valve is set too high.
C— The hand pump outport check valve is sticking open.

8455. What type of valve in an aircraft hydraulic system permits fluid to flow freely in one direction, but restricts the rate at which fluid is allowed to flow in the other direction?

A— Check valve.
B— Orifice restrictor.
C— Orifice check valve.

8456. The main system pressure relief valve in a simple hydraulic system equipped with a power control valve should be adjusted

A— with the power control valve held in the CLOSED position.
B— while one or more actuating units are in operation.
C— with the power control valve in the OPEN position.

8457. A hydraulic accumulator is charged with an air preload of 1,000 PSI. When a hydraulic system pressure of 3,000 PSI is developed, the pressure on the air side of the accumulator will be

A— 1,000 PSI.
B— 3,000 PSI.
C— 4,000 PSI.

8458. How is the air in a hydraulic accumulator prevented from entering the fluid system?

A— By forcing the oil/air mixture through a centrifugal separating chamber that prevents the air from leaving the accumulator.
B— By physically separating the air chamber from the oil chamber with a flexible or movable separator.
C— By including a valve that automatically closes when the fluid level lowers to a preset amount.

8459. After a hydraulic accumulator has been installed and air chamber charged, the main system hydraulic pressure gauge will not show a hydraulic pressure reading until

A— at least one selector valve has been actuated to allow fluid to flow into the fluid side of the accumulator.
B— the air pressure has become equal to the fluid pressure.
C— the fluid side of the accumulator has been charged.

8460. Which must be done before adjusting the relief valve of a main hydraulic system incorporating a pressure regulator?

A— Eliminate the action of the unloading valve.
B— Adjust all other system relief valves which have a lower pressure setting.
C— Manually unseat all system check valves to allow unrestricted flow in both directions.

8461. Which seals are used with vegetable base hydraulic fluids?

A— Silicone rubber.
B— Butyl rubber.
C— Natural rubber.

8462. The air that is expended and no longer needed when an actuating unit is operated in a pneumatic system is

A— exhausted or dumped, usually overboard.
B— returned to the compressor.
C— charged or pressurized for use during the next operating cycle.

8463. Some hydraulic systems incorporate a device which is designed to remain open to allow a normal fluid flow in the line, but closed if the fluid flow increases above an established rate. This device is generally referred to as a

A— hydraulic fuse.
B— flow regulator.
C— metering check valve.

8464. When hydraulic system pressure control and relief units fail to function properly, how are most systems protected against overpressure?

A— A shear section on the main hydraulic pump drive shaft.
B— One or more hydraulic fuses installed in the pressure and return lines.
C— A shuttle valve interconnecting the main and emergency systems.

8465. A worn hydraulic pump shaft seal can normally be detected by

A— hydraulic fluid flowing from the pump drain line.
B— evidence of hydraulic fluid combined in the engine oil.
C— the presence of hydraulic fluid around the pump mounting pad.

8466. If an engine-driven hydraulic pump of the correct capacity fails to maintain normal system pressure during the operation of a cowl flap actuating unit, the probable cause is

A— mechanical interference to the movement of the cowl flap.
B— a partial restriction in the inport of the selector valve.
C— restriction in the pump outlet.

8467. Before removing the filler cap of a pressurized hydraulic reservoir,

A— relieve the hydraulic system pressure.
B— actuate several components in the system.
C— relieve the air pressure.

8468. What happens to the output of a constant-displacement hydraulic pump when the hydraulic system pressure regulator diverts the fluid from the system to the reservoir?

A— The output pressure remains the same, but the volume reduces.
B— The output pressure reduces, but the volume remains the same.
C— The output pressure and volume remain the same.

8469. Hydraulic system accumulators serve which of the following functions?

1. Dampen pressure surges.
2. Supplement the system pump when demand is beyond the pump's capacity.
3. Store power for limited operation of components if the pump is not operating.
4. Ensure a continuous supply of fluid to the pump.

A— 2, 3.
B— 1, 2, 3, 4.
C— 1, 2, 3.

8470. Chattering of the hydraulic pump during operation is an indication

A— of low accumulator preload.
B— that the main system relief valve is sticking open.
C— that air is entering the pump.

8471. Quick-disconnect couplings in hydraulic systems provide a means of

A— easily replacing hydraulic lines in areas where leaks are common.
B— quickly connecting and disconnecting hydraulic lines and eliminate the possibility of contaminates entering the system.
C— quickly connecting and disconnecting hydraulic lines without loss of fluid or entrance of air into the system.

8472. Which seal/material is used with phosphate ester base hydraulic fluids?

A— Silicone rubber.
B— Butyl rubber.
C— Neoprene rubber.

8473. A hydraulic pump is a constant-displacement type if it

A— produces an unregulated constant pressure.
B— produces a continuous positive pressure.
C— delivers a uniform rate of fluid flow.

8474. A hydraulic motor converts fluid pressure to

A— linear motion.
B— rotary motion.
C— angular motion.

8475. A crossflow valve which is designed to bypass fluid from one side of an actuating cylinder to the other side, under certain conditions, may be found in some aircraft installed in the

A— engine cowl flap system.
B— landing gear system.
C— flap overload system.

8476. Hydraulic fluid filtering elements constructed of porous paper are normally

A— cleaned and reused.
B— discarded at regular intervals and replaced with new filtering elements.
C— not approved for use in certificated aircraft.

8477. A pilot reports that when the hydraulic pump is running, the pressure is normal. However, when the pump is stopped, no hydraulic pressure is available. This is an indication of a

A— leaking selector valve.
B— low accumulator fluid preload.
C— leaking accumulator air valve.

8478. If fluid is added to a nonpressurized reservoir in a constant-pressure hydraulic system while the system is pressurized,

A— fluid will spray violently out of the reservoir when the filler neck cap is removed.
B— the fluid level will increase when system pressure is reduced.
C— air will be drawn into the system, when the filler neck cap is removed, resulting in pump chattering and possible damage.

8479. In a hydraulic system that has a reservoir pressurized with turbine-engine compressor bleed air, which unit reduces the air pressure between the engine and reservoir?

A— Relief valve.
B— Air bleed relief valve.
C— Air pressure regulator.

8480. What is the main purpose of a pressurized reservoir in a hydraulic system?

A— Prevent tank collapse at altitude.
B— Prevent hydraulic pump cavitation.
C— Prevent hydraulic fluid from foaming.

8481. Two hydraulic actuating cylinders which have the same length of stroke but different diameters are connected to the same source. The two cylinders will operate at

A— the same internal pressure but will exert different amounts of force.
B— different internal pressures and will exert different amounts of force.
C— the same internal pressure and will exert equal force.

8482. Hydraulic fluid reservoirs are sometimes designed with a standpipe in one of the outlet ports in order to assure emergency supply of fluid. The outlet port with the standpipe in it furnishes fluid to the

A— emergency pump when the fluid supply to the normal system has been depleted.
B— emergency pump at any time it is required.
C— normal system power pump.

8483. An emergency supply of fluid is often retained in the main hydraulic system reservoir by the use of a standpipe located in the

A— inlet from the main hydraulic system.
B— outlet to the emergency pump.
C— outlet to the main system pump.

8484. To check the air charge in a hydraulic accumulator,

A— reduce all hydraulic pressure, then observe the reading on the accumulator air gauge.
B— observe the first reading on the hydraulic system gauge while operating a component in the system.
C— read it directly from the auxiliary pressure gauge.

8485. How would the air pressure charge in the accumulator be determined if the engine is inoperative, but the system still has hydraulic pressure?

A— Read it directly from the main system pressure gauge with all actuators inoperative.
B— Build up system pressure with the emergency pump and then read the pressure on a gauge attached to the air side of the accumulator.
C— Operate a hydraulic unit slowly and note the pressure at which a rapid pressure drop begins as it goes toward zero.

8486. How many of these seals are used with petroleum base hydraulic fluids?

1. Synthetic rubber.
2. Natural rubber.
3. Neoprene rubber.

A— One.
B— Two.
C— Three.

8487. Hydraulic system thermal relief valves are set to open at a

A— lower pressure than the system relief valve.
B— higher pressure than the system relief valve.
C— lower pressure than the system pressure regulator.

8488. Chatter in a hydraulic system is caused by

A— excessive system pressure.
B— insufficient system pressure.
C— air in the system.

8489. If hydraulic fluid is released when the air valve core of the accumulator is depressed, it is evidence of

A— excessive accumulator air pressure.
B— a leaking check valve.
C— a ruptured diaphragm or leaking seals.

8490. Although dents in the heel of a bend are not permissible, they are acceptable in the remainder of the hydraulic tube providing they do not exceed what percent of the tube diameter?

A— 5.
B— 10.
C— 20.

8491. If the hydraulic system pressure is normal while the engine-driven pump is running, but there is no pressure after the engine has been shut off, it indicates

A— the system relief valve setting is too high.
B— no air pressure in the accumulator.
C— the pressure regulator is set too high.

8492. The purpose of restrictors in hydraulic systems is to

A— control the rate of movement of hydraulically operated mechanisms.
B— allow the flow of fluid in one direction only.
C— lower the operating pressure of selected components.

8493. A common cause of slow actuation of hydraulic components is

A— cold fluid.
B— restricted orifices.
C— internal leakage in the actuating unit.

8494. A loud hammering noise in a hydraulic system having an accumulator usually indicates

A— air in the fluid.
B— too much preload in the accumulator.
C— too low or no preload in the accumulator.

8495. Teflon hose that has developed a permanent set from being exposed to high pressure or temperature should

A— not be straightened or bent further.
B— not be reinstalled once removed.
C— be immediately replaced.

8496. In a typical high-pressure pneumatic system, if the moisture separator does not vent accumulated water when the compressor shuts down, a likely cause is a

A— saturated chemical dryer.
B— malfunctioning pressure transmitter.
C— malfunctioning solenoid dump valve.

8497. Which section of a turbojet engine provides air for the pressurization and air-conditioning systems of a jet aircraft?

A— Compressor.
B— Turbine.
C— Intake.

8498. At which component in an air-cycle cooling system does air undergo a pressure and temperature drop?

A— Expansion turbine.
B— Primary heat exchanger.
C— Refrigeration bypass valve.

8499. In a freon vapor-cycle cooling system, where is cooling air obtained for the condenser?

A— Turbine engine compressor.
B— Ambient air.
C— Pressurized cabin air.

8500. What is ventilating air used for on a combustion heater?

A— Provides combustion air for ground blower.
B— Carries heat to the places where needed.
C— Provides air required to support the flame.

8501. Turbine engine air used for air-conditioning and pressurization is generally called

A— compressed air.
B— ram air.
C— bleed air.

8502. In the combustion heater, combustion air system, what prevents too much air from entering the heaters as air pressure increases?

A— Either a combustion air relief valve or a differential pressure regulator.
B— Only a differential pressure regulator can be used.
C— Only a combustion air relief valve can be used.

8503. The cabin pressure of an aircraft in flight is maintained at the selected altitude by

A— controlling the air inflow rate.
B— inflating door seals and recirculating conditioned cabin air.
C— controlling the rate at which air leaves the cabin.

8504. What controls the operation of the cabin pressure regulator?

A— Cabin altitude.
B— Bleed air pressure.
C— Compression air pressure.

8505. The basic air-cycle cooling system consists of

A— a source of compressed air, heat exchangers, and a turbine.
B— heaters, coolers, and compressors.
C— ram air source, compressors, and engine bleeds.

8506. The purpose of the dump valve in a pressurized aircraft is to relieve

A— all positive pressure from the cabin.
B— a negative pressure differential.
C— pressure in excess of the maximum differential.

8507. What component might possibly be damaged if liquid refrigerant is introduced into the low side of a vapor-cycle cooling system when the pressure is too high or the outside air temperature is too low?

A— Compressor.
B— Condenser.
C— Evaporator.

8508. How can it be determined that a vapor-cycle cooling system is charged with the proper amount of freon?

A— Air bubbles in the sight glass disappear.
B— The compressor loads up and RPM decreases.
C— Air bubbles appear in the sight glass.

8509. When charging a vapor-cycle cooling system after evacuation, the low-pressure gauge fails to come out of a vacuum. What is indicated?

A— Blockage in the system.
B— The expansion valve failed to close.
C— The compressor is not engaging.

8510. What component in a vapor-cycle cooling system would most likely be at fault if a system would not take a freon charge?

A— Expansion valve.
B— Condenser.
C— Receiver-dryer.

8511. Frost or ice buildup on a vapor-cycle cooling system evaporator would most likely be caused by

A— the mixing valve sticking closed.
B— moisture in the evaporator.
C— inadequate airflow through the evaporator.

8512. What test is used to determine the serviceability of an oxygen cylinder?

A— Pressure test with compressed air.
B— Pressure test with oxygen.
C— Pressure test with water.

8513. How often should standard weight high-pressure oxygen cylinders be hydrostatically tested?

A— Every 5 years.
B— Every 4 years.
C— Every 3 years.

8514. To be eligible for recharging, a DOT 3HT oxygen cylinder must have been hydrostatically tested within

A— 3 years.
B— 5 years.
C— 7 years.

8515. What type of oxygen system uses the rebreather bag-type mask?

A— Diluter demand.
B— Continuous flow.
C— Demand.

8516. The positive-displacement type compressor that is most suitable for use in large piston-engined aircraft for cabin pressurization is the

A— reciprocating type.
B— vane type.
C— roots blower.

8517. In large aircraft that use positive-displacement compressors for cabin pressurization, how is oil-free delivery of air achieved?

A— By using separate bearing chambers, rubber seals, and labyrinth seals in the compressor.
B— By the use of air-oil separators.
C— By the use of a dry lubricant in the compressor.

8518. For emergency or backup use in pressurized aircraft, which is generally the least complicated and requires the least maintenance?

A— Chemical oxygen candle systems.
B— High-pressure oxygen systems.
C— Low-pressure oxygen systems.

8519. The main cause of contamination in gaseous oxygen systems is

A— moisture.
B— dust and other airborne particulates.
C— other atmospheric gases.

8520. Where does the last stage of cooling in an air-cycle air-conditioning system occur?

A— Refrigeration unit compressor.
B— Secondary heat exchanger.
C— Expansion turbine.

8521. The point at which freon flowing through a vapor-cycle cooling system gives up heat and changes from a gas to a liquid is the

A— condenser.
B— evaporator.
C— expansion valve.

8522. The point at which freon flowing through a vapor-cycle cooling system absorbs heat and changes from a liquid to a gas is the

A— condenser.
B— evaporator.
C— expansion valve.

8523. How is the cabin pressure of a pressurized aircraft usually controlled?

A— By a pressure-sensitive switch that causes the pressurization pump to turn on or off as required.
B— By an automatic outflow valve that dumps all the pressure in excess of the amount for which it is set.
C— By a pressure-sensitive valve that controls the output pressure of the pressurization pump.

8524. Which is considered a good practice concerning the inspection of heating and exhaust systems of aircraft utilizing a jacket around the engine exhaust as a heat source?

A— Supplement physical inspections with periodic operational carbon monoxide detection tests.
B— All exhaust system components should be removed periodically, and their condition determined by the magnetic-particle inspection method.
C— All exhaust system components should be removed and replaced at each 100-hour inspection period.

8525. What will result if auxiliary (ambient) ventilation is selected during pressurized flight while at cruising altitude?

A— An increase in cabin pressure.
B— Decreased cabin altitude.
C— Increased cabin altitude.

8526. The cabin pressure control setting has a direct influence upon the

A— outflow valve opening.
B— pneumatic system pressure.
C— inflow valve opening.

8527. The function of the evaporator in a freon cooling system is to

A— liquefy freon in the line between the compressor and the condenser.
B— lower the temperature of the cabin air.
C— transfer heat from the freon gas to ambient air.

8528. What is the purpose of a mixing valve in an air-conditioning system?

A— Control the supply of hot, cool, and cold air.
B— Distribute conditioned air evenly to all parts of the cabin.
C— Combine ram air with conditioned air.

8529. What component of a pressurization system prevents the cabin altitude from becoming higher than airplane altitude?

A— Cabin rate-of-descent control.
B— Negative pressure relief valve.
C— Positive pressure relief valve.

8530. If the liquid level gauge in a vapor-cycle cooling system indicates a low freon charge, the system should

A— be operated and a pressure check performed.
B— be operated for a period of time to reach a stable condition and then the freon level rechecked.
C— not be operated until freon and oil have been added.

8531. If the cabin rate of climb is too great, the controls should be adjusted to cause the

A— outflow valve to close slower.
B— outflow valve to close faster.
C— cabin compressor speed to decrease.

8532. The position of the thermostatic expansion valve in a vapor-cycle cooling system is determined by temperature and pressure of the

A— freon entering the evaporator.
B— air in the outlet of the condenser.
C— freon in the outlet of the evaporator.

8533. The function of the condenser in a freon cooling system is to

A— transfer heat from the freon gas to ambient air.
B— change liquid freon into a gas before it enters the compressor.
C— transfer heat from the cabin air to the liquid freon.

8534. The function of an expansion valve in a freon cooling system is to act as a metering device and to

A— reduce the pressure of the gaseous freon.
B— increase the pressure of the liquid freon.
C— reduce the pressure of the liquid freon.

8535. Which prevents a loss of pressurization through a disengaged cabin air compressor?

A— Firewall shutoff valve.
B— Cabin pressure outflow valve.
C— Delivery air duct check valve.

8536. When servicing an air-conditioning system that has lost all of its freon, it is necessary to

A— check oil and add as necessary, evacuate the system, relieve vacuum, and add freon.
B— check oil and add as necessary, evacuate the system, and add freon.
C— check oil and add as necessary, and add freon.

8537. The primary function of the cabin pressurization system outflow valve is to

A— provide protection against overpressurization.
B— maintain the desired cabin pressure.
C— maintain the same cabin air pressure at all altitudes.

8538. One purpose of a jet pump in a pressurization and air-conditioning system is to

A— produce a high pressure for operation of the outflow valve.
B— provide for augmentation of airflow in some areas of the aircraft.
C— assist in the circulation of freon.

8539. After cleaning or replacing the filtering element in a combustion heater fuel system, the system should be pressurized and

A— all connections checked for leaks.
B— the fuel filter bypass valve reset to the filter position.
C— a sample of fuel taken downstream from the filter to ensure proper operation of the new filtering element.

8540. The operation of an aircraft combustion heater is usually controlled by a thermostat circuit which

A— alternately turns the fuel on and off, a process known as cycling.
B— meters the amount of fuel continuously entering the heater and therefore regulates the heater's BTU output.
C— regulates the voltage applied to the heater's ignition transformer.

8541. The air-cycle cooling system produces cold air by

A— extracting heat energy across a compressor.
B— passing air through cooling coils that contain a refrigerant.
C— extracting heat energy across an expansion turbine.

8542. (Refer to figure 13.) Determine what unit is located immediately downstream of the expansion valve in a freon refrigeration system.

A— Condenser.
B— Compressor.
C— Evaporator coils.

8543. When checking a freon system, a steady stream of bubbles in the sight gauge indicates the charge is

A— high.
B— correct.
C— low.

8544. An aircraft fuselage is subjected to five major stresses. Pressurization is classified as a

A— tension stress.
B— compression stress.
C— torsion stress.

8545.
Which best describes cabin differential pressure?

A— Difference between cabin flight altitude pressure and Mean Sea Level pressure.
B— Difference between the ambient and internal air pressure.
C— Difference between cabin pressure controller setting and actual cabin pressure.

8546. The purpose of evacuating a freon air-conditioner system is to

A— check for leaks.
B— assure that all check valves are sealed.
C— remove moisture.

8547. The cabin pressurization modes of operation are

A— isobaric, differential, and maximum differential.
B— differential, unpressurized, and isobaric.
C— ambient, unpressurized, and isobaric.

8548. (1) Usually the air bled from a gas-turbine engine compressor is free from contamination and can be used safely for cabin pressurization.

(2) Independent cabin compressors can be engine-driven through accessory drive gearing or can be powered by bleed air from a turbine engine compressor.

Regarding the above statements,

A— only No. 1 is true.
B— only No. 2 is true.
C— both No. 1 and No. 2 are true.

8549. A pressurization controller uses

A— bleed air pressure, outside air temperature, and cabin rate of climb.
B— barometric pressure, cabin altitude, and cabin rate of change.
C— cabin rate of climb, bleed air volume, and cabin pressure.

8550. What unit in a vapor-cycle cooling system serves as a reservoir for the refrigerant?

A— Receiver-dryer.
B— Evaporator.
C— Condenser.

8551. What is the condition of the refrigerant as it enters the condenser of a vapor-cycle cooling system?

A— High-pressure liquid.
B— Low-pressure liquid.
C— High-pressure vapor.

8552. What is the condition of the refrigerant as it enters the evaporator of a vapor-cycle cooling system?

A— High-pressure liquid.
B— Low-pressure liquid.
C— High-pressure vapor.

8553. The evacuation of a vapor-cycle cooling system removes any water that may be present by

A— drawing out the liquid.
B— raising the boiling point of the water and drawing out the vapor.
C— lowering the boiling point of the water and drawing out the vapor.

8554. What is the condition of the refrigerant as it leaves the evaporator of a vapor-cycle cooling system?

A— Low-pressure liquid.
B— Low-pressure vapor.
C— High-pressure vapor.

8555. What is the condition of the refrigerant as it leaves the condenser of a vapor-cycle cooling system?

A— Low-pressure liquid.
B— High-pressure liquid.
C— High-pressure vapor.

8556. In what position should the bottle be placed when adding liquid freon to a vapor-cycle cooling system?

A— Vertical with the outlet at the top.
B— Horizontal with the outlet to the side.
C— Vertical with the outlet at the bottom.

8557. When purging a freon air-conditioning system, it is important to release the charge at a slow rate. What is the reason for the slow-rate discharge?

A— Prevent the large amount of freon from contaminating the surrounding atmosphere.
B— Prevent excessive loss of refrigerant oil.
C— Prevent condensation from forming and contaminating the system.

8558. When a vapor-cycle cooling system is not in operation, what is an indication that the system is leaking freon?

A— Oil seepage.
B— Bubbles in the sight glass.
C— An ozone-like odor in the immediate area.

8559. In an operating vapor-cycle cooling system, if the two lines connected to the expansion valve are essentially the same temperature, what does this indicate?

A— The system is functioning normally.
B— The expansion valve is not metering freon properly.
C— The compressor is pumping too much refrigerant.

8560. The purpose of a subcooler in a vapor-cycle cooling system is to

A— augment the cooling capacity during periods of peak demand.
B— aid in quick cooling a hot aircraft interior.
C— cool the freon to prevent premature vaporization.

8561. (1) A small amount of water in a vapor-cycle cooling system can freeze in the receiver-dryer and stop the entire system operation.

(2) Water in a vapor-cycle cooling system will react with refrigerant to form hydrochloric acid which is highly corrosive to the metal in the system.

Regarding the above statements,

A— only No. 1 is true.
B— only No. 2 is true.
C— both No. 1 and No. 2 are true.

8562. When Refrigerant-12 is passed over an open flame, it

A— changes to methane gas.
B— is broken down into its basic chemical elements.
C— changes to phosgene gas.

8563. What type of oil is suitable for use in vapor-cycle cooling systems?

A— Low viscosity engine oil.
B— Special refrigeration oil only.
C— Highly refined synthetic oil, free from impurities with special water absorbing additives.

8564. When an aircraft's oxygen system has developed a leak, the lines and fittings should be

A— removed and replaced.
B— inspected using a special oxygen system dye penetrant.
C— bubble tested with a special soap solution manufactured specifically for this purpose.

8565. If oxygen bottle pressure is allowed to drop below a specified minimum, it may cause

A— the pressure reducer to fail.
B— the automatic altitude control valve to open.
C— moisture to collect in the bottle.

8566. What controls the amount of oxygen delivered to a mask in a continuous-flow oxygen system?

A— Calibrated orifice.
B— Pressure reducing valve.
C— Pilot's regulator.

8567. In the diluter demand oxygen regulator, when does the demand valve operate?

A— When the diluter control is set at normal.
B— When the user demands 100 percent oxygen.
C— When the user breathes.

8568. The primary difference between aviation breathing oxygen and most other types of commercially available compressed oxygen is that

A— the other types are usually somewhat less than 99.5 percent pure oxygen.
B— aviation breathing oxygen has had all the water vapor removed.
C— aviation breathing oxygen has a higher percentage of water vapor to help prevent drying of a person's breathing passages and possible dehydration.

8569. What is used in some oxygen systems to change high cylinder pressure to low system pressure?

A— Pressure reducer valve.
B— Calibrated fixed orifice.
C— Diluter demand regulator.

8570. In a high-pressure oxygen system, if the pressure reducer fails, what prevents high-pressure oxygen from entering the system downstream?

A— Check valve.
B— Pressure relief valve.
C— Manifold control valve.

8571. High-pressure cylinders containing oxygen for aviation use can be identified by their

A— green color and the words "BREATHING OXYGEN" stenciled in 1-inch white letters.
B— yellow color and the words "AVIATOR'S BREATHING OXYGEN" stenciled in 1-inch white letters.
C— green color and the words "AVIATOR'S BREATHING OXYGEN" stenciled in 1-inch white letters.

8572. (Refer to figure 14.) One hour after an oxygen system was charged for a leakage check, the oxygen pressure gauge read 460 PSI at 63 °F; 6 hours later the temperature was 50 °F. (A 5 PSI change is the maximum allowable in a 6-hour period.) What pressure gauge readings would be acceptable to remain within the allowable limits?

A— 455 to 460 PSI.
B— 446 to 450 PSI.
C— 445 to 450 PSI.

8573. The greatest danger from hypoxia during long exposure at 10,000 feet MSL altitude is

A— headache and fatigue.
B— increased pulse, respiration, and sweating with possibility of dehydration.
C— impaired vision and judgment.

8574. In a gaseous oxygen system, which of the following are vented to blow out plugs in the fuselage skin?

A— Pressure relief valves.
B— Filler shutoff valves.
C— Pressure reducer valves.

8575. The purpose of pressurizing aircraft cabins is to

(1) create the proper environment for prevention of hypoxia.

(2) permit operation at high altitudes.

Regarding the above statements,

A— only No. 1 is true.
B— only No. 2 is true.
C— both No. 1 and No. 2 are true.

8576. (1) Oxygen used in aircraft systems is at least 99.5 percent pure and is practically water free.

(2) Oxygen used in aircraft systems is 99.5 percent pure and is hospital quality.

Regarding the above statements,

A— only No. 1 is true.
B— both No. 1 and No. 2 are true.
C— neither No. 1 nor No. 2 is true.

8577. Oxygen systems in unpressurized aircraft are generally of the

A— continuous-flow and pressure-demand types.
B— pressure-demand type only.
C— portable-bottle type only.

8578. The purpose of the airflow metering aneroid assembly found in oxygen diluter demand regulators is to

A— regulate airflow in relation to oxygen flow when operating in emergency or diluter demand positions.
B— regulate airflow in relation to cabin altitude when in diluter demand position.
C— automatically put the regulator in emergency position if the demand valve diaphragm ruptures.

8579. If a high-pressure oxygen cylinder is to be installed in an airplane, it must meet the specifications of the

A— aircraft manufacturer or the cylinder manufacturer.
B— Department of Transportation.
C— National Transportation Safety Board or the Standards of Compressed Gas Cylinders.

8580. Before a high-pressure oxygen cylinder is serviced, it must be the correct type and have been

A— hydrostatically tested within the proper time interval.
B— approved by the National Transportation Safety Board.
C— inspected by a certificated airframe mechanic.

8581. A contaminated oxygen system is normally purged with

A— oxygen.
B— compressed air.
C— nitrogen.

8582. How should you determine the amount of oxygen in a portable, high-pressure cylinder?

A— Weigh the cylinder and its contents.
B— Read the pressure gauge mounted on the cylinder.
C— Measure the pressure at the mask.

8583. What may be used as a lubricant on oxygen system tapered pipe thread connections?

A— Silicone dielectric compound.
B— Glycerine.
C— Teflon tape.

8584. Which oxygen system uses a molecular sieve as part of the system?

A— Mechanically-separated oxygen system.
B— Liquid oxygen system.
C— Chemical oxygen system.

8585. How many of the following are characteristic of a chemical or solid state oxygen system?

1. An adjustable oxygen release rate.
2. A volume storage capacity about three times that of compressed oxygen.
3. The system generators are inert below 400 °F even under severe impact.
4. A distribution and regulating system similar to gaseous oxygen systems.

A— Two.
B— Three.
C— Four.

8586. An aircraft magnetic compass is swung at specified operating intervals in order to determine the

A— accuracy of the lubber line.
B— compass precession.
C— compass deviation.

8587. The operating mechanism of most hydraulic pressure gauges is

A— a Bourdon tube.
B— an airtight diaphragm.
C— an evacuated bellows filled with an inert gas to which suitable arms, levers, and gears are attached.

8588. What is the fixed line mark attached to the compass bowl of a magnetic compass called?

A— Reeder line.
B— Lubber line.
C— Reference line.

8589. (1) Aircraft instruments are color-coded to direct attention to approaching operating difficulties.

(2) Aircraft instruments range markings are not specified by Federal Aviation Regulations but are standardized by aircraft manufacturer agreement.

Regarding the above statements,

A— only No. 1 is true.
B— only No. 2 is true.
C— both No. 1 and No. 2 are true.

8590. When swinging a magnetic compass, the compensators are adjusted to correct for

A— magnetic influence deviation.
B— small magnetic disturbances.
C— magnetic variations.

8591. What will be the result if the instrument static pressure line becomes disconnected inside a pressurized cabin during cruising flight?

A— The altimeter and airspeed indicator will both read low.
B— The altimeter and airspeed indicator will both read high.
C— The altimeter will read low and the airspeed indicator will read high.

8592. The maximum deviation (during level flight) permitted in a magnetic direction indicator installed in an aircraft is

A— 6°.
B— 8°.
C— 10°.

8593. Magnetic compass bowls are filled with a liquid to

A— retard precession of the float.
B— reduce deviation errors.
C— dampen the oscillation of the float.

8594. Instrument static system leakage can be detected by observing the rate of change in indication of the

A— airspeed indicator after suction has been applied to the static system to cause a prescribed equivalent airspeed to be indicated.
B— altimeter after pressure has been applied to the static system to cause a prescribed equivalent altitude to be indicated.
C— altimeter after suction has been applied to the static system to cause a prescribed equivalent altitude to be indicated.

8595. The maximum altitude loss permitted during an unpressurized aircraft instrument static pressure system integrity check is

A— 50 feet in 1 minute.
B— 200 feet in 1 minute.
C— 100 feet in 1 minute.

8596. Which statement regarding an aircraft instrument vacuum system is true?

A— Dry-type vacuum pumps with carbon vanes are very susceptible to damage from solid airborne particles and must take in only filtered air.
B— Vacuum systems are generally more effective at high altitudes than positive pressure systems.
C— If the air inlet to each vacuum instrument is connected to a common atmospheric pressure manifold, the system generally will be equipped with individual instrument filters only.

8597. When an aircraft altimeter is set at 29.92" Hg on the ground, the altimeter will read

A— pressure altitude.
B— density altitude.
C— field elevation.

8598. Which of the following instrument discrepancies could be corrected by an aviation mechanic?

1. Red line missing.
2. Case leaking.
3. Glass cracked.
4. Mounting screws loose.
5. Case paint chipped.
6. Leaking at line B nut.
7. Will not zero out.
8. Fogged.

A— 1, 4, 6.
B— 3, 4, 5, 6.
C— 1, 4, 5, 6.

8599. Which of the following instrument discrepancies would require replacement of the instrument?

1. Red line missing.
2. Case leaking.
3. Glass cracked.
4. Mounting screws loose.
5. Case paint chipped.
6. Leaking at line B nut.
7. Will not zero out.
8. Fogged.

A— 2, 3, 7, 8.
B— 1, 4, 6, 7.
C— 1, 3, 5, 8.

8600. Which of the following instrument conditions is acceptable and would not require correction?

1. Red line missing.
2. Case leaking.
3. Glass cracked.
4. Mounting screws loose.
5. Case paint chipped.
6. Leaking at line B nut.
7. Will not zero out.
8. Fogged.

A— 1.
B— 5.
C— None.

8601. A barometric altimeter indicates pressure altitude when the barometric scale is set at

A— 29.92" Hg.
B— 14.7" Hg.
C— field elevation.

8602. A Bourdon tube instrument may be used to indicate

1. pressure.
2. temperature.
3. position.

A— 1 and 2.
B— 1.
C— 2 and 3.

8603. A turn-and-bank instrument indicates

A— the longitudinal attitude of the aircraft during climb and descent.
B— trim and serves as an emergency source of bank information in case the attitude indication fails.
C— the need for corrections in pitch and bank anytime the aircraft deviates from a preselected attitude.

8604. Thermocouple leads

A— are designed for a specific installation and may not be altered.
B— may be installed with either lead to either post of the indicator.
C— may be repaired using solderless connectors.

8605. A synchro transmitter is connected to a synchro receiver

A— mechanically through linkage.
B— electromagnetically without wires.
C— electrically with wires.

8606. The operation of an angle-of-attack indicating system is based on detection of differential pressure at a point where the airstream flows in a direction

A— not parallel to the true angle of attack of the aircraft.
B— parallel to the angle of attack of the aircraft.
C— parallel to the longitudinal axis of the aircraft.

8607. Turbine engine exhaust gas temperatures are measured by using

A— iron/constantan thermocouples.
B— chromel/alumel thermocouples.
C— ratiometer electrical resistance thermometers.

8608. Fuel flow transmitters are designed to transmit data

A— mechanically.
B— electrically.
C— utilizing fluid power.

8609. Which of the following causes of aircraft magnetic compass inaccuracies may be compensated for by mechanics?

A— Deviation.
B— Magnetic compass current.
C— Variation.

8610. Who is authorized to repair an aircraft instrument?

1. A certificated mechanic with an airframe rating.
2. A certificated repairman with an airframe rating.
3. A certificated repair station approved for that class instrument.
4. A certificated airframe repair station.

A— 1, 2, 3, and 4.
B— 3 and 4.
C— 3.

8611. What does a reciprocating engine manifold pressure gauge indicate when the engine is not operating?

A— Zero pressure.
B— The differential between the manifold pressure and the atmospheric pressure.
C— The existing atmospheric pressure.

8612. The minimum requirements for testing and inspection of instrument static pressure systems required by FAR Section 91.411 are contained in

A— Type Certificate Data Sheets.
B— AC 43.13-1A.
C— FAR Part 43, appendix E.

8613. Which condition would be most likely to cause excessive vacuum in a vacuum system?

A— Vacuum pump overspeed.
B— Vacuum relief valve improperly adjusted.
C— Vacuum relief valve spring weak.

8614. Data transmitted between components in an EFIS are converted into

A— digital signals.
B— analog signals.
C— carrier wave signals.

8615. The function of a CRT in an EFIS is to

A— allow the pilot to select the appropriate system configuration for the current flight situation.
B— display alphanumeric data and representations of aircraft instruments.
C— receive and process input signals from aircraft and engine sensors and send the data to the appropriate display.

8616. The function of a symbol generator (SG) in an EFIS is to

A— display alphanumeric data and representations of aircraft instruments.
B— allow the pilot to select the appropriate system configuration for the current flight situation.
C— receive and process input signals from aircraft and engine sensors and send the data to the appropriate display.

8617. The function of a display controller in an EFIS is to

A— display alphanumeric data and representations of aircraft instruments.
B— allow the pilot to select the appropriate system configuration for the current flight situation.
C— receive and process input signals from aircraft and engine sensors and send the data to the appropriate display.

8618. A radar altimeter determines altitude by

A— transmitting a signal and receiving back a reflected signal.
B— receiving signals transmitted from ground radar stations.
C— means of transponder interrogation.

8619. A radar altimeter indicates

A— flight level (pressure) altitude.
B— altitude above sea level.
C— altitude above ground level.

8620. Resistance-type temperature indicators using Wheatstone bridge or ratiometer circuits may be used to indicate the temperatures of which of the following?

1. Free air.
2. Exhaust gas temperature.
3. Carburetor air.
4. Coolant (engine).
5. Oil temperature.
6. Cylinder head temperature.

A— 1, 2, 3, 4, 5, and 6.
B— 1, 3, 4, and 5.
C— 1, 2, 3, and 6.

8621. When flags such as NAV, HDG, or GS are displayed on an HSI, the indication is

A— that function is inoperative.
B— that function is operating.
C— to call attention to deviation from the desired setting, or flight path, or heading, etc.

8622. The number of shock mounts required for an original instrument panel installation is determined by the

A— size of the panel.
B— weight of the complete panel unit.
C— number of instruments to be installed.

8623. Which procedure should you use if you find a vacuum-operated instrument glass loose?

A— Mark the case and glass with a slippage mark.
B— Replace the glass.
C— Install another instrument.

8624. Which instruments are connected to an aircraft's pitot-static system?

1. Vertical speed indicator.
2. Cabin altimeter.
3. Altimeter.
4. Cabin rate-of-change indicator.
5. Airspeed indicator.

A— 1, 2, 3, 4, and 5.
B— 1, 2, and 4.
C— 1, 3, and 5.

8625. How many of the following instruments will normally have range markings?

1. Airspeed indicator.
2. Altimeter.
3. Cylinder head temperature gauge.

A— One.
B— Two.
C— Three.

8626. How would an airspeed indicator be marked to show the best rate-of-climb speed (one engine inoperative)?

A— A red radial line.
B— A blue radial line.
C— A green arc.

8627. The green arc on an aircraft temperature gauge indicates

A— the instrument is not calibrated.
B— the desirable temperature range.
C— a low, unsafe temperature range.

8628. What marking on an instrument is used to indicate whether the glass has slipped?

A— White index mark.
B— Blue radial line.
C— Red radial line.

8629. What marking color is used to indicate if a cover glass has slipped?

A— Red.
B— White.
C— Yellow.

8630. Aircraft instruments should be marked and graduated in accordance with

A— the instrument manufacturer's specifications.
B— both the aircraft and engine manufacturers' specifications.
C— the specific aircraft maintenance or flight manual.

8631. Aircraft instrument panels are generally shock-mounted to absorb

A— all vibration.
B— low-frequency, high-amplitude shocks.
C— high-frequency, high-amplitude shocks.

8632. The method of mounting aircraft instruments in their respective panels depends on the

A— instrument manufacturer.
B— design of the instrument case.
C— design of the instrument panel.

8633. How is a flangeless instrument case mounted in an instrument panel?

A— By four machine screws which extend through the instrument panel.
B— By an expanding-type clamp secured to the back of the panel and tightened by a screw from the front of the instrument panel.
C— By a metal shelf separate from and located behind the instrument panel.

8634. Why are most electrical instruments mounted in iron or steel cases?

A— To avoid damage to the instrument during maintenance.
B— To facilitate removal or installation.
C— To prevent interference from outside magnetic fields.

8635. When installing an instrument in an aircraft, who is responsible for making sure it is properly marked?

A— The aircraft owner.
B— The instrument installer.
C— The instrument manufacturer.

8636. Where may a person look for the information necessary to determine the required markings on an engine instrument?

1. Engine manufacturer's specifications.
2. Aircraft flight manual.
3. Instrument manufacturer's specifications.
4. Aircraft maintenance manual.

A— 2 or 4.
B— 1 or 4.
C— 2 or 3.

8637. A certificated mechanic with airframe and powerplant ratings may

A— perform minor repairs to engine instruments.
B— perform minor repairs and minor alterations to engine instruments.
C— not perform repairs to engine instruments.

8638. A red radial line on the face of an engine instrument indicates

A— normal operating range.
B— caution range.
C— maximum or minimum safe operating limits.

8639. A certificated mechanic may perform

A— minor repairs to instruments.
B— 100-hour inspections of instruments.
C— instrument overhaul.

8640. An aircraft instrument panel is electrically bonded to the aircraft structure to

A— act as a restraint strap.
B— provide current return paths.
C— aid in the panel installation.

8641. How many of the following are controlled by gyroscopes?

1. Attitude indicator.
2. Heading indicator.
3. Turn needle of the turn-and-slip indicator.

A— Three.
B— Two.
C— One.

8642. The lubber line on a directional gyro is used to

A— represent the nose of the aircraft.
B— align the instrument glass in the case.
C— represent the wings of the aircraft.

8643. Which instruments are connected to an aircraft's static pressure system only?

1. Vertical speed indicator.
2. Cabin altimeter.
3. Altimeter.
4. Cabin rate-of-change indicator.
5. Airspeed indicator.

A— 1 and 3.
B— 2, 4, and 5.
C— 2 and 4.

8644. When an unpressurized aircraft's static pressure system is leak checked to comply with the requirements of FAR Section 91.411, what aircraft instrument(s) may be used in lieu of a pitot-static system tester?

1. Vertical speed indicator.
2. Cabin altimeter.
3. Altimeter.
4. Cabin rate-of-change indicator.
5. Airspeed indicator.

A— 1 or 5.
B— 2 or 4.
C— 3.

8645. If a static pressure system check reveals excessive leakage, the leak(s) may be located by

A— pressurizing the system and adding leak detection dye.
B— isolating portions of the line and testing each portion systematically, starting at the instrument connections.
C— removing and visually inspecting the line segments.

8646. When performing the static system leakage check required by FAR Section 91.411, the technician utilizes

A— static pressure.
B— positive pressure.
C— negative pressure.

8647. What is the primary purpose of an autopilot?

A— To relieve the pilot of control of the aircraft during long periods of flight.
B— To fly a more precise course for the pilot.
C— To obtain the navigational aid necessary for extended overwater flights.

8648. Which of the following provides manual maneuverability of the aircraft while the autopilot is engaged?

A— Servo-amplifier.
B— Directional gyro indicator.
C— Flight controller.

8649. In an autopilot, which signal nullifies the input signal to the ailerons?

A— Displacement signal.
B— Course signal.
C— Followup signal.

8650. In which control element of an autopilot system is an attitude indicator?

A— Command.
B— Sensing.
C— Input.

8651. What is the operating principle of the sensing device used in an autopilot system?

A— The reaction of the force 90° away from the applied force in the direction of gyro rotation.
B— The relative motion between a gyro and its supporting system.
C— The rate of change of motion between the gyro gimbal rings and the aircraft.

8652. What will occur if an aircraft attitude or heading is changed by its autopilot system in order to correct a deviation and the involved control surfaces are returned to neutral immediately after the aircraft has reached its correct position?

A— Normal operation.
B— Undershoot and oscillation.
C— Overshoot and oscillation.

8653. What component of an autopilot system applies torque to the control surfaces of an aircraft?

A— Servo.
B— Controller.
C— Gyro.

8654. What is the main purpose of a servo in an autopilot system?

A— Correct for displacement of the aircraft about its axis.
B— Change mechanical energy to electrical energy.
C— Move the control surface as commanded.

8655. Which channel of an autopilot detects changes in pitch attitude of an aircraft?

A— Elevator.
B— Aileron.
C— Rudder.

8656. The elevator channel of an autopilot controls the aircraft about which axis of rotation?

A— Roll.
B— Longitudinal.
C— Lateral.

8657. What component is the sensing device in an electromechanical autopilot system?

A— Servo.
B— Gyro.
C— Controller.

8658. A fully integrated autopilot controls the aircraft around how many axes?

A— Two.
B— Three.
C— Four.

8659. Dutch roll, a combination yawing and rolling oscillation that affects many sweptwing aircraft, is counteracted with

A— a flight director system.
B— an aileron damper system.
C— a yaw damper system.

8660. When operationally checking an autopilot system on the ground, after the aircraft's main power has been switched on, the autopilot should be engaged

A— only after the gyros come up to speed and the amplifier warms up.
B— whenever the operator desires.
C— for only a few minutes at a time.

8661. Installed radio equipment is protected from damage due to jolts and vibration by

A— shock mounts.
B— spring and/or viscous damper mounted racks.
C— rubber or foam cushioning material between circuit chassis and case.

8662. (1) Use solder to attach bonding jumpers on radio equipment.

(2) Radio equipment is bonded to the aircraft in order to provide a low-impedance ground and to minimize radio interference from static electrical charges.

Regarding the above statements,

A— only No. 1 is true.
B— both No. 1 and No. 2 are true.
C— only No. 2 is true.

8663. Which of the following must be displayed in an aircraft equipped with a two-way radio?

A— Radio station license.
B— Restricted radiotelephone operator's permit.
C— Certificate of minimum performance standards.

8664. Which of the following issues the license required to operate an aircraft radio transmitter?

A— National Transportation Safety Board.
B— Federal Aviation Administration.
C— Federal Communications Commission.

8665. Part of the ADF systems used on aircraft includes

A— VHF transmitters.
B— marker beacons.
C— sense and loop antennas.

8666. When installing coaxial cable, it should be secured firmly along its entire length

A— at 1-foot intervals.
B— wherever the cable sags.
C— at 2-foot intervals.

8667. The emergency locator transmitter (ELT) battery

A— replacement date must be marked on the outside of the transmitter.
B— must be replaced annually.
C— replacement date must be computed from the date of installation.

8668. An emergency locator transmitter (ELT) battery must be capable of furnishing power for signal transmission for at least

A— 36 hours.
B— 48 hours.
C— 72 hours.

8669. The preferred location of an ELT is

A— where it is readily accessible to the pilot or a member of the flightcrew while the aircraft is in flight.
B— as far aft as possible.
C— as far aft as possible, but forward of the vertical fin.

8670. An emergency locator transmitter (ELT) is normally activated by an inertial switch or equivalent mechanism if subjected to a force of a prescribed intensity and duration. It must activate when the force is applied

A— parallel to the longitudinal axis of the aircraft.
B— parallel to the vertical axis of the aircraft.
C— in any direction relative to the aircraft axes.

8671. How may the battery replacement date for an emergency locator transmitter (ELT) be determined?

A— By removing the batteries and testing them under a measured load to determine if 50 percent of the useful life remains.
B— By observing the battery replacement date marked on the outside of the transmitter.
C— By activating the transmitter and measuring the signal strength.

8672. How may the operation of an installed emergency locator transmitter (ELT) be verified during aircraft inspection?

A— By moving the deactivating switch from the DISARM position to the ARM position while monitoring the civil emergency frequency with a communications receiver.
B— By visually inspecting the transmitter installation and checking its power supply.
C— By tuning a communications receiver to the civil emergency frequency and activating the ELT momentarily.

8673. Static dischargers help eliminate radio interference by dissipating static electricity into the atmosphere at

A— low current levels.
B— high voltage level.
C— high current levels.

8674. Long Range Navigation (LORAN) systems determine aircraft location by

A— measuring the inertial forces acting on the aircraft.
B— means of pulsed signals transmitted from ground stations.
C— means of signals transmitted to and from navigation satellites.

8675. Typically, a prerecorded emergency announcement may be automatically initiated over a large aircraft's passenger address system in the event of

A— cabin decompression (during flight).
B— a hard landing.
C— an engine failure.

8676. VHF radio signals are commonly used in

A— ATC communications.
B— VOR navigation.
C— both VOR navigation and ATC communications.

8677. On modern large aircraft, what electronic device typically monitors flight parameters and performs autopilot functions?

A— Flight management computer.
B— Transponder.
C— Control/display unit.

8678. During operation, a GPWS typically monitors the radio (radar) altimeter; air data computer; instrument landing system; and

A— aileron, rudder, and elevator positions.
B— landing gear and flap positions.
C— spoiler, slat, and stabilizer positions.

8679. In general, the purpose of an aircraft transponder is to

A— continually transmit heading, speed, and rate of climb/decent etc. information to ATC.
B— monitor aircraft speed, heading, altitude, and attitude whenever the autopilot system is engaged.
C— receive an interrogation signal from a ground station and automatically send a reply back.

8680. When an antenna is installed, it should be fastened

A— to the primary structure at the approximate intersection of the three aircraft axes.
B— with a reinforcing doubler on each side of the aircraft skin.
C— so that loads imposed are transmitted to the aircraft structure.

8681. After an automatic direction finding antenna has been installed, the

A— antenna must be grounded.
B— loop must be calibrated.
C— transceiver must be compensated.

8682. Doublers are used when antennas are installed to

A— eliminate antenna vibration.
B— prevent oil canning of the skin.
C— reinstate the structural strength of the aircraft skin.

8683. One antenna can be used for the radio range and standard broadcast bands in light aircraft because the

A— two ranges are close together.
B— antenna is omnidirectional.
C— antenna length may be electronically adjusted.

8684. What characteristics of the installation of a rigid antenna on a vertical stabilizer should be evaluated?

A— Polarization and impedance.
B— Impedance and interference.
C— Flutter and vibration.

8685. A gasket or sealant is used between the antenna mast and fuselage skin

A— to prevent the entry of moisture.
B— for aircraft pressurization only.
C— to prevent abrasion between the antenna mast and fuselage skin.

8686. The preferred location of a VOR antenna on light aircraft is on

A— the bottom of the fuselage and as far forward as possible.
B— top of the cabin with the apex of the V pointing forward.
C— top of the vertical stabilizer.

8687. The purpose of a localizer is to

A— set the airplane on the proper approach angle to the runway.
B— indicate the distance the airplane is from the end of the runway.
C— align the airplane with the center of the runway.

8688. (Refer to figure 15.) What is the approximate drag load on an antenna with a frontal area of .125 square feet installed on an aircraft with a speed of 225 MPH?

A— 2.069 pounds.
B— 2.073 pounds.
C— 2.080 pounds.

8689. (Refer to figure 15.) What is the approximate drag load on an antenna with a frontal area of .137 square feet installed on an aircraft with a speed of 275 MPH?

A— 3.387 pounds.
B— 3.741 pounds.
C— 3.592 pounds.

8690. A DME antenna should be located in a position on the aircraft that will

A— not be blanked by the wing when the aircraft is banked.
B— permit interruptions in DME operation.
C— eliminate the possibility of the DME locking on a station.

8691. When bending coaxial cable, the bend radius should be at least

A— 10 times the diameter of the cable.
B— 15 times the diameter of the cable.
C— 20 times the diameter of the cable.

8692. When installing a DME antenna, it should be aligned with the

A— null position.
B— angle of incidence.
C— centerline on the airplane.

8693. (Refer to figure 16.) Which of the antennas shown is a typical DME antenna?

A— 1.
B— 2.
C— 4.

8694. (Refer to figure 16.) Which of the antennas shown is a typical glideslope antenna?

A— 2.
B— 3.
C— 4.

8695. The addition of avionics and associated antenna systems forward of the CG limit will affect

A— empty weight and useful load.
B— CG limits and useful load.
C— useful load and maximum gross weight.

8696. How much clearance from the seat bottom is required when installing radio equipment under a seat?

A— 3 inches with the seat unoccupied.
B— No set minimum as long as the equipment receives adequate cooling and damage protection.
C— 1 inch with the seat occupied and subjected to maximum downward seat spring deflection.

8697. The purpose of a glideslope system is to

A— provide for automatic altitude reporting to air traffic control.
B— indicate the distance the airplane is from the end of the runway.
C— assist the pilot in making a correct angle of descent to the runway.

8698. Fuel jettisoning is usually accomplished

A— through a common manifold and outlet in each wing.
B— by gravity flow into the outboard wing tanks and overboard through a common outlet in each wing.
C— through individual outlets for each tank.

8699. The primary purpose of an aircraft's fuel jettison system is to quickly achieve a

A— lower landing weight.
B— balanced fuel load.
C— reduced fire hazard.

8700. (1) The fuel jettison valve must be designed to allow flight personnel to close the valve during any part of the jettisoning operation.

(2) During the fuel jettisoning operation, the fuel must discharge clear of any part of the airplane.

Regarding the above statements,

A— both No. 1 and No. 2 are true.
B— only No. 2 is true.
C— neither No. 1 nor No. 2 is true.

8701. Which of the following is employed to maintain lateral stability when jettisoning fuel?

A— Two separate independent systems.
B— Crossfeed system.
C— Two interconnected systems.

8702. A fuel jettison system is required if the maximum takeoff weight exceeds the maximum landing weight on

A— transport category aircraft only.
B— both transport category and general aviation aircraft.
C— general aviation aircraft only.

8703. Fuel is moved overboard in most fuel jettison systems by

A— boost pumps.
B— gravity.
C— gravity and engine-driven fuel pumps.

8704. Fuel jettisoning past the limits prescribed by Federal Aviation Regulations is usually prevented by

A— closely monitoring the fuel quantity and turning off the fuel dump switch(es).
B— dump limit valves or a low-level circuit.
C— standpipes in the fuel tanks.

8705. Which procedure must be followed when defueling aircraft with sweptback wings?

A— Defuel all the tanks at one time.
B— Defuel the inboard wing tanks first.
C— Defuel the outboard wing tanks first.

8706. (Refer to figure 17.) What is the purpose of the pump crossfeed valve?

A— Balance the fuel in the tanks.
B— Allow operation of engines from one tank.
C— Allow operation of the left engine when the right fuel-boost pump is inoperative.

8707. Normal fuel crossfeed system operation in multiengine aircraft

A— calls for jettisoning of fuel overboard to correct lateral instability.
B— reduces contamination and/or fire hazards during fueling or defueling operations.
C— provides a means to maintain a balanced fuel load condition.

8708. What is the primary purpose of the crossfeed system?

A— To allow the feeding of any engine from any tank.
B— To allow the feeding of fuel from one tank for defueling.
C— To provide automatic refueling of a tank to any desired level.

8709. Fuel system components must be bonded and grounded in order to

A— drain off static charges.
B— prevent stray currents.
C— retard galvanic corrosion.

8710. A typical large transport aircraft fuel manifold system allows how many of the following?

1. All tanks can be serviced through a single connection.
2. Any engine can be fed from any tank.
3. All engines can be fed from all tanks simultaneously.
4. A damaged tank can be isolated from the rest of the fuel system.

A— Two.
B— Three.
C— Four.

8711. The use of turbine fuels in aircraft has resulted in some problems not normally associated with aviation gasolines. One of these problems is

A— increasing viscosity of fuel as fuel temperature lowers at altitude.
B— higher vapor pressure.
C— microbial contaminants.

8712. What is used in many aircraft to prevent bubbles in the fuel after it leaves the tank when atmospheric pressure is lower than fuel vapor pressure?

A— Air-fuel separators.
B— Anti-foaming additives.
C— Boost pumps.

8713. Which of the following precautions is most important during refueling operations?

A— All outside electrical sources must be disconnected from the aircraft.
B— Fuel to be used must be appropriately identified.
C— All electrical switches must be in OFF position.

8714. Before fueling an aircraft by using the pressure fueling method, what important precaution should be observed?

A— The truck pump pressure must be correct for that refueling system.
B— The truck pump pressure must be adjusted for minimum filter pressure.
C— The aircraft's electrical system must be on to indicate quantity gauge readings.

8715. What flight safety-related advantage does a pressure fueling system provide?

A— Keeps the aircraft within weight and balance limitations.
B— Reduces the chances for fuel contamination.
C— Reduces the time required for fueling.

8716. Aircraft pressure fueling systems instructional procedures are normally placarded on the

A— fuel control panel access door.
B— lower wing surface adjacent to the access door.
C— aircraft ground connection point.

8717. Pressure fueling of aircraft is usually accomplished through

A— pressure connections on individual fuel tanks.
B— at least one single point connection.
C— individual fuel tank overwing and/or fuselage access points.

8718. Which of the following may be used to repair most integral fuel tanks?

A— Welding.
B— Brazing.
C— Riveting.

8719. How is the outlet fuel pressure regulated on a submerged, single-speed, centrifugal-type fuel pump?

A— By the engine-driven pump's design and internal clearance.
B— By the first check valve downstream from the pump.
C— By the pump's design and internal clearances.

8720. What is one purpose of a fuel tank vent?

A— To maintain atmospheric pressure.
B— To decrease fuel vapor pressure.
C— To decrease tank internal air pressure.

8721. What publication defines the test pressures to be followed when testing a repaired fuel tank of a small aircraft?

A— Federal Aviation Regulations.
B— Manufacturer's service letters.
C— AC 43.13-1A.

8722. If it is necessary to enter an aircraft's fuel tank, which procedure should be avoided?

A— Continue purging the tank during the entire work period.
B— Station an assistant outside the fuel tank access to perform rescue operations if required.
C— Conduct the defueling and tank purging operation in an air-conditioned building.

8723. What is the recommended practice for cleaning a fuel tank before welding?

A— Purge the tank with air.
B— Flush the inside of the tank with clean water.
C— Steam clean the tank interior.

8724. An aircraft's integral fuel tank is

A— usually located in the bottom of the fuselage.
B— a part of the aircraft structure.
C— a self-sealing tank.

8725. Which gas is used for purging an aircraft fuel tank?

A— Helium or argon.
B— Carbon dioxide.
C— Carbon monoxide.

8726. Why is the main fuel strainer located at the lowest point in the fuel system?

A— It traps any small amount of water that may be present in the fuel system.
B— It provides a drain for residual fuel.
C— It filters and traps all micro-organisms that may be present in the fuel system.

8727. The purpose of a diaphragm in a vane-type fuel pump is to

A— equalize fuel pressure at all speeds.
B— vary fuel pressure according to throttle setting.
C— compensate fuel pressures to altitude changes.

8728. When moving the mixture control on a normally operating engine into the idle cutoff position, engine RPM should

A— slightly increase before the engine starts to die.
B— slightly decrease and then drop rapidly.
C— remain the same until the cutoff is effected, then drop rapidly.

8729. Entrained water in aviation turbine fuel is a hazard because of its susceptibility to freezing as it passes through the filters. What is a common method of preventing this hazard?

A— Micromesh fuel strainers.
B— High-velocity fuel pumps.
C— Anti-icing fuel additives.

8730. Fuel leaks are usually classified as a stain, a seep, a heavy seep, or a running leak. As a general rule,

A— stains, seeps, and heavy seeps are not flight hazards.
B— all fuel leaks regardless of location or severity are considered a hazard to flight.
C— stains, seeps, and heavy seeps, (in addition to running leaks) are considered flight hazards when located in unvented areas of the aircraft.

8731. The presence of fuel stains around a fuel nozzle would indicate

A— too much fuel pressure.
B— excessive airflow across the venturi.
C— clogged fuel nozzle.

8732. What should be used to inert an integral fuel tank before attempting repairs?

A— CO_2.
B— Water.
C— Steam.

8733. What should be used to remove flux from an aluminum tank after welded repairs?

A— Soft brush and warm water.
B— 5 percent solution of nitric or sulfuric acid.
C— Mild solution of soap and warm water.

8734. What method would be used to check for internal leakage of a fuel valve without removing the valve from the aircraft?

A— Place the valve in the OFF position, drain the strainer bowl, and with boost pump on, watch to see if fuel flows to the strainer bowl.
B— Remove fuel cap(s), turn boost pump(s) on, and watch for bubbling in the tanks.
C— Apply regulated air pressure on the downstream side of the fuel pump and listen for air passing through the valve.

8735. Why are jet fuels more susceptible to water contamination than aviation gasoline?

A— Jet fuel has a higher viscosity than gasoline.
B— Jet fuel is lighter than gasoline; therefore, water is more easily suspended.
C— Condensation is greater because of the higher volatility of jet fuels.

8736. When installing a rigid fuel line, 1/2 inch in diameter, at what intervals should the line be supported?

A— 24 inches.
B— 12 inches.
C— 16 inches.

8737. The probe of a capacitance-type fuel level gauge is essentially a

A— float-actuated variable capacitor.
B— capacitor with fuel and air acting as one plate.
C— capacitor with fuel and air acting as a dielectric.

8738. The capacitance-type (electronic) fuel quantity indicator

A— has no moving parts in the tank.
B— has two tubes separated by a mica dielectric in the tank.
C— utilizes a float operated variable capacitor.

8739. What type of remote-reading fuel quantity indicating system has several probes installed in each fuel tank?

A— Electromechanical.
B— Electronic.
C— Direct reading.

8740. Which aircraft fuel quantity indicating system incorporates a signal amplifier?

A— Electronic.
B— Sight glass.
C— Electrical.

8741. A drip gauge may be used to measure

A— the amount of fuel in the tank.
B— system leakage with the system shut down.
C— fuel pump diaphragm leakage.

8742. The electronic-type fuel quantity indicating system consists of a bridge circuit,

A— an amplifier, an indicator, and a tank unit.
B— a tank, an amplifier, and an indicator.
C— a tank unit, a tank, and an amplifier.

8743. A probe or a series of probes is used in what kind of fuel quantity indicating system?

A— Selsyn.
B— Capacitor.
C— Synchro.

8744. The electronic-type fluid quantity indicating system is more accurate in measuring fuel level because

A— only one probe and one indicator are necessary for multiple tank configurations.
B— aircraft attitude has no effect on fluid quantity indication.
C— it measures by weight instead of volume.

8745. One advantage of electrical and electronic fuel quantity indicating systems is that

A— the indicators are calibrated in gallons; therefore, no conversion is necessary.
B— only one transmitter and one indicator are needed regardless of the number of tanks.
C— several fuel tank levels can be read on one indicator.

8746. A fuel totalizer is a component which indicates the

A— total amount of fuel being consumed by all engines.
B— amount of fuel in any given tank.
C— amount of fuel in all tanks.

8747. What is the dielectric (nonconducting material) in a capacitance-type fuel quantity indicating system?

A— Outer shell of the capacitor.
B— Fuel in the tank.
C— Fuel and air in the tank.

8748. A capacitance-type fuel quantity indicating system measures fuel in

A— pounds.
B— pounds per hour.
C— gallons.

8749. What are the four general types of fuel quantity gauges?

1. Sight glass.
2. Mechanical.
3. Electrical.
4. Electronic.
5. Bourdon tube.
6. Vane-type transmitter.
7. Litmus indicator.
8. Direct-reading static pressure type.

A— 1, 2, 3, 4.
B— 1, 3, 6, 8.
C— 2, 3, 5, 7.

8750. How does temperature affect fuel weight?

A— Cold fuel is heavier per gallon.
B— Warm fuel is heavier per gallon.
C— Temperature has no effect.

8751. One advantage of electrical and electronic fuel quantity indicating systems is that the indicator

A— can be located any distance from the tank(s).
B— has no movable devices.
C— always measures volume instead of mass.

8752. When fuel quantity is measured in pounds instead of gallons, the measurement will be more accurate because fuel volume

A— varies with temperature change.
B— increases when temperature decreases.
C— varies with changes in atmospheric pressure.

8753. An electrical-type fuel quantity indicating system consists of an indicator in the cockpit and a

A— float-operated transmitter installed in the tank.
B— float resting on the surface of the tank.
C— float-operated receiver installed in the tank.

8754. What is the purpose of a float-operated transmitter installed in a fuel tank?

A— It sends an electric signal to the fuel quantity indicator.
B— It senses the total amount of fuel density.
C— It senses the dielectric qualities of fuel and air in the tank.

8755. In an electronic-type fuel quantity indicating system, the tank sensing unit is a

A— capacitor.
B— variable resistor.
C— variable inductor.

8756. What must each fuel quantity indicator be calibrated to read during level flight when the quantity of fuel remaining is equal to the unusable fuel supply?

A— The total unusable fuel quantity.
B— Both the total unusable fuel quantity and the unusable fuel quantity in each tank.
C— Zero.

8757. What unit would be adjusted to change the fuel pressure warning limits?

A— Fuel flowmeter bypass valve.
B— Pressure-sensitive mechanism.
C— Fuel pressure relief valve.

8758. Select one means of controlling the fuel temperature on turbine-powered aircraft.

A— Engine lubricating oil at the fuel filter.
B— Engine bleed air around the fuel tank.
C— Engine bleed air to a heat exchanger.

8759. What is the purpose of flapper-type check valves in integral fuel tanks?

A— To allow defueling of the tanks by suction.
B— To prevent fuel from flowing away from the boost pumps.
C— To allow the engine-driven pumps to draw fuel directly from the tank if the boost pump fails.

8760. What unit is generally used to actuate the fuel pressure warning system?

A— Fuel flowmeter.
B— Pressure-sensitive mechanism.
C— Fuel pressure gauge.

8761. What method is used on turbine-powered aircraft to determine when the condition of the fuel is approaching the danger of forming ice crystals?

A— Fuel pressure warning.
B— Fuel pressure gauge.
C— Fuel temperature indicator.

8762. Which of the following would give the first positive indication that a change-over from one fuel tank to another is needed?

A— Fuel pressure warning.
B— Fuel pressure gauge.
C— Fuel quantity indicator.

8763. A fuel pressure warning switch contacts close and warning light is turned on when

A— a measured quantity of fuel has passed through it.
B— the fuel flow stops.
C— the fuel pressure drops below specified limits.

8764. A transmitter in a fuel pressure warning system serves what function?

A— Transmits an electrical signal to fluid pressure.
B— Converts fluid pressure to an electrical signal.
C— Transmits fluid pressure directly to the indicator.

8765. Where is fuel pressure taken for the pressure warning signal on most aircraft engines?

A— Outlet side of the boost pump.
B— Fuel pressure line of the carburetor.
C— Between the fuel pump and the strainer.

8766. Which of the following is necessary to effectively troubleshoot a fuel pressure warning system?

A— The manufacturer's maintenance manuals.
B— AC 43.13-1A, Acceptable Methods, Techniques, and Practices — Aircraft Inspection and Repair.
C— A set of Federal Aviation Regulations.

8767. Which of the following would be most useful to locate and troubleshoot an internal fuel leak in a large aircraft fuel system?

A— AC 43.13-1A, Acceptable Methods, Techniques, and Practices — Aircraft Inspection and Repair.
B— Federal Aviation Regulations.
C— A fuel system schematic.

8768. In some aircraft with several fuel tanks, the possible danger of allowing the fuel supply in one tank to become exhausted before the selector valve is switched to another tank is prevented by the installation of

A— a fuel pressure warning signal system.
B— a fuel pressure relief valve.
C— an engine fuel pump bypass valve.

8769. (1) The function of a fuel heater is to protect the engine fuel system from ice formation.

(2) An aircraft fuel heater cannot be used to thaw ice in the fuel screen.

Regarding the above statements,

A— only No. 1 is true.
B— only No. 2 is true.
C— both No. 1 and No. 2 are true.

8770. (1) Gas-turbine-engine fuel systems are very susceptible to the formation of ice in the fuel filters.

(2) A fuel heater operates as a heat exchanger to warm the fuel.

Regarding the above statements,

A— only No. 1 is true.
B— only No. 2 is true.
C— both No. 1 and No. 2 are true.

8771. (1) A fuel heater can use engine bleed air as a source of heat.

(2) A fuel heater can use engine lubricating oil as a source of heat.

Regarding the above statements,

A— only No. 1 is true.
B— both No. 1 and No. 2 are true.
C— neither No. 1 nor No. 2 is true.

8772. (1) A fuel pressure gauge is a differential pressure indicator.

(2) A fuel pressure gauge indicates the pressure of the fuel entering the carburetor.

Regarding the above statements,

A— only No. 2 is true.
B— both No. 1 and No. 2 are true.
C— neither No. 1 nor No. 2 is true.

8773. (1) A fuel pressure relief valve is required on an aircraft positive-displacement fuel pump.

(2) A fuel pressure relief valve is required on an aircraft centrifugal fuel boost pump.

Regarding the above statements,

A— only No. 1 is true.
B— only No. 2 is true.
C— both No. 1 and No. 2 are true.

8774. The primary purpose of a fuel tank sump is to provide a

A— positive system of maintaining the design minimum fuel supply for safe operation.
B— place where water and dirt accumulations in the tank can collect and be drained.
C— reserve supply of fuel to enable the aircraft to land safely in the event of fuel exhaustion.

8775. Why are integral fuel tanks used in many large aircraft?

A— To reduce fire hazards.
B— To facilitate servicing.
C— To reduce weight.

8776. If an aircraft is fueled from a truck or storage tank which is known to be uncontaminated with dirt or water, periodic checks of the aircraft's fuel tank sumps and system strainers

A— can be eliminated except for the strainer check before the first flight of the day and the fuel tank sump check during 100-hour or annual inspections.
B— are still necessary due to the possibility of contamination from other sources.
C— can be sharply reduced since contamination from other sources is relatively unlikely and of little consequence in modern aircraft fuel systems.

8777. Aircraft defueling should be accomplished

A— with the aircraft's communication equipment on and in contact with the tower in case of fire.
B— in a hangar where activities can be controlled.
C— in the open air for good ventilation.

8778. Integral fuel tanks are

A— usually constructed of nonmetallic material.
B— readily removed from the aircraft.
C— formed by the aircraft structure.

8779. What precautions must be observed if a gravity-feed fuel system is permitted to supply fuel to an engine from more than one tank at a time?

A— The tank airspaces must be interconnected.
B— The fuel outlet ports of each tank must have the same cross-sectional area.
C— Each tank must have a valve in its outlet that automatically shuts off the line when the tank is empty.

8780. The purpose of the baffle plate in a fuel tank is to

A— provide an expansion space for the fuel.
B— resist fuel surging within the fuel tank.
C— provide internal structural integrity.

8781. What minimum required markings must be placed on or near each appropriate fuel filler cover on transport category aircraft?

A— The word "fuel," the minimum fuel grade or designation for the engines, and the total fuel tank capacity.
B— The word "fuel," the minimum fuel grade or designation for the engines, and the usable fuel tank capacity.
C— The word "fuel," the minimum fuel grade or permissible fuel designations, and the maximum permissible fueling and defueling pressure for pressure fueling systems.

8782. What is one disadvantage of using aromatic aviation fuels?

A— A fuel intercooler is required.
B— Deteriorates rubber parts.
C— Results in low fuel volatility.

8783. Fuel-boost pumps are operated

A— to provide a positive flow of fuel to the engine.
B— primarily for fuel transfer.
C— automatically from fuel pressure.

8784. Flapper valves are used in fuel tanks to

A— reduce pressure.
B— prevent a negative pressure.
C— act as check valves.

8785. Why are centrifugal-type boost pumps used in fuel systems of aircraft operating at high altitude?

A— Because they are positive displacement pumps.
B— To supply fuel under pressure to engine-driven pumps.
C— To permit cooling air to circulate around the motor.

8786. Why is it necessary to vent all aircraft fuel tanks?

A— To ensure a positive head pressure for a submerged boost pump.
B— To exhaust fuel vapors.
C— To limit pressure differential between the tank and atmosphere.

8787.
According to FAR Part 23, what minimum required markings must be placed at or near each appropriate fuel filler cover for reciprocating engine-powered airplanes?

A— The word "Avgas" and the minimum fuel grade.
B— The word "Fuel" and usable fuel capacity.
C— The word "Avgas" and the total fuel capacity.

8788. The location of leaks and defects within the internal portions of the fuel system can usually be determined by

A— visual inspection for evidence of wet spots and stains, and feeling for unusually warm components.
B— performing a fuel flow check.
C— observing the pressure gauge and operating the selector valves.

8789. What type of fuel-booster pump requires a pressure relief valve?

A— Concentric.
B— Sliding vane.
C— Centrifugal.

8790. To prevent vapor lock in fuel lines at high altitude, some aircraft are equipped with

A— vapor separators.
B— direct-injection-type carburetors.
C— booster pumps.

8791. A fuel temperature indicator is located in the fuel tanks on some turbine-powered airplanes to tell when the fuel may be

A— getting cold enough to form hard ice.
B— in danger of forming ice crystals.
C— about to form rime ice.

8792. When inspecting a fuel system, you should check all valves located downstream of boost pumps with the pumps

A— at idle.
B— dormant.
C— operating.

8793. The type of fuel-boost pump that separates air and vapor from the fuel before it enters the line to the carburetor is the

A— gear-type pump.
B— centrifugal-type pump.
C— sliding vane-type pump.

8794. (1) Most large aircraft pressure refueling systems consist of a pressure refueling hose receptacle and a panel of controls and gauges that permit one person to fuel or defuel any or all fuel tanks of an aircraft.

(2) Because of the fuel tank area, there are more advantages to a pressure fueling system in light aircraft.

Regarding the above statements,

A— only No. 1 is true.
B— only No. 2 is true.
C— both No. 1 and No. 2 are true.

8795. Which of the following could result if aviation gasoline does not vaporize readily enough?

1. Hard starting.
2. Poor acceleration.
3. Vapor locking.
4. Excessive gum deposits.
5. High cylinder temperatures.
6. Uneven fuel distribution.
7. Excessive crankcase dilution.

A— 1, 2, 6, 7.
B— 2, 3, 4, 5.
C— 3, 4, 5, 6.

8796. (1) If aviation gasoline vaporizes too readily, fuel lines may become filled with vapor and cause increased fuel flow.

(2) A measure of a gasoline's tendency to vapor lock is obtained from the Reid vapor pressure test.

Regarding the above statements,

A— only No. 2 is true.
B— both No. 1 and No. 2 are true.
C— neither No. 1 nor No. 2 is true.

8797. Microbial growth is produced by various forms of micro-organisms that live and multiply in the water interfaces of jet fuels. Which of the following could result if microbial growth exists in a jet fuel tank and is not corrected?

1. Interference with fuel flow.
2. Interference with fuel quantity indicators.
3. Engine seizure.
4. Electrolytic corrosive action in a metal tank.
5. Lower grade rating of the fuel.
6. Electrolytic corrosive action in a rubber tank.

A— 1, 2, 4.
B— 2, 3, 5.
C— 1, 5, 6.

8798. The vapor pressure of aviation gasoline is

A— lower than the vapor pressure of automotive gasoline.
B— higher than the vapor pressure of automotive gasoline.
C— approximately 20 PSI at 100 °F.

8799. What can be done to eliminate or minimize the microbial growth problem in an aircraft jet fuel tank?

A— Use anti-icing and antibacterial additives.
B— Add CO_2 as a purgative.
C— Keep the fuel tank topped off.

8800. What is the maximum vapor pressure allowable for an aircraft fuel?

A— 7 PSI.
B— 5 PSI.
C— 3 PSI.

8801. If a bladder-type fuel tank is to be left empty for an extended period of time, the inside of the tank should be coated with a film of

A— engine oil.
B— linseed oil.
C— ethylene glycol.

8802. How may the antiknock characteristics of a fuel be improved?

A— By adding a knock inhibitor.
B— By adding a knock enhancer.
C— By adding a fungicide agent.

8803. Some electric motors have two sets of field windings wound in opposite directions so that the

A— speed of the motor can be more closely controlled.
B— power output of the motor can be more closely controlled.
C— motor can be operated in either direction.

8804. One purpose of a growler test is to determine the presence of

A— an out-of-round commutator.
B— a broken field lead.
C— a shorted armature.

8805. Electric wire terminals for most aircraft applications must be what type?

A— Slotted.
B— Hook.
C— Ring.

8806. What is the principal advantage of the series-wound dc motor?

A— High starting torque.
B— Suitable for constant speed use.
C— Low starting torque.

8807. If a generator is equipped with a vibrator-type voltage regulator, the actual time the voltage regulator points remain open

A— depends on the load carried by the generator.
B— is controlled by the reverse-current cutout relay point clearance.
C— is increased when the external load is greater than the generator output.

8808. What is a cause of generator brush arcing?

A— Seating brushes with No. 000 sandpaper.
B— Carbon dust particles.
C— Low spring tension.

8809. When ac generators are operated in parallel, the

A— amperes and frequency must both be equal.
B— frequency and voltage must both be equal.
C— amperes and voltage must both be equal.

8810. The starting current of a series-wound dc motor, in passing through both the field and armature windings, produces a

A— low starting torque.
B— speed slightly higher when unloaded.
C— high starting torque.

8811. Which motor would be most likely to have an armature brake?

A— Starter motor.
B— Landing light retraction motor.
C— Inverter drive motor.

8812. The method most often used in overcoming the effect of armature reaction is through the use of

A— interpoles.
B— shaded poles.
C— drum-wound armatures in combination with a negatively connected series field.

8813. The only practical method of maintaining a constant voltage output from an aircraft generator under varying conditions of speed and load is to vary the

A— strength of the magnetic field.
B— number of conductors in the armature.
C— speed at which the armature rotates.

8814. The pole pieces or shoes used in a dc generator are a part of the

A— armature assembly.
B— field assembly.
C— brush assembly.

8815. How many cycles of ac voltage are produced in a six-pole alternator of the revolving-field type for each revolution of the rotor?

A— Four.
B— Three.
C— Six.

8816. If the reverse current cutout relay contact points fail to open after the generator output has dropped below battery potential, current will flow through the generator armature

A— in the normal direction and through the shunt field opposite the normal direction.
B— and the shunt field opposite the normal direction.
C— opposite the normal direction and through the shunt field in the normal direction.

8817. How does the magnetic brake used to stop rotation of an electric motor armature operate?

A— Centrifugal force releases a rotating brake cog from a stationary notch when the armature reaches a certain speed and magnetic force re-engages the cog when the electrical power is turned off.
B— A friction brake is applied by a magnet and released by a spring.
C— A friction brake is applied by a spring and released by a magnet.

8818. In a generator, what eliminates any possible sparking to the brush guides caused by the movement of the brushes within the holder?

A— The brush pigtail.
B— Brush spring tension.
C— Undercutting the mica on the commutator.

8819. A series-wound dc electric motor will normally require

A— more current at high RPM than at low RPM.
B— approximately the same current throughout its operating range of speed.
C— more current at low RPM than at high RPM.

8820. The type of electric wire terminals used for most aircraft applications, in addition to providing good current carrying capabilities, are designed primarily

A— to prevent circuit failure due to terminal disconnection.
B— for uncomplicated and rapid circuit connection and disconnection.
C— for permanent connection to the circuit.

8821. Aluminum wire must be stripped very carefully because

A— high resistance will develop in stripping nicks.
B— stripping nicks can cause short circuits.
C— individual strands will break easily after being nicked.

8822. The commutator of a generator

A— changes direct current produced in the armature into alternating current as it is taken from the armature.
B— changes alternating current produced in the armature into direct current as it is taken from the armature.
C— reverses the current in the field coils at the proper time in order to produce direct current.

8823. An ammeter in a battery charging system is for the purpose of indicating the

A— amperage available for use.
B— total amperes being used in the airplane.
C— rate of current used to charge the battery.

8824. Which of the following is not one of the purposes of interpoles in a generator?

A— Reduce field strength.
B— Overcome armature reaction.
C— Reduce arcing at the brushes.

8825. To test generator or motor armature windings for opens,

A— place armature in a growler and connect a 110V test light on adjacent segments; light should light.
B— check adjacent segments on commutator with an ohmmeter on the high resistance scale.
C— use a 12/24V test light between the armature core segments and the shaft.

8826. Where are position lights for navigation located on all civil airplanes?

A— Aft (L & R) and forward.
B— Forward (L & R) and aft.
C— Upper, lower, forward, and aft.

8827. To what depth is the mica insulation between the commutator bars of a dc generator undercut?

A— One-half the width of the mica.
B— Equal to twice the width of the mica.
C— Equal to the width of the mica.

8828. A voltage regulator controls generator output by

A— introducing a resistance in generator-to-battery lead in the event of overload.
B— shorting out field coil in the event of overload.
C— varying current flow to generator field coil.

8829. Which type of dc generator is not used as an airplane generator?

A— Externally grounded.
B— Series wound.
C— Compound wound.

8830. What is the most accurate type of frequency-measuring instrument?

A— Integrated circuit chip having a clock circuit.
B— Electrodynamometers using electromagnetic fields.
C— Electromagnets using one permanent magnet.

8831. During ground operation, aircraft generator cooling is usually accomplished by

A— auxiliary air cooled through an air/fuel heat exchanger.
B— an integral fan.
C— an external motor-driven fan.

8832. What does a rectifier do?

A— Changes direct current into alternating current.
B— Changes alternating current into direct current.
C— Reduces voltage.

8833. What type of instrument is used for measuring very high values of resistance?

A— Megohmmeter.
B— Shunt-type ohmmeter.
C— Multimeter.

8834. When a diode is checked for an open circuit or a short circuit, it should be

A— in the circuit.
B— checked with a milliamp ammeter.
C— disconnected from the circuit.

8835. When handling a high voltage capacitor in an electrical circuit, be sure it

A— has a full charge before removing it from the circuit.
B— has at least a residual charge before removing it from the circuit.
C— is fully discharged before removing it from the circuit.

8836. Which of the following is most likely to cause thermal runaway in a nickel-cadmium battery?

A— A high internal resistance condition.
B— Excessive current draw from the battery.
C— Constant current charging of the battery to more than 100 percent of its capacity.

8837. How can it be determined if a transformer winding has some of its turns shorted together?

A— Measure the input voltage with an ohmmeter.
B— The output voltage will be high.
C— The transformer will get hot in normal operation.

8838. Which of the following are the major parts of a dc motor?

1. Armature assembly.
2. Field assembly.
3. Brush assembly.
4. Commutator.
5. Pole piece.
6. Rheostat.
7. End frame.

A— 1, 2, 3, 7.
B— 2, 3, 4, 5.
C— 3, 5, 6, 7.

8839. (1) There are three basic types of dc motors; series, shunt, and compound.

(2) In the series motor, the field windings, consisting of relatively few turns of heavy wire, are connected in series with the armature winding.

Regarding the above statements,

A— only No. 1 is true.
B— only No. 2 is true.
C— both No. 1 and No. 2 are true.

8840. For general electrical use in aircraft, the acceptable method of attaching a terminal to a wire is by

A— crimping.
B— soldering.
C— crimping and soldering.

8841. Which of the following factors must be taken into consideration when determining the wire size to use for an aircraft installation?

1. Mechanical strength.
2. Allowable power loss.
3. Ease of installation.
4. Resistance of current return path through the aircraft structure.
5. Permissible voltage drop.
6. Current carrying capability of the conductor.
7. Type of load (continuous or intermittent).

A— 2, 5, 6, 7.
B— 1, 2, 4, 5.
C— 2, 4, 6, 7.

8842. When selecting hardware for attaching bonding connections to an aircraft structure, which of the following should be considered?

1. Mechanical strength.
2. Allowable power loss.
3. Ease of installation.
4. Permissible voltage drop.
5. Amount of current to be carried.
6. Type of load (continuous or intermittent).

A— 1, 3, 5.
B— 4, 5, 6.
C— 1, 2, 3.

8843. How should the splices be arranged if several are to be located in an electrical wire bundle?

A— Staggered along the length of the bundle.
B— Grouped together to facilitate inspection.
C— Enclosed in a conduit.

8844. What is the minimum bend radii for an electrical wire bundle?

A— Ten times the outside diameter of the bundle.
B— Five times the outside diameter of the bundle.
C— Fifteen times the outside diameter of the bundle.

8845. When approved, splices may be used to repair manufactured harnesses or installed wiring. The maximum number of splices permitted between any two connectors is

A— one.
B— two.
C— three.

8846. AN/MS electrical connectors are specifically designed to meet

A— Technical Standard Order (TSO) specifications.
B— military specifications.
C— International Civil Aviation Organization (ICAO) standards.

8847. A newer method of attaching a pin or socket to an individual wire in an MS electrical connector is by

A— crimping.
B— soldering.
C— crimping and soldering.

8848. The pin section of an AN/MS connector is normally installed on

A— the power supply side of a circuit.
B— the ground side of a circuit.
C— either side of a circuit (makes no difference).

8849. The voltage output of an alternator may be regulated by controlling the

A— speed of the alternator.
B— voltage output of the dc exciter.
C— resistance in the rotor windings.

8850. If several long lengths of electrical cable are to be installed in rigid conduit, the possibility of damage to the cable as it is pulled through the conduit will be reduced by

A— dusting the cable with powdered graphite.
B— dusting the cable with powdered soapstone.
C— applying a light coat of dielectric grease.

8851. Grounding is electrically connecting a conductive object to the primary structure. One purpose of grounding is to

A— prevent current return paths.
B— allow static charge accumulation.
C— prevent development of radio frequency potentials.

8852. What is normally used to bond noncontinuous stainless steel aircraft components?

A— Stainless steel jumpers.
B— Copper jumpers.
C— Aluminum jumpers.

8853. Aircraft fuse capacity is rated in

A— volts.
B— ohms.
C— amperes.

8854. When adding a rheostat to a light circuit to control the light intensity, it should be connected in

A— parallel with the light.
B— series with the light.
C— series parallel with the light switch.

8855. Circuits that must be operated only in an emergency or whose inadvertent activation could endanger a system frequently employ

A— guarded switches.
B— push-pull-type circuit breakers only (no switches).
C— spring-loaded to off toggle or rocker switches.

8856. If one switch is used to control all navigation lights, the lights are most likely connected

A— in series with each other and parallel to the switch.
B— in series with each other and in series with the switch.
C— parallel to each other and in series with the switch.

8857. Oil canning of the sides of aluminum or steel electrical junction boxes is considered to be

A— normal operation in vibration prone areas.
B— a shorting hazard.
C— acceptable operation.

8858. Electric wiring installed in aircraft without special enclosing means (open wiring) offers the advantages of ease of installation, simple maintenance, and reduced weight. When bundling open wiring, the bundles should

A— be limited as to the number of cables to minimize damage from a single electrical fault.
B— include at least one shielded cable to provide good bonding of the bundle to the airframe.
C— be limited to a minimum bend radius of five times the bundle diameter to avoid excessive stresses on the cable insulation.

8859. During inspection of the terminal strips of an aircraft electrical system, it should be determined that

A— only locknuts have been used for terminal attachment to the studs.
B— the terminal studs are anchored against rotation.
C— only plain nuts and lockwashers have been used for terminal attachment to the studs.

8860. What protection to wires and cables does conduit provide when used in aircraft installations?

A— Electromagnetic.
B— Mechanical.
C— Structural.

8861. Which of the following should be avoided in conduit installation?

A— Support the conduit to prevent chafing against the structure.
B— Provide drainholes at the lowest point in a conduit run.
C— Locate conduit to provide a footstep or handhold.

8862. If the (+) terminal of a voltmeter is connected to the (–) terminal of the source voltage and the (–) terminal of the meter is connected to the (+) terminal of the source voltage, the voltmeter will read

A— correctly.
B— low voltage.
C— backwards.

8863. When using the voltage drop method of checking circuit resistance, the

A— input voltage must be maintained at a constant value.
B— output voltage must be maintained at a constant value.
C— input voltage must be varied.

8864. The nominal rating of electrical switches refers to continuous

A— current rating with the contacts open.
B— voltage rating with the contacts closed.
C— current rating with the contacts closed.

8865. Aircraft electrical junction boxes located in a fire zone are usually constructed of

A— asbestos.
B— cadmium-plated steel.
C— stainless steel.

8866. Which conduit is normally used where necessary to minimize radio interference?

A— Flexible brass.
B— Flexible aluminum.
C— Rigid steel.

8867. The primary considerations when selecting electric cable size are

A— current-carrying capacity and allowable voltage drop.
B— the voltage and amperage of the load it must carry.
C— the system voltage and cable length.

8868. The navigation lights of some aircraft consist of a single circuit controlled by a single switch which has an ON position and an OFF position, with no additional positions possible. This switch is referred to as a

A— double-pole, single-throw (DPST), two-position switch.
B— single-pole, double-throw (SPDT), two-position switch.
C— single-pole, single-throw (SPST), two-position switch.

8869. Electric circuits are protected from overheating by means of

A— thermocouples.
B— shunts.
C— fuses.

8870. How does the routing of coaxial cables differ from the routing of electrical wiring?

A— Coaxial cables are routed parallel with stringers or ribs.
B— Coaxial cables are routed at right angles to stringers or ribs.
C— Coaxial cables are routed as directly as possible.

8871. Which of the following aluminum electrical cable sizes should be selected to replace a No. 10 copper electrical cable?

A— No. 4.
B— No. 6.
C— No. 8.

8872. In installations where the ammeter is in the generator or alternator lead, and the regulator system does not limit the maximum current that the generator or alternator can deliver, the ammeter can be redlined at what percent of the generator or alternator rating?

A— 50.
B— 75.
C— 100.

8873. Which statement relating to electric wiring is true?

A— When attaching a terminal to the end of an electric cable, it should be determined that the strength of the cable-to-terminal joint is at least twice the tensile strength of the cable.
B— When attaching a terminal to the end of an electric cable, it should be determined that the strength of the cable-to-terminal joint is at least equal to the tensile strength of the cable itself.
C— All electric cable splices should be covered with soft insulating tubing (spaghetti) for mechanical protection against external abrasion.

8874. Bonding connections should be tested for

A— resistance value.
B— amperage value.
C— reactance.

8875. What kind of switch should you install in a single wire circuit that required the switch to be manually held in the ON position?

A— Single-pole, single-throw (SPST), two-position normally open (NO).
B— Single-pole, single-throw (SPST), single-position.
C— Single-pole, double-throw (SPDT), single-position normally open (NO).

8876. A circuit breaker is installed in an aircraft electrical system primarily to protect the

A— circuit and should be located as close to the source as possible.
B— circuit and should be located as close to the unit as possible.
C— electrical unit in the circuit and should be located as close to the source as possible.

8877. How should a voltmeter be connected?

A— In series with the source.
B— In parallel with the load.
C— In series with the load.

8878. If aluminum electrical cable is to be substituted for No. 12 copper electrical cable, which gauge should be selected?

A— No. 8.
B— No. 6.
C— No. 10.

8879. If it is necessary to use an electrical connector where it may be exposed to moisture, the mechanic should

A— coat the connector with grease.
B— use a special moisture-proof type.
C— spray the connector with varnish or zinc-chromate.

8880. The two kinds of fuse holders used most commonly in aircraft circuits are

A— clip-on, screw-in types.
B— plug-in, clip types.
C— plug-in, mechanical reset types.

8881. If a wire is installed so that it comes in contact with some moving parts, what protection should be given the wire?

A— Wrap with soft wire solder into a shield.
B— Wrap with friction tape.
C— Pass through conduit.

8882. In the American Wire Gauge (AWG) system of numbers used to designate electrical wire sizes, the number assigned to a size is related to its

A— combined resistance and current-carrying capacity.
B— current-carrying capacity.
C— cross-sectional area.

8883. What is the allowable voltage drop for a No. 18 copper wire 50 feet long to carry 12.5 amperes, continuous operation?

Use the formula VD = RLA

VD = Voltage drop
R = Resistance per ft = .00644
L = Length of wire
A = Amperes

A— 1/2V.
B— 1V.
C— 4V.

8884. What is the purpose of the selection of derated switches for known continuous load current applications?

A— To calculate the voltage drop across the circuit.
B— To prevent short circuits in the motor field windings.
C— To obtain reasonable switch efficiency and service life.

8885. What is the advantage of a circuit breaker when compared to a fuse?

A— Never needs replacing.
B— Always eliminates the need of a switch.
C— Resettable and reusable.

8886. What is the advantage of a current limiter?

A— It breaks circuit quickly.
B— It can be reset easily.
C— It will take overload for a short period.

8887. Where electric cables must pass through holes in bulkheads, formers, ribs, firewalls, etc., the wires should be protected from chafing by

A— wrapping with electrical tape.
B— using a rubber grommet.
C— wrapping with plastic.

8888. In aircraft electrical systems, automatic reset circuit breakers

A— should not be used as circuit protective devices.
B— are useful where only temporary overloads are normally encountered.
C— must be used in all circuits essential to safe operation of the aircraft.

8889. A certain switch is described as a single-pole, double-throw switch (SPDT). The throw of a switch indicates the number of

A— circuits each pole can complete through the switch.
B— terminals at which current can enter or leave the switch.
C— places at which the operating device (toggle, plunger, etc.) will come to rest and at the same time open or close a circuit.

8890. When considering an alteration, the criteria upon which the selection of electric cable size should be based are

A— applied voltage and allowable voltage drop.
B— current-carrying capacity and allowable voltage drop.
C— current-carrying capacity and applied voltage.

8891. What is an important factor in selecting aircraft fuses?

A— The current exceeds a predetermined value.
B— The voltage rating should be lower than the maximum circuit voltage.
C— Capacity matches the needs of the circuit.

8892. The circuit breaker in the instrument lighting system protects the

A— lights from too much current.
B— wiring from too much current.
C— wiring from too much voltage.

8893. One advantage of using ac electrical power in aircraft is

A— that ac electrical motors can be reversed while dc motors cannot.
B— greater ease in stepping the voltage up or down.
C— that the effective voltage is 1.41 times the maximum instantaneous voltage; therefore, less power input is required.

8894. Why are the iron cores of most induction coils laminated?

A— To reduce the core reluctance.
B— To increase the core permeability.
C— To reduce the effects of eddy currents.

8895. Certain transport aircraft use ac electrical power for all normal operation and battery furnished dc electrical power for standby emergency use. In aircraft of this type that operate no dc generators, the batteries are kept charged by

A— inverters which use the aircraft's ac generators as a source of power.
B— alternators which use the aircraft's generators as a source of power.
C— rectifiers which use the aircraft's ac generators as a source of power.

8896. The voltage in an ac transformer secondary that contains twice as many loops as the primary will be

A— greater and the amperage less than in the primary.
B— greater and the amperage greater than in the primary.
C— less and the amperage greater than in the primary.

8897. If the positive field lead between a generator and a generator control panel breaks and is shorted while the engine is running, a voltmeter connected to generator output would indicate

A— zero voltage.
B— residual voltage.
C— normal voltage.

8898. What is a method used for restoring generator field residual magnetism?

A— Flash the fields.
B— Reseat the brushes.
C— Energize the armature.

8899. One of the chief advantages of alternating current is that it can be transmitted at a high voltage with a low power loss; the voltage can then be changed to any desired value of

A— dc by means of inverters.
B— dc by means of transformers.
C— ac by means of transformers.

8900. Which of the following must be accomplished when installing an anticollision light?

A— Install a switch independent of the position light switch.
B— Use shielded electrical cable to assure fail-safe operation.
C— Connect the anticollision light to the aircraft position light switch.

8901. The inductor-type inverter output voltage is controlled by the

A— number of poles and the speed of the motor.
B— voltage regulator.
C— dc stator field current.

8902. When using an ohmmeter to check the continuity of a generator field coil, the coil should

A— be removed from the generator housing.
B— show high resistance when the meter prods are connected to the terminals of the coil.
C— show very low resistance if it is a series field coil.

8903. The strength of the core of an electromagnet depends upon the material from which it is constructed and which of the following?

A— The number of turns of wire in the coil and the applied voltage.
B— The number of turns of wire in the coil and the amount of current (amperes) passing through the coil.
C— The size (cross section) and the number of turns of wire in the coil and the applied voltage.

8904. A voltage regulator controls generator voltage by changing the

A— resistance in the generator output circuit.
B— current in the generator output circuit.
C— resistance of the generator field circuit.

8905. The overvoltage control automatically protects the generator system when excessive voltage is present by

A— opening the shunt field circuit.
B— opening and resetting the field control relay.
C— breaking a circuit to the trip coil of the field control relay.

8906. When dc generators are operated in parallel to supply power for a single load, their controls include an equalizer circuit to assure that all generators share the load equally. The equalizer circuit operates by

A— increasing the output of the low generator to equal the output of the high generator.
B— decreasing the output of the high generator to equal the output of the low generator.
C— increasing the output of the low generator and decreasing the output of the high generator until they are equal.

8907. Which of the following is considered to be an intermittent duty circuit?

A— Anticollision light circuit.
B— Landing light circuit.
C— Instrument panel light circuit.

8908. The most common method of regulating the voltage output of a compound dc generator is to vary the

A— current flowing through the shunt field coils.
B— total effective field strength by changing the reluctance of the magnetic circuit.
C— resistance of the series field circuit.

8909. (Refer to figure 18.) Which of the batteries are connected together incorrectly?

A— 1.
B— 2.
C— 3.

8910. (Refer to figure 19.) Upon completion of the landing gear extension cycle, the green light illuminated and the red light remained lit. What is the probable cause?

A— Short in the down limit switch.
B— Short in the gear safety switch.
C— Short in the up limit switch.

8911. If any one generator in a 24-volt dc system shows low voltage, the most likely cause is

A— an out-of-adjustment voltage regulator.
B— shorted or grounded wiring.
C— a defective reverse current cutout relay.

8912. How can the direction of rotation of a dc electric motor be changed?

A— Interchange the wires which connect the motor to the external power source.
B— Reverse the electrical connections to either the field or armature windings.
C— Rotate the positive brush one commutator segment.

8913. Aircraft which operate only ac generators (alternators) as a primary source of electrical power normally provide current suitable for battery charging through the use of

A— a stepdown transformer and a rectifier.
B— an inverter and a voltage-dropping resistor.
C— a dynamotor with a half-wave dc output.

8914. During inspection of an anticollision light installation for condition and proper operation, it should be determined that

A— electrical or mechanical interconnections are provided so that the anticollision light will operate at all times that the position light switch is in the ON position.
B— an appropriately rated fuse is in position at the light to protect the connecting wiring against electrical faults.
C— the anticollision light can be operated independently of the position lights.

8915. Major adjustments on equipment such as regulators, contactors, and inverters are best accomplished outside the airplane on test benches with necessary instruments and equipment. Adjustment procedure should be as outlined by

A— the equipment manufacturer.
B— the FAA.
C— aircraft technical orders.

8916. A battery-generator system provides direct current. On installations requiring alternating current from the battery-generator system, it is necessary to have

A— a transformer.
B— an inverter.
C— a variable resistor between the battery and generator.

8917. A relay is

A— a magnetically operated switch.
B— a device which converts electrical energy to kinetic energy.
C— any conductor which receives electrical energy and passes it on with little or no resistance.

8918. The purpose of a rectifier in an electrical system is to change

A— the frequency of alternating current.
B— direct current to alternating current.
C— alternating current to direct current.

8919. What is the ratio of turns between the primary coil winding and the secondary coil winding of a transformer designed to triple its input voltage?

A— Primary will have one-third as many turns as its secondary.
B— Primary will have twice as many turns as its secondary.
C— Primary will have three times as many turns as its secondary.

8920. In an ac circuit with no phase lead or lag, which is true?

A— Real power is zero.
B— Real power is greater than apparent power.
C— Real power equals apparent power.

8921. How are generators rated?

A— Watts at rated voltage.
B— Amperes at rated voltage.
C— The impedance at rated voltage.

8922. How is a shunt-wound dc generator connected?

A— One field is shunted across the other.
B— Both fields are shunted across the armature.
C— The field and armature are shunted with a capacitor.

8923. The poles of a generator are laminated to

A— reduce flux losses.
B— increase flux concentration.
C— reduce eddy current losses.

8924. What is the frequency of an alternator dependent upon?

A— Voltage.
B— RPM.
C— Current.

8925. The generator rating is usually found stamped on the

A— firewall.
B— generator.
C— engine.

8926. Residual voltage is a result of magnetism in the

A— field windings.
B— field shoes.
C— armature.

8927. In troubleshooting an electrical circuit, if an ohmmeter is properly connected across a circuit component and some value of resistance is read,

A— the component has continuity and is open.
B— either the component or the circuit is shorted.
C— the component has continuity and is not open.

8928. CSD driven generators are usually cooled by

A— oil spray.
B— an integral fan.
C— both ram air and an integral fan.

8929. A CSD unit drives a generator through the use of

A— a synchronous electric motor.
B— an infinitely variable mechanical gearing system.
C— a variable hydraulic pump and hydraulic motor.

8930. Integrated drive generators (IDG) employ a type of high output ac generator that utilizes

A— brushes and slip rings to carry generated dc exciter current to the rotating field.
B— battery current to excite the field.
C— a brushless system to produce current.

8931. The Built In Test Equipment (BITE) employed in automated ac power systems typically indicates the defective component to the technician via

A— digital multimeter readings.
B— light emitting diode (LED) coded readouts on the control units.
C— plug-in test circuits that are easily installed and removed.

8932. When necessary during operation, CSD disconnect is usually accomplished by

A— a switch in the cockpit.
B— circuit breaker activation.
C— a shear section in the input shaft.

8933. A CSD unit that is disconnected in flight, due to a malfunction such as overtemperature, may be reconnected

A— automatically if the temperature falls back into the normal operating range.
B— manually by the flightcrew.
C— only on the ground.

8934. The purpose of antiskid generators is to

A— monitor hydraulic pressure applied to brakes.
B— indicate when a tire skid occurs.
C— measure wheel rotational speed and any speed changes.

8935. In a brake antiskid system, when an approaching skid is sensed, an electrical signal is sent to the skid control valve which

A— acts as a bypass for the debooster cylinders.
B— relieves the hydraulic pressure on the brake.
C— equalizes the hydraulic pressure in adjacent brakes.

8936. An antiskid system is

A— a hydraulic system.
B— an electrohydraulic system.
C— an electrical system.

8937. Antiskid braking systems are generally armed by

A— a centrifugal switch.
B— a switch in the cockpit.
C— the rotation of the wheels above a certain speed.

8938. A typical takeoff warning indication system, in addition to throttle setting, monitors the position of which of the following?

A— Ailerons, elevators, speed brake, and steerable fuselage landing gear.
B— Elevators, speed brake, flaps, and steerable fuselage landing gear.
C— Aerodynamically actuated slats, elevators, flaps, and speed brake.

8939. The primary purpose of a takeoff warning system is to alert the crew that a flight control is not properly set prior to takeoff. It is actuated by the closing of the

A— landing gear retract switch.
B— ground proximity switch (squat switch).
C— thrust lever switch.

8940. (1) An airspeed indicator measures the differential between pitot and static air pressures surrounding the aircraft at any moment of flight.

(2) An airspeed indicator measures the differential between pitot and cabin air pressures at any moment of flight.

Regarding the above statements,

A— both No. 1 and No. 2 are true.
B— only No. 2 is true.
C— only No. 1 is true.

8941. The angle-of-attack detector operates from differential pressure when the airstream

A— is parallel to the longitudinal axis of the aircraft.
B— is not parallel to the true angle of attack of the aircraft.
C— is parallel to the angle of attack of the aircraft.

8942. (1) When an airplane is slowed below approximately 20 MPH, the antiskid system automatically deactivates to give the pilot full control of the brakes for maneuvering and parking.

(2) An antiskid system consists basically of three components; wheel speed sensors, control box, and control valves.

Regarding the above statements,

A— only No. 1 is true.
B— only No. 2 is true.
C— both No. 1 and No. 2 are true.

8943. In an antiskid system, wheel skid is detected by

A— an electrical sensor.
B— a discriminator.
C— a sudden rise in brake pressure.

8944. Which of the following functions does a skid control system perform?

1. Normal skid control.
2. Normal braking.
3. Fail safe protection.
4. Locked wheel skid control.
5. Touchdown protection.
6. Takeoff protection.

A— 1, 2, 3, 4.
B— 1, 3, 4, 5.
C— 1, 2, 5, 6.

8945. What effect, if any, will a wheel going into a skid have on the aircraft's braking value?

A— Greatly reduced.
B— Slightly increased.
C— No effect.

8946. At what point in the landing operation does normal skid control perform its function?

A— When wheel rotation has slowed but not come to a stop.
B— When wheel rotation has come to a stop.
C— Anytime the wheel is rotating.

8947. (1) An antiskid system is designed to apply enough force to operate just below the skid point.

(2) A warning lamp lights in the cockpit when the antiskid system is turned off or if there is a system failure.

Regarding the above statements,

A— only No. 1 is true.
B— only No. 2 is true.
C— both No. 1 and No. 2 are true.

8948. When an airplane's primary flight control surfaces are set for a particular phase of flight, such as landing or takeoff, the corresponding control-surface indicating system will show

A— flap/slat position.
B— speed break position.
C— trim position.

8949. The pneumatic (reed) type stall warning system installed in some light aircraft is activated by

A— static air pressure.
B— positive air pressure.
C— negative air pressure.

8950. Stall warning systems are generally designed to begin warning the pilot when a stall

A— is imminent.
B— is starting to occur.
C— first affects the outboard portions of the wings.

8951.
(Refer to figure 19) What is the indication of the red landing gear position light under the following conditions?

Aircraft on jacks.
Landing gear in transit.
Warning horn sounding.

A— extinguished.
B— flashing.
C— illuminated.

8952. (Refer to figure 19). Which repair should be made if the gear switch was placed in UP position and the gear does not retract?

A— Replace electrical wire No. 15.
B— Replace the down limit switch.
C— Replace electrical wire No. 12.

8953. Which of the following conditions is most likely to cause the landing gear warning signal to sound?

A— Landing gear locked down and throttle advanced.
B— Landing gear locked down and throttle retarded.
C— Landing gear not locked down and throttle retarded.

8954. (Refer to figure 20.) What will illuminate the amber indicator light?

A— Closing the nosewheel gear full retract switch.
B— Retarding one throttle and closing the left wheel gear locked down switch.
C— Closing the nose, left and right wheel gear full retract switches.

8955. (Refer to figure 20.) What is the minimum circumstance that will cause the landing gear warning horn to indicate an unsafe condition?

A— All gears up and one throttle retarded.
B— Any gear up and both throttles retarded.
C— Any gear not down and locked, and one throttle retarded.

8956. Where is the landing gear safety switch usually located?

A— On the main gear shock strut.
B— On the landing gear drag brace.
C— On the pilot's control pedestal.

8957. What safety device is actuated by the compression and extension of a landing gear strut?

A— Uplock switch.
B— Downlock switch.
C— Ground safety switch.

8958. Which repair would require a landing gear retraction test?

A— Landing gear safety switch.
B— Red warning light bulb.
C— Gear downlock microswitch.

8959. Landing gear warning systems usually provide which of the following indications?

A— Red light for unsafe gear, no light for gear down, green light for gear up.
B— Green light for gear up and down, red light for unsafe gear.
C— Red light for unsafe gear, green light for gear down, no light for gear up.

8960. In most modern hydraulically actuated landing gear systems, the order of gear and fairing door operation is controlled by

A— sequence valves.
B— shuttle valves.
C— microswitches.

8961. What landing gear warning device(s) is/are incorporated on retractable landing gear aircraft?

A— A visual indicator showing gear position.
B— A light which comes on when the gear is fully down and locked.
C— A horn or other aural device and a red warning light.

8962. When a landing gear safety switch on a main gear strut closes at liftoff, which system is deactivated?

A— Landing gear position system.
B— Antiskid system.
C— Aural warning system.

8963. The rotor in an autosyn remote indicating system uses

A— an electromagnet.
B— a permanent magnet.
C— neither an electromagnet nor a permanent magnet.

8964. The basic difference between an autosyn and a magnesyn indicating system is the

A— rotor.
B— transmitter.
C— receiver.

8965. The rotor in a magnesyn remote indicating system uses

A— a permanent magnet.
B— an electromagnet.
C— an electromagnet and a permanent magnet.

8966. Microswitches are used primarily as limit switches to

A— limit generator output.
B— control electrical units automatically.
C— prevent overcharging of a battery.

8967. Which of the following are some uses for a dc selsyn system?

1. Indicates position of retractable landing gear.
2. Indicates the angle of incidence of an aircraft.
3. Indicates the altitude of an aircraft.
4. Indicates cowl flaps or oil cooler door position.
5. Indicates fuel quantity.
6. Indicates the rate of climb of an aircraft.
7. Indicates position of wing flaps.

A— 1, 4, 5, 7.
B— 2, 3, 4, 5.
C— 2, 3, 5, 6.

8968. (1) A dc selsyn system is a widely used electrical method of indicating a remote mechanical movement or position.

(2) A synchro-type indicating system is an electrical system used for transmitting information from one point to another.

Regarding the above statements,

A— only No. 1 is true.
B— only No. 2 is true.
C— both No. 1 and No. 2 are true.

8969. When installing pneumatic surface-bonded type deicer boots,

A— remove all paint from the area to be covered by the deicer boot.
B— apply a solution of glycerin and water between the rubber and the wing skin.
C— apply a silastic compound between the rubber and the wing skin.

8970. Which of the following are found in a laminated integral electrically heated windshield system?

1. Autotransformer.
2. Heat control relay.
3. Heat control toggle switch.
4. 24V dc power supply.
5. Indicating light.

A— 1, 2, 4, 5.
B— 2, 3, 4, 5.
C— 1, 2, 3, 5.

8971. What is one check for proper operation of a pitot/static tube heater after replacement?

A— Ammeter reading.
B— Voltmeter reading.
C— Continuity check of system.

8972. What controls the inflation sequence in a pneumatic deicer boot system?

A— Shuttle valve.
B— Vacuum pump.
C— Distributor valve.

8973. What is the source of pressure for inflating deicer boots on reciprocating engine aircraft?

A— Vane-type pump.
B— Gear-type pump.
C— Piston-type pump.

8974. Which of the following regulates the vacuum of the air pump to hold the deicing boots deflated when the pneumatic deicing system is off?

A— Distributor valve.
B— Pressure regulator.
C— Suction relief valve.

8975. What may be used to clean deicer boots?

A— Unleaded gasoline or Jet A fuel.
B— Naphtha.
C— Soap and water.

8976. Some aircraft are protected against airframe icing by heating the leading edges of the airfoils and intake ducts. When is this type of anti-ice system usually operated during flight?

A— Continuously while the aircraft is in flight.
B— In symmetric cycles during icing conditions to remove ice as it accumulates.
C— Whenever icing conditions are first encountered or expected to occur.

8977. Which of the following indications occur during a normal operational check of a pneumatic deicer system?

A— Relatively steady readings on the pressure gauge and fluctuating readings on the vacuum gauge.
B— Fluctuating readings on the pressure gauge and relatively steady readings on the vacuum gauge.
C— Pressure and vacuum gauges will fluctuate as the deicer boots inflate and deflate.

8978. What method is usually employed to control the temperature of an anti-icing system using surface combustion heaters?

A— Thermo-cycling switches.
B— Thermostats in the cockpit.
C— Heater fuel shutoff valves.

8979. What is the purpose of the distributor valve in a deicing system utilizing deicer boots?

A— To equalize the air pressure to the left and right wings.
B— To sequence the deicer boot inflations.
C— To distribute anti-icing fluid to the deicer boots.

8980. What is the purpose of the oil separator in the pneumatic deicing system?

A— To protect the deicer boots from oil deterioration.
B— To remove oil from air exhausted from the deicer boots.
C— To prevent an accumulation of oil in the vacuum system.

8981. Where are the heat sensors located on most aircraft with electrically heated windshields?

A— Imbedded in the glass.
B— Attached to the glass.
C— Around the glass.

8982. Three possible sources of hot air for the operation of a wing thermal anti-icing system are

A— engine bleed air, vacuum pump, and compressed air tank.
B— engine bleed air, combustion heaters, and augmenter tubes.
C— combustion heaters, augmenter tubes, and exhaust gases.

8983. What maintains normal windshield temperature control in an electrically heated windshield system?

A— Thermal overheat switches.
B— Thermistors.
C— Electronic amplifiers.

8984. Arcing in an electrically heated windshield panel usually indicates a breakdown in the

A— temperature-sensing elements.
B— autotransformers.
C— conductive coating.

8985. Which of the following directs vacuum to the deicer boots for hold down in flight?

A— Vacuum relief valve.
B— Ejector.
C— Distributor valve.

8986. How do deicer boots help remove ice accumulations?

A— By preventing the formation of ice.
B— By breaking up ice formations.
C— By allowing only a thin layer of ice to build up.

8987. Why are the tubes in deicer boots alternately inflated?

A— Alternate inflation of deicer boot tubes keeps disturbance of the airflow to a minimum.
B— Alternate inflation of deicer boot tubes does not disturb airflow.
C— Alternate inflation of deicer boot tubes relieves the load on the air pump.

8988. Carburetor icing may be eliminated by which of the following methods?

A— Alcohol spray and heated induction air.
B— Ethylene glycol spray and heated induction air.
C— Electrically heating air intake, ethylene glycol spray, or alcohol spray.

8989. Why should a chemical rain repellant not be used on a dry windshield?

A— It will etch the glass.
B— It will restrict visibility.
C— It will cause glass crazing.

8990. What is the principle of a windshield pneumatic rain removal system?

A— An air blast spreads a liquid rain repellant evenly over the windshield that prevents raindrops from clinging to the glass surface.
B— An air blast forms a barrier that prevents raindrops from striking the windshield surface.
C— A pneumatic rain removal system is simply a mechanical windshield wiper system that is powered by pneumatic system pressure.

8991. What mixture may be used as a deicing fluid to remove frost from an aircraft surface?

A— Ethylene glycol and isopropyl alcohol.
B— Methyl ethyl ketone and ethylene glycol.
C— Naphtha and isopropyl alcohol.

8992. Which of the following is the best means to use when removing wet snow from an aircraft?

A— A brush or a squeegee.
B— Hot air.
C— Warm water.

8993. What are three methods of anti-icing aircraft windshields?

1. Blanket-type heating system.
2. An electric heating element in the windshield.
3. Heated air circulating system.
4. Hot water system.
5. Windshield wipers and anti-icing fluid.
6. Ribbon-type heating system.

A— 2, 3, 5.
B— 1, 2, 6.
C— 2, 3, 4.

8994. What icing condition may occur in warm weather when there is no visible moisture present?

A— Glaze ice.
B— Rime ice.
C— Carburetor ice.

8995. What should be used to melt the ice in a turbine engine if the compressor is immobile because of ice?

A— Deicing fluid.
B— Anti-icing fluid.
C— Hot air.

8996. What is used as a temperature-sensing element in an electrically heated windshield?

A— Thermocouple.
B— Thermistor.
C— Thermometer.

8997. In what area of an aircraft would you find a carbon monoxide detector?

A— Surface combustion heater compartment.
B— Cockpit and/or cabin.
C— Engine and/or nacelle.

8998. What occurs when a visual smoke detector is activated?

A— A warning bell within the indicator alarms automatically.
B— A lamp within the indicator illuminates automatically.
C— The test lamp illuminates and an alarm is provided automatically.

8999. The types of fire-extinguishing agents for aircraft interior fires are

A— water, carbon dioxide, dry chemical, and halogenated hydrocarbons.
B— water, dry chemical, methyl bromide, and chlorobromomethane.
C— water, carbon tetrachloride, carbon dioxide, and dry chemical.

9000. When air samples contain carbon monoxide, portable carbon monoxide detectors containing yellow silica gel will turn which color?

A— Blue.
B— Green.
C— Red.

9001. Smoke detection instruments are classified by their method of

A— construction.
B— maintenance.
C— detection.

9002. Smoke detectors which use a measurement of light transmissibility in the air are called

A— electromechanical devices.
B— photoelectrical devices.
C— visual devices.

9003. A contaminated carbon monoxide portable test unit would be returned to service by

A— heating the indicating element to 300 °F to reactivate the chemical.
B— installing a new indicating element.
C— evacuating the indicating element with CO_2.

9004. Which fire-detection system measures temperature rise compared to a reference temperature?

A— Fenwal continuous loop.
B— Lindberg continuous element.
C— Thermocouple.

9005. A carbon dioxide (CO_2) hand-held fire extinguisher may be used on an electrical fire if the

A— horn is nonmetallic.
B— handle is insulated.
C— horn is nonmagnetic.

9006. The proper fire-extinguishing agent to use on an aircraft brake fire is

A— water.
B— carbon dioxide.
C— dry powder chemical.

9007. Smoke in the cargo and/or baggage compartment of an aircraft is commonly detected by which instrument?

A— Chemical reactor.
B— Photoelectric cell.
C— Sniffer.

9008. Light refraction smoke detectors

A— measure a reduction in the amount of visible or infrared light in the surrounding area.
B— sense light reflected from smoke particles passing through a chamber.
C— use radiation induced ionization to detect the presence of smoke.

9009. Why does the Fenwal fire-detection system use spot detectors wired parallel between two separate circuits?

A— A control unit is used to isolate the bad system in case of malfunction.
B— This installation is equal to two systems: a main system and a reserve system.
C— A short may exist in either circuit without causing a false fire warning.

9010. A fire-extinguisher container can be checked to determine its charge by

A— attaching a remote pressure gauge.
B— weighing the container and its contents.
C— a hydrostatic test.

9011. What is the color code for fire-extinguisher lines?

A— Brown.
B— Yellow.
C— Red and green.

9012. The most common cause of false fire warnings in continuous-loop fire-detection systems is

A— improper routing or clamping of loops.
B— moisture.
C— dents, kinks, or crushed sensor sections.

9013. A thermocouple in a fire-detection system causes the warning system to operate because

A— it generates a small current when heated.
B— heat decreases its electrical resistance.
C— it expands when heated and forms a ground for the warning system.

9014. The thermocouple fire-warning system is activated by a

A— certain temperature.
B— core resistance drop.
C— rate-of-temperature rise.

9015. When used in fire-detection systems having a single indicator light, thermal switches are wired in

A— parallel with each other and in series with the light.
B— series with each other and the light.
C— series with each other and parallel with the light.

9016. Built-in aircraft fire-extinguishing systems are ordinarily charged with

A— carbon monoxide and nitrogen.
B— freon and nitrogen.
C— sodium bicarbonate.

9017. In reference to aircraft fire-extinguishing systems,

(1) during removal or installation, the terminals of discharge cartridges should be grounded or shorted.

(2) before connecting cartridge terminals to the electrical system, the system should be checked with a voltmeter to see that no voltage exists at the terminal connections.

Regarding the above statements,

A— only No. 2 is true.
B— both No. 1 and No. 2 are true.
C— neither No. 1 nor No. 2 is true.

9018. What method is used to detect the thermal discharge of a built-in fire-extinguisher system?

A— A discoloring of the yellow plastic disk in the thermal discharge line.
B— A rupture of the red plastic disk in the thermal discharge line.
C— The thermal plug missing from the side of the bottle.

9019. The thermal switches of a bimetallic thermal-switch type fire-detection system are heat-sensitive units that complete circuits at a certain temperature. They are connected in

A— parallel with each other, and in parallel with the indicator lights.
B— parallel with each other, but in series with the indicator lights.
C— series with each other, but in parallel with the indicator lights.

9020. (Refer to figure 21.) Using the chart, determine the temperature range for a fire-extinguishing agent storage container with a pressure of 330 PSIG. (Consider 330 PSIG for both minimum and maximum pressure.)

A— 47 to 73 °F.
B— 47 to 71 °F.
C— 45 to 73 °F.

9021. (Refer to figure 21.) Determine what pressure is acceptable for a fire extinguisher when the surrounding area temperature is 33 °F.

A— 215 to 302 PSIG.
B— 214 to 301 PSIG.
C— 215 to 301 PSIG.

9022. On a periodic check of fire-extinguisher containers, the pressure was not between minimum and maximum limits. What procedure should be followed?

A— Release pressure if above limits.
B— Replace the extinguisher container.
C— Increase pressure if below limits.

9023. In some fire-extinguishing systems, evidence that the system has been intentionally discharged is indicated by the absence of a

A— red disk on the side of the fuselage.
B— green disk on the side of the fuselage.
C— yellow disk on the side of the fuselage.

9024. If a fire-extinguisher cartridge is removed from a discharge valve for any reason, it

A— must be pressure checked.
B— is recommended that the cartridge be used only on the original discharge valve assembly.
C— cannot be used again.

9025. Which of the following are fire precautions which must be observed when working on an oxygen system?

1. Display "No Smoking" placards.
2. Provide adequate fire-fighting equipment.
3. Keep all tools and oxygen servicing equipment free from oil or grease.
4. Avoid checking aircraft radio or electrical systems.

A— 1, 3, and 4.
B— 1, 2, and 4.
C— 1, 2, 3, and 4.

9026. Which fire-extinguishing agent is considered to be the least toxic?

A— Carbon dioxide.
B— Bromotrifluoromethane (Halon 1301).
C— Bromochloromethane (Halon 1011).

9027. Maintenance of fire-detection systems includes the

A— repair of damaged sensing elements.
B— removal of excessive loop or element material.
C— replacement of damaged sensing elements.

9028. A squib, as used in a fire-protection system, is a

A— temperature-sensing device.
B— device for causing the fire-extinguishing agent to be released.
C— probe used for installing frangible disks in extinguisher bottles.

APPENDIX 1

DEPARTMENT OF TRANSPORTATION
FEDERAL AVIATION ADMINISTRATION

AVIATION MECHANIC AIRFRAME
SUBJECT MATTER KNOWLEDGE CODES

To determine the knowledge area in which a particular question was incorrectly answered, compare the subject matter code(s) on AC Form 8080-2, Airmen Written Test Report, to the subject matter outline that follows. The total number of test items missed may differ from the number of subject matter codes shown on the AC Form 8080-2, since you may have missed more than one question in a certain subject matter code.

Wood Structures

A01 Service and repair wood structures
A02 Identify wood defects
A03 Inspect wood structures

Aircraft Covering

B01 Select and apply fabric and fiberglass covering materials
B02 Inspect, test, and repair fabric and fiberglass

Aircraft Finishes

C01 Apply trim, letters, and touchup paint
C02 Identify and select aircraft finishing materials
C03 Apply finishing materials
C04 Inspect finishes and identify defects

Sheet Metal and Non-Metallic Structures

D01 Select, install, and remove special fasteners for metallic, bonded, and composite structures
D02 Inspect bonded structures
D03 Inspect, test, and repair fiberglass, plastics, honeycomb, composite, and laminated primary and secondary structures
D04 Inspect, check, service, and repair windows, doors, and interior furnishings
D05 Inspect and repair sheet-metal structures
D06 Install conventional rivets
D07 Form, lay out, and bend sheet metal

Welding

E01 Weld magnesium and titanium
E02 Solder stainless steel
E03 Fabricate tubular structures
E04 Solder, braze, gas-, and arc-weld steel
E05 Weld aluminum and stainless steel

Assembly and Rigging

F01 Rig rotary-wing aircraft
F02 Rig fixed-wing aircraft
F03 Check alignment of structures
F04 Assemble aircraft components, including flight control surfaces
F05 Balance, rig, and inspect movable primary and secondary flight control surfaces
F06 Jack aircraft

Airframe Inspection

G01 Perform airframe conformity and airworthiness inspections
HXX Reserved
IXX Reserved
JXX Reserved

Aircraft Landing Gear Systems

K01 Inspect, check, service, and repair landing gear, retraction systems, shock struts, brakes, wheels, tires, and steering systems

Hydraulic and Pneumatic Power Systems

L01 Repair hydraulic and pneumatic power system components

L02 Identify and select hydraulic fluids

L03 Inspect, check, service, troubleshoot, and repair hydraulic and pneumatic power systems

Cabin Atmosphere Control Systems

M01 Inspect, check, service, troubleshoot, and repair heating, cooling, air-conditioning, pressurization, and air cycle machines

M02 Inspect, check, troubleshoot, service, and repair oxygen systems

Aircraft Instrument Systems

N01 Inspect, check, service, troubleshoot, and repair electronic flight instrument systems and both mechanical and electrical heading, speed, altitude, temperature, pressure, and position indicating systems to include the use of built-in test equipment

N02 Install instruments and perform a static pressure system leak test

Communication and Navigation Systems

O01 Inspect, check, and troubleshoot autopilot, servos and approach coupling systems

O02 Inspect, check, and service aircraft electronic communication and navigation systems, including VHF, passenger address interphones and static discharge devices, aircraft VOR, ILS, LORAN, radar beacon transponders, flight management computers, and GPWS

O03 Inspect and repair antenna and electronic equipment installations

Aircraft Fuel Systems

P01 Check and service fuel dump systems

P02 Perform fuel management, transfer, and defueling

P03 Inspect, check, and repair pressure fueling systems

P04 Repair aircraft fuel system components

P05 Inspect and repair fluid quantity indicating systems

P06 Troubleshoot, service, and repair fluid pressure and temperature warning systems

P07 Inspect, check, service, troubleshoot, and repair aircraft fuel systems

Aircraft Electrical Systems

Q01 Repair and inspect aircraft electrical system components; crimp and splice wiring to manufacturer's specifications; and repair pins and sockets of aircraft connectors

Q02 Install, check, and service airframe electrical wiring, controls, switches, indicators, and protective devices

Q03 Inspect, check, troubleshoot, service, and repair alternating and direct current electrical systems

Q04 Inspect, check, and troubleshoot constant speed and integrated speed drive generators

Position and Warning Systems

R01 Inspect, check, and service speed and configuration warning systems, electrical brake controls, and antiskid systems

R02 Inspect, check, troubleshoot, and service landing gear position indicating and warning systems

Ice and Rain Control Systems

S01 Inspect, check, troubleshoot, service, and repair airframe ice and rain control systems

Fire Protection Systems

T01 Inspect, check, and service smoke and carbon monoxide detection systems

T02 Inspect, check, service, troubleshoot, and repair aircraft fire detection and extinguishing systems

NOTE: AC 00-2, Advisory Circular Checklist, transmits the status of all FAA advisory circulars (AC's), as well as FAA internal publications and miscellaneous flight information such as AIM, Airport/Facility Directory, written test question books, practical test standards, and other material directly related to a certificate or rating. To obtain a free copy of the AC 00-2, send your request to:

U.S. Department of Transportation
Utilization and Storage Section, M-443.2
Washington, DC 20590

Appendix 1

ABBREVIATIONS AND REFERENCES

The following abbreviations are used to identify the reference associated with each test question listed in appendix 1.

AC	—	Advisory Circular
AEE	—	Aircraft Electricity and Electronics - Glencoe Division, Macmillan/McGraw-Hill Publ. Co.
AMR	—	Aircraft Maintenance and Repair - Glencoe Division, Macmillan/McGraw-Hill Publishing Co.
AP	—	Aircraft Powerplants - Glencoe Division, Macmillan/McGraw-Hill Publishing Co.
DAT	—	Dictionary of Aeronautical Terms - Aviation Supplies and Academics (ASA) Publications
EA-AAC-1	—	Aircraft Air-conditioning (Vapor Cycle) - International Aviation Publishers (IAP), Inc.
EA-FMS	—	Aircraft Fuel Metering Systems - IAP, Inc.
EA-AH-1	—	Aircraft Hydraulic System - IAP, Inc.
EA-AIS	—	Aircraft Instrument Systems - IAP, Inc.
EA-AOS-1	—	Aircraft Oxygen System - Aviation Maintenance Publishers (AMP) 1975
EA-ITP-A2	—	Airframe Section Textbook - IAP, Inc.
EA-ITP-G2	—	General Section Textbook - IAP, Inc.
EA-NMR	—	Aircraft Bonded Structure - IAP, Inc.
EA-WB-1	—	Welding Guidelines with Aircraft Supplement - IAP, Inc.
EA-356	—	Aircraft Radio Systems - IAP, Inc.
EA-358	—	Advanced Composites - IAP, Inc.
FAR	—	Federal Aviation Regulations
MBM	—	Marathon Battery Manual
MMM	—	Manufacturer's Maintenance Manual
TSO	—	Technical Standard Order

QUESTIONS AND REFERENCES

A01-A03:
8001. AMR
8002. AC 43.13-1A
8003. AC 43.13-1A
8004. AC 43.13-1A
8005. AC 43.13-1A
8006. AC 43.13-1A
8007. AC 43.13-1A
8008. AC 65-15A
8009. AC 43.13-1A
8010. AC 65-15A
8011. AC 65-15A
8012. AC 65-15A
8013. AMR
8014. AC 43.13-1A
B01-B02:
8015. AC 43.13-1A
8016. AC 65-15A
8017. AC 43.13-1A
8018. AC 43.13-1A
8019. AC 65-15A
8020. AC 65-15A
8021. AMR
8022. AMR
8023. AC 65-15A
8024. AC 43.13-1A
8025. AC 43.13-1A
C01-C04:
8026. AMR
8027. EA-ITP-A2
8028. DAT
8029. AMR
8030. EA-ITP-A2
8031. AMR
8032. AC 65-15A
8033. EA-ITP-A2
8034. EA-ITP-A2
8035. AC 43.13-1A
8036. AC 65-15A
8037. EA-ITP-A2
8038. AC 65-15A
8039. AMR
8040. AC 65-15A
D01:
8041. AC 65-15A
8042. EA-ITP-A2
8043. AC 65-9A
8044. AC 65-15A
8045. AC 65-15A
8046. EA-ITP-A2
8047. AC 65-9A
8048. AC 65-9A
8049. AC 65-9A
8050. EA-ITP-A2
8051. AMR
8052. EA-ITP-A2
8053. EA-ITP-A2
D02:
8054. AC 65-15A
8055. AMR
8056. AMR
8057. EA-NMR
8058. EA-NMR
8059. EA-ITP-A2
8060. AMR
8061. EA-NMR
8062. EA-NMR
8063. AMR
8064. AC 43.13-1A
D03:
8065. EA-ITP-A2
8066. EA-ITP-A2
8067. EA-ITP-A2
8068. AC 65-15A
8069. AC 43.13-1A
8070. EA-ITP-A2
8071. AC 65-15A
8072. EA-ITP-A2
8073. EA-ITP-A2
8074. EA-ITP-A2
8075. EA-ITP-A2
8076. EA-ITP-A2
8077. EA-ITP-A2
8078. EA-NMR
8079. EA-NMR
8080. EA-358
8081. AC 43.13-1A
8082. AC 43.13-1A
8083. AC 43.13-1A
8084. EA-358
8085. EA-ITP-A2
8086. EA-ITP-A2
8087. EA-ITP-A2
8088. EA-ITP-A2
8089. EA-ITP-A2
D04:
8090. AC 65-15A
8091. AC 43.13-1A
8092. TSO
8093. AC 65-15A
8094. AC 65-15A
8095. AC 65-15A
8096. FAR 23.853
8097. AC 65-15A
8098. AC 65-15A
D05:
8099. AC 65-15A
8100. AC 65-9A &

EA-ITP-G2
8101. AC 43.13-1A
8102. AC 65-15A
8103. AC 65-15A
8104. AC 65-9A
8105. AC 65-15A
8106. AC 65-15A
8107. AC 65-15A
8108. AC 43.13-1A
8109. AC 65-15A
8110. AC 65-15A
8111. AC 43.13-1A
8112. AC 65-9A
8113. AC 65-15A
8114. AC 65-15A
8115. AC 65-9A
8116. AC 43.13-1A
8117. AMR
8118. AC 65-9A
8119. AC 65-15A
8120. AMR
8121. AC 65-15A
8122. AC 65-9A
8123. AC 65-15A
8124. AC 65-15A
8125. AC 65-15A
8126. AMR
8127. AC 65-15A
8128. AC 65-15A
8129. EA-ITP-G2
D06:
8130. AC 65-9A
8131. AC 65-9A
8132. AC 43.13-1A
8133. AMR
8134. AC 65-9A
8135. AC 43.13-1A
8136. EA-ITP-G2
8137. AC 65-9A
8138. AC 65-9A
8139. AC 65-9A
8140. AC 43.13-1A
8141. AC 65-9A
8142. AC 65-9A
8143. AC 43.13-1A
8144. AC 65-15A
8145. AC 65-9A
8146. AC 43.13-1A
8147. AC 65-9A
8148. AC 43.13-1A
8149. AC 65-9A
8150. AC 65-15A
8151. AMR
8152. AC 65-15A
8153. AC 65-15A
8154. AC 65-9A
8155. AC 65-9A
8156. AC 65-15A
D07:
8157. AC 65-15A
8158. AC 65-15A
8159. AC 65-15A
8160. AC 65-15A
8161. AC 65-15A
8162. AC 65-15A
8163. AC 65-15A
8164. AC 65-15A
8165. AC 65-15A
8166. AC 65-15A
8167. AC 65-15A
8168. AC 65-15A
8169. AC 65-15A
8170. AC 65-9A
8171. AC 65-15A
8172. AC 65-15A
8173. AC 65-15A
8174. AC 65-15A
8175. AMR
8176. AC 65-15A
8177. AC 65-15A
8178. AC 65-9A
E01-E03:
8179. EA-ITP-A2
8180. AC 65-15A
8181. AC 65-15A
8182. AC 65-15A
8183. AC 65-15A
8184. AC 65-15A
8185. AC 43.13-1A
E04:
8186. AC 43.13-1A
8187. AC 65-15A
8188. AC 65-15A
8189. AMR
8190. AC 65-15A
8191. AC 65-15A
8192. AC 43.13-1A & EA-ITP-A2
8193. AMR
8194. AC 65-15A
8195. AMR
8196. AC 65-15A
8197. AC 65-15A
8198. AC 65-15A
E05:
8199. AC 65-15A
8200. AC 65-15A
8201. AC 65-15A
8202. AC 65-15A
8203. AC 65-15A
8204. AC 65-15A
8205. AC 65-15A
8206. AC 65-15A
8207. AC 65-15A
8208. AC 43.13-1A
8209. AC 65-15A
8210. AC 65-15A
8211. AMR & EA-WB-1
F01:
8212. AC 65-15A
8213. AC 65-15A
8214. AC 65-15A
8215. AC 43.13-2A
8216. AC 65-15A
8217. AC 65-15A
8218. AC 65-15A
8219. AC 65-15A
8220. AC 65-15A
8221. AC 65-15A
8222. AC 61-13B
8223. AC 61-13B
8224. AC 61-13B
8225. AC 65-15A
8226. AC 65-15A
F02:
8227. AC 65-15A
8228. AC 43.13-1A
8229. EA-ITP-A2
8230. AC 65-15A
8231. AC 65-15A
8232. AC 65-15A
8233. AC 65-15A
8234. AC 65-15A
8235. AC 65-15A
8236. AC 65-15A
8237. AC 65-15A
8238. AC 65-15A
8239. AC 65-15A
8240. AC 65-15A
8241. AC 65-15A
8242. AC 65-15A
8243. AC 65-15A
8244. AC 65-15A
8245. AC 65-15A
8246. AC 65-15A
8247. AC 65-15A
8248. AC 65-15A
8249. AC 65-15A
8250. AC 65-15A
F03-F04:
8251. AC 65-15A
8252. AC 65-15A
8253. AC 65-15A
8254. AC 65-15A
8255. AC 65-15A
8256. AC 65-15A
8257. AC 43.13-1A
8258. AC 43.13-1A
8259. AC 43.13-1A
8260. AC 43.13-1A
8261. AC 43.13-1A
8262. AC 65-9A
8263. AC 43.13-1A
8264. AC 65-15A
F05:
8265. AC 65-15A
8266. FAR 23.689(a)(1)
8267. AC 65-15A
8268. EA-ITP-A2
8269. AC 65-15A
8270. AC 65-15A
8271. AC 43.13-1A
8272. AC 43.13-1A
8273. AC 43.13-1A
8274. AC 65-15A
8275. AC 65-9A
8276. AC 65-15A
8277. FAR 23.677(a)
8278. AC 65-15A
8279. AC 65-15A
8280. AC 65-15A
8281. AC 65-15A
8282. AC 65-15A
8283. AC 65-15A
8284. AMR
8285. AC 65-15A
8286. AC 65-15A
8287. AC 65-9A
8288. AC 65-15A
8289. AC 65-15A
8290. AC 65-15A
F06-G01:
8291. AC 65-9A
8292. AC 65-9A
8293. AC 65-9A
8294. AC 65-9A
8295. FAR 43.7
8296. FAR 43
8297. FAR 91.409
8298. FAR 43.11
8299. FAR 91.409
8300. FAR 43.7(b)
8301. FAR 91.409
8302. FAR 65
8303. FAR 91.409
8304. FAR 91.409
K01:
8305. AMR & AC 65-9A
8306. AC 65-15A
8307. AC 65-15A
8308. AC 65-15A
8309. AC 65-15A
8310. AC 65-15A
8311. AC 43.13-1A
8312. AC 65-15A
8313. AC 65-15A & EA-ITP-A2
8314. AC 65-15A

8315. AMR
8316. AC 65-15A
8317. AC 65-15A
8318. AC 65-15A
8319. AC 43.13-1A
8320. AC 65-15A
8321. AC 65-15A
8322. AC 65-15A
8323. AC 65-15A
8324. EA-ITP-A2
8325. AC 65-15A
8326. AC 65-15A
8327. AC 43.13-1A
8328. AC 65-15A
8329. AC 65-15A
8330. FAR 43.3 & APP A
8331. AC 65-15A
8332. AC 65-15A
8333. AC 65-15A
8334. AC 65-15A
8335. AC 65-15A
8336. AMR
8337. AC 65-15A
8338. AC 65-15A
8339. AC 65-15A
8340. AC 65-15A
8341. AC 65-15A
8342. AC 65-15A
8343. AC 65-15A
8344. EA-ITP-A2
8345. AC 65-15A
8346. AC 65-15A
8347. AC 65-15A
8348. AC 65-15A
8349. AC 65-15A
8350. AC-65-15A
8351. AC 43.13-1A
8352. AC 65-15A
8353. AC 65-15A
8354. AC 65-15A
8355. AC 65-15A
8356. AC 65-15A
8357. AC 65-15A
8358. AC 65-15A
8359. AC 65-15A
8360. EA-ITP-A2
8361. AC 65-15A
8362. AC 65-15A
8363. AC 65-15A
8364. AC 65-15A
8365. AC 65-15A
8366. AC 65-15A
8367. AC 65-15A
8368. AC 65-15A
8369. AMR
8370. AC 65-15A
8371. AMR
8372. AC 65-15A
8373. AC 65-9A
8374. EA-AH-1
8375. AC 65-15A
8376. AC 65-15A
8377. EA-AH-1
8378. EA-AH-1
8379. EA-AH-1
8380. EA-AH-1
8381. AMR
8382. EA-ITP-A2
8383. AC 65-15A
8384. AC 65-9A
8385. AC 65-9A
L01:
8386. AC 43.13-1A
8387. AC 65-9A
8388. AC 65-15A
8389. AC 65-15A
8390. AC 65-15A
8391. AC 65-15A
8392. AC 65-15A
8393. AC 65-15A
8394. AC 65-15A
8395. AC 65-15A
8396. EA-ITP-A2
8397. AC 65-15A
8398. AC 65-15A
8399. AC 65-15A
8400. AC 65-15A
8401. AC 65-15A
8402. AC 65-9A
8403. AC 65-15A
8404. AC 65-9A
8405. AC 65-9A
8406. EA-ITP-A2
8407. EA-AH-1
8408. AMR
8409. AC 65-9A
8410. EA-ITP-A2
L02:
8411. EA-ITP-A2
8412. AC 65-15A
8413. AC 65-15A
8414. AC 65-15A
8415. AC 65-15A
8416. AC 65-15A
8417. EA-ITP-A2
8418. AC 65-15A
8419. AC 65-15A
8420. AC 65-15A
8421. AC 65-15A
8422. AC 65-15A
8423. AC 65-15A
8424. AC 65-15A
8425. AC 65-15A
8426. AC 65-15A
8427. EA-AH-1
8428. EA-AH-1
8429. EA-AH-1
8430. EA-ITP-A2
8431. AC 65-15A
L03:
8432. AC 65-15A
8433. EA-ITP-A2
8434. AMR
8435. AC 65-15A
8436. AC 65-15A
8437. AMR
8438. AC 65-15A
8439. AC 65-15A
8440. AMR
8441. AC 65-15A
8442. AC 65-15A
8443. AC 65-15A
8444. AC 65-15A
8445. AC 65-9A
8446. AC 65-9A
8447. AC 65-9A
8448. AMR
8449. AC 65-15A
8450. AC 65-15A
8451. AC 65-15A
8452. AC 65-15A
8453. AC 65-15A
8454. AMR
8455. AC 65-15A
8456. AC 65-15A
8457. AC 65-15A
8458. AC 65-15A
8459. AC 65-15A
8460. AC 65-15A
8461. AC 65-15A
8462. AC 65-15A
8463. AMR
8464. AC 65-15A
8465. AC 65-15A
8466. AC 65-15A
8467. AC 65-15A
8468. AC 65-15A
8469. AMR
8470. AC 65-15A
8471. EA-ITP-A2
8472. AC 65-15A
8473. AC 65-15A
8474. AC 65-15A
8475. AMR
8476. AC 65-15A
8477. AC 65-15A
8478. AC 65-15A
8479. AC 65-15A
8480. AC 65-15A
8481. AC 65-9A
8482. AC 65-15A
8483. AC 65-15A
8484. AC 65-15A
8485. AC 65-15A
8486. AC 65-15A
8487. AC 65-15A
8488. AC 65-15A
8489. AC 65-15A
8490. AC 65-15A
8491. AC 65-15A
8492. AMR
8493. EA-AH-1
8494. EA-AH-1
8495. AC 43.13-1A
8496. AC 65-15A
M01:
8497. AC 65-15A
8498. AC 65-15A
8499. AC 65-15A
8500. AC 65-15A
8501. AC 65-15A
8502. AC 65-15A
8503. EA-ITP-A2
8504. AC 65-15A
8505. AC 65-15A
8506. AC 65-15A
8507. EA-AAC-1
8508. EA-AAC-1
8509. EA-AAC-1
8510. AC 65-15A
8511. EA-AAC-1
8512. AC 43.13-1A
8513. AC 43.13-1A
8514. AC 43.13-1A
8515. EA-ITP-A2
8516. AC 65-15A
8517. AC 65-15A
8518. AC 65-15A
8519. AC 65-15A
8520. AC 65-15A
8521. AC 65-15A
8522. AC 65-15A
8523. AC 65-15A
8524. AC 65-15A
8525. AC 65-15A
8526. AC 65-15A
8527. AC 65-15A
8528. AC 65-15A
8529. AC 65-15A
8530. AC 65-15A
8531. AC 65-15A
8532. AC 65-15A
8533. AC 65-15A
8534. AC 65-15A
8535. AC 65-15A
8536. AC 65-15A
8537. AMR

8538. AMR
8539. AC 65-15A
8540. AC 65-15A
8541. AC 65-15A
8542. AC 65-15A
8543. AC 65-15A
8544. AC 65-15A
8545. AC 65-15A
8546. AC 65-15A
8547. AC 65-15A
8548. AC 65-15A
8549. AC 65-15A
8550. EA-AAC-1
8551. EA-AAC-1
8552. EA-AAC-1
8553. EA-ITP-A2
8554. EA-AAC-1
8555. EA-AAC-1
8556. EA-AAC-1
8557. EA-AAC-1
8558. EA-ITP-A2
8559. EA-AAC-1
8560. AC 65-15A
8561. EA-ITP-A2
8562. EA-ITP-A2
8563. EA-ITP-A2
M02:
8564. AC 65-15A
8565. AC 65-15A
8566. AC 65-15A
8567. AC 65-15A
8568. AC 65-15A
8569. AC 65-15A
8570. AC 65-15A
8571. AC 65-15A
8572. AC 65-15A
8573. AC 65-15A
8574. AC 65-15A
8575. AC 65-15A
8576. AC 65-15A
8577. AC 65-15A
8578. AC 65-15A
8579. EA-ITP-A2
8580. EA-ITP-A2
8581. AC 65-15A
8582. EA-ITP-A2 & EA-AOS-1
8583. EA-ITP-A2
8584. EA-ITP-A2
8585. AC 65-15A
N01:
8586. AC 65-15A
8587. AC 65-15A
8588. AC 65-15A
8589. AC 65-15A & FAR 23
8590. AC 65-15A
8591. AC 65-15A
8592. FAR 23.1327
8593. AC 65-15A
8594. AC 65-15A
8595. FAR 23.1325
8596. AC 65-15A
8597. AC 65-9A
8598. FAR 65.81
8599. FAR 65.81
8600. AC 65-15A
8601. DAT
8602. AC 65-15A
8603. AC 65-15A
8604. AC 65-15A
8605. AC 65-15A
8606. AC 65-15A
8607. AC 65-15A
8608. AC 65-15A
8609. AC 65-15A
8610. AC 65-15A
8611. AC 65-15A
8612. FAR 91.411
8613. AC 65-15A
8614. AEE
8615. AEE
8616. AEE
8617. EA-ITP-A2
8618. AEE
8619. AEE
8620. AEE
8621. AMR
N02:
8622. AC 65-15A
8623. AC 65-15A
8624. AC 65-15A
8625. AC 65-15A
8626. FAR 23.1545
8627. AC 65-15A
8628. AC 65-15A
8629. AC 65-15A
8630. AC 65-15A
8631. AC 65-15A
8632. AC 65-15A
8633. AC 65-15A
8634. AC 65-15A
8635. AC 65-15A
8636. AC 65-15A
8637. FAR 65.81
8638. AC 65-15A
8639. FAR 65.81(a)
8640. AC 65-15A
8641. AC 65-15A
8642. EA-AIS
8643. AC 65-15A
8644. EA-ITP-A2
8645. EA-ITP-A2
8646. EA-ITP-A2
O01:
8647. AC 65-15A
8648. AC 65-15A
8649. AC 65-15A
8650. AC 65-15A
8651. AEE
8652. EA-ITP-A2
8653. AC 65-15A
8654. AC 65-15A
8655. AC 65-15A
8656. AP
8657. AEE
8658. AEE
8659. EA-ITP-A2
8660. AC 65-15A
O02:
8661. AC 65-15A
8662. AC 43.13-2A
8663. AC 65-15A
8664. AC 65-15A
8665. AC 65-15A
8666. AC 65-15A
8667. AC 65-15A
8668. AC 65-15A
8669. EA-ITP-A2
8670. AC 91-44A
8671. AC 65-15A
8672. AC 65-15A
8673. EA-ITP-A2
8674. EA-ITP-A2
8675. EA-356
8676. AEE
8677. AEE
8678. AEE
8679. AEE
O03:
8680. AC 43.13-2A
8681. AC 65-15A
8682. AC 65-9A
8683. AC 65-15A
8684. AC 43.13-2A
8685. AC 65-15A
8686. AC 43.13-2A
8687. AC 65-15A
8688. AC 43.13-2A
8689. AC 43.13-2A
8690. AC 65-15A
8691. AC 65-15A
8692. AC 65-15A
8693. AC 65-15A
8694. AC 65-15A
8695. AC 65-15A
8696. AC 43.13-2A
8697. AC 65-15A
P01-P03:
8698. AC 65-9A
8699. AMR
8700. FAR 23.1001
8701. AC 65-9A
8702. AC 65-9A
8703. EA-ITP-A2
8704. EA-ITP-A2
8705. MMM
8706. AC 65-9A
8707. AC 65-9A
8708. AC 65-9A
8709. AC 65-9A
8710. AC 65-9A
8711. AC 43.13-1A
8712. AC 65-9A
8713. AC 65-9A
8714. AC 65-9A
8715. AC 65-9A
8716. AC 65-9A
8717. AC 65-9A
P04:
8718. AC 65-9A
8719. AC 65-9A
8720. AC 65-9A
8721. FAR 23.965(a)(1)
8722. AC 65-9A
8723. AC 65-9A
8724. AC 65-9A
8725. AC 43.13-1A
8726. AC 65-9A
8727. AC 65-9A
8728. EA-FMS
8729. AC 43.13-1A
8730. EA-ITP-A2
8731. EA-FMS
8732. AC 43.13-1A
8733. AC 43.13-1A
8734. AC 43.13-1A
8735. EA-ITP-G2
8736. AC 43.13-1A
P05:
8737. AC 65-9A
8738. AC 65-9A
8739. AC 65-9A
8740. AC 65-9A
8741. AC 65-9A
8742. AC 65-9A
8743. AC 65-15A
8744. AC 65-9A
8745. AC 65-9A
8746. AC 65-9A
8747. EA-ITP-A2
8748. AC 65-9A
8749. AC 65-9A
8750. AC 65-9A
8751. AC 65-9A
8752. AC 65-9A
8753. AC 65-9A
8754. AC 65-9A

8755. EA-ITP-A2
8756. FAR 23.1337
P06:
8757. AC 65-9A
8758. AC 65-9A
8759. AC 65-9A
8760. AC 65-9A
8761. AC 65-9A
8762. AC 65-9A
8763. AC 65-9A
8764. AC 65-9A
8765. AC 65-9A
8766. AC 65-9A
8767. AC 65-9A
8768. AC 65-9A
8769. AC 65-12A
8770. AC 65-12A
8771. AC 65-12A
8772. AC 65-9A
8773. AC 65-9A
P07:
8774. AC 65-9A
8775. AC 65-9A
8776. AC 65-9A
8777. AC 43.13-2A
8778. AC 65-9A
8779. FAR 23.951(b)
8780. AC 65-9A
8781. FAR 25.1557
8782. AC 65-9A
8783. AC 65-9A
8784. AC 65-9A
8785. AC 65-9A
8786. AC 65-9A
8787. FAR 23.1557
8788. AC 65-9A
8789. AC 65-9A
8790. AC 65-9A
8791. AC 65-9A
8792. AC 43.13-1A
8793. AC 65-9A
8794. AC 65-9A
8795. AC 65-9A
8796. AC 65-9A
8797. AC 65-9A
8798. AC 65-9A
8799. AC 43.13-1A
8800. AC 65-9A
8801. EA-ITP-A2
8802. AC 65-9A
Q01:
8803. AC 65-9A
8804. AEE
8805. AEE
8806. AC 65-9A
8807. AC 65-9A
8808. AC 65-9A
8809. AC 65-9A
8810. AC 65-9A
8811. AC 65-9A
8812. AC 65-9A
8813. AC 65-9A
8814. AC 65-9A
8815. AC 65-9A
8816. AC 65-9A
8817. AC 65-9A
8818. AC 65-9A
8819. AC 65-9A
8820. AEE
8821. AC 65-15A
8822. AC 65-9A
8823. AC 65-9A
8824. AC 65-9A
8825. AEE
8826. FAR 23.1385
8827. AC 65-9A
8828. AC 65-9A
8829. AC 65-9A
8830. EA-ITP-G2
8831. AEE
8832. EA-ITP-G2
8833. EA-ITP-G2
8834. EA-ITP-G2
8835. EA-ITP-G2
8836. MBM
8837. EA-ITP-G2
8838. AC 65-9A
8839. AC 65-9A
8840. AC 65-9A
8841. AC 65-15A
8842. AC 65-15A
8843. AC 43.13-1A
8844. AC 43.13-1A
8845. AC 43.13-1A
8846. AEE
8847. AEE
8848. EA-ITP-A2
8849. AC 65-9A
Q02:
8850. AC 65-15A
8851. AC 65-15A
8852. AC 65-15A
8853. AC 65-9A
8854. AC 65-9A
8855. EA-ITP-A2
8856. AC 65-9A
8857. AC 43.13-1A
8858. AC 43.13-1A
8859. AC 43.13-1A
8860. AC 65-15A
8861. AC 65-15A
8862. AC 65-9A
8863. AC 43.13-1A
8864. AC 43.13-1A
8865. AC 43.13-1A
8866. AC 43.13-1A
8867. AC 65-15A
8868. AC 65-15A
8869. AC 65-9A
8870. AC 65-15A
8871. AC 43.13-1A
8872. AC 43.13-1A
8873. AC 43.13-1A
8874. AC 43.13-1A
8875. AC 65-9A
8876. AC 65-15A
8877. AC 65-9A
8878. AC 43.13-1A
8879. AC 65-15A
8880. AC 65-9A
8881. AC 43.13-1A
8882. AC 43.13-1A
8883. AC 43.13-1A
8884. AC 65-15A
8885. AC 43.13-1A
8886. AC 65-9A
8887. AC 43.13-1A
8888. AC 65-15A
8889. AC 65-9A
8890. AC 65-15A
8891. AC 65-9A
8892. AC 65-9A
Q03:
8893. AC 65-9A
8894. AEE
8895. AC 65-9A
8896. AC 65-9A
8897. AEE
8898. AC 65-9A
8899. AC 65-9A
8900. AC 43.13-2A
8901. AC 65-9A
8902. AC 65-9A
8903. AC 65-9A
8904. AC 65-9A
8905. AC 65-9A
8906. AC 65-9A
8907. AC 43.13-1A
8908. AC 65-9A
8909. AEE
8910. AC 65-9A
8911. AC 65-9A
8912. AC 65-9A
8913. AC 43.13-2A
8914. AC 65-9A
8915. AC 65-9A
8916. AC 65-9A
8917. AC 65-9A
8918. AC 65-9A
8919. AC 65-9A
8920. AC 65-9A
8921. AC 65-9A
8922. AC 65-9A
8923. AC 65-9A
8924. AC 65-9A
8925. AC 65-9A
8926. AC 65-15A
8927. AC 65-9A
Q04:
8928. AEE
8929. AEE
8930. AEE
8931. EA-ITP-A2
8932. AEE
8933. EA-ITP-A2
R01:
8934. AC 65-15A
8935. AC 65-15A
8936. AC 65-15A
8937. AC 65-15A
8938. AMR
8939. AC 65-15A
8940. FAR 23.1323
8941. AC 65-15A
8942. EA-ITP-A2
8943. AC 65-15A
8944. AC 65-15A
8945. AC 65-15A
8946. AC 65-15A
8947. AC 65-15A
8948. AMR
8949. AMR
8950. AMR
R02:
8951. AC 65-15A
8952. AC 65-15A
8953. AC 65-15A
8954. AC 65-15A
8955. AC 65-15A
8956. AC 65-15A
8957. AC 43.13-1A
8958. AC 65-15A
8959. AC 65-15A
8960. AC 65-15A
8961. AC 65-15A
8962. AC 65-15A
8963. EA-AIS
8964. EA-AIS
8965. EA-AIS
8966. AC 65-9A
8967. AC 65-15A
8968. AC 65-15A
S01:
8969. AC 65-15A
8970. AC 65-15A
8971. AC 43.13-1A
8972. AC 65-15A
8973. AC 65-15A

8974. AC 65-15A
8975. AC 65-15A
8976. AC 65-15A
8977. AC 65-15A
8978. AC 65-15A
8979. AC 65-15A
8980. AC 65-15A
8981. AC 65-15A
8982. AC 65-15A
8983. AC 65-15A
8984. AC 65-15A
8985. AC 65-15A
8986. AC 65-15A
8987. AC 65-15A
8988. AC 65-15A
8989. AC 65-15A
8990. AC 65-15A
8991. AC 65-15A
8992. AC 65-15A
8993. AC 65-15A
8994. AC 65-15A
8995. AC 65-15A
8996. AC 65-15A
T01:
8997. AC 65-15A
8998. AC 65-15A
8999. AC 65-15A
9000. AC 65-15A
9001. AC 65-15A
9002. AC 65-15A
9003. AC 65-15A
9004. AC 65-15A
9005. AC 65-15A
9006. AC 65-9A
9007. AC 65-15A
9008. AC 65-15A
9009. AC 65-15A
T02:
9010. AC 65-15A
9011. AC 65-15A
9012. AC 65-15A
9013. AC 65-15A
9014. AC 65-15A
9015. AC 65-15A
9016. EA-ITP-A2
9017. AP
9018. AC 65-15A
9019. AC 65-15A
9020. AC 65-15A
9021. AC 65-15A
9022. AC 65-15A
9023. AC 65-15A
9024. AC 65-15A
9025. AC 65-15A
9026. AC 65-15A
9027. AC 65-15A
9028. AC 65-15A

APPENDIX 2

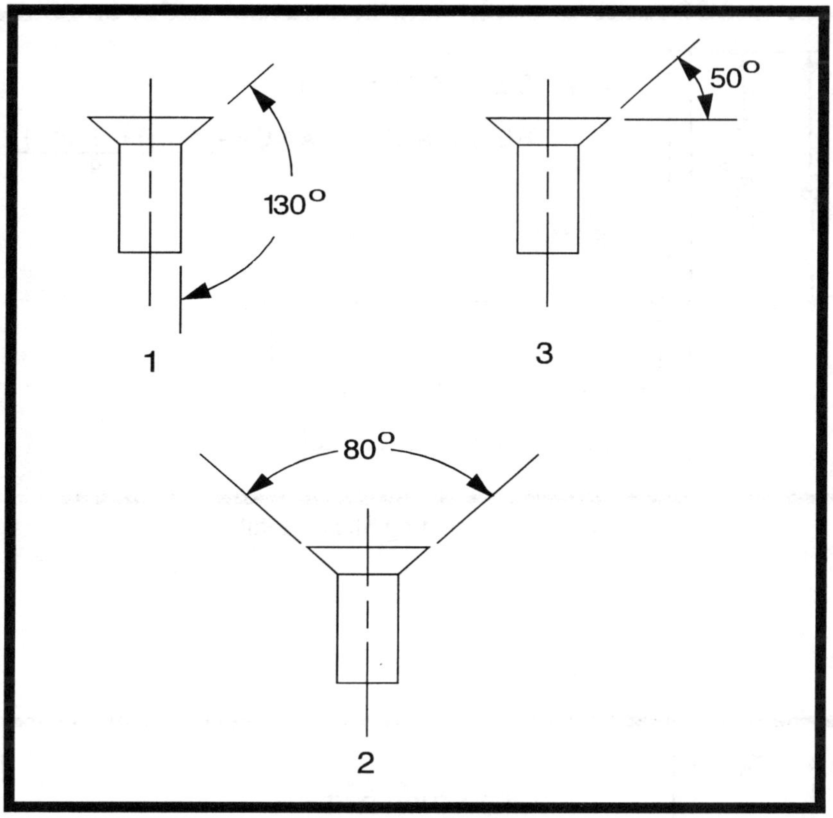

FIGURE 1.—Rivets.

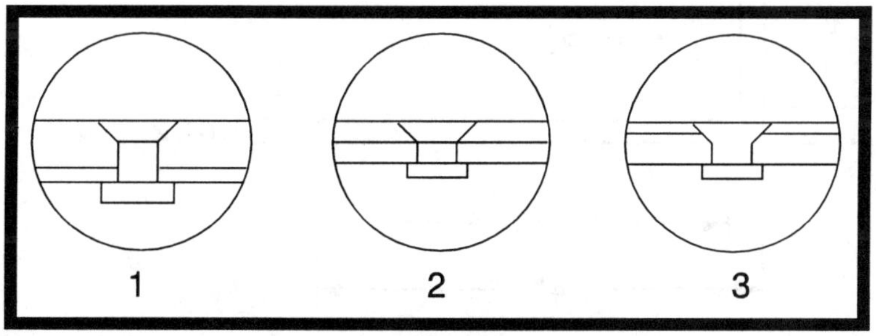

FIGURE 2.—Countersinking.

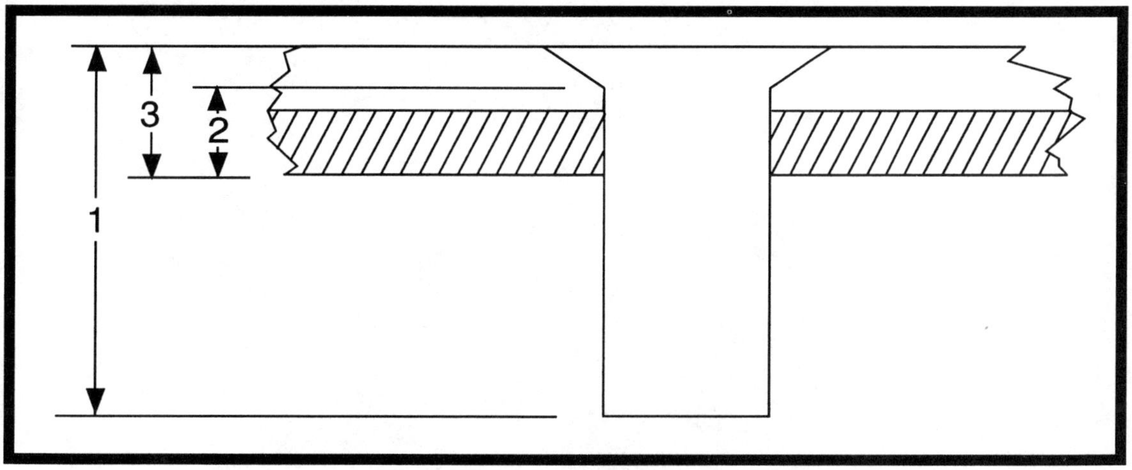

FIGURE 3.—Grip Length.

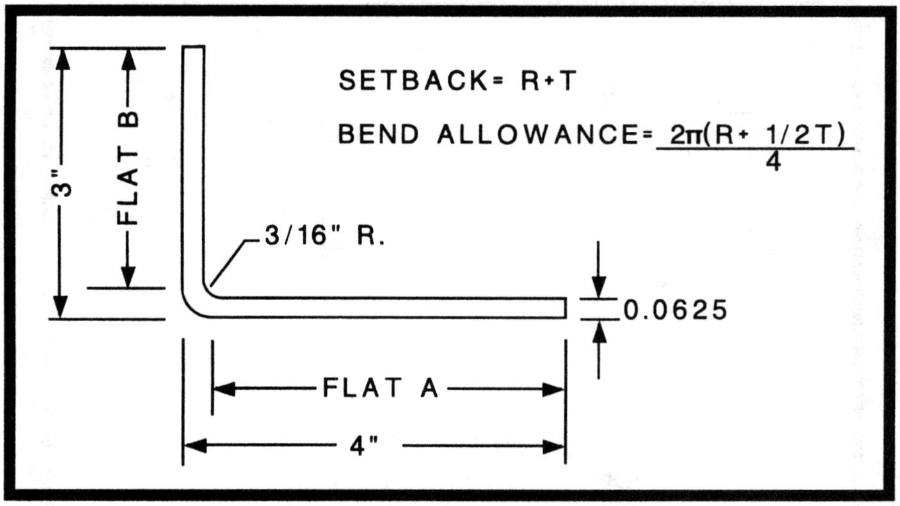

FIGURE 4.—Bending Sheet Metal.

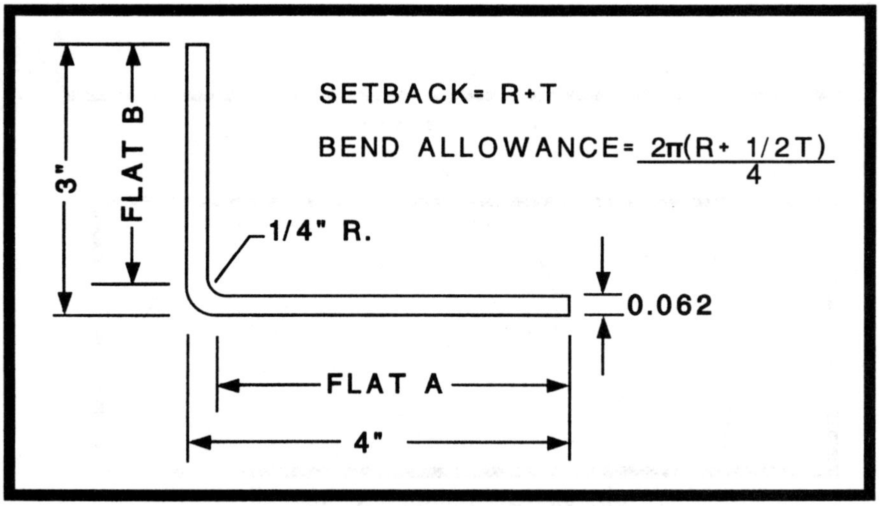

FIGURE 5.—Sheet Metal Layout.

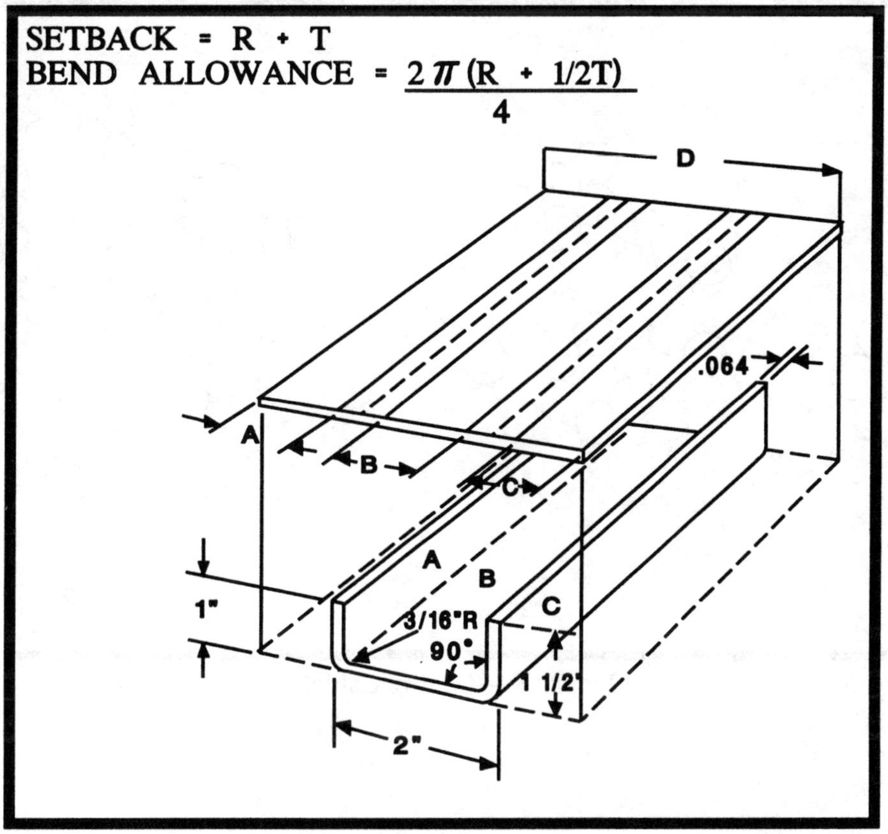

FIGURE 6.—Sheet Metal Layout.

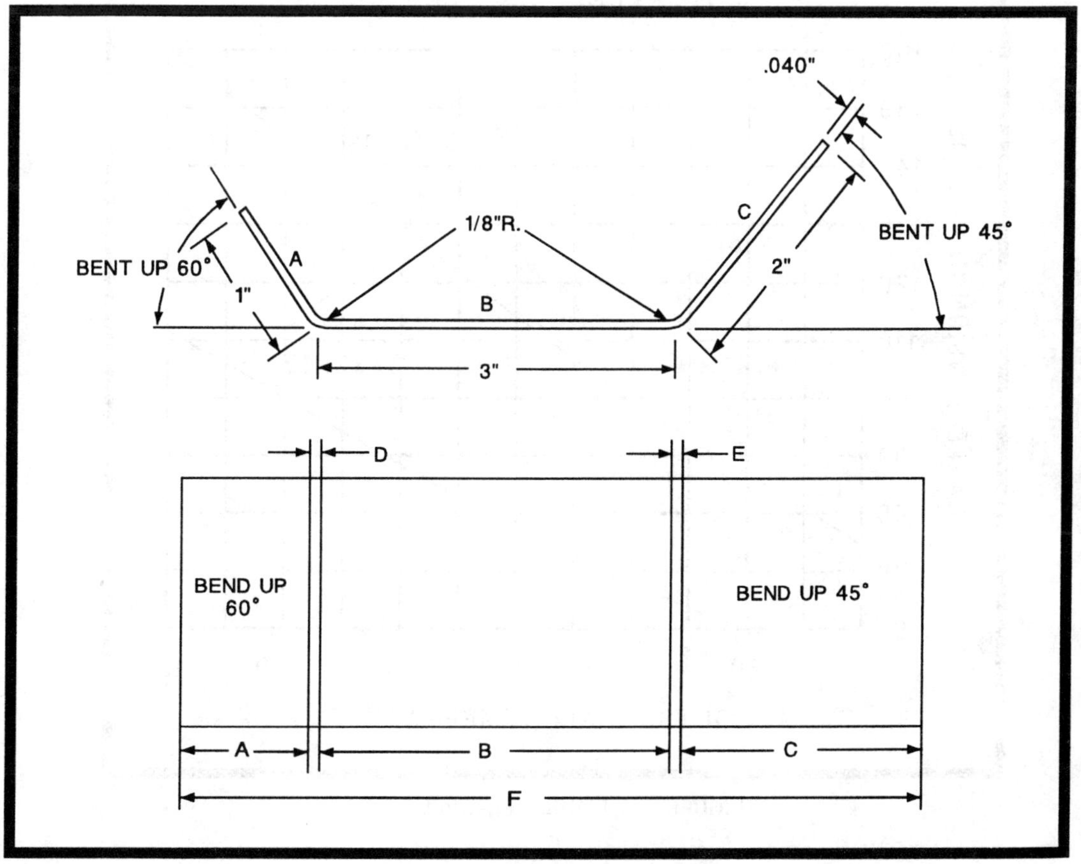

FIGURE 7.—Sheet Metal Layout.

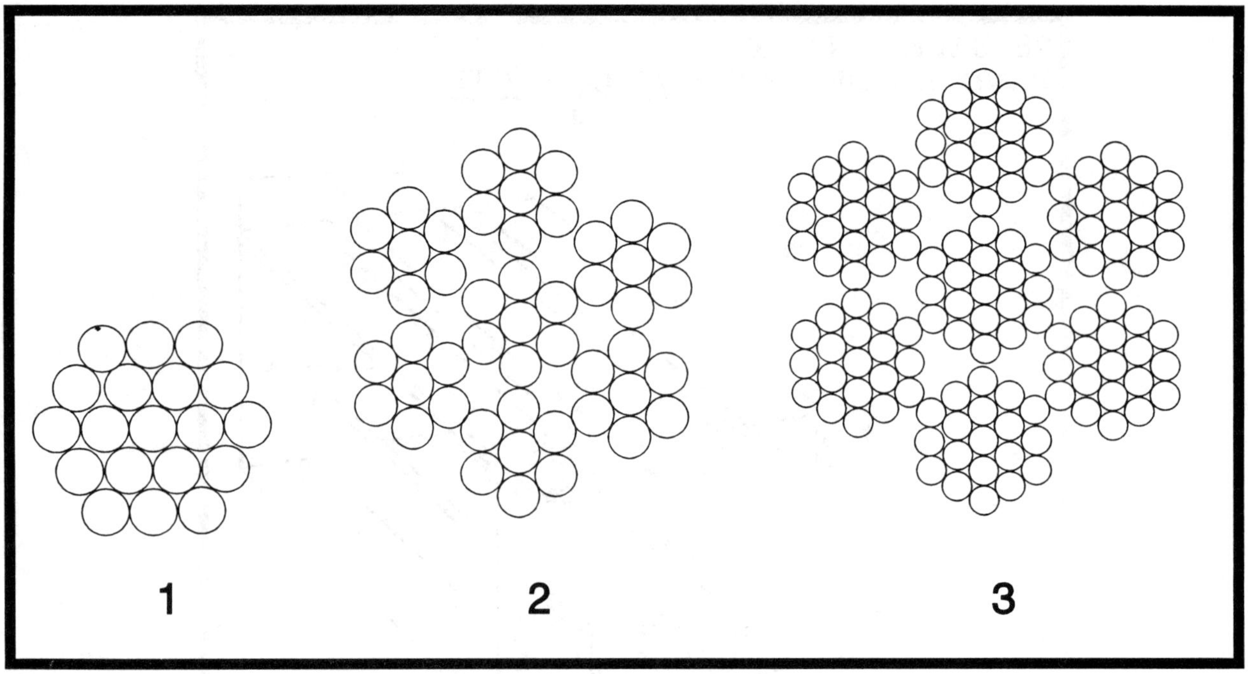

FIGURE 8.—Control Cable.

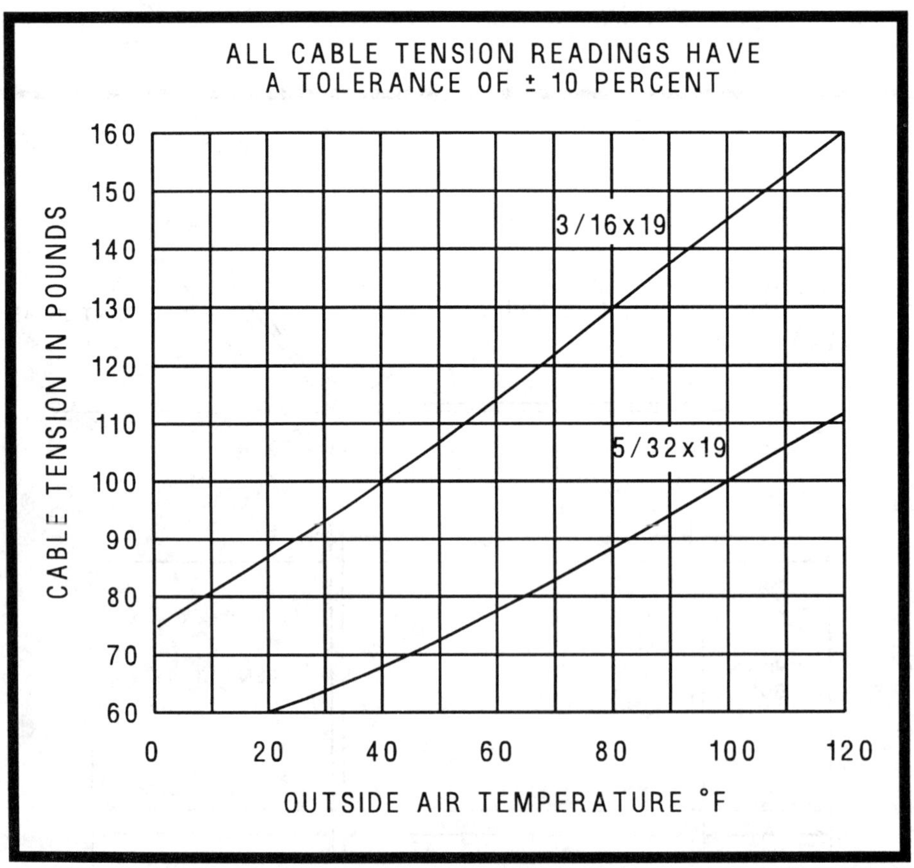

FIGURE 9.—Cable Tension Chart.

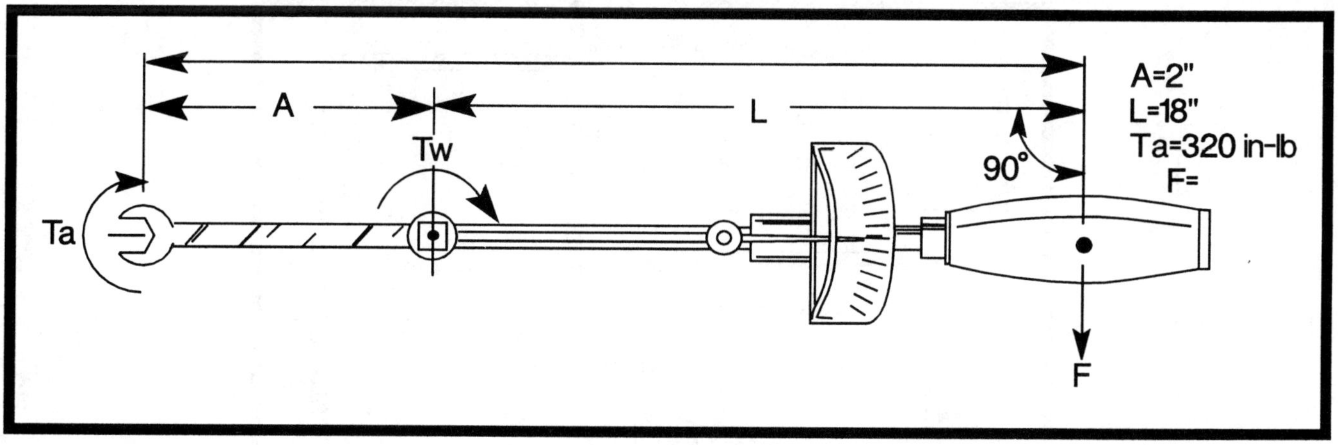

FIGURE 10.—Torque Value.

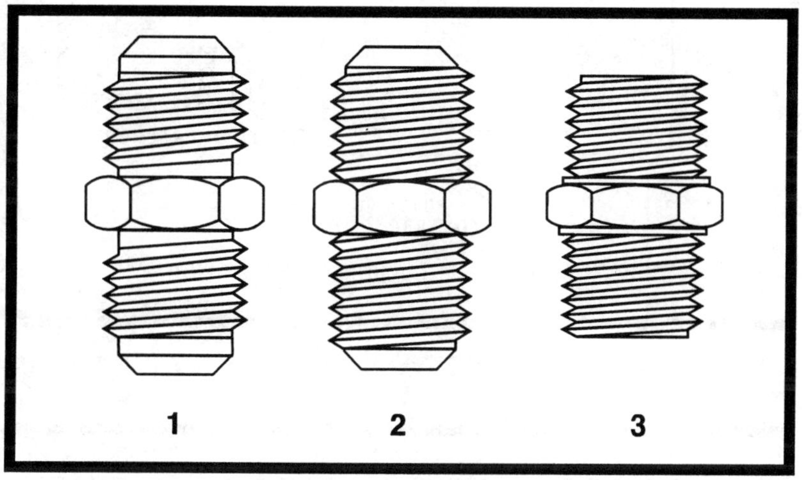

FIGURE 11.—Fittings.

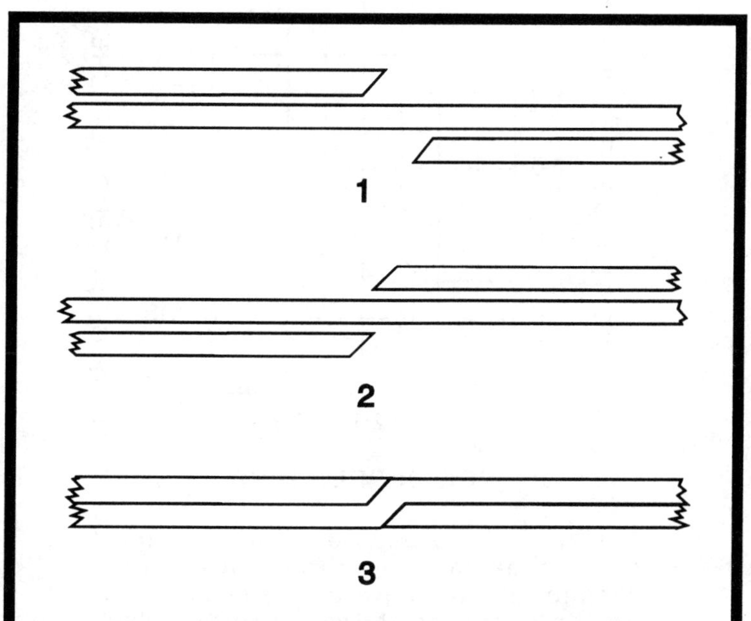

FIGURE 12.—Backup Rings.

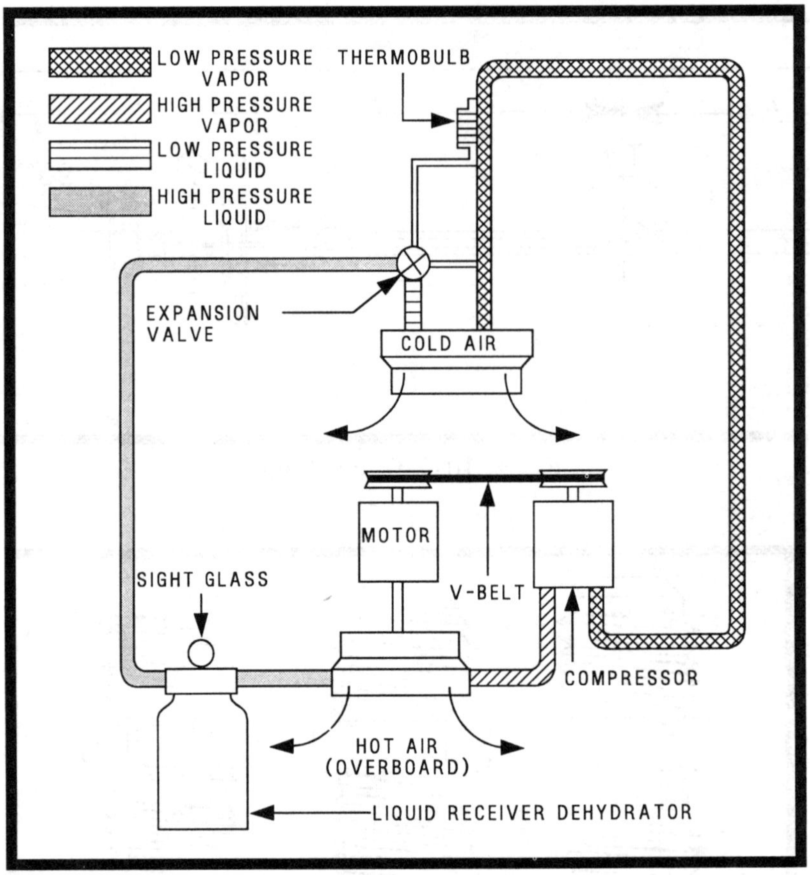

FIGURE 13.—Cooling System.

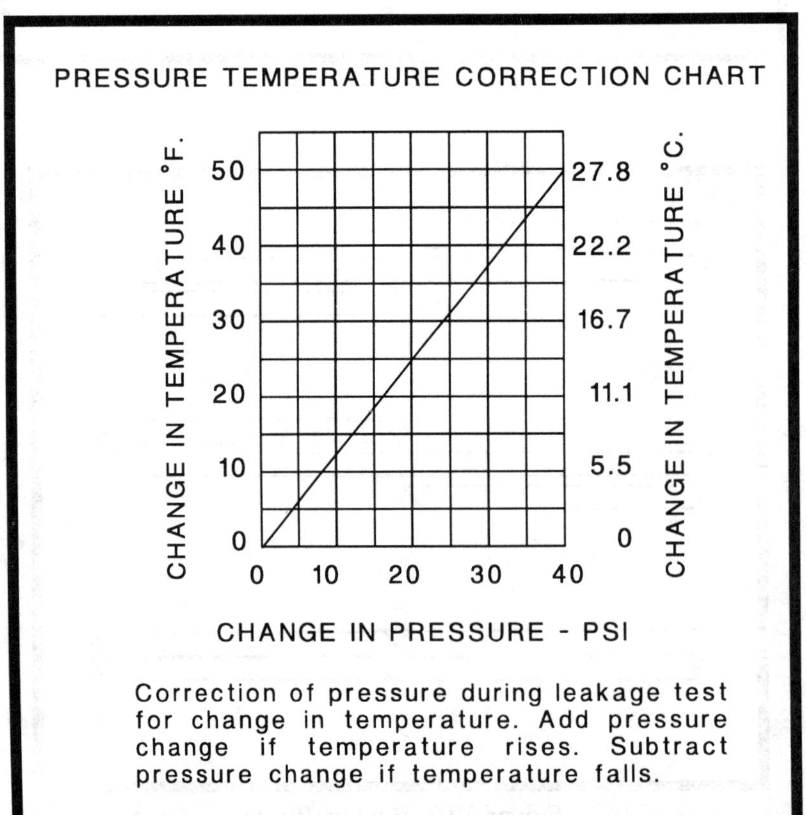

FIGURE 14.—Pressure Temperature Correction Chart.

$$D = .000327AV^2$$

FIGURE 15.—Formula.

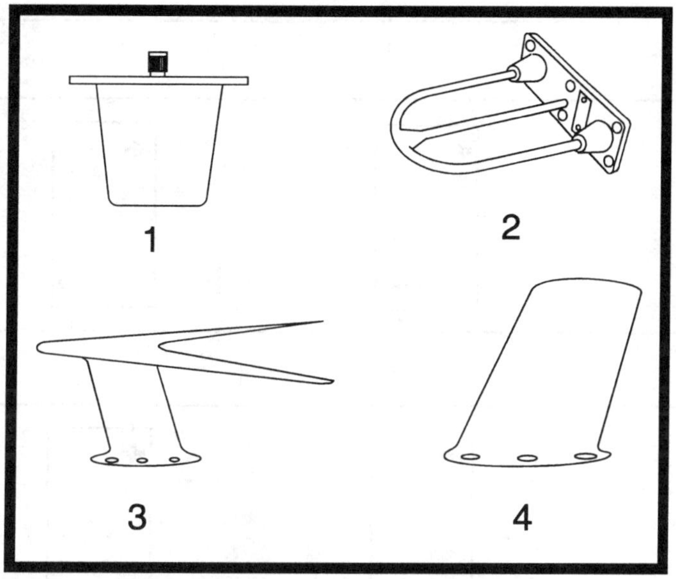

FIGURE 16.—Antennas.

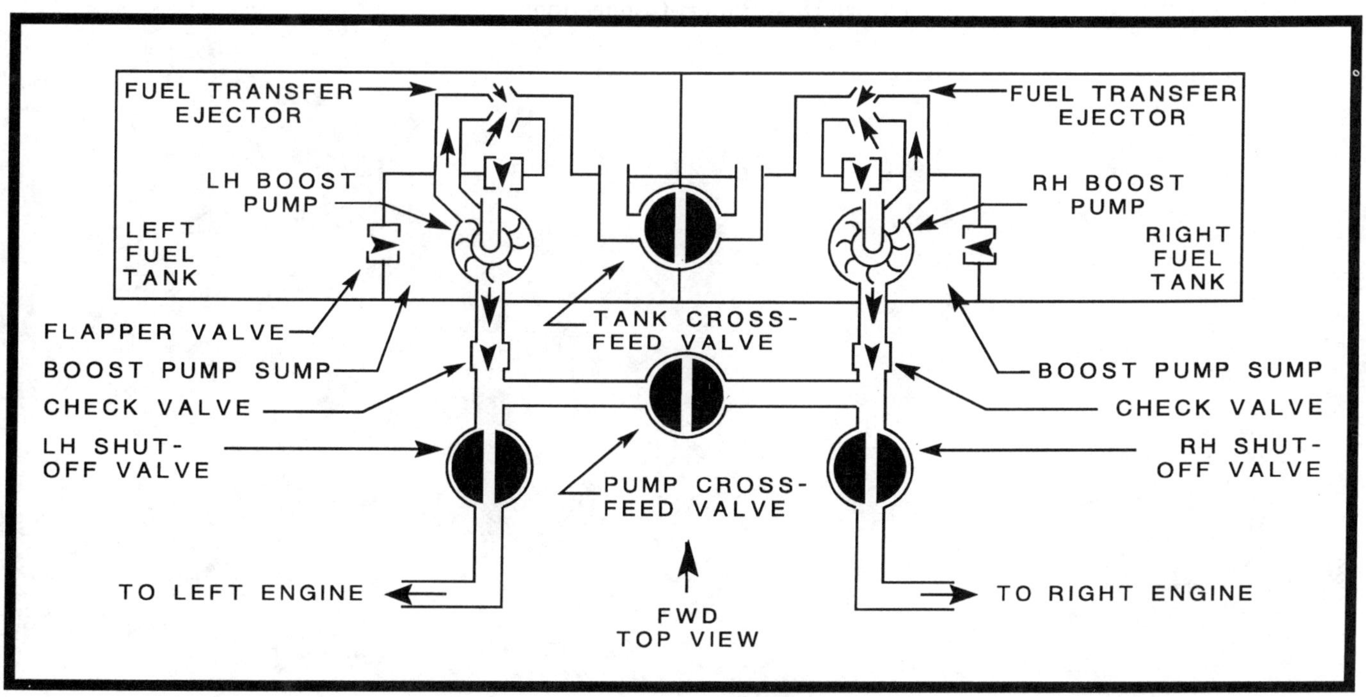

FIGURE 17.—Fuel System.

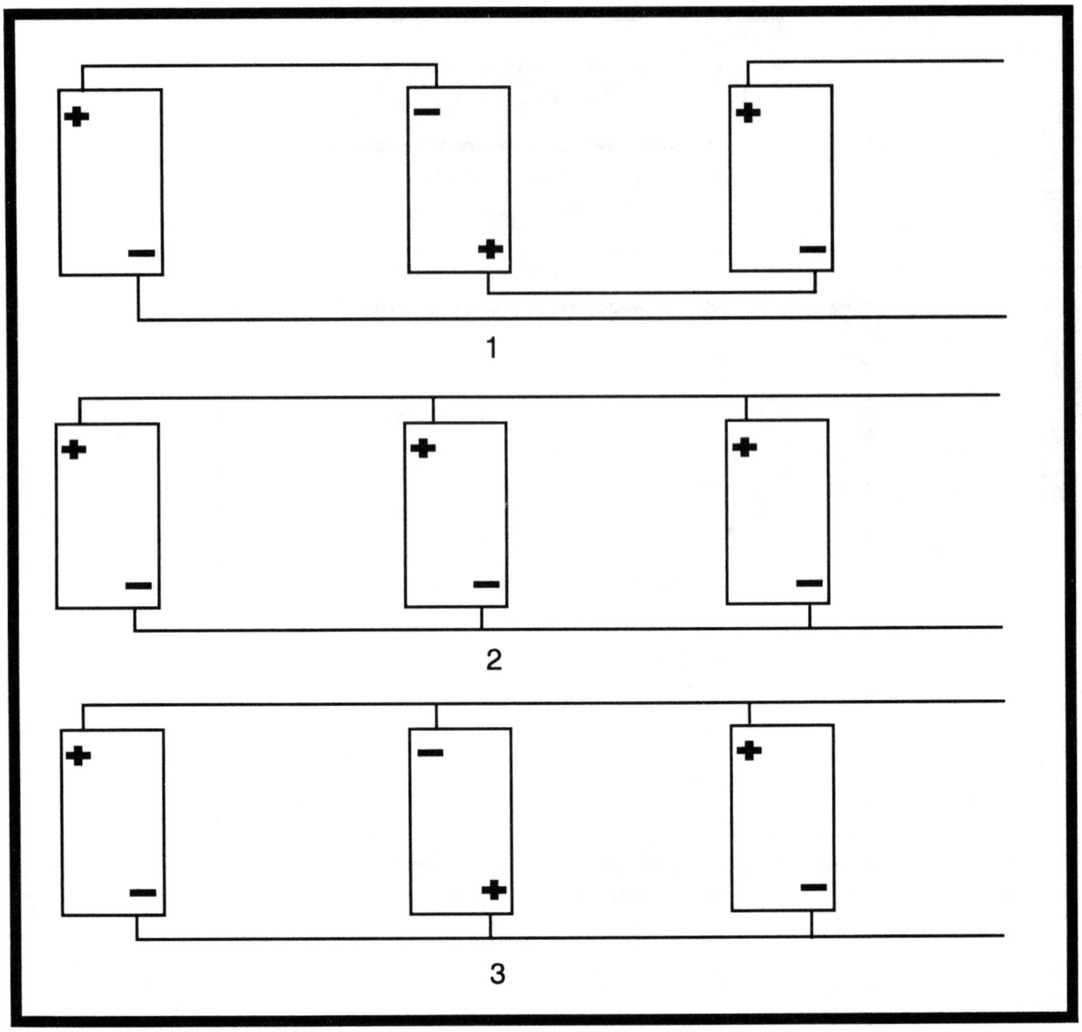

FIGURE 18.—Battery Connections.

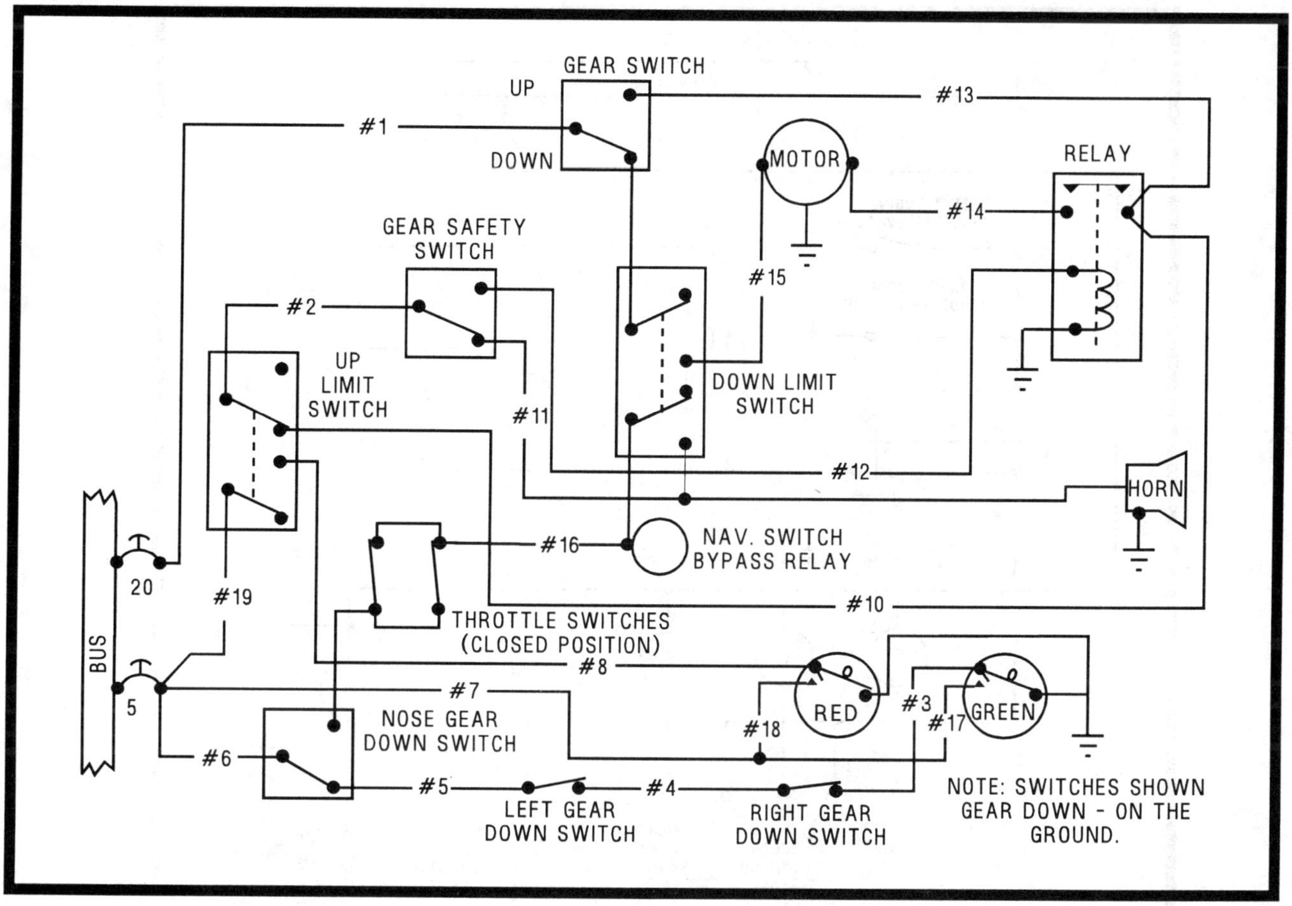

FIGURE 19.—Landing Gear Circuit.

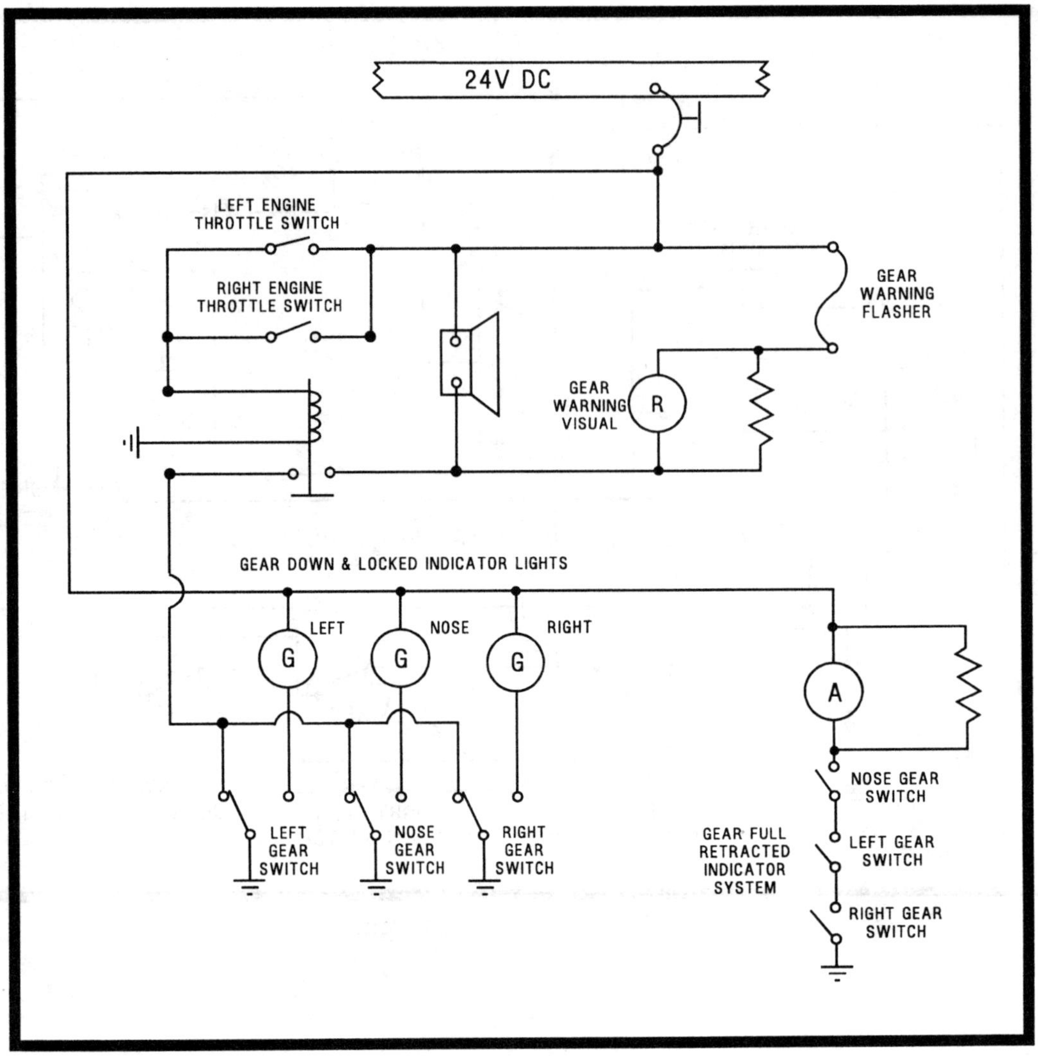

FIGURE 20.—Landing Gear Circuit.

CONTAINER PRESSURE VERSUS TEMPERATURE		
TEMPERATURE °F	CONTAINER PRESSURE (PSIG)	
	MINIMUM	MAXIMUM
-40	60	145
-30	83	165
-20	105	188
-10	125	210
0	145	230
10	167	252
20	188	275
30	209	295
40	230	317
50	255	342
60	284	370
70	319	405
80	356	443
90	395	483
100	438	523

FIGURE 21.—Fire Extinguisher Chart.

APPENDIX 3

1994-96 AIRFRAME MECHANICS FAA EXAM BOOK
ANSWERS, EXPLANATIONS AND REFERENCES

Question numbers, along with the correct answers are in bold type, then the references in *italic* type, followed by the explanation.

8001. (B) *EA-ITP-ASG2E, Page 1-3; EA-43.13-1A/3, Page 6*
Solid spruce spars may be replaced with laminated ones or vice versa, provided the material is of the same high quality.

8002. (B) *EA-ITP-ASG2E, Page 1-3; EA-43.13-1A/3, Page 3*
When a wood splice is made, the bond between the two pieces of wood consists only of a layer of glue. How strong the joint will be depends on how well the gluing operation was performed.

8003. (C) *EA-ITP-A2, Pages 250-251; EA-ITP-ASG2E, Page 1-3; EA-43.13-1A/3, Page 1*
The species used to repair a part should be the same as that of the original whenever possible; however, permissible substitutes are given in EA-43.13-1A/3.

8004. (C) *EA-ITP-ASG2E, Page 1-4; EA-43.13-1A/3, Page 7*
In cases of elongated boltholes in a spar, or cracks in the vicinity of boltholes, a new spar section should be spliced in, or the spar should be replaced.

8005. (A) *EA-ITP-ASG2E, Page 1-3; EA-ITP-AB, Page 34; EA-43.13-1A/3, Page 3*
Compression failures are characterized by a buckling of the fibers that appear as streaks on the surface of the piece, substantially at right angles to the grain.

8006. (C) *EA-ITP-A2, Page 250; EA-ITP-ASG2E, Page 1-3; EA-43.13-1A/3, Page 3.*
Decay. All pieces must be free from rot, dote, red heart, purple heart, and all other forms of decay.

8007. (A) *EA-ITP-ASG2E, Page 1-3; EA-ITP-AB, Page 34; EA-43.13-1A/3, Page 3*
Mineral streaks are not acceptable in wood if they are accompanied by any decay.

8008. (C) *EA-ITP-ASG2E, Page 1-4; EA-AC65-15A, Page 8*
I-beam spars are externally routed on both sides to reduce weight while retaining adequate strength.

8009. (A) *EA-ITP-ASG2E, Page 1-3; EA-43.13-1A/3, Page 3*
Pin knot clusters. Small clusters are acceptable, provided they produce only a small effect on grain direction.

8010. (B) *EA-ITP-A2, Page 9; EA-ITP-ASG2E, Page 1-4; EA-AC65-15A, Page 13*
A cantilever wing is one that uses no external bracing.

8011. (A) *EA-ITP-ASG2E, Page 1-3; EA-AC65-15A, Page 228*
Laminated wood is an assembly of two or more layers of wood which have been glued together with the grain of all layers or laminations approximately parallel.

8012. (A) *EA-ITP-A2, Pages 259-262; EA-ITP-ASG2E, Page 1-4; EA-ITP-AB, Pages 43-47; EA-AC65-15A, Pages 235-239*
Circular and elliptical patches are good to use when repairing plywood skin because they don't have any sharp corners which could set up a stress in the skin.

8013. (A) *EA-ITP-ASG2E, Page 1-3; Aircraft Maintenance and Repair, Page 104*
If the glue is in good condition any separation will occur in the wood and not in the glue.

8014. (C) *EA-ITP-ASG2E, Page 1-3; EA-43.13-1A/3, Page 3*
This defect is caused from the wood being overstressed in compression due to natural forces during the growth of the tree, felling trees on rough or irregular ground, or rough handling of logs or lumber.

8015. (C) *EA-ITP-A2, Pages 136-137; EA-ITP-ASG2E, Page 1-5; EA-ITP-GB, Pages 153-154; EA-43.13-1A/3, Page 87*
Finishing tape is used to cover all the lacing, and it must be of at least the same width and quality as that used on the original airplane. It is applied with dope, after the first or second coat of dope has dried.

8016. (C) *EA-ITP-A2, Page 125; EA-ITP-ASG2E, Page 1-4; EA-ITP-AB, Page 141; EA-AC65-15A, Page 85*
The quality and strength of fabric, surface tape, lacing cord, and thread, etc. are determined by the aircraft's never-exceed speed and the pounds per square foot of wing loading.

8017. (C) *EA-ITP-A2, Page 133; EA-ITP-ASG2E, Page 1-4; EA-43.13-1A/3, Page 85*
The French-fell seam is illustrated in the above references.

8018. (C) *EA-ITP-A2, Page 138; EA-ITP-ASG2E, Page 1-5; EA-ITP-AB, Page 156; EA-43.13-1A/3, Page 87*
Pinked surface tape is sometimes applied over the trailing edges of control surfaces and airfoils. If separation of the tape from the trailing edge begins, it will tear at a notched section, and thereby prevent loosening of the entire strip, which could seriously affect the controllability of the aircraft.

8019. (A) *EA-ITP-A2, Page 141; EA-ITP-ASG2E, Page 1-5; EA-AC65-15A, Page 88*
Glass cloth is not affected by moisture, mildew, chemicals, or acids.

8020. (C) *EA-ITP-A2, Pages 141-142; EA-ITP-ASG2E, Page 1-5; EA-ITP-AB, Page 160; EA-AC65-15A, Pages 99-100*
Tears in fabric should be repaired by sewing the torn edges together using a baseball stitch, and doping a piece of pinked-edge fabric over the tear. If this damaged area were to exceed 16″ in any direction, a new panel would need to be sewn and doped in.

8021. (C) *EA-ITP-A2, Page 125; EA-ITP-ASG2E, Page 1-4*
The stength of aircraft fabric is classified by determining its tensil strength, in lb. per inch of width. Do not confuse this with lb. per square inch.

8022. (A) *EA-ITP-A2, Page 128; EA-ITP-ASG2E, Page 1-6*
Rejuvenator is a mixture of very potent solvents and plasticizers. It may be used to restore the resilience of doped surfaces.

8023. (B) *EA-ITP-A2, Page 132; EA-ITP-ASG2E, Page 1-4; EA-ITP-AB, Page 150; EA-AC65-15A, Page 90*
Machine sewn seams should be of the folded fell or french fell types. A plain lapped seam is satisfactory where selvaged edges or pinked edges are joined.

8024. (B) *EA-ITP-A2, Page 129; EA-ITP-ASG2E, Page 1-4*
Fabric is considered airworthy until it deteriorates to a strength of less than 70% of the strength of new fabric. If Grade A Cotton is used on an aircraft that requires only Intermediate Grade fabric, it can deteriorate to 46 lb. (70% of the strength of Intermediate Grade fabric) before it must be replaced.

8025. (B) *EA-ITP-ASG2E, Page 1-4; EA-AC65-15A, Page 91*
Treat all parts of the structure that come into contact with doped fabric with a protective coating, such as aluminum foil, dope-proof paint, or cellulose tape. Clad aluminum and stainless steel parts need not be dope proofed.

8026. (A) *EA-ITP-A2, Page 160; EA-ITP-ASG2E, Page 1-7*
For 12-inch characters the width will be ⅔ 12=8 inches. The 1 will be 2 inches wide, and the spacing will be ¼ 8 inches, or 2 inches. This will give us 8 +2 +2 +2 +8 +2 +8 +2 +8 +2 +8 or 52-inches.

8027. (C) *EA-ITP-A2, Page 169; EA-ITP-ASG2E, Page 1-5*
Masking tape should not be allowed to remain on any longer than necessary. If it is difficult to remove, try to soak the tape free with aliphatic naptha.

8028. (B) *EA-ITP-A2, Page 156; EA-ITP-ASG2E, Page 1-6*
Retarder is a special type of thinner having rich solvents. These dry very slowly and prevent the temperature drop at the surface which causes blush.

8029. (A) *EA-ITP-A2, Page 152; EA-ITP-ASG2E, Page 1-7*
As one component of their system, wash primers use phosphoric acid to create a phosphate film on the surface of the aluminum.

8030. (C) *EA-ITP-A2, Page 150; EA-ITP-ASG2E, Page 1-7*
Polyurethane is a two-part chemically cured finish that has a solids content of up to 60%. The high gloss that is characteristic of these finishes is due to the slow-flowing resins used.

8031. (C) *EA-ITP-A2, Page 146; EA-ITP-ASG2E, Page 1-7*
Aluminum pigment is mixed with clear dope and applied to fabric surfaces to help combat the deterioration caused by sunlight.

8032. (B) *EA-ITP-ASG2E, Page 1-7; EA-AC65-15A, Page 113*
Acetone is a fast-evaporating dope solvent that is suitable for removing grease from fabric prior to doping.

8033. **(B)** *EA-ITP-A2, Page 172; EA-ITP-ASG2E, Page 1-7*
When dry sanding a fabric covered aircraft, be sure to electrically ground the structure. Static electricity from sanding can cause a spark to jump inside the structure and ignite volatile fumes.

8034. **(A)** *EA-ITP-A2, Page 152; EA-ITP-ASG2E, Page 1-7*
One of the most critical aspects of applying a wash primer is the necessity to have sufficient moisture in the air. The moisture is required to convert the acid in the primer into a phosphate film. This conversion process takes about 30 minutes. If sufficient moisture is not present, the finish coat will trap the acid against the surface and cause corrosion.

8035. **(A)** *EA-ITP-A2, Page 134; EA-ITP-ASG2E, Page 1-4; EA-AC65-15A, Page 111*
Fungicidal dope normally is used as the first coat for fabrics to prevent rotting. The first coat is applied extremely thin so it can thoroughly saturate both sides of the fabric.

8036. **(C)** *EA-ITP-A2, Page 157; EA-ITP-ASG2E, Page 1-7*
Moisture, skin oils and other contaminants may be left on a surface when you touch it. To prevent problems with the finishing system you should avoid touching surfaces after the final cleaning or wipe-down with M.E.K.

8037. **(B)** *EA-ITP-A2, Page 152; EA-ITP-ASG2E, Page 1-7*
About the most critical aspect of the application of wash primer is the necessity of having sufficient moisture in the air. This moisture is required to properly convert the acid into a phosphate film. This process should take about 30-minutes.

8038. **(A)** *EA-ITP-A2, Page 147; EA-ITP-ASG2E, Page 1-6; EA-ITP-AB, Page 166; EA-AC65-15A, Page 124*
Sags and runs result from too much paint being applied, causing the film of wet paint to move by gravity, and presenting a sagging appearance.

8039. **(B)** *EA-ITP-A2, Page 167; EA-ITP-ASG2E, Page 1-6*
Blushing occurs because of the effects of high humidity and the temperature drop during the evaporation of the solvents in the dope. Anything which affects either the humidity or the speed of solvent evaporation could cause blushing.

8040. **(C)** *EA-ITP-ASG2E, Page 1-7; EA-AC65-15A, Page 122*
Old wash primer coats may be overcoated directly with epoxy finishes. A new second coat of wash primer must be applied if an acrylic finish is to be applied.

8041. **(B)** *EA-ITP-A2, Page 175; EA-ITP-ASG2E, Page 1-7; EA-ITP-AB, Page 77; EA-AC65-15A, Page 132*
Rivets hold pieces of aircraft skin together, and in a properly designed riveted joint, the rivets support shear loads only.

8042. **(A)** *EA-ITP-A2, Pages 208-209; EA-ITP-ASG2E, Page 1-8*
Huckbolts and Lockbolts are manufactured to the same Federal Standard. In each case, the tension-type has 4 to 5 locking grooves and the shear-type has 2.

8043. **(A)** *EA-ITP-A2, Page 200; EA-ITP-ASG2E, Page 1-7; EA-ITP-GB, Page 300; EA-AC65-9A, Page 151*
The 2117-T rivet, known as the field rivet, is used more than any other for riveting aluminum alloy structures. The field rivet is in wide demand because it is ready for use as received, and needs no further heat treating or annealing.

8044. **(C)** *EA-ITP-A2, Page 211; EA-ITP-ASG2E, Page 1-7; EA-ITP-AB, Page 103; EA-AC65-15A, Page 165*
As a general rule, the rivet diameter should be not less than three times the thickness of the thickest sheet being riveted.

8045. **(A)** *EA-ITP-A2, Page 212; EA-ITP-ASG2E, Page 1-7; EA-ITP-AB, Page 114*
A properly formed shop head is one-half the shank diameter high, its diameter is one-and-one-half times that of the shank, and it is concentric with the hole.

8046. **(C)** *EA-ITP-A2, Page 207; EA-ITP-ASG2E, Page 1-8*
The hand tools used for installation of Hi-Loks are an open end wrench and hex key (allen wrench).

8047. **(A)** *EA-ITP-ASG2E, Page 1-10; EA-AC65-9A, Page 141*
When working with Dzus fasteners, it is worthwhile to know that the body diameter, length, and head type may be identified by the markings found on the head of the stud.

8048. **(A)** *EA-ITP-ASG2E, Page 1-10; EA-AC65-9A, Page 141*
The length of a Dzus fastener is measured in hundredths of an inch and is the distance from the head of the stud to the bottom of the spring hole.

8049. (C) *EA-ITP-ASG2E, Page 1-10; EA-AC65-9A, Page 141*
The diameter of a Dzus fastener is always measured in sixteenths of an inch.

8050. (B) *EA-ITP-A2, Page 211; EA-ITP-ASG2E, Page 1-10*
Rivnuts are a combination of rivet and nut. A hole is drilled through the skin and the Rivnut, which resembles a hollow rivet with threads inside its shank, is slipped into the hole and upset with a special puller.

8051. (B) *EA-ITP-A2, Page 205; EA-ITP-ASG2E, Page 1-9*
Both Cherrymax and Olympic Lock rivets require special pulling type tools for installation.

8052. (A) *EA-ITP-A2, Pages 333-334; EA-ITP-ASG2E, Page 1-11*
If the metal of a solid rivet were expanded to completely fill the hole in a composite structure, the rivet would expand against the sides of the laminate and could cause delamination at the edges of the hole.

8053. (B) *EA-ITP-A2, Page 334; EA-ITP-ASG2E, Page 1-11*
Aluminum fasteners should not be used with carbon/graphite material because of its tendency to corrode the aluminum.

8054. (C) *EA-ITP-A2, Page 191; EA-ITP-ASG2E, Page 1-10; EA-ITP-AB, Pages 55-56; EA-AC65-15A, Page 201*
Aircraft structure must not only be strong, but rigid as well. Very thin aluminum skins normally provide adequate strength, but lack the rigidity needed. A structural material that provides both the strength and rigidity with light weight is the honeycomb structure.

8055. (B) *EA-ITP-A2, Page 284; EA-ITP-ASG2E, Page 1-10*
A change in sound when performing a ring (coin-tap) test may indicate an area of delamination. The change in sound could also be due to the transition to a different internal structure. More sophisticated testing would be necessary to determine the extent of the delamination, if present.

8056. (A) *EA-ITP-A2, Pages 284-285; EA-ITP-ASG2E, Page 1-11*
X-ray inspection allows you to "see" moisture trapped in the structure, and acoustic emission testing allows you to hear the noise associated with the bubbles generated by the corrosion process.

8057. (C) *EA-ITP-A2, Page 291; EA-ITP-ASG2E, Page 1-11; EA-ITP-AB, Page 62*
After damaged honeycomb core has been removed, either a balsa wood or honeycomb replacement plug is installed. If balsa is used, the plug should be cut so the grain is perpendicular to the skin.

8058. (B) *EA-ITP-A2, Page 290; EA-ITP-ASG2E, Page 1-11; EA-ITP-AB, Page 61*
A doubler should be cut from a piece of aluminum the same thickness as the original face or one and one-half times as thick, and it should be tapered at a ratio of about 100:1.

8059. (B) *EA-ITP-A2, Page 276; EA-ITP-ASG2E, Page 1-12*
Mix enough extra resin in each batch to make an identical lay-up on a piece of scrap aluminum. This test sample will give a good indication of the cure time needed, as well as the physical characteristics of the lay-up.

8060. (A) *EA-ITP-ASG2E, Page 1-11*
Acoustical emission monitoring is a system using a very sensitive microphone and an amplifier. With it you can pick up the sound of the bubbles associated with active corrosion.

8061. (A) *EA-ITP-A2, Page 287; EA-ITP-ASG2E, Page 1-12; EA-ITP-AB, Page 57*
When metal is used as a facing for laminated structure, it must be protected from corrosion by using a corrosion-inhibiting primer and a finish that will completely seal the atmosphere away from the repaired area.

8062. (A) *EA-ITP-A2, Page 284; EA-ITP-ASG2E, Page 1-11; EA-ITP-AB, Page 57*
If there is surface indication of damage to a fiberglass honeycomb or fiberglass laminated sheet, remove the paint from both sides of the sheet in the suspect area and hold a strong light so it will shine through.

8063. (B) *EA-ITP-A2, Page 284; EA-ITP-ASG2E, Page 1-11; EA-ITP-AB, Page 57; EA-AC65-15A, Page 201*
If honeycomb structure is faced with metal, the way to find delamination is by tapping on the surface with a coin. A good bond will produce a clear ringing sound. A delamination will produce a sound which is dull and lifeless.

8064. (B) *EA-ITP-A2, Page 281; EA-ITP-ASG2E, Page 1-12; EA-43.13-1A/3, Page 77*
Phenolic microballoons are added to lower the density and give greater flexibility, thus lowering stress concentrations in the repair area.

8065. **(A)** *EA-ITP-A2, Page 278; EA-ITP-ASG2E, Page 1-12*
The pot life of a catalyzed resin is the amount of time that the mixed resins will be workable. Some resin systems have a very short (45 minutes) pot life, others have a long pot life (4 hours).

8066. **(A)** *EA-ITP-A2, Page 264; EA-ITP-ASG2E, Page 1-12*
A thermoplastic material is hard in its original state, but will soften and become pliable when it is heated. It can be molded or shaped and will retain its shape when it cools.

8067. **(B)** *EA-ITP-A2, Page 266; EA-ITP-ASG2E, Page 1-11*
S-glass is a magnesia-alumina-silicate glass. It is used where a very high tensile strength fiberglass is needed.

8068. **(A)** *EA-ITP-ASG2E, Page 1-12; EA-AC65-15A, Page 214*
To identify a type of plastic as acrylic or acetate, put a drop of zinc chloride on the plastic. If the zinc chloride has no effect, the plastic is acrylic.

8069. **(B)** *EA-ITP-A2, Page 288; EA-ITP-ASG2E, Page 1-11; EA-ITP-AB, Page 61; EA-43.13-1A/3, Page 74*
Superficial scars, scratches, surface abrasion, or rain erosion can generally be repaired by applying one or more coats of a suitable resin, catalyzed to cure at room temperature.

8070. **(A)** *EA-ITP-A2, Page 265; EA-ITP-ASG2E, Page 1-11*
E-glass is a borosilicate glass, which is the most common type of fiberglass used for reinforcement. E-glass is known as electric glass because of its high resistivity.

8071. **(C)** *EA-ITP-ASG2E, Page 1-11; EA-AC65-15A, Page 206*
Bonded honeycomb structure damage up to 1" in diameter can be repaired by a hole filling technique known as the "potted compound repair".

8072. **(B)** *EA-ITP-A2, Page 266; EA-ITP-ASG2E, Page 1-12*
In a woven application, there are typically more threads woven into the warp than the fill direction. This material will be stronger in the warp direction. Warp direction may be indicated by inserting another type of thread at periodic intervals.

8073. **(C)** *EA-ITP-A2, Page 303; EA-ITP-ASG2E, Page 1-12*
A warp compass is a tool which can be used to reference the orientation of the warp of the fabric.

8074. **(B)** *EA-ITP-A2, Page 303; EA-ITP-ASG2E, Page 1-12*
Structural composite parts are engineered and manufactured to endure specific stress loads. Their ability to endure these stress loads is dependent, in great measure, to the way in which the fibers are oriented. The fiber orientation of patches must be the same as in the original structure.

8075. **(C)** *EA-ITP-A2, Page 304; EA-ITP-ASG2E, Page 1-12*
A fabric/resin mixture should be about 60/40. A resin rich repair is more susceptible to cracking due to a lack of adequate fiber for support. A resin starved repair is weak in those areas where sufficient resin does not provide stiffness, or because fibers are not held together.

8076. **(A)** *EA-ITP-A2, Page 323; EA-ITP-ASG2E, Page 1-12*
Bleeders are absorbent materials which are used to soak up the excess resins. Always use bleeders in conjunction with a release fabric, peel ply or release film.

8077. **(B)** *EA-ITP-A2, Page 305; EA-ITP-ASG2E, Page 1-12*
Pre-impregnated (Pre-preg) materials simplify work with composites. For proper curing, vacuum bagging is often used to apply pressure, and heat used to produce a proper cure.

8078. **(B)** *EA-ITP-A2, Page 276; EA-ITP-ASG2E, Page 1-12; EA-ITP-AB, Page 54*
When hinges are installed on top of a large flat surface, using long strips of glass fiber and resin, the shrinkage will often warp the surface out of shape. One way to handle this kind of situation is to use short strips of fiberglass to bond the fasteners to the panel.

8079. **(A)** *EA-ITP-A2, Page 288; EA-ITP-ASG2E, Page 1-11; EA-ITP-AB, Page 59*
The surface to be repaired should first be cleaned with soap and water, then scrubbed with methyl-ethyl-ketone. The next step is to sand the area and, finally, it is cleaned again with methyl-ethyl-ketone to remove any sanding residue and moisture.

8080. **(C)** *EA-ITP-A2, Page 333; EA-ITP-ASG2E, Page 1-12*
Generally coolant or cutting fluid is not used with composites, for a variety of reasons. Many of these materials will absorb liquids. If water is used, the area should be thoroughly dried before bonding.

8081. (A) *EA-ITP-ASG2E, Page 1-11; EA-43.13-1A/3, Page 74*
Damage, not exceeding the first layer or ply of fiberglass laminate, can be repaired by filling with a putty consisting of a compatible room-temperature-setting resin and clean short glass fibers.

8082. (A) *EA-ITP-ASG2E, Page 1-11; EA-43.13-1A/3, Page 75*
Damage that is completely through a laminate sandwich structure may be repaired by such methods as scarfed joint, stepped joint, etc.

8083. (B) *EA-ITP-ASG2E, Page 1-11; EA-43.13-1A/3, Page 75*
Damage that extends completely through one face of a laminate structure and into the core requires the replacement of the damaged core and facing.

8084. (C) *EA-ITP-A2, Page 285; EA-358, Page 130; EA-ITP-ASG2E, Page 1-10*
The older type of fiberglass repairs cannot be used on advanced composite structures. Old style fiberglass repair methods were not necessarily intended to be used to restore full structural strength.

8085. (C) *EA-ITP-A2, Page 306; EA-ITP-ASG2E, Page 1-10*
When the proper facilities, curing, and bagging equipment are not available, a pre-cured patch inserted with blind fasteners may be used. This type of repair may be considered temporary until the damage can be scarfed down and patches correctly laminated on with heat or pressure.

8086. (A) *EA-ITP-A2, Page 281; EA-ITP-ASG2E, Page 1-11*
Fillers are materials that, when added to resins, control viscosity and weight, increase pot life and strength, and make resin application easier. Although all items listed are fillers, only flox and chopped fibers (answer A) add strength.

8087. (B) *EA-ITP-A2, Page 282; EA-ITP-ASG2E, Page 1-11*
It is important, when doing a repair, to line up the ribbon direction of the replacement honeycomb core with the ribbon direction of the original part.

8088. (A) *EA-ITP-A2, Page 266; EA-ITP-ASG2E, Page 1-10*
Arimid is the name given to aromatic polymide fibers. They are usually characterized by their yellow color, light weight, excellent tensile strength and remarkable flexibility.

8089. (B) *EA-ITP-A2, Page 267; EA-ITP-ASG2E, Page 1-10*
Carbon/graphite fiber is very strong, stiff and has a higher compressive strength than Kevlar. Special corrosion control techniques must be employed when carbon/graphite materials are in contact with aluminum.

8090. (C) *EA-ITP-A2, Page 345; EA-ITP-ASG2E, Page 1-12; EA-AC65-15A, Page 218*
Crazing of plastic enclosures appears as a network of cracks running in all directions over the surface of the plastic. It can also occur within the plastic, at or near cemented joints.

8091. (B) *EA-ITP-A2, Page 345; EA-ITP-ASG2E, Page 1-13; EA-ITP-AB, Page 74; EA-43.13-1A/3, Page 159*
Screws or bolts that go through the windshield should be tightened down snug and then backed out a full turn, so the plastic can shift as it expands and contracts.

8092. (C) *EA-ITP-ASG2E, Page 1-13*
The standards which safety belts must meet are contained in Technical Standard Order C22. The Technical Standard Order System is covered in FAR 21, Subpart O, TSO, C22.

8093. (B) *EA-ITP-ASG2E, Page 1-13; EA-43.13-1A/3, Page 159*
When rivets are used in the installation of windshields, the installer should provide adequate spacers, or other satisfactory means, to prevent excessive tightening of the frames to the plastic.

8094. (A) *EA-ITP-ASG2E, Page 1-13; EA-43.13-1A/3, Page 159*
Plastics expand and contract considerably more than the metal channels in which they are mounted. To accommodate this, windshields are mounted to a sufficient depth in the channel to prevent them from falling out when the panel contracts at low temperatures, or deforms under load.

8095. (C) *EA-ITP-A2, Page 344; EA-ITP-ASG2E, Page 1-12; EA-ITP-AB, Page 73; EA-AC65-15A, Page 219*
After a windshield is cleaned, a thin coating of wax should be applied to fill any minute scratches that may be present.

8096. (B) *EA-ITP-ASG2E, Page 1-13*
The upholstery materials which are installed in the cabin of standard category airplanes must meet the requirements of FAR 23.

8097. (B) *EA-ITP-A2, Page 342; EA-ITP-ASG2E, Page 1-13; EA-ITP-AB, Page 71; EA-AC65-9A, Page 217*
There are two methods for cementing plastics — soaking and gluing. The most common method used is soaking method.

8098. (B) *EA-ITP-A2, Page 342; EA-ITP-ASG2E, Page 1-13; EA-ITP-AB, Page 70; EA-AC65-15A, Page 216*
When drilling plastic, the drill bit used should be carefully ground, and be free from nicks and burrs, which would affect the surface finish. The drill bit should be ground with a greater included angle than would be used for drilling soft metal.

8099. (C) *EA-ITP-ASG2E, Page 1-17; EA-AC65-15A, Pages 194-195, 242*
Gussets are used in the repair of aircraft structures to join and reinforce intersecting structural members. They are used extensively in wood structures.

8100. (C) *EA-ITP-G2, Page 360; EA-ITP-ASG2E, Page 1-17; EA-ITP-GB, Page 349; EA-AC65-9A, Page 538*
A twist drill has a conical point with cutting edges formed by the ends of the flutes.

8101. (A) *EA-ITP-A2, Pages 212-213; EA-ITP-ASG2E, Page 1-16; EA-ITP-AB, Pages 103-104; EA-43.13-1A/3, Page 51*
The rivets identified in this question are $4/32''$ in diameter. Minimum edge distance, according to the EA-43.13-1A/3 is two times the diameter. This gives an edge distance of $1/4''$. One-fourth of an inch taken away from each side of the plate leaves the dimensions as $4\frac{1}{2}$ by $9\frac{1}{2}$. With four times the rivet diameter spacing, the spacing will be $1/2''$. Because the plate has two $4\frac{1}{2}''$ sides and two $9\frac{1}{2}''$ sides (in terms of rivet installation), there are 28″ total to be covered, with $1/2''$ spacing, or 56 rivets needed.

8102. (B) *EA-ITP-A2, Page 28; EA-ITP-ASG2E, Page 1-13; EA-AC65-15A, Page 3*
Primary bending loads in the semimonocoque fuselage are taken up by longitudinal members called longerons. These are supplemented by other longitudinal members called stringers.

8103. (A) *EA-ITP-A2, Page 237; EA-ITP-ASG2E, Page 1-17; EA-ITP-AB, Page 129*
Scratches in aluminum alloy skins encourage corrosion to form, and concentrate the stresses enough that they may cause the part to crack. If the scratch is not too deep, it can be burnished with a smooth, round-end piece of steel to force the metal back into the scratch.

8104. (B) *EA-ITP-G2, Pages 362-363; EA-ITP-ASG2E, Page 1-16; EA-ITP-GB, Page 350; EA-AC65-9A, Page 539*
When drilling soft metals a cutting angle of 90° may be more efficient.

8105. (B) *EA-ITP-G2, Page 363; EA-ITP-ASG2E, Page 1-17; EA-ITP-GB, Page 350; EA-AC65-15A, Page 169*
When drilling hard metals, the twist drill should have an included angle of 118°, and should be operated at low speeds.

8106. (C) *EA-ITP-A2, Page 213; EA-ITP-ASG2E, Page 1-16; EA-ITP-AB, Pages 103-104; EA-AC65-15A, Pages 164-165*
The minimum edge distance when riveting is two times the diameter of the rivet, which in this case is $1/4''$. Because this is an overlap splice, there are two edge distances involved, or a total of $1/2''$. The minimum pitch, or distance between rivets, should be no less than three times the diameter. The transverse pitch, or distance between rows, should be about 75% of the pitch. In this case, 75% of $3/8$ is $9/32$. The edge distance total of $1/2''$, and the transverse pitch of $9/32$, gives a total of $25/32$ minimum overlap. The closest answer to this is 1″.

8107. (C) *EA-ITP-ASG2E, Page 1-17; EA-ITP-AB, Page 128; EA-AC65-15A, Page 176*
The presence of chipped or cracked paint around the heads of rivets may indicate shifted or loose rivets. If tipped or loose rivet heads are noticed, they will generally show up in groups of several consecutive rivets, and will probably be tipped in the same direction.

8108. (A) *EA-ITP-A2, Page 213; EA-ITP-ASG2E, Page 1-16; EA-ITP-AB, Page 103; EA-AC65-15A, Page 165*
If rivets are too close to the edge, the sheet is likely to tear, but if they are too far back, the edge of the sheet will lift. Accepted practice is to have an edge distance of no less than two times the diameter of the rivet.

8109. (C) *EA-ITP-G2, Page 363; EA-ITP-ASG2E, Page 1-17; EA-ITP-GB, Page 350*
When drilling stainless steel, angles as flat as 150° can be used. Drilling of hard metals should always be done with the drill turning at a slow speed.

8110. (C) *EA-ITP-A2, Page 213; EA-ITP-ASG2E, Page 1-16; EA-ITP-AB, Page 104; EA-AC65-15A, Page 165*
The minimum spacing, or pitch, for a single row of rivets is three times the diameter of the rivet shank.

8111. (A) *EA-ITP-A2, Page 215; EA-ITP-ASG2E, Page 1-14; EA-ITP-AB, Page 108*
When a rivet has a 100° countersunk head, it has a head whose sides form a 100° angle. In choice "A" for this question, there are 260° of angle shown, if you don't include the rivet's head. This leaves 100° for the head of the rivet.

8112. (A) *EA-ITP-G2, Page 356; EA-ITP-ASG2E, Page 1-17; EA-ITP-GB, Pages 344-345; EA-AC65-9A, Pages 536-537*
Files which have one row of teeth are known as "single-cut", and those with two rows of teeth as "double-cut". The coarseness of the teeth are rated from coarse to fine: Coarse cut, bastard cut, second cut, smooth cut, and dead smooth cut. Generally speaking, double cut files are used for removing the most material. When cutting with the file, one should apply pressure only on the forward cut, unless the material is soft. In this case the file may be permitted to remain in contact with the material on the return to help remove chips from the teeth.

8113. (C) *EA-ITP-A2, Page 216; EA-ITP-ASG2E, Page 1-15; EA-AC65-15A, Page 170*
As a general rule, use the drill or machine countersink method when the thickness of the material is greater than the thickness of the rivet head. If the material is too thin, drill or machine countersinking will grind away too much of the material. In this case, dimpling should be used.

8114. (A) *EA-ITP-A2, Pages 175-176, 239; EA-ITP-ASG2E, Page 1-17; EA-ITP-AB, Pages 78, 131-132*
When repairing a small hole on a metal stressed skin, the riveted joint needs to be designed so that the rivets take the stress, and not the skin, so if there is a failure, it will be the rivets that shear and not the skin that tears.

8115. (B) *EA-ITP-G2, Page 365; EA-ITP-ASG2E, Page 1-17; EA-ITP-GB, Page 352; EA-AC65-9A, Page 541*
When a reamer is being used, it should be rotated in the cutting direction only. The reamer should be turned steadily and evenly to prevent chattering or marking and scoring of the hole walls.

8116. (C) *EA-ITP-A2, Page 243; EA-ITP-ASG2E, Page 1-13; EA-ITP-AB, Pages 133-134; EA-43.13-1A/3, Pages 60 and 64*
Repairs or splices involving stringers are allowed, but it is especially important that care be taken when splicing stringers on the lower surface of stressed skin wings. It is in this area of the wing that very high tension stresses exist.

8117. (A) *EA-ITP-ASG2E, Page 1-17; EA-43.13-1A/3, Page 60*
Aluminum members which are slightly bent may be straightened cold and examined with a magnifying glass for injury to the material. Reinforce the straightened parts to an extent depending on the condition of the material.

8118. (B) *EA-ITP-A2, Page 180; EA-ITP-ASG2E, Page 1-14; EA-AC65-9A, Page 201*
Clad aluminum alloys consist of an aluminum alloy core coated with a layer of pure aluminum, to a depth of approximately 5½% on each side. The pure aluminum coating affords a dual protection for the core, preventing contact with any corrosive agents, and protecting the core electrolytically by preventing any attack caused by scratching, or from other abrasions.

8119. (B) *EA-ITP-A2, Pages 8-9; EA-ITP-ASG2E, Page 1-17; EA-AC65-15A, Pages 7 and 13*
Wings of the cantilever design are built so that no external bracing is needed.

8120. (A) *EA-ITP-A2, Page 238; EA-ITP-ASG2E, Page 1-17; EA-ITP-AB, Page 129*
Aircraft structural units, which are made from sheet metal, are generally repairable if the repair procedures used bring the aircraft back to an airworthy condition, and meet the requirements of the Federal Aviation Administration.

8121. (B) *EA-ITP-A2, Page 213; EA-ITP-ASG2E, Page 1-16; EA-ITP-AB, Page 104; EA-AC65-15A, Page 165*
Rivet spacing, known as pitch, is determined by the diameter of the rivets being used.

8122. (A) *EA-ITP-G2, Page 362; EA-ITP-ASG2E, Page 1-16; EA-ITP-GB, Page 350; EA-AC65-9A, Page 539*
For most drilling, a twist drill with a cutting angle of 118° will be sufficient; however, when drilling soft metals, a cutting angle of 90° may be more efficient.

8123. (C) *EA-ITP-A2, Page 181; EA-ITP-ASG2E, Page 1-19; EA-ITP-AB, Pages 83-84; EA-AC65-9A, Page 201*
Alcad 2024-T3 aluminum has a pure aluminum coating over the aluminum alloy core. If the coating is damaged, the aluminum alloy will be subject to outside corrosive elements.

8124. (C) *EA-ITP-A2, Pages 25-28; EA-ITP-ASG2E, Page 1-14; EA-ITP-AB, Page 26; EA-AC65-15A, Page 3*
The monocoque fuselage relies largely on the strength of the skin or covering to carry the primary stresses.

8125. (B) *EA-ITP-A2, Page 28; EA-ITP-ASG2E, Page 1-14; EA-AC65-15A, Page 4*
Stringers and longerons prevent tension and compression from bending the fuselage.

8126. (A) *EA-ITP-A2, Page 213; EA-ITP-ASG2E, Page 1-16*
The distance between the rows of rivets in a multi-row layout should be about 75% of the pitch, and the rivets in adjacent rows should be staggered.

8127. (B) *EA-ITP-A2, Page 213; EA-ITP-ASG2E, Page 1-16*
The distance between adjacent rivets in a row is called the pitch of the rivet.

8128. (C) *EA-ITP-A2, Page 217; EA-ITP-ASG2E, Page 1-14; EA-AC65-15A, Page 170; EA-SM, Page 31*
See Figure 5-47; EA-AC65-15A. Sometimes the maximum diameter rivet that can be used will be governed by the top skin thickness. This is why some skins are dimpled.

8129. (C) *EA-ITP-G2, Page 424; EA-ITP-ASG2E, Page 1-17*
Fretting corrosion may occur around rivets in a skin, and will be indicated by dark deposits around the rivet heads streaming out behind, give the appearance of "rivet smoking".

8130. (C) *EA-ITP-A2, Page 201; EA-ITP-ASG2E, Page 1-15; EA-ITP-GB, Page 299; EA-AC65-9A, Pages 152-153*
Markings on the heads of rivets are used to classify their characteristics. The different markings indicate the composition of the rivet stock.

8131. (A) *EA-ITPOA2, Page 200; EA-ITP-ASG2E, Page 1-15*
Type D or DD rivets are known as "ice box" rivets because they must be stored at low temperatures after having been annealed. Within munutes after being removed from refrigeration they will begin to age harden. Only after the age hardening process will they achieve their full strength.

8132. (B) *EA-ITP-A2, Page 213; EA-ITP-ASG2E, Page 1-16; EA-ITP-AB, Pages 102-103; EA-AC65-15A, Page 165*
The length of the rivet has no effect on the rivet spacing.

8133. (C) *EA-ITP-A2, Page 200; EA-ITP-ASG2E, Page 1-15; EA-ITP-GB, Page 300; EA-AC65-9A, Page 152*
When an aluminum alloy is heated to a specific temperature and quenched in water or in brine, it will not get hard immediately as steel does, but will gradually become hard and gain its full strength. Rivets made of an alloy such as 2017 can have their aging process retarded by storing them in a refrigerator.

8134. (A) *EA-ITP-A2, Page 200; EA-ITP-ASG2E, Page 1-15; EA-ITP-GB, Page 299; EA-AC65-9A, Page 151*
The 1100 rivet, which is composed of 99.45% pure aluminum, is very soft. It is generally used for riveting non-structural parts.

8135. (B) *EA-ITP-A2, Page 220; EA-ITP-ASG2E, Page 1-17; EA-ITP-AB, Pages 111-112*
The radius of the cup of the rivet set must be slightly larger than the radius of the rivet head. If it is too small, it will produce a small indentation on the head of the rivet. If it is too large, it will produce the same indentation on the skin around the rivet.

8136. (A) *EA-ITP-A2, Page 200; EA-ITP-ASG2E, Page 1-15; EA-ITP-GB, Page 300; EA-AC65-9A, Page 152*
Rivets classed as "D" and "DD" become hard within a few hours of being heat treated. If this happens, they must be heat treated again before they can be used.

8137. (B) *EA-ITP-A2, Page 201; EA-ITP-ASG2E, Page 1-15*
The first number following the letter (material code) indicates the diameter of the rivet in $\frac{1}{32}$-inch increments. $\frac{4}{32}$ = $\frac{1}{8}$-inch diameter. The number following the dash is the length in $\frac{1}{16}$-inch increments $\frac{8}{16}$ = $\frac{1}{2}$ inch.

8138. (A) *EA-ITP-G2, Page 391; EA-ITP-ASG2E, Page 1-14; EA-ITP-GB, Pages 379-380; EA-AC65-9A, Page 199*
The first digit in this aluminum alloy 2024 number indicates what the primary alloying agent is. 1000 series would indicate almost pure aluminum. 2000 series would indicate copper as the primary alloying agent. 3000 series would indicate manganese as the primary alloying agent, and so on.

8139. (A) *EA-ITP-G2, Page 391; EA-ITP-ASG2E, Page 1-14; EA-AC65-9A, Page 199; EA-SM, Page 22*
In the 2xxx through 8xxx groups, the first digit indicates the major alloying agent used in the formation of the alloy. A head is the only way to tell what alloy the rivet is actually made of.

8140. (B) *EA-ITP-A2, Page 212; EA-ITP-ASG2E, Page 1-15; EA-ITP-GB, Page 302; EA-43.13-1A/3, Page 53*
To properly form the shop head, a rivet should be equal to the thickness of the material being joined, plus one and one-half times the rivet diameter. Two pieces of .040 material plus 1.5 × .094 will give us a rivet length of .221. Standard rivet lengths are in 1⁄32″ increments, and the closest standard length to .221 would be 1⁄4″.

8141. (A) *EA-ITP-A2, Page 200; EA-ITP-ASG2E, Page 1-15; EA-ITP-GB, Page 300; EA-AC65-9A, Page 151*
There are more rivets made of 2117 alloy than any other. These rivets are identified by a single dimple on their head.

8142. (C) *EA-ITP-A2, Page 201; EA-ITP-ASG2E, Page 1-15; EA-ITP-GB, Page 302; EA-AC65-9A, Page 153*
The second dash number indicates the rivet's length in sixteenths of an inch. In this case, it is 5⁄16″.

8143. (B) *EA-ITP-A2, Page 200; EA-ITP-ASG2E, Page 1-15; EA-ITP-GB, Page 300; EA-AC65-9A, Page 151*
The reason the 2117-T rivet is so popular is that it is ready for use as received, and needs no further heat treating or annealing. It also has a high resistance to corrosion.

8144. (C) *EA-ITP-A2, Page 201; EA-ITP-ASG2E, Page 1-15; EA-AC65-15A, Page 165*
The grip length of a rivet is that portion of the rivet which is passing through the material.

8145. (B) *EA-ITP-A2, Page 212; EA-ITP-ASG2E, Page 1-15; EA-ITP-AB, Page 103; EA-AC65-15A, Page 165*
To join the two pieces of aluminum together, the rivet will need to be .064″ long plus 1½ times its diameter. The AN470AD-4-4 rivet is 1⁄8″ in diameter and 1⁄4″ long. This rivet meets the length requirement.

8146. (C) *EA-ITP-A2, Page 212; EA-ITP-ASG2E, Page 1-15; EA-ITP-AB, Pages 102-103; EA-43.13-1A/3, Page 59 (Figure 2-24); EA-SM, Page 28*
To find the rivet shank length, add the material thickness and then add 1½ times the rivet diameter. The material required to form the shop head is .3125″ or 5⁄16″ total shank length.

8147. (C) *EA-ITP-ASG2E, Page 1-15; EA-AC65-9A, Page 151*
Mild steel rivets are used for riveting steel parts.

8148. (C) *EA-ITP-A2, Page 200; EA-ITP-ASG2E, Page 1-15; EA-ITP-GB, Pages 300-301; EA-AC65-9A, Page 151*
A "DD" rivet is made from 2024-T aluminum alloy. These rivets are stored in a refrigerator until used. They may be heat treated just before use to soften them, which makes the riveting easier.

8149. (B) *EA-ITP-G2, Pages 442-443; EA-ITP-ASG2E, Pages 1-15; EA-AC65-9A, Page 180*
To prevent corrosion between dissimilar metal joints in which magnesium alloy is involved, two coats of zinc chromate are applied to each surface. Then a layer of pressure-sensitive vinyl tape is applied smoothly and firmly enough to prevent air bubbles and wrinkles. The separating material may be metal primer, aluminum tape, washers, grease, or sealant, depending on the metals involved.

8150. (B) *EA-ITP-A2, Page 212; EA-ITP-ASG2E, Page 1-15; EA-ITP-AB, Pages 102-103; EA-SM, Page 28; EA-AC65-15A, Page 165*
When determining total rivet length, the combined thickness of the materials to be joined must be known. This measurement is grip length. The total length of the rivet must be grip length plus 1½ times the diameter of the rivet shank (necessary to form a shop head on the rivet).

8151. (B) *EA-ITP-A2, Page 222; EA-ITP-ASG2E, Page 1-17*
Using a drill one size smaller than the shank of the rivet, and drilling only through the head should prevent enlarging the rivet hole.

8152. (C) *EA-ITP-ASG2E, Page 1-18; EA-AC65-15A, Page 176*
When rivets are to be inspected which have been critically loaded, but show no visible signs of distortion, some of the rivet heads should be drilled off. If, upon examination, the shank appears joggled and the holes in the sheet are misaligned, the rivet has failed in shear.

8153. (A) *EA-ITP-ASG2E, Page 1-17; EA-AC65-15A, Page 176*
Shear failure is perhaps the most common of rivet failures. It is simply a breakdown of the rivet shank by forces acting along the plane of two adjacent sheets, causing a slipping action which may be severe enough to cut the rivet shank in two. If the shank becomes loaded beyond the yield point of the material and remains overloaded, a permanent shift is established in the sheets and the rivet shank may become joggled.

8154. (B) *EA-ITP-A2, Page 200; EA-ITP-ASG2E, Page 1-15; EA-ITP-AB, Page 99; EA-SM, Page 22; EA-AC65-9A, Page 151*
The 5056 rivet is used for riveting magnesium alloy structures because of its corrosion resistant qualities in combination with magnesium.

8155. (C) *EA-ITP-G2, Page 307; EA-ITP-ASG2E, Page 1-15; EA-ITP-GB, Page 301; EA-AC65-9A, Page 151*
Monel rivets are used for riveting nickel/steel alloys. They can be substituted for those made of corrosion-resistant steel in some cases.

8156. (B) *EA-ITP-A2, Page 200; EA-ITP-ASG2E, Page 1-16; EA-ITP-AB, Page 103*
The length of rivet to use is determined by adding together the thickness of the material, and then adding 1½ times the diameter of the rivets to this value. The thickness of the material in this case is .096 of an inch. A No. 30 drill is .1285″ in diameter, which would take a ⅛″ diameter rivet. One and one-half times the rivet diameter, or 3⁄16 of an inch, plus the thickness of the material, would call for a rivet 5⁄16 of an inch in length.

8157. (A) *EA-ITP-A2, Pages 226-227; EA-ITP-ASG2E, Page 1-21; EA-ITP-AB, Page 118; EA-AC65-15A, Pages 147-148*
The length of flat "A" is equal to the total length from the mold point to the end of the flat, or 4″, minus the setback. The setback for a 90° bend is equal to the metal thickness plus the radius of the bend. The thickness of the metal, .0625 or 1⁄16″, plus the bend radius of 3⁄16″, is equal to ¼ of an inch. Four inches minus ¼″ leaves a flat of 3.75″.

8158. (A) *EA-ITP-A2, Pages 226-227; EA-ITP-ASG2E, Page 1-21; EA-ITP-AB, Page 120; EA-AC65-15A, Pages 147-149*
The amount of material required to make a bend is known as bend allowance. The bend allowance for a 90° bend may be calculated using the formula given in the drawing. Simply substitute 3.14 for pi, .1875 for R, and .0625 for T. Complete the calculations.

8159. (A) *EA-ITP-A2, Pages 226-227; EA-ITP-ASG2E, Page 1-21; EA-ITP-AB, Page 118; EA-AC65-15A, Pages 148-149; EA-SM, Page 65*
Calculating setback formulas is accomplished using the value obtained from a "K" chart.

8160. (B) *EA-ITP-A2, Pages 226-227; EA-ITP-ASG2E, Page 1-21; EA-ITP-AB, Page 121; EA-AC65-15A, Page 152*
The flat pattern layout for a 90° bend should be accomplished by the following steps. First calculate the setback using the formula given (.312). Then deduct the setback from the first overall dimension of flat A (2.688). Calculate the bend allowance using the formula given (.441). Deduct the setback from the overall dimension of the flat B (3.688). Add these three values to get the length of the flat pattern layout.

8161. (B) *EA-ITP-A2, Pages 226-227; EA-ITP-ASG2E, Page 1-18; EA-ITP-AB, Pages 126-127; EA-AC65-15A, Page 161*
With a bumping block clamped in a bench vise, a soft-faced mallet or hard-wood drive block and suitable mallet should be used to start the bumping action. With light blows of the mallet, work the material down gradually from the edges. The object of the bumping process is to work the material into shape by stretching it, rather than by forcing it into the form with heavy blows. Always start bumping near the edge of the form; never start near the center of the blister.

8162. (A) *EA-ITP-A2, Page 229; EA-ITP-ASG2E, Page 1-19; EA-ITP-AB, Page 120; EA-AC65-15A, Page 151*
When a closed angle is referred to, it means an angle that is formed by taking a flat piece of stock and bending it past 90°, to where the two ends of the stock start coming together. This is shown in Chapter 2 of the ITP Airframe Textbook. Because there are 180° in a piece of flat stock, to end up with a closed angle of 15° would require a bend of 165°.

8163. (B) *EA-ITP-A2, Page 225; EA-ITP-ASG2E, Page 1-18; EA-ITP-AB, Page 117; EA-AC65-15A, Page 146*
The sheet metal we use for aircraft construction and repair was formed from an ingot of aluminum alloy passed through a series of rollers until it was reduced to the thickness needed. In the process of rolling, the metal assumes a grain structure which is easy to see in a piece of sheet aluminum alloy. When laying out a pattern, the bends in the metal should be made across the grain as much as possible.

8164. (A) *EA-ITP-A2, Page 226; EA-ITP-ASG2E, Page 1-21; EA-ITP-AB, Pages 118 and 120; EA-AC65-15A, Pages 147-149*
The smaller the bend radius in an L-shaped bracket, the greater the amount of material needed to make the bracket. If the bracket is made from 1/8" sheet metal, with a bend radius of 1/8", the flat layout would need to be 3.742" long. As the bend radius becomes greater, the length of the flat layout becomes less.

8165. (B) *EA-ITP-A2, Page 226; EA-ITP-ASG2E, Page 1-19; EA-ITP-AB, Page 117; EA-AC65-15A, Page 146*
The line along which the portion of sheet being bent does not shrink or stretch is called the neutral axis of the metal. It is not located exactly in the center of the sheet, but is actually about 44.5% of the sheet thickness from the inside of the bend. For practical purposes, we can consider this to be in the center.

8166. (A) *EA-ITP-A2, Page 226; EA-ITP-ASG2E, Page 1-19; EA-ITP-AB, Page 120; EA-SM, Page 63; EA-AC65-15A, Page 146*
The radius of bend on a sheet of material is the radius of the bend as measured on the inside of the curved material.

8167. (B) *EA-ITP-A2, Page 226; EA-ITP-ASG2E, Page 1-19; EA-ITP-AB, Page 117; EA-AC65-15A, Page 146*
The radius of bend of a sheet of material is the radius of the bend as measured on the inside of the curved material. The minimum radius of bend of a sheet of material is the sharpest curve, or bend, to which the sheet can be bent without critically weakening the metal at the bend.

8168. (B) *EA-ITP-A2, Page 227; EA-ITP-ASG2E, Page 1-21; EA-ITP-AB, Page 121; EA-AC65-15A, Page 152*
In order to make a flat pattern layout, the technician must know the radius of the bend, because this must be accounted for when calculating bend allowance. The thickness of the material is needed so the technician can calculate the setback, and thickness is also accounted for in calculating bend allowance. The degree of bend is also needed when calculating bend allowance.

8169. (C) *EA-ITP-A2, Page 226; EA-ITP-ASG2E, Page 1-19; EA-ITP-AB, Page 117; EA-AC65-15A, Page 150*
The bend radius for a piece of sheet metal is always to the inside of the metal being formed, unless otherwise stated.

8170. (C) *EA-ITP-ASG2E, Page 1-18; EA-AC65-9A, Page 201*
Aluminum alloy sheets are marked with the specification number on approximately every square foot of material. If for any reason this identification is not on the material, it is possible to separate the heat-treatable alloys from the non-heat-treatable alloys by immersing a sample of the material in a 10% solution of caustic soda. The heat-treatable alloys will turn black due to the copper content.

8171. (A) *EA-ITP-A2, Page 236; EA-ITP-ASG2E, Page 1-19; EA-ITP-AB, Page 127; EA-AC65-15A, Page 162*
A joggle is an offset formed on an angle strip to allow clearance for a sheet or an extrusion.

8172. (C) *EA-ITP-A2, Page 226; EA-ITP-ASG2E, Page 1-19; EA-ITP-AB, Page 117; EA-AC65-15A, Page 146*
Bending a strip compresses the material on the inside of the curve and stretches the material on the outside of the curve. However, at some distance between these two extremes lies a space which is not affected by either force. This is known as the neutral line or neutral axis.

8173. (B) *EA-ITP-A2, Pages 227-228; EA-ITP-ASG2E, Page 1-21; EA-ITP-AB, Pages 121-122; EA-AC65-15A, Pages 152-153*
The dimension for flat "A" is its mold line of 1" minus the setback. The setback is given in this question as .252, so the dimension of flat "A" is .748". The dimension for flat "B" is its mold line of 2" minus two setbacks (one at each end), or 1.496". The dimension of flat "C" is its mold line of 1.5" minus the setback, or 1.248".

8174. (B) *EA-ITP-A2, Pages 227-228; EA-ITP-ASG2E, Page 1-21; EA-ITP-AB, Pages 121-122; EA-AC65-15A, Pages 152-153*
Dimension "D" in this figure is the total flat layout dimension. This dimension would be equal to the three flats "A", "B", and "C", plus two bend allowances for the two bends. The values for the three flats was calculated earlier, and the bend allowance is given in the question as .345. The three flats plus two of the bend allowances would give a total dimension of 4.182".

8175. (C) *EA-ITP-A2, Page 229; EA-ITP-ASG2E, Page 1-21*
The sight line is the mark on a flat sheet which is set even with the nose of the radius bar of the cornice brake and serves as a guide in bending. This may also be referred to as the brake line.

8176. (B) *EA-ITP-A2, Pages 227-228; EA-ITP-ASG2E, Page 1-21; EA-ITP-AB, Pages 121-122; EA-AC65-15A, Pages 152-153*
The dimension "F" is the total flat layout for the object. This dimension is going to equal the individual flats "A", plus "B", plus "C", plus the bend allowances at "D" and "E". Keeping in mind that the dimension of a flat is equal to the mold line minus the setback, the following dimensions can be calculated: Flat "A" = 1" – .095 or .905" Flat "B" = 3" – .095 and .068 or 2.837" Flat "C" = 2" – .068 or 1.932" Bend Allowance at "D" = .150 Bend Allowance at "E" = .112 The total of the five dimensions above is 5.936".

8177. (C) *EA-ITP-A2, Page 229; EA-ITP-ASG2E, Page 1-21; EA-ITP-AB, Pages 118 and 120; EA-AC65-15A, Pages 146-148*
When accounting for the total flat layout dimension when the object has a single bend, the two flats forming the bend must both have the setback subtracted from their mold line dimension, and then the bend allowance is added in.

8178. (C) *EA-ITP-G2, Page 391; EA-ITP-ASG2E, Page 1-19; EA-ITP-GB, Page 378; EA-AC65-9A, Page 205*
Heat treatment is a series of operations involving the heating and cooling of metals in the solid state. Its purpose is to change a mechanical property or combination of mechanical properties so that the metal will be more useful, serviceable, and safe for a definite purpose. By heat treating, a metal can be made harder, stronger, and more resistant to impact.

8179. (A) *EA-ITP-A2, Page 348; EA-ITP-ASG2E, Page 1-25*
A stream of inert gas such as argon or helium flows out of the torch and envelopes the arc to exclude oxygen from the area and prevent the formation of oxides.

8180. (C) *EA-ITP-A2, Page 365; EA-ITP-ASG2E, Page 1-25; EA-ITP-AB, Page 204; EA-AC65-15A, Page 262*
The filler rod should be of the same composition as the base metal.

8181. (A) *EA-ITP-ASG2E, Page 1-25; EA-AC65-15A, Page 261*
Repair of magnesium by welding is limited by the fact that if it is used as a structural member, it has probably been heat treated. As is true with heat treated aluminum, welding heat treated magnesium does not leave as much strength as the original metal had.

8182. (B) *EA-ITP-A2, Pages 362-363; EA-ITP-ASG2E, Page 1-25; EA-AC65-15A, Page 264*
The oxyacetylene flame for silver soldering should be neutral, but may have a slight excess of acetylene. It must be soft, not harsh.

8183. (A) *EA-ITP-A2, Pages 362-363; EA-ITP-ASG2E, Page 1-25; EA-ITP-AB, Page 201; EA-AC65-15A, Page 264*
It is necessary to use flux in all silver soldering operations because of the necessity for having the base metal chemically clean without the slightest film of oxide to prevent the silver solder from coming into intimate contact with the base metal.

8184. (B) *EA-ITP-ASG2E, Page 1-25; EA-AC65-15A, Page 279*
All welding on an engine mount should be of the highest quality, since vibration tends to accentuate any minor defect. Engine-mount members should preferably be repaired by using a larger diameter replacement tube telescoped over the stub of the original member, using fishmouth and rosette welds.

8185. (C) *EA-ITP-A2, Page 367; EA-ITP-ASG2E, Page 1-25; EA-ITP-AB, Page 206; EA-43.13-1A/3, Pages 32-33*
A dent at a cluster should be repaired by welding a specially formed steel patch plate over the dented area and surrounding tubes, as shown in the EA-43.13-1A/3.

8186. (A) *EA-ITP-ASG2E, Page 1-25; EA-43.13-1A/3, Page 32*
Do not weld brazed or soldered parts as the brazing mixture or solder will penetrate the hot steel and weaken it.

8187. (A) *EA-ITP-ASG2E, Page 1-25; EA-ITP-AB, Page 208; EA-AC65-15A, Page 265*
Soft soldering is used in combination with mechanical seams; that is, seams that are riveted, bolted or folded. Soldering is also used to join electrical connections.

8188. (C) *EA-ITP-ASG2E, Page 1-25; EA-AC65-15A, Page 265*
Soft soldering is used chiefly for copper, brass, and coated iron in combination with mechanical seams; that is, seams that are riveted, bolted, or folded. This process is used to join electrical connections, but it must not be relied on to withstand any stress.

8189. (C) *EA-ITP-A2, Page 366; EA-ITP-ASG2E, Page 1-25*
You must be very careful about using welding as a repair procedure. Generally materials, or weldments, which have been heat treated after fabrication are not candidates for field welded repairs. 4130 steel tubing is commonly used both during fabrication and repair.

8190. (A) *EA-ITP-A2, Page 355; EA-ITP-ASG2E, Page 1-24*
The size of the tip used on an oxyacetylene torch will determine the amount of heat applied to the work. The thickness of the metal will determine which tip size is required.

8191. (B) *EA-ITP-ASG2E, Page 1-24; EA-AC65-15A, Pages 258-259*
Care must be taken when welding steel, to ensure that a carburizing (reducing) flame is not used. This flame is one with an insufficient supply of oxygen to ensure complete combustion of the acetylene, and it allows carbon to escape unburned. This carbon will combine with the steel during the welding operation and cause it to become hard and brittle.

8192. (B) *EA-ITP-A2, Page 354; EA-ITP-ASG2E, Page 1-24; EA-43.13-1A/3, Page 30*
When selecting a welding rod, ensure that it is compatible with the materials to be welded.

8193. (A) *EA-ITP-A2, Page 359; EA-ITP-ASG2E, Page 1-25*
The flame use for welding aluminum should be soft and neutral or slightly carburizing, to be sure that there will be no possibility of the metal being oxidized from the welding.

8194. (C) *EA-ITP-ASG2E, Page 1-25; EA-AC65-15A, Page 265*
The tip on a soldering copper should be blunt and the entire head should be fairly wide. These features allow the soldering copper to retain its heat for a long enough period of time to perform its intended function.

8195. (B) *EA-ITP-A2, Page 359; EA-ITP-ASG2E, Page 1-24*
Filing or grinding of a weld will reduce its strength.

8196. (A) *EA-ITP-A2, Page 350; EA-ITP-ASG2E, Page 1-23*
Acetylene becomes unstable at pressures of 15 PSI or higher. At 30 PSI it becomes so unstable that the slightest amount of heat generated by the friction of acetylene against the atmosphere will cause it to explode.

8197. (C) *EA-ITP-A2, Page 350; EA-ITP-ASG2E, Page 1-23; EA-ITP-AB, Page 191; EA-AC65-15A, Page 248*
Under low pressure at normal temperatures, acetylene is a stable compound. But when compressed in a container to pressures greater than 15 PSI, it becomes dangerously unstable. For this reason, manufacturers fill the acetylene storage cylinders with a porous substance and saturate this substance with acetone. Since acetone is capable of absorbing approximately 25 times its own volume of acetylene gas, a cylinder containing the correct amount of acetone can be pressurized to 250 PSI.

8198. (A) *EA-ITP-ASG2E, Page 1-23; EA-AC65-15A, Page 253*
A backfire may be caused by touching the tip against the work, overheating the tip, a loose tip or head, or dirt or slag in the end of the tip.

8199. (A) *EA-ITP-A2, Page 348; EA-ITP-ASG2E, Page 1-25; EA-ITP-AB, Page 202; EA-AC65-15A, Page 266*
In gas-shielded arc welding, a gas is used as a covering shield around the arc to prevent the atmosphere from contaminating the weld.

8200. (B) *EA-ITP-A2, Page 359; EA-ITP-ASG2E, Page 1-25; EA-ITP-AB, Page 199; EA-AC65-15A, Page 260*
Aluminum welding flux is designed to remove the aluminum oxide by chemically combining with it. The flux can be painted directly on the top and bottom of the joint if no filler rod is required; if filler rod is used, it can be coated.

8201. (B) *EA-ITP-A2, Page 359; EA-ITP-ASG2E, Page 1-25; EA-ITP-AB, Page 199; EA-AC65-15A, Page 260*
Using the proper flux when welding aluminum is extremely important. Aluminum welding flux is designed to remove the aluminum oxide by chemically combining with it.

8202. (A) *EA-ITP-ASG2E, Page 1-25; EA-AC65-15A, Page 260*
Preheating of aluminum before welding lessens the strain caused by the large coefficient of expansion of this metal. Preheating also assures a more complete weld penetration.

8203. (A) *EA-ITP-ASG2E, Page 1-24; EA-AC65-15A, Page 259*
A slightly carburizing flame is recommended for welding stainless steel. The flame should be adjusted so that a feather of excess acetylene, about $\frac{1}{16}$" long, forms around the inner cone.

8204. **(C)** *EA-ITP-A2, Page 359; EA-ITP-ASG2E, Page 1-25; EA-AC65-15A, Page 260*
Using the proper flux when welding is extremely important. Aluminum welding flux is designed to remove the aluminum oxide by chemically combining with it.

8205. **(B)** *EA-ITP-A2, Page 355; EA-ITP-ASG2E, Page 1-24; EA-ITP-AB, Page 195; EA-AC65-15A, Page 250*
The nature of the weld, the material, the experience of the welder, and the position in which the weld is to be made, all determine the correct size of the tip opening. The size of the tip opening, in turn, determines the amount of heat applied to the work.

8206. **(B)** *EA-ITP-ASG2E, Page 1-23; EA-ITP-AB, Pages 191-192; EA-AC65-15A, Page 248*
Oxygen and acetylene cylinders used in welding operations are made of seamless steel.

8207. **(B)** *EA-ITP-A2, Page 359; EA-ITP-ASG2E, Page 1-24; EA-AC65-9A, Page 486*
If it is properly done, the penetration in a butt weld should be 100% of the thickness of the base metal.

8208. **(C)** *EA-ITP-G2, Page 393; EA-ITP-ASG2E, Page 1-25; EA-ITP-GB, Page 380; EA-AC65-9A, Page 216*
When aluminum alloy is being formed, it will become strain hardened, and often before the forming can be completed, the metal must be annealed to relieve the stress.

8209. **(C)** *EA-ITP-ASG2E, Page 1-25; EA-AC65-15A, Page 261*
Edge notching is recommended in aluminum welding because it aids in getting full penetration and also prevents local distortion. All butt welds in material over .135″ thick are generally notched in some manner.

8210. **(B)** *EA-ITP-ASG2E, Page 1-25; EA-ITP-AB, Page 196; EA-AC65-15A, Page 259*
Too much acetylene will add carbon to the metal and cause it to lose its resistance to corrosion.

8211. **(C)** *EA-ITP-A2, Page 348; EA-ITP-ASG2E, Page 1-25*
Argon or helium are used to exclude oxygen from the area of the molten metal to prevent oxidation.

8212. **(C)** *EA-ITP-A2, Page 84; EA-ITP-ASG2E, Page 1-26*
The tail rotor, controlled by pedals operated by the pilot, compensates for the torque of the main rotor and provides directional control.

8213. **(A)** *EA-ITP-A2, Page 81; EA-ITP-ASG2E, Page 1-27; EA-ITP-AB, Page 260; EA-AC65-15A, Page 56*
The collective pitch control varies the lift of the main rotor by increasing or decreasing the pitch of all blades at the same time. This changing of lift on the blades controls the vertical flight of the helicopter.

8214. **(B)** *EA-ITP-A2, Pages 83-84; EA-ITP-ASG2E, Page 1-26; EA-ITP-AB, Pages 262-263; EA-AC65-15A, Page 50*
The tail rotor blades on a helicopter counteract the effect of the torque produced when the engine drives the main rotor blades. If the pitch angle on the tail rotor blades is decreased, the helicopter's tail will spin in the direction of torque.

8215. **(C)** *EA-AC-43.13-1A/3; EA-ITP-ASG2E, Page 1-28*
The FAR's specify the requirements for quick-release devices used in conjunction with helicopter external load operations. The ideal location of the cargo release would allow the line of action to always pass through the helicopter's center-of-gravity.

8216. **(C)** *EA-ITP-A2, Page 78; EA-ITP-ASG2E, Page 1-26; EA-AC65-15A, Page 31*
Angle of attack is defined as the angle between the chord line of the wing and the direction of the relative wind.

8217. **(C)** *EA-ITP-A2, Page 79; EA-ITP-ASG2E, Page 1-27; EA-ITP-AB, Pages 260-261; EA-AC65-15A, Pages 55-56*
The rapidly rotating rotor blades create a disk area that can be tilted in any direction with respect to the supporting rotor mast. Horizontal movement is controlled by changing the direction of tilt of the main rotor to produce a force in the desired direction.

8218. **(B)** *EA-ITP-A2, Page 87; EA-ITP-ASG2E, Page 1-28*
Rotor blade tracking determines that each blade follows directly in the path of the blade in front of it.

8219. **(C)** *EA-ITP-A2, Pages 83-84; EA-ITP-ASG2E, Page 1-26; EA-ITP-AB, Pages 262-263; EA-AC65-15A, Page 50*
In addition to being used to counteract the torque of the main rotor, the tail rotor also provides directional control when the helicopter is in a hover.

8220. (C) *EA-ITP-A2, Page 78; EA-ITP-ASG2E, Page 1-28; EA-ITP-AB, Pages 255-256; EA-AC65-15A, Pages 51-52*
When a helicopter is moving forward, the blade on the right side of the helicopter is moving forward, against airflow, and the blade on the left side is moving aft, with the airflow. This causes the advancing blade to produce more lift, and therefore rise. The retreating blade produces less lift and will drop. The dropping of the retreating blade increases its angle of attack and helps solve the potential problem of dissymmetry of lift.

8221. (C) *EA-ITP-A2, Page 87; EA-ITP-ASG2E, Page 1-28; EA-AC65-15A, Page 78*
When the main rotor blades do not "cone" by the same amount during rotation, it is referred to as "out of track".

8222. (A) *EA-ITP-A2, Page 90; EA-ITP-ASG2E, Page 1-28; EA-ITP-AB, Page 269*
The drag imposed by the main rotor blades is tremendous. One feature of the clutch in the helicopter is that it allows the engine to be started with the rotor system disengaged. This relieves the load on the starter.

8223. (C) *EA-ITP-A2, Page 83; EA-ITP-ASG2E, Page 1-26; EA-ITP-AB, Page 253; EA-AC65-15A, Page 50*
Newton's Third Law of Motion states, "For every action there is an equal and opposite reaction." As the main rotor of a helicopter turns in one direction, the fuselage tends to rotate in the opposite direction. This tendency for the fuselage to rotate is called torque.

8224. (A) *EA-ITP-A2, Page 90; EA-ITP-ASG2E, Page 1-28; EA-ITP-AB, Page 270*
There must be some form of freewheeling device in a helicopter to release the engine from the rotor any time the speed of the engine drops below that required to drive the rotor. In the case of an engine failure, it is essential that the engine be disconnected immediately lest it slow the rotor and prevent the rotor being turned by the autorotative forces.

8225. (B) *EA-ITP-A2, Page 81; EA-ITP-ASG2E, Page 1-26; EA-ITP-AB, Page 260; EA-AC65-15A, Page 55*
Movement about the longitudinal axis is called roll. This is effected by moving the cyclic pitch control to the right or left.

8226. (B) *EA-ITP-A2, Page 81; EA-ITP-ASG2E, Page 1-26; EA-ITP-AB, Page 260; EA-AC65-15A, Page 56*
Movement about the lateral axis produces a nose up or nose down attitude. This is effected by moving the cyclic pitch control fore and aft.

8227. (A) *EA-ITP-A2, Page 49; EA-ITP-ASG2E, Page 1-32; EA-ITP-AB, Page 226; EA-AC65-15A, Pages 38-39*
Lateral stability (stability about the longitudinal axis), or roll stability, is provided primarily by dihedral in the wings. Dihedral is the positive acute angle between the wing and the lateral axis of the airplane.

8228. (A) *EA-ITP-A2, Page 91; EA-ITP-ASG2E, Page 1-35; EA-ITP-AB, Page 274; EA-AC65-15A, Pages 72-73*
The position of the wing attachment fuselage fittings will determine the angle of incidence for the wing. These fittings must be positioned exactly, according to specifications, or the aerodynamics of the aircraft will be altered.

8229. (A) *EA-ITP-A2, Page 91; EA-ITP-ASG2E, Page 1-36; EA-ITP-AB, Page 274*
Increasing the angle of incidence, that angle between the chord line of the wing and the longitudinal axis of the airplane, is called "washing the wing in", and it increases the lift. If the left wing of an airplane is flying heavy, washing it in will increase its lift and probably correct the problem.

8230. (B) *EA-ITP-A2, Page 50; EA-ITP-ASG2E, Page 1-36; EA-AC65-15A, Page 75*
On most single engine, propeller driven airplanes, the leading edge of the vertical fin is offset to the longitudinal center line to counteract engine torque.

8231. (B) *EA-ITP-A2, Page 49; EA-ITP-ASG2E, Page 1-30; EA-ITP-AB, Page 226; EA-AC65-15A, Page 37*
When an aircraft has a tendency to keep a constant angle of attack with reference to the relative wind — that is, when it does not tend to put its nose down and dive, or lift its nose and stall — it is said to have longitudinal stability. Longitudinal stability refers to motion in pitch.

8232. (C) *EA-ITP-A2, Page 45; EA-ITP-ASG2E, Page 1-29; EA-ITP-AB, Page 224; EA-AC65-15A, Page 31*
On an asymmetrical airfoil, center of pressure moves forward as the angle of attack increases. On a symmetrical airfoil, center of pressure does not change with changes in angle of attack.

8233. (B) *EA-ITP-A2, Page 91; EA-ITP-ASG2E, Page 1-29; EA-ITP-AB, Page 274; EA-AC65-15A, Page 32*
The acute angle which the wing chord makes with the longitudinal axis of the aircraft is called the angle of incidence, or the angle of wing setting.

8234. (B) *EA-ITP-A2, Page 49; EA-ITP-ASG2E, Page 1-29; EA-ITP-AB, Page 226; EA-AC65-15A, Page 33*
The longitudinal, or pitch, stability of an airplane determines its ability to be flown hands-off at any airspeed. The center of gravity is located ahead of the aerodynamic center of the wing, and in straight and level flight, the wing produces a nose-down moment.

8235. (A) *EA-ITP-A2, Page 53; EA-ITP-ASG2E, Page 1-30; EA-ITP-AB, Page 231; EA-AC65-15A, Page 4*
Turning the nose of the aircraft causes the aircraft to rotate about its vertical axis. Rotation of the aircraft about the vertical axis is called yawing. This motion is controlled by using the rudder.

8236. (B) *EA-ITP-A2, Pages 51-52; EA-ITP-ASG2E, Page 1-30; EA-ITP-AB, Page 229; EA-AC65-15A, Page 44*
When the nose of an aircraft is raised or lowered, it is rotated about its lateral axis. Elevators are the movable control surfaces that cause this rotation.

8237. (B) *EA-ITP-A2, Page 91; EA-ITP-ASG2E, Page 1-36; EA-ITP-AB, Page 274*
Increasing the angle of incidence, that angle between the chord line of the wing and the longitudinal axis of the airplane, is called "washing the wing in" and it increases the lift. An increase in lift also means an increase in drag because of the increased angle of attack.

8238. (A) *EA-ITP-A2, Page 59; EA-ITP-ASG2E, Page 1-34; EA-ITP-AB, Page 237; EA-AC65-15A, Page 47*
The use of flaps increases the camber of a wing, and therefore the lift of the wing, making it possible for the speed of the aircraft to be decreased without stalling.

8239. (A) *EA-ITP-A2, Page 91; EA-ITP-ASG2E, Page 1-35; EA-ITP-AB, Page 274; EA-AC65-15A, Page 38*
When a wing is improperly rigged, so that it has a greater angle of incidence than the manufacturer calls for, the wing will produce more lift than it should. An aircraft in this condition will not be laterally, or directionally, stable.

8240. (C) *EA-ITP-A2, Page 39; EA-ITP-ASG2E, Page 1-29; EA-ITP-AB, Page 216; EA-AC65-15A, Page 31*
The chord of an airfoil or wing section is an imaginary straight line which passes through the section from the leading edge to the trailing edge.

8241. (B) *EA-ITP-A2, Page 45; EA-ITP-ASG2E, Page 1-29; EA-ITP-AB, Page 220; EA-AC65-15A, Page 35*
The same factors that enter into the production of lift also produce drag. At low angles of attack, there is very little drag, but as the angle of attack increases, so does the drag.

8242. (B) *EA-ITP-G2, Page 11; EA-ITP-ASG2E, Page 1-29; EA-AC65-15A, Page 33*
The larger the wingspan is, as compared to the chord, the greater the lift obtained from the wing. This comparison of wingspan and chord is called the wing's aspect ratio.

8243. (A) *EA-ITP-A2, Page 57; EA-ITP-ASG2E, Page 1-34; EA-ITP-AB, Page 226; EA-AC65-15A, Page 37*
By design, airplanes are nose heavy. The nose heavy tendency of airplanes is balanced out by a downward force on the horizontal stabilizer and elevator. If an elevator trim tab is improperly rigged, the pilot will need to hold pressure on the control column to keep the airplane flying straight and level.

8244. (A) *EA-ITP-A2, Page 49; EA-ITP-ASG2E, Page 1-30; EA-ITP-AB, Page 226; EA-AC65-15A, Page 37*
When an aircraft has a tendency to keep a constant angle of attack with reference to the relative wind — that is, when it does not tend to put its nose down and dive, or lift its nose and stall — it is said to have longitudinal stability. When an aircraft has a tendency to increase its pitching movement once it has been set in motion, it has very poor longitudinal stability.

8245. (A) *EA-ITP-A2, Page 61; EA-ITP-ASG2E, Page 1-35; EA-ITP-AB, Pages 239-240; EA-AC65-15A, Page 49*
Many high performance airplanes have a portion of the wing leading edge mounted on tracks so it can extend outward and create a duct to direct high energy air down over the surface and delay separation to a very high angle of attack. These devices are known as slats, and they serve to reduce the stall speed of the airplane.

8246. (C) *EA-ITP-A2, Page 91; EA-ITP-ASG2E, Page 1-36; EA-ITP-AB, Page 274; EA-AC65-15A, Page 32*
The angle of incidence for an airplane's wing is the angle between the wing chord and the longitudinal axis of the airplane. Setting up this angle is part of rigging the wing, and once it is set it does not change (for most wings).

8247. (B) *EA-ITP-ASG2E, Page 1-29; EA-ITP-AB, Page 224; EA-AC65-15A, Page 32*
Turbulence, or erratic movement of air currents, can cause buffeting in an aircraft. If the aircraft has proper stability, as turbulence ceases, the aircraft will return to its original flight condition.

8248. **(C)** *EA-ITP-A2, Page 53; EA-ITP-ASG2E, Page 1-30; EA-ITP-AB, Pages 226-227; EA-AC65-15A, Pages 37-38*
When an aircraft is moving along its lateral axis, it means that it is moving in a rolling direction (one wing dropping and the other one rising). Movement along the lateral axis is also movement about the longitudinal axis, or the nose to tail axis, and this is controlled by the ailerons.

8249. **(B)** *EA-ITP-A2, Page 62; EA-ITP-ASG2E, Page 1-35; EA-ITP-AB, Page 240*
Stall strips provide a progressive stall and help maintain control during a stall. Should they be removed for maintenance, stall strips must be re-installed in the correct location.

8250. **(C)** *EA-ITP-ASG2E, Page 1-35; EA-AC65-15A, Page 72*
Normally, rigging and alignment checks should not be done in the open. If this cannot be avoided, the aircraft should be positioned with the nose into the wind.

8251. **(C)** *EA-ITP-A2, Page 91; EA-ITP-ASG2E, Page 1-35; EA-ITP-AB, Page 274; EA-AC65-15A, Page 74*
After installing a wing, the wing is adjusted to get the proper dihedral. This is determined by using a dihedral board that has a specific taper. It is held against the main spar on the bottom of the wing, at the location specified by the manufacturer. It is essential that the airplane be level, both longitudinally and laterally, when this check is made. A bubble level is used to check this.

8252. **(A)** *EA-ITP-A2, Page 91; EA-ITP-ASG2E, Page 1-35*
See question above for an explanation and reference.

8253. **(C)** *EA-ITP-A2, Page 91; EA-ITP-ASG2E, Page 1-35; EA-AC65-15A, Page 75*
The precise figures, tolerances and checkpoints for a particular aircraft's symmetry check will be found in the applicable service or maintenance manual.

8254. **(C)** *EA-ITP-G2, Page 238; EA-ITP-ASG2E, Page 1-35; EA-ITP-GB, Page 238; EA-AC65-15A, Page 6*
The buttock line, or butt line, is a width measurement left or right of, and parallel to, the vertical center line.

8255. **(B)** *EA-ITP-G2, Page 237; EA-ITP-ASG2E, Page 1-35; EA-ITP-GB, Page 238; EA-AC65-15A, Page 6*
Fuselage stations are numbered in inches from a reference, or zero point, known as the reference datum. The distance to a given point is measured in inches parallel to a center line, extending through the aircraft from the nose through the center of the tail cone.

8256. **(B)** *EA-ITP-A2, Page 91; EA-ITP-ASG2E, Page 1-35; EA-ITP-AB, Page 274; EA-AC65-15A, Pages 74-75*
Incidence is usually checked on the surface of the wing at two specific locations to ensure that the wing is free from twist. A variety of incidence boards are used, along with a bubble level, to check for the proper angle of incidence.

8257. **(C)** *EA-ITP-A2, Page 98; EA-ITP-ASG2E, Page 1-37; EA-43.13-1A/3, Page 102*
After a control cable has been swaged, the junction of the swaged fitting and the cable is painted with red paint. At all subsequent service inspections of the swaged fittings, the painted section is checked for any cable slippage.

8258. **(B)** *EA-ITP-A2, Page 98; EA-ITP-ASG2E, Page 1-37; EA-ITP-AB, Page 278; EA-43.13-1A/3, Page 104*
After the compression of the fitting is completed, it should be checked with a special gage. If the compressions have been properly made, the fitting will exactly fit the gage.

8259. **(C)** *EA-ITP-A2, Pages 102-103; EA-ITP-ASG2E, Page 1-38; EA-ITP-AB, Page 283; EA-43.13-1A/3, Page 112*
In the single and double wrap methods of safetying turnbuckles, there is an option of straight or spiral wiring. In either method, the wiring must be terminated with at least four wraps around the shank of the turnbuckle.

8260. **(A)** *EA-ITP-A2, Page 98; EA-ITP-ASG2E, Page 1-37; EA-43.13-1A/3, Page 100*
When swaging tools are used, it is important that all the manufacturer's instructions, including "go and no go" dimensions, be followed in detail to avoid defective and inferior swaging. Observance of all instructions should result in a terminal developing the full rated strength of the cable.

8261. **(C)** *EA-ITP-ASG2E, Page 1-39; EA-43.13-1A/3, Pages 120-121*
In normal practice, a cotter pin is the only safety device used on a castle nut and is the correct choice. In some special installations safety wire will be used.

8262. (A) *EA-ITP-ASG2E, Page 1-39; EA-AC65-9A, Page 145*
The AN381 corrosion-resistant steel cotter pin is used in locations where nonmagnetic material is required, or in locations where resistance to corrosion is desired.

8263. (B) *EA-ITP-G2, Page 330; EA-ITP-ASG2E, Page 1-39; EA-ITP-AB, Page 322; EA-43.13-1A/3, Page 120*
When fiber-type self-locking nuts are reused, check the fiber carefully to make sure it has not lost its locking friction, or become brittle. Do not reuse locknuts if they can be run up finger tight.

8264. (A) *EA-ITP-A2, Page 15; EA-ITP-ASG2E, Page 1-36; EA-ITP-AB, Page 227; EA-AC65-15A, Page 38*
Stability about the vertical axis is referred to as directional stability. The vertical stabilizer is the primary surface which controls directional stability.

8265. (A) *EA-ITP-A2, Page 99; EA-ITP-ASG2E, Page 1-38; EA-ITP-AB, Pages 279-280*
Aircraft manufacturers generally route the control cables in the most direct manner possible. At points where a change in direction is needed, a pulley is used. These pulleys must be carefully aligned so the cable rides squarely in the center of the pulley's groove.

8266. (C) *EA-ITP-ASG2E, Page 1-37*
Each cable, cable fitting, turnbuckle, splice, and pulley used in a small airplane must meet approved specifications. In addition, no cable smaller than 1/8" diameter may be used in primary control systems.

8267. (C) *EA-ITP-ASG2E, Page 1-34; EA-AC65-15A, Page 82*
Any time repairs on a control surface add weight fore or aft of the hinge center line, the control surface must be re-balanced. Any control surface that is out of balance will be unstable and will not remain in a streamlined position during normal flight.

8268. (B) *EA-ITP-A2, Page 99; EA-ITP-ASG2E, Page 1-37*
Using a piece of cloth to check cables for broken strands protects you hands from injury.

8269. (A) *EA-ITP-A2, Pages 100-102; EA-ITP-ASG2E, Page 1-38; EA-ITP-AB, Page 282; EA-AC65-15A, Page 71 (Figure 2-65)*
The answer requires some simple logic. Aluminum expands greater than steel, therefore a hot airplane is longer. The cables expand also, but not to the same extent. Cable tension is a specified average. When hot they are too tight; cold, too loose.

8270. (B) *EA-ITP-A2, Page 13; EA-ITP-ASG2E, Page 1-34*
To eliminate flutter it is extremely important that control surfaces be balanced so that their center of gravity does not fall behind their hinge line.

8271. (A) *EA-ITP-A2, Page 101; EA-ITP-ASG2E, Page 1-38*
If both sides of a control cable pulley are worn excessively, it is an indication that the pulley is misaligned.

8272. (C) *EA-ITP-ASG2E, Page 1-38; EA-43.13-1A/3, Page 110*
When inspecting fairleads for wear and alignment, the technician must ensure that the cable deflection is no greater than 3°.

8273. (B) *EA-ITP-ASG2E, Page 1-38; EA-43.13-1A/3, Page 99*
Locate splices so that no portion comes closer than 2" to any fair-lead or pulley.

8274. (B) *EA-ITP-ASG2E, Page 1-32; EA-AC65-15A, Page 41*
Differential aileron travel provides more aileron up travel than down travel for a given movement of the control stick or wheel in the cockpit.

8275. (C) *EA-ITP-A2, Pages 103-104; EA-ITP-ASG2E, Page 1-39; EA-ITP-AB, Pages 283-284*
When inspecting a push/pull tube-type control system, the technician should check to be sure that the rod ends are screwed far enough into the fitting. This is accomplished by checking the small hole drilled into the fitting. If a piece of safety wire can pass through the hole, the rod end is not screwed in far enough.

8276. (A) *EA-ITP-ASG2E, Page 1-38; EA-43.13-1A/3, Page 27*
If a control surface is properly balanced, rigged, and adjusted, the most likely cause of vibration is too much play in the attachment fittings due to wear. To correct this, the fittings would need to be replaced.

8277. (A) *EA-ITP-ASG2E, Page 1-34;EA-FAR-M, See FAR 23.677(a).*

8278. (B) *EA-ITP-A2, Page 49; EA-ITP-ASG2E, Page 1-30; EA-ITP-AB, Pages 226-227; EA-AC65-15A, Page 38*
Movement of an aircraft about its longitudinal (fore and aft) axis is a lateral, or rolling, motion. The tendency to return to the original attitude from such motion is called lateral stability.

8279. (A) *EA-ITP-A2, Page 59; EA-ITP-ASG2E, Page 1-33; EA-ITP-AB, Page 236; EA-AC65-15A, Page 46*
Servo tabs aid in moving the control surface and holding it in position. Spring tabs are used to aid in moving a primary control surface.

8280. (C) *EA-ITP-A2, Pages 54-57; EA-ITP-ASG2E, Page 1-32; EA-ITP-AB, Pages 230-232; EA-AC65-15A, Pages 41 and 44*
Moving the control stick back and to the left is telling the airplane to climb and bank to the left. Climbing is accomplished by moving the elevator up, creating a downward force on the tail to raise the nose. Banking to the left is accomplished by moving the right wing aileron down to increase the lift on the right wing, and moving the left wing aileron up, to decrease the lift on the left wing.

8281. (B) *EA-ITP-A2, Page 57; EA-ITP-ASG2E, Page 1-32; EA-ITP-AB, Page 236; EA-AC65-15A, Page 46*
Moving the cockpit control for the elevator trim tabs toward the nose down position is telling the airplane that you want the elevator to move down, because this will cause the nose of the airplane to come down. For a trim tab to work, it needs to move in the opposite direction of the way you want the flight control to move.

8282. (C) *EA-ITP-A2, Pages 54-57; EA-ITP-ASG2E, Page 1-32; EA-ITP-AB, Pages 230-232; EA-AC65-15A, Pages 41 and 44*
Moving the control stick forward and to the right is telling the airplane to descend and to bank to the right. To drop the nose of the airplane the elevator needs to move down to put an upward force on the tail. To bank the airplane to the right, the aileron on the right wing needs to come up to decrease the lift on that wing, and the aileron on the left wing needs to move down to increase the lift on that wing.

8283. (B) *EA-ITP-A2, Pages 100-102; EA-ITP-ASG2E, Page 1-38; EA-ITP-AB, Pages 282-283; EA-AC65-15A, Pages 70-71*
When cables are rigged excessively tight, it creates a great deal of strain on the system. Not only does premature wearing of the pulleys take place, but the flight controls will also feel very heavy and difficult to move because of the strain.

8284. (A) *EA-ITP-ASG2E, Page 1-32; EA-AC65-15A, Page 41*
Differential aileron control on an airplane provides more aileron up travel than down travel for a given movement of the control column.

8285. (A) *EA-ITP-A2, Page 95; EA-ITP-ASG2E, Page 1-36; EA-ITP-AB, Page 275; EA-AC65-15A, Page 73*
When using the universal propeller protractor to measure control surface travel, the control surface must be moved to neutral before the protractor is placed on the surface, and the ring adjuster turned to center the bubble in the spirit level.

8286. (C) *EA-ITP-A2, Page 95; EA-ITP-ASG2E, Page 1-37; EA-ITP-AB, Page 275; EA-AC65-15A, Page 73*
The universal propeller protractor can be used to measure the travel of a control surface that has up and down movement. The flaps have such a movement.

8287. (C) *EA-ITP-A2, Pages 96-97; EA-ITP-ASG2E, Page 1-37; EA-ITP-AB, Page 276; EA-AC65-9A, Page 143*
The 7 ×19 cable is made up of seven strands of 19 wires each. It is extra flexible, and is used in primary control systems.

8288. (C) *EA-ITP-A2, Page 102; EA-ITP-ASG2E, Page 1-38; EA-AC65-15A, Page 70*
Cable tension regulators are used in some flight control systems because there is considerable difference in temperature expansion of the aluminum aircraft structure and the steel control cables. Some large aircraft incorporate tension regulators in the control cable systems to automatically maintain a given cable tension.

8289. (B) *EA-ITP-A2, Page 102; EA-ITP-ASG2E, Page 1-38; EA-ITP-AB, Pages 282-283; EA-AC65-15A, Page 71*
The use of this chart is explained in the EA-AC65-15A.

8290. (B) *EA-ITP-ASG2E, Pages 1-32; EA-AC65-15A, Page 41*
Differential aileron travel provides more aileron up travel than down travel.

8291. (B) *EA-ITP-ASG2E, Page 1-40; EA-AC65-9A, Page 61*
If possible, aircraft should be weighed in a closed building where there are no air currents to cause incorrect scale readings.

8292. (A) *EA-ITP-ASG2E, Page 1-40; EA-AC65-9A, Page 525*
Prior to jacking an aircraft, it should be determined if the aircraft configuration will permit it. On some aircraft, the stress panels or plates must be in place when the aircraft is jacked to avoid structural damage.

8293. (C) *EA-ITP-G2, Page 526; EA-ITP-ASG2E, Page 1-40; EA-AC65-9A, Page 465; EA-AC39-7A, Item 8, Item 12(e)*
Airworthiness Directives are Federal Aviation Regulations and must be complied with, unless specific exemption is granted. Compliance with Airworthiness Directives, like any other maintenance action, must be recorded in the maintenance record.

8294. (B) *EA-ITP-ASG2E, Page 1-40; EA-AC65-9A, Page 464*
When doing work on electrical equipment, whether it involves repair or overhaul, the manufacturer's technical data should always be used. This is the source of data which will give specific information on the piece of equipment you are working on.

8295. (A) *EA-ITP-A2, Page 107; EA-ITP-ASG2E, Page 1-40*
When an annual inspection is performed and the aircraft is not approved for return to service, the owner is supplied with a list of the discrepancies. An A&P mechanic, with the necessary experience, tools, and tech data, may correct the discrepancies provided that they do not involve instrument repair or major repair to propellers. If the corrective action does involve a major repair to some other part of the aircraft, however, the A&P cannot approve the aircraft for return to service. This must be done by an authorized inspector.

8296. (C) *EA-ITP-ASG2E, Page 1-40*
Supplemental Type Certificates are issued by the FAA when someone proposes and gets approval for a change to a Type Certificated Product (aircraft, aircraft engine, propeller). When a change is made in accordance with a Supplemental Type Certificate, it is considered to be a major alteration, and an FAA Form 337 must be filled out.

8297. (A) *EA-ITP-ASG2E, Page 1-40*
The 100-hour inspection may be exceeded by not more than 10 hours if necessary to reach a place at which the inspection can be done. The excess time, however, is included in computing the next 100 hours of time in service.

8298. (B) *EA-ITP-A2, Page 107; EA-ITP-ASG2E, Page 1-40*
In FAR 43, paragraph 43.11, the suggested wording is given for the approval or disapproval for return to service after a 100-hour inspection.

8299. (A) *EA-ITP-ASG2E, Page 1-40*
The 100-hour inspection may be exceeded by not more than 10 hours if necessary to reach a place at which the inspection can be done. The excess time, however, is included in computing the next 100 hours of time in service.

8300. (B) *EA-ITP-A2, Page 107; EA-ITP-ASG2E, Page 1-40*
If an annual inspection is performed and not approved for return to service, the owner of the airplane is supplied with a list of the discrepancies. Providing the discrepancies are minor repair items, an A&P mechanic with the necessary experience, tools, and technical data is authorized to correct the discrepancies and approve the aircraft for return to service.

8301. (A) *EA-ITP-ASG2E, Page 1-39*
An aircraft that is due for an annual inspection may not be operated unless a special flight permit has been issued. This certificate, often called a ferry permit, may be issued so that the aircraft can be flown to an appropriate maintenance facility where the annual inspection will be performed.

8302. (B) *EA-ITP-A2, Page 107; EA-ITP-ASG2E, Page 1-39*
If an aircraft is operated for compensation or hire, it must be given an inspection of the same scope and detail as the annual inspection every 100 hours of operation. The A&P Technician returning the aircraft to service after a 100-hour inspection is not required to have an Inspection Authorization.

8303. (A) *EA-ITP-A2, Page 107; EA-ITP-ASG2E, Page 1-39;EA-FAR-M, See FAR 91.409(b).*

8304. (A) *EA-ITP-ASG2E, Page 1-39*
FAR 91, Subpart F is titled "Large and Turbine-Powered Multi-Engine Airplanes." It used to be under this portion of FAR 91 that the inspection requirements were given for large and turbine-powered multi-engine airplanes. This requirement now is identified in 91.409.

8305. (C) *EA-ITP-A2, Page 494; EA-ITP-ASG2E, Page 1-48*
Tires should be stored in a cool, dry area, out of direct sunlight and away from electrical machinery (which may generate ozone). Special care should be taken to avoid contact with hydrocarbon compounds (petroleum products).

8306. (C) *EA-ITP-A2, Page 467; EA-ITP-ASG2E, Page 1-44; EA-ITP-AB, Page 453; EA-AC65-15A, Page 361*
When the brake pedal is released, the master cylinder piston is returned to the OFF position by a return spring. If the spring were broken, the piston would not return and some pressure could stay in the system. This would cause the brakes to drag.

8307. (B) *EA-ITP-A2, Page 472; EA-ITP-ASG2E, Page 1-46; EA-ITP-AB, Page 446; EA-AC65-15A, Page 372*
Bleeding brakes is for the purpose of getting air out of the system which could cause spongy and ineffective brakes. Bleeding is accomplished by forcing fluid out of the system under pressure, which takes the air out of the system at the same time.

8308. (A) *EA-ITP-A2, Pages 445-446; EA-ITP-ASG2E, Page 1-41; EA-ITP-AB, Pages 418-419; EA-AC65-15A, Page 343*
Some shock struts are equipped with a damping or snubbing device consisting of a recoil valve on the piston or recoil tube, to reduce the rebound during the extension stroke, and to prevent too rapid an extension of the shock strut.

8309. (B) *EA-ITP-ASG2E, Page 1-46; EA-ITP-AB, Page 448; Aircraft Maintenance and Repair, Page 368*
The most likely cause of a brake grabbing (not applying consistent force) when it is applied is brake fluid on the linings.

8310. (A) *EA-ITP-A2, Page 466; EA-ITP-ASG2E, Page 1-45; EA-ITP-AB, Page 440; EA-AC65-15A*
A common type of parking brake is one that operates by depressing the brake pedals and pulling a parking brake handle, which locks the brakes in a depressed position with a ratchet mechanism. If the master cylinder piston has a faulty seal and allows fluid to pass by it, the brakes will release even though the pedals are still held depressed.

8311. (A) *EA-ITP-A2, Page 497; EA-ITP-ASG2E, Page 1-49; EA-ITP-AB, Page 473; EA-AC65-15A, Page 387*
When inflating a tire in which a new tube has just been installed, put the tire in a safety cage and gradually bring the air pressure up to the recommended value to seat the beads and then deflate the tire. Then reinflate the tire to the correct pressure. This inflation, deflation, and re-inflation procedure allows the tube to straighten itself out inside the tire.

8312. (B) *EA-ITP-A2, Pages 445-446; EA-ITP-ASG2E, Page 1-42; EA-ITP-AB, Page 418; EA-AC65-15A, Page 343*
Most shock struts employ a metering pin for controlling the rate of fluid flow from the lower chamber into the upper chamber. During the compression stroke, the rate of fluid flow is not constant, but is controlled automatically by the variable shape of the metering pin as it passes through the orifice.

8313. (C) *EA-ITP-A2, Pages 452-453; EA-ITP-ASG2E, Page 1-41; EA-ITP-AB, Pages 423 and 425; EA-AC65-15A, Page 404; EA-FAR Part 43.13*
After working on the gear retraction system of an aircraft in such a way that the proper operation of the system could be affected, it is absolutely essential that a retraction test be performed on the gear. One can rest assured that the aircraft maintenance manual will require a retraction test in such an instance.

8314. (B) *EA-ITP-A2, Page 455; EA-ITP-ASG2E, Page 1-49; EA-ITP-AB, Page 427; EA-AC65-15A, Page 387*
A maintenance technician should always be sure to deflate tires completely before demounting. For even a better practice, it is recommended that tires be deflated before wheels are removed from the aircraft. With split rim wheels, there is always a possibility of the rim separating when the tire is under pressure.

8315. (B) *EA-ITP-A2, Page 460; EA-ITP-ASG2E, Page 1-45; EA-ITP-AB, Pages 433-434; EA-AC65-15A, Page 368*
In a single-disc brake system, when the pedal is pushed, hydraulic pressure from the brake control unit enters the brake cylinders and forces the pistons and their pucks against the rotating disk. The rotating disk is keyed to the landing gear wheel so that it is free to move laterally within the brake cavity of the wheel.

8316. (C) *EA-ITP-ASG2E, Page 1-47; EA-AC65-15A, Page 372; Aircraft Maintenance and Repair, Page 368*
Although the most likely cause of spongy brakes is air in the system, another good candidate is a flexible brake line which is deteriorated. This deteriorated brake line will flex and expand when pressure is applied, acting just like air being compressed in a brake system.

8317. (C) *EA-ITP-A2, Page 465; EA-ITP-ASG2E, Page 1-46; EA-AC65-15A, Page 371*
The expander tube type of brake uses a tube made of neoprene reinforced with fabric, which fluid enters under pressure to expand the brakes and apply a stopping force. This system needs fluid pressure to operate.

8318. (A) *EA-ITP-ASG2E, Page 1-46; EA-AC65-15A, Pages 371-372*
For brakes equipped with adjusters, clearance between the brake blocks and drum is usually set to a minimum value of .002 to .015 of an inch.

8319. (A) *EA-ITP-ASG2E, Page 1-49; EA-AC65-15A, Page 380; EA-43.13-1A/3, Page 150*
A stripe or mark that extends from the rim of the wheel onto the tire is used to detect slippage of the tire on the rim. Slippage can occur if the tire is under-inflated, or if it is subjected to severe stress when landing or braking.

8320. (C) *EA-ITP-A2, Pages 472-473; EA-ITP-ASG2E, Pages 1-47; EA-ITP-AB, Page 447*
Bleeding of brakes should continue until no more air bubbles come through the bleeder hose into the container.

8321. (C) *EA-ITP-A2, Page 457; EA-ITP-ASG2E, Page 1-48; EA-ITP-AB, Page 429*
The most difficult area of an aircraft wheel to inspect is the bead seat region. This area, which is highly stressed by the inflated tire, can be distorted or cracked by a hard landing or a seriously over-inflated tire.

8322. (C) *EA-ITP-A2, Page 469; EA-ITP-ASG2E, Page 1-45; EA-ITP-AB, Pages 442-443; EA-AC65-15A, Pages 364-365*
Debooster units are generally used on aircraft equipped with a high pressure hydraulic system and low pressure brakes. Brake debooster cylinders reduce the pressure to the brakes and increase the volume of fluid flow. When the brake pedals are released, the debooster causes a suction in the line to the brake assembly, resulting in faster release of the brakes.

8323. (B) *EA-ITP-A2, Page 448; EA-ITP-ASG2E, Page 1-42; EA-ITP-AB, Page 420; EA-AC65-15A, Page 348*
Toe-in is adjusted on landing gear using an oleo-type shock absorber by adding or removing washers from between the torque links.

8324. (C) *EA-ITP-A2, Page 422; EA-ITP-ASG2E, Page 1-42; EA-ITP-AB, Page 397*
A raised "H" on the stem of an air valve core stands for high pressure. This type of air valve core is used on shock struts or accumulators, not tires.

8325. (A) *EA-ITP-A2, Pages 497-498; EA-ITP-ASG2E, Page 1-49; EA-ITP-AB, Page 113; EA-AC65-15A, Page 385*
Balance in an aircraft wheel assembly is very important. From a wear standpoint, when the wheels are in landing position, a heavy spot in a wheel assembly will have a tendency to remain at the bottom and thus will always strike the ground or runway first. This results in severe wear at one area of the tire tread. In addition, unbalanced tires can cause severe vibration.

8326. (A) *EA-ITP-ASG2E, Page 1-46; EA-AC65-15A, Page 371; Aircraft Maintenance and Repair, Page 368*
If pressure is applied to an expander tube brake system without the wheel installed, the brake blocks will be pushed out too far, and the brake assembly will be subject to damage.

8327. (B) *EA-ITP-A2, Page 449; EA-ITP-ASG2E, Page 1-41; EA-ITP-AB, Page 423; EA-43.13-1A/3, Page 145*
All retractable landing gear systems must have some means of lowering the gear in the event the main extension system should fail. Most emergency systems employ either mechanical, pressure-bottle or free-fall extension capabilities.

8328. (B) *EA-ITP-A2, Page 453; EA-ITP-ASG2E, Page 1-43; EA-ITP-AB, Pages 421-422; EA-AC65-15A, Page 360*
The steer damper accomplishes two separate functions; one is steering the nose wheel and the other is shimmy damping. The steer damper automatically reverts to damping when, for any reason, the flow of high-pressure fluid is removed from the inlet of the steer damper.

8329. (C) *EA-ITP-A2, Page 446; EA-ITP-ASG2E, Page 1-42; EA-ITP-AB, Page 420; EA-AC65-15A, Page 348*
Torque links keep the landing gear pointed in a straight ahead direction; one torque link connects to the shock strut cylinder, while the other connects to the piston.

8330. (C) *EA-ITP-ASG2E, Page 1-48*
According to FAR 43, pilots are allowed to remove, install, and repair landing gear tires on aircraft owned and operated by them.

8331. (C) *EA-ITP-A2, Page 469; EA-ITP-ASG2E, Page 1-45; EA-ITP-AB, Pages 441-442; EA-AC65-15A, Page 362*
Power brake control valve systems are used on aircraft requiring a large volume of fluid to operate the brakes. Because larger aircraft require larger brakes with greater fluid displacement and pressures, an independent master cylinder system is not practical.

8332. (C) *EA-ITP-A2, Page 467; EA-ITP-ASG2E, Page 1-45; EA-AC65-15A, Page 361*
The typical master cylinder has a compensating port or valve that permits fluid to flow from the brake chamber back to the reservoir when excessive pressure is developed in the brake line due to temperature changes. If this port is blocked, the brakes will be slow to release.

8333. (B) *EA-ITP-A2, Pages 445-446; EA-ITP-ASG2E, Page 1-41; EA-ITP-AB, Pages 418-419; EA-AC65-15A, Page 343*
Most shock struts employ a metering pin for controlling the rate of fluid flow from the lower chamber of the strut into the upper chamber. During the compression stroke, the rate of fluid flow is not constant, but is controlled by the variable shape of the metering pin as it passes through the orifice.

8334. (B) *EA-ITP-A2, Page 467; EA-ITP-ASG2E, Page 1-45; EA-ITP-AB, Pages 439-440; EA-AC65-15A, Page 361*
Internal leakage in a master cylinder means that the fluid is passing by the piston seal rather than being forced out to the system under pressure. If this happens in a brake master cylinder, the brakes will not hold pressure, and will fade.

8335. (B) *EA-ITP-A2, Pages 445-446; EA-ITP-ASG2E, Page 1-42; EA-AC65-15A, Page 345*
The extension stroke of an oleo shock strut occurs at the end of the compression stroke, as the energy stored in the compressed air causes the aircraft to start moving upward in relation to the ground and wheels. A snubbing device is used to prevent the strut from rebounding too rapidly, and a sleeve, spacer, or bumper ring is incorporated to limit the extension stroke of the strut.

8336. (C) *EA-ITP-A2, Page 450; EA-ITP-ASG2E, Page 1-41; EA-ITP-AB, Page 425; EA-AC65-15A, Page 349*
In a retractable landing gear system, sequence valves are used to ensure that the landing gear doors open and close at the proper time.

8337. (A) *EA-ITP-A2, Page 467; EA-ITP-ASG2E, Page 1-45; EA-ITP-AB, Page 442; EA-AC65-15A, Page 362*
In a power brake system, a line is tapped off from the main hydraulic system pressure line. A check valve is installed in this line to prevent loss of brake system pressure in case of main system failure.

8338. (B) *EA-ITP-ASG2E, Page 1-46; EA-ITP-AB, Pages 437-438; EA-AC65-15A, Page 368*
If the parking brake is set when the brakes are hot, there is a chance the brake disks will warp. This is not as serious a problem as it once was because of the higher quality alloys in use today, and because the disks are thicker than they used to be.

8339. (C) *EA-ITP-A2, Page 470; EA-ITP-ASG2E, Page 1-45; EA-AC65-15A, Page 363*
Each brake actuating line in a power brake system incorporates a shuttle valve for the purpose of isolating the emergency brake system from the normal brake system.

8340. (A) *EA-ITP-ASG2E, Page 1-42; EA-AC65-15A, Pages 343-344*
A packing gland designed to seal the sliding joint between the upper and lower telescoping cylinders is installed in the open end of the outer cylinder. A packing nut holds the gland in place. When the nut is being tightened, the aircraft should be on jacks, and the strut checked for freedom of movement. The nut should not be tightened more than the manufacturer calls for, or it will cause the strut to bind.

8341. (C) *EA-ITP-A2, Page 446; EA-ITP-ASG2E, Page 1-42; EA-AC65-15A, Pages 346-347*
Proper methods of servicing shock struts with fluid and nitrogen will be found in aircraft manufacturers service manuals.

8342. (C) *EA-ITP-ASG2E, Page 1-45; EA-AC65-15A, Page 363*
The purpose of a relief valve in a power brake system is to bypass fluid into the return line, if the pressure in the system gets too high. If the temperature of the brake fluid in a power brake system were to get too high, the relief valve would unseat due to the thermal expansion of the fluid.

8343. (B) *EA-ITP-A2, Page 490, EA-AC65-15A, Page 379; EA-ITP-ASG2E, Page 1-48*
Proper inflation is undoubtedly the most necessary maintenance function for safe, long service from aircraft tires.

8344. (A) *EA-ITP-A2, Pages 446-448; EA-ITP-ASG2E, Page 1-42; EA-ITP-AB, Page 421*
Toe in or toe out is the amount the wheels deviate from a straight ahead condition. If the front of the wheels try to move together, they are toed in. If the front of the wheels try to move apart, they are toed out. A toed out condition will cause the longitudinal axis of the wheels, if extended, to meet somewhere behind the aircraft.

8245. (C) *EA-ITP-A2, Page 467; EA-ITP-ASG2E, Page 1-45; EA-AC65-15A, Page 361*
The typical master cylinder has a compensating port or valve that permits fluid to flow from the brake chamber back to the reservoir when excessive pressure is developed in the brake line, due to temperature changes. During normal operation, the valve allows reservoir fluid to fill the master cylinder when the brakes are released.

8346. (A) *EA-ITP-A2, Pages 445-446; EA-ITP-ASG2E, Page 1-42; EA-ITP-AB, Page 419*
The air/oil shock strut absorbs the landing impact by letting oil transfer from the oil chamber into the air chamber around a tapered metering pin. Taxi shocks are taken up by the compressed air in the upper chamber. Bottoming out on landing is caused by low fluid level.

8347. (B) *EA-ITP-A2, Page 452; EA-ITP-ASG2E, Page 1-43; EA-AC65-15A, Page 353*
Centering devices include such units as internal centering cams to center the nose wheel as it retracts into the wheel well.

8348. (C) *EA-ITP-A2, Page 446; EA-ITP-ASG2E, Page 1-42; EA-AC65-15A, Page 347*
On some shock struts the correct amount of inflation is determined by measuring the amount of extension (in inches) between two given points on the strut.

8349. (C) *EA-ITP-A2, Page 469; EA-ITP-ASG2E, Page 1-45; EA-ITP-AB, Pages 442-443; EA-AC65-15A, Page 364*
In some power brake control valve systems, debooster cylinders are used in conjunction with the power brake control valves. Brake deboosters reduce the pressure to the brake and increase the volume of fluid flow.

8350. (C) *EA-ITP-A2, Page 446; EA-ITP-ASG2E, Page 1-42; EA-ITP-AB, Pages 418-419; EA-AC65-15A, Pages 342-343*
A properly serviced shock strut should not bottom unless it is subjected to an excessive force during a hard landing. If a strut does bottom during normal operation, something is wrong inside the strut, and it must be taken apart and inspected. Increasing air pressure or fluid level is definitely not the answer.

8351. (C) *EA-ITP-A2, Page 494; EA-ITP-ASG2E, Page 1-49; EA-43.13-1A/3, Page 147*
The wide variation in tire operating environments which may affect total carcass life and serviceability makes it inadvisable to prescribe arbitrarily the maximum number of times a high speed tire should be retreaded. This aspect, therefore, is controlled by thorough inspection of the carcass before retreading.

8352. (B) *EA-ITP-A2, Page 462; EA-ITP-ASG2E, Page 1-45; EA-ITP-AB, Pages 436-437; EA-AC65-15A, Page 368*
Of the answers provided with this question, the only one that is remotely correct is the worn brake linings. A leakage in the master cylinder could cause the excessive pedal travel, but not if the brakes are hard and effective. The push rod between the brake pedal and the master cylinder being too long would not affect the pedal travel. A worn brake disk could cause the problem if it were worn on the faces, but the distractor identifies wear on the perimeter of the disk. Single-disk brakes will typically have an adjusting pin that keeps the clearance between the brake linings and the disk constant, so the pedal travel will not change as the linings wear. If this feature were not present, worn brake linings would cause increased pedal travel.

8353. (B) *EA-ITP-A2, Page 490; EA-ITP-ASG2E, Page 1-48; EA-ITP-AB, Page 465; EA-AC65-15A, Page 379*
The proper inflation pressure is that specified by the aircraft manufacturer in his service manuals or operator's manuals. Even for the same tire, this pressure will vary from one aircraft to another.

8354. (C) *EA-ITP-A2, Pages 445-446; EA-ITP-ASG2E, Page 1-42; EA-ITP-AB, Page 419*
The most likely cause of a shock strut bottoming on landing is low fluid level.

8355. (B) *EA-ITP-A2, Page 472; EA-ITP-ASG2E, Page 1-47; EA-ITP-AB, Pages 446-447; EA-AC65-15A, Pages 372-373*
A sure sign of air in a brake system which uses an independent master cylinder is spongy brakes. A good firm brake pedal is a good sign that all the air has been purged from the system.

8356. (A) *EA-ITP-A2, Page 467; EA-ITP-ASG2E, Page 1-45; EA-AC65-15A, Page 361*
The compensating port in a master cylinder allows fluid to flow back to the reservoir when the brakes are released. If the compensating port is blocked, pressure will be held on the system when the brakes are released and they will drag.

8357. (B) *EA-ITP-A2, Page 469; EA-ITP-ASG2E, Page 1-45; EA-ITP-AB, Page 442 (Figure 6-133); EA-AC65-15A, Page 364*
A typical mounting location for a debooster cylinder would be on the landing gear shock strut, in line between the brake control valve and the brake actuating cylinder.

8358. (C) *EA-ITP-A2, Page 392; EA-ITP-ASG2E, Page 1-42*
Due to the difference in composition vegetable base, petroleum base and phosphate ester base fluids will not mix. Neither are the type of seals for any one fluid usable with, or tolerant of, any of the other fluids.

8359. (A) *EA-ITP-A2, Page 469; EA-ITP-ASG2E, Page 1-45*
Lockout deboosters employ a spring-loaded valve to prevent fluid from entering the lower chamber until the reset handle is lifted.

8360. (A) *EA-ITP-A2, Page 493; EA-ITP-ASG2E, Page 1-48*
If the tire has been exposed to enough heat to melt the fusible plug, it should be replaced. This excessive heat will cause damage to the tire that may not be obvious.

8361. (C) *EA-ITP-ASG2E, Page 1-48; EA-AC65-15A, Page 380*
Nylon tires will develop temporary flat spots when they sit under a static load. Aircraft that are to remain idle for more than 3 days should either be moved every 48 hours or be blocked up to take the load off the tires.

8362. (B) *EA-ITP-A2, Pages 489-490; EA-ITP-ASG2E, Page 1-48; EA-ITP-AB, Pages 465-466; EA-AC65-15A, Page 378*
Tires are as vital to the operation of aircraft as they are to the operation of an automobile. During ground operation, tires can be considered as ground control surfaces. The same rules of safe driving and careful inspection apply on the runways as on the highway. These rules include control of speed, braking, cornering, inspection for proper inflation, checking for cuts and bruises, and signs of tread wear. The toughest demand on tires is the heat buildup during lengthy ground operations.

8363. (C) *EA-ITP-A2, Page 455; EA-ITP-ASG2E, Page 1-48; EA-ITP-AB, Page 426; EA-AC65-15A, Page 383*
The fusible plug in an aircraft wheel is designed to melt and let the air escape from the tire if the temperature gets too high. This is done so the tire will not blow out because of the excessive temperature.

8364. (B) *EA-ITP-ASG2E, Page 1-48; EA-AC65-15A, Page 379*
Differences of air pressure in tires mounted as duals, whether main or nose, should be cause for concern, as it ordinarily means that one tire is carrying more of the load than the other. If there is a difference of more than 5 lbs., it should be noted in the aircraft log.

8365. (A) *EA-ITP-A2, Page 490; EA-ITP-ASG2E, Page 1-48; EA-AC65-15A, Page 379*
Wait at least two hours after a flight before checking pressures (three hours in hot weather).

8366. (C) *EA-ITP-A2, Page 491; EA-ITP-ASG2E, Page 1-48; EA-ITP-AB, Page 467; EA-AC65-15A, Page 380*
Under-inflation results in harmful and potentially dangerous effects. One of the problems it can cause is rapid or uneven wear at or near the edge of the tread.

8367. (C) *EA-ITP-A2, Page 491; EA-ITP-ASG2E, Page 1-48; EA-ITP-AB, Page 466; EA-AC65-15A, Page 397*
If the center ribs are worn away while the shoulder ribs still have an appreciable depth, the tire has been operated in an over-inflated condition.

8368. (C) *EA-ITP-A2, Page 446; EA-ITP-ASG2E, Page 1-42; EA-AC65-15A, Page 347*
After an empty shock strut is serviced, it must be bled to get the air out of the fluid cavity. This is done by extending and compressing the strut until there is no sign of air bubbles coming from the strut bleed nose. This same operation also ensures the proper seating of the seals in the strut.

8369. (B) *EA-ITP-ASG2E, Page 1-56; Aircraft Maintenance and Repair, Page 330*
Chevron seals in shock struts are used to seal the oil in the strut. They are located between the inner and outer cylinders at the lower end of the shock strut. They are held in place between packing spacers, which compress the chevron seals and force them to form a tight seal.

8370. (C) *EA-ITP-A2, Page 446; EA-ITP-ASG2E, Page 1-42; EA-AC65-15A, Page 347*
When servicing a shock strut, the air pressure should be released and the strut fully compressed. The strut should then be filled to the level of the air valve opening with an approved type of hydraulic fluid.

8371. (B) *EA-ITP-ASG2E, Page 1-46*
We are unable to find another reference for this question. It is assumed that the McGraw Hill text from which the question was drawn is not a current edition. From years of experience with absolute braking systems, answer #2 seems the best.

8372. (A) *EA-ITP-A2, Page 452; EA-ITP-ASG2E, Page 1-41; EA-AC65-15A, Page 353*
Gear warning devices are incorporated on all retractable gear aircraft, and usually consist of a horn or some other aural device, and a red warning light. The horn blows and the light comes on when one or more throttles are retarded and the landing gear is in any position other than down and locked.

8373. (C) *EA-ITP-A2, Page 579; EA-ITP-ASG2E, Page 1-41; EA-ITP-AB, Page 355; EA-AC65-9A, Page 448*
Because a motor to operate the landing gear needs to be able to reverse its direction of rotation, a split field series wound motor would be used. By using a single-pole, double-throw switch, it is a possible to change the direction of current flow through the motor, and thereby change the direction of rotation.

8374. (C) *EA-ITP-A2, Page 426; EA-ITP-ASG2E, Page 1-41; EA-ITP-AB, Page 401; EA-AC65-9A, Page 168*
When a chevron seal is installed, it must have its open end facing the direction from which the pressure is applied. This causes the seal to open up and exert outward pressure against the surface it is trying to seal.

8375. (C) *EA-ITP-A2, Page 452; EA-ITP-ASG2E, Page 1-41; EA-AC65-15A, Page 353*
Centering devices include such units as internal centering cams to center the nose wheel as it retracts into the wheel well.

8376. (C) *EA-ITP-A2, Pages 418-419; EA-ITP-ASG2E, Page 1-52; EA-ITP-AB, Page 393; EA-AC65-15A, Pages 319 and 323*
When a constant-delivery pump is used in a hydraulic system in which the pressure must be kept at a constant value, a pressure regulator is required. This pressure regulator discharges fluid from the pressure line into a reservoir return line when the pressure exceeds the predetermined maximum for which the valve is adjusted. This valve also unloads the pump when the pressure in the system is within normal operating parameters.

8377. (B) *EA-ITP-A2, Pages 420-421; EA-ITP-ASG2E, Page 1-53; EA-ITP-AB, Page 395; EA-AC65-15A, Page 324*
The function of an accumulator is to dampen pressure surges, to aid or supplement the power pump when several units are operating, to store power for the limited operation of the system when the pump is not running, and to supply fluid under pressure to compensate for small internal or external leaks.

8378. (B) *EA-ITP-A2, Page 399; EA-ITP-ASG2E, Page 1-51; EA-ITP-AB, Page 372*
To continue the simplification of the hydraulic systems, many manufacturers use an electric motor to drive the hydraulic pump and incorporate it with the necessary valves and reservoir into a single unit. This unit is typically called a "power pack".

8379. (A) *EA-ITP-G2, Page 290; EA-ITP-GB, Page 286; EA-AH-1, Page 41; EA-ITP-ASG2E, Page 1-50*
Aircraft flexible hose is marked with a yellow, red, or white stripe running the length of the hose, called a lay line. This lay line is not only for identification of the hose, but it serves to indicate whether or not the line was twisted when it was installed.

8380. (A) *EA-ITP-A2, Page 428; EA-ITP-ASG2E, Page 1-50; EA-ITP-AB, Page 403*
The part number is the only sure way of knowing that the correct O-ring is being used, but most rings are also marked with a series of colored dots or stripes to indicate the type of fluid with which they are compatible. A blue dot or stripes indicates compatibility with MIL-H-5606 hydraulic fluid.

8381. (B) *EA-ITP-ASG2E, Page 1-53; Aircraft Maintenance and Repair, Page 324*
Excessive fluctuation of hydraulic pressure is a sure sign of inadequate supply of fluid.

8382. (C) *EA-ITP-A2, Page 403; EA-ITP-ASG2E, Page 1-51; EA-ITP-AB, Page 378; EA-AC65-15A, Page 314*
One of the more efficient types of filters used in aircraft hydraulic systems is made of specially treated paper folded into pleats to increase its surface area. This pleated paper micronic element, as it is called, is wrapped around a spring steel wire coil to prevent its collapsing.

8383. (B) *EA-ITP-A2, Page 414; EA-ITP-ASG2E, Page 1-53; EA-ITP-AB, Pages 388-389; EA-AC65-15A, Page 327*
There are many instances in an aircraft hydraulic system when it is desirable to allow fluid to flow in one direction but prevent its flow in the opposite direction. This can be done by using check valves.

8384. (A) *EA-ITP-G2, Page 376; EA-ITP-GB, Page 365 (Figure 2-89) ; EA-ITP-ASG2E, Figure Reference*
Adapter length can be figured by using the formula indicated.

8385. (C) *EA-ITP-G2, Page 374; EA-ITP-GB, Page 365; EA-43.13-1A/3, Page 119; EA-ITP-ASG2E, Figure Reference*
Because there are 12 inches in a foot, there are 12 inch-pounds in one foot-pound. To convert inch-pounds to foot-pounds, divide the inch-pounds by 12. In this case, 440 divided by 12 equals 36.6 ft.-lbs.

8386. (C) *EA-ITP-A2, Page 429; EA-ITP-ASG2E, Page 1-56; EA-ITP-AB, Page 404; EA-43.13-1A/3, Page 171*
When installing an O-ring over a sharp edge or threads, you should cover the sharp portion with paper, aluminum foil, brass shim stock, or with a piece of plastic.

8387. (A) *EA-ITP-A2, Page 426; EA-ITP-ASG2E, Page 1-56; EA-ITP-AB, Page 402; EA-AC65-9A, Page 167*
The most commonly used two way seal is the O-ring, which may be used as either a gasket or a packing.

8388. (A) *EA-ITP-A2, Page 413; EA-ITP-ASG2E, Page 1-53; EA-ITP-AB, Page 388; EA-AC65-15A, Page 327*
There are many instances in an aircraft hydraulic system where it is desirable to allow fluid to flow in one direction but prevent its flow in the opposite direction. This can be done by using check valves.

8389. (C) *EA-ITP-A2, Page 410; EA-ITP-ASG2E, Page 1-55; EA-ITP-AB, Page 387; EA-AC65-15A, Page 329*
Selector valves are used to control the direction of movement of an actuating unit. A selector valve provides a pathway for the simultaneous flow of hydraulic fluid into and out of a connected actuating unit.

8390. (C) *EA-ITP-ASG2E, Page 1-58; EA-AC65-15A, Pages 338-339*
The compressor inlet air for a pneumatic system is filtered through a high temperature, 10 micron filter and the air pressure is regulated by an absolute pressure regulator to provide a stabilized source of air for the compressor.

8391. (A) *EA-ITP-A2, Pages 435-437; EA-ITP-ASG2E, Page 1-52; EA-ITP-AB, Page 410; EA-AC65-15A, Pages 334-335*
Relief valves are used in pneumatic systems to prevent damage. They act as pressure-limiting units and prevent excessive pressures from bursting lines and blowing out seals. Check valves are used in both pneumatic and hydraulic systems to limit the flow to one direction.

8392. (A) *EA-ITP-A2, Page 412; EA-ITP-ASG2E, Page 1-55; EA-ITP-AB, Page 387*
In an open center-type of selector valve, when the valve is in the OFF or neutral position, fluid flows straight through the valve from the pump and on to the next selector valve or back to the reservoir.

8393. (B) *EA-ITP-A2, Pages 392-393; EA-ITP-ASG2E, Page 1-50; EA-ITP-AB, Page 376; EA-AC65-15A, Page 311*
Skydrol® hydraulic fluid is a phosphate ester base fluid, so the packing materials used with this type of fluid must be compatible with an ester base.

8394. (C) *EA-ITP-A2, Page 418; EA-ITP-ASG2E, Page 1-52; EA-ITP-AB, Pages 409-410; EA-AC65-15A, Page 334*
Relief valves are used in pneumatic systems to prevent damage. They act as pressure-limiting units and prevent excessive pressures from bursting lines and blowing out seals.

8395. (C) *EA-ITP-A2, Page 439; EA-ITP-ASG2E, Page 1-58; EA-ITP-AB, Page 409; EA-AC65-15A, Page 339*
Moisture in a compressed air system will condense and freeze when the pressure of the air is dropped for actuation and, for this reason, every bit of water must be removed from the air. A moisture separator collects the water that is in the air on a baffle and holds it until the system is shut down. The separator will remove approximately 98% of the water in the air.

8396. (B) *EA-ITP-A2, Page 402; EA-ITP-ASG2E, Page 1-56*
The easiest way to remove air from an aircraft hydraulic system is to operate various hydraulic components (often the flaps) until air is no longer heard in the system.

8397. (B) *EA-ITP-A2, Page 435; EA-ITP-ASG2E, Page 1-57; EA-ITP-AB, Page 409; EA-AC65-15A, Page 334*
Of the four items mentioned in this question, the only one that pneumatic systems utilize is a relief valve.

8398. (C) *EA-ITP-A2, Page 410; EA-ITP-ASG2E, Page 1-55; EA-ITP-AB, Page 387; EA-AC65-15A, Page 329*
Selector valves are used to control the direction of movement of an actuating unit. A selector valve provides a pathway for the simultaneous flow of hydraulic fluid into and out of a connected actuating unit.

8399. (A) *EA-ITP-A2, Page 412; EA-ITP-ASG2E, Page 1-55; EA-AC65-15A, Page 330*
The four-way closed-center selector valve is one of the most commonly used selector valves in an aircraft hydraulic system. In this type of valve, when the selector is OFF, all of the valve ports are blocked and fluid cannot flow into or out of the valve.

8400. (C) *EA-ITP-A2, Page 427; EA-ITP-ASG2E, Page 1-56; EA-ITP-AB, Page 402; EA-AC65-9A, Page 167*
An O-ring of the appropriate size can withstand pressures up to about 1500 PSI without distortion, but beyond this, there is a tendency for the ring to extrude into the groove between the two mating surfaces. To prevent this, an anti-extrusion or backup ring should be used.

8401. (A) *EA-ITP-A2, Pages 418-419; EA-ITP-ASG2E, Page 1-52; EA-ITP-AB, Page 393; EA-AC65-15A, Page 324*
One purpose of a pressure regulator is to manage the output of the pump to maintain system operating pressure within a predetermined range. The other purpose is to permit the pump to turn without resistance (termed unloading the pump) at times when pressure in the system is within normal operating range.

8402. (B) *EA-ITP-ASG2E, Page 1-56; EA-ITP-AB, Page 401; EA-AC65-9A, Page 166*
Packings are made of synthetic or natural rubber. They are generally used as "running" seals, that is, in units that contain moving parts, such as actuating cylinders, pumps, selector valves, etc.

8403. (C) *EA-ITP-A2, Page 387; EA-ITP-ASG2E, Page 1-50; EA-AC65-15A, Page 309*
Hydraulic systems combine the advantages of light weight, ease of installation, simplification of inspection, and minimum maintenance requirements. Hydraulic operations are nearly 100% efficient.

8404. (B) *EA-ITP-G2, Page 200; EA-ITP-ASG2E, Page 1-57*
Rigid tubing should have at least one bend, and should be fabricated so that it is not necessary to "pull" the tube into alignment.

8405. (C) *EA-ITP-G2, Page 299; EA-ITP-GB, Page 295; EA-ITP-ASG2E, Page 1-57*
All tubing should have at least one bend when it is installed to absorb the strains from vibration and from the dimensional changes of the aircraft structure due to temperature changes.

8406. (C) *EA-ITP-A2, Page 427; EA-ITP-ASG2E, Page 1-56; EA-ITP-AB, Page 402*
Extrusion of O-ring material under high pressure can be effectively stopped by using a backup ring. Check the angle cuts on the backup ring for proper overlap.

8407. (B) *EA-ITP-ASG2E, Page 1-55; EA-AC65-15A, Page 321*
Pump drive couplings are designed to serve as safety devices.

8408. (A) *EA-ITP-ASG2E, Page 1-53; EA-ITP-AB, Page 397; Manufacturer's Maintenance Manual*
Before removing any component in an hydraulic system, the system pressure must be relieved. Some systems utilize a pressurized reservoir. Always check the manufacturer's maintenance manual instructions for the removal of components. Before disassembling an accumulator, make sure that the preload pressure has been discharged.

8409. (A) *EA-ITP-G2, Page 296; EA-ITP-AB, Page 289 (Figure 3-29); EA-AC65-9A, Page 103 (Figure 5-6); EA-ITP-ASG2E, Illustration*
These figures compare the differences between fittings.

8410. (C) *EA-ITP-A2, Page 428; EA-ITP-ASG2E, Page 1-57; EA-ITP-AB, Page 403*
Illustration B would cause a "hump" in the compressed "O" ring and a shearing tendency in the backup ring.

8411. (A) *EA-ITP-A2, Page 392; EA-ITP-ASG2E, Page 1-50; EA-ITP-AB, Page 376; EA-AC65-15A, Page 310*
Natural rubber seals should be used with vegetable base fluids.

8412. (B) *EA-ITP-A2, Page 391; EA-ITP-ASG2E, Page 1-50; EA-AC65-15A, Page 309*
One of the most important properties of any hydraulic fluid is its viscosity. Viscosity is internal resistance to flow.

8413. (C) *EA-ITP-A2, Page 391; EA-ITP-ASG2E, Page 1-50; EA-AC65-15A, Page 309*
Viscosity is internal resistance to flow.

8414. (A) *EA-ITP-A2, Page 392; EA-ITP-ASG2E, Page 1-50; EA-AC65-15A, Page 311*
Mineral base hydraulic fluid is processed from petroleum. This type of fluid is flammable.

8415. (C) *EA-ITP-A2, Page 392; EA-ITP-ASG2E, Page 1-50; EA-ITP-AB, Page 376; EA-AC65-15A, Pages 310-311*
To assure proper system operation and to avoid damage to nonmetallic components of the hydraulic system, the correct fluid must be used. When adding fluid to a system, use the type specified in the aircraft manufacturer's maintenance manual or on the instruction plate affixed to the reservoir or unit being serviced. To avoid servicing with the wrong type of fluid, each type of fluid is dyed a certain color.

8416. (C) *EA-ITP-A2, Page 392; EA-ITP-ASG2E, Page 1-50; EA-ITP-AB, Page 376; EA-AC65-15A, Page 311*
Mineral based hydraulic fluid is processed from petroleum. It has an odor similar to penetrating oil and is dyed red.

8417. (C) *EA-ITP-A2, Page 292; EA-ITP-ASG2E, Page 1-50*
Skydrol is quite susceptible to contamination by water from the atmosphere, and must be kept tightly sealed.

8418. (B) *EA-ITP-A2, Page 392; EA-ITP-ASG2E, Page 1-50; EA-ITP-AB, Page 376; EA-AC65-15A, Page 311*
Skydrol® fluid has a very high flash point. Even though it might flash at exceedingly high temperatures, Skydrol fluid will not tend to spread fire.

8419. (C) *EA-ITP-A2, Page 387; EA-ITP-ASG2E, Page 1-50; EA-ITP-GB, Pages 78-79; EA-AC65-9A, Page 222*
Gases and liquids are both fluids. Although there are many differences between the characteristics of the two, one characteristic that all fluids have is their ability to readily transmit pressure. Liquids do a better job of transmitting pressure, however, because they are for all practical purposes, incompressible.

8420. (A) *EA-ITP-A2, Page 392; EA-ITP-ASG2E, Page 1-50; EA-ITP-AB, Page 376; EA-AC65-15A, Pages 310-311*
There are three types of hydraulic fluid currently being used in civil aircraft — vegetable, mineral, and phosphate ester base.

8421. (C) *EA-ITP-A2, Pages 390-392; EA-ITP-ASG2E, Page 1-50; EA-AC65-15A, Pages 309-310*
A good hydraulic fluid has a low viscosity, good chemical stability, high flash point, and high fire point.

8422. (B) *EA-ITP-A2, Page 392; EA-ITP-ASG2E, Page 1-50; EA-ITP-AB, Page 376; EA-AC65-15A, Page 311*
Skydrol® 500 A and B hydraulic fluids are phosphate ester base synthetics, light purple in color, fire resistant, and are used with butyl rubber seals.

8423. (A) *EA-ITP-A2, Page 392; EA-ITP-ASG2E, Page 1-50; EA-ITP-AB, Page 376; EA-AC65-15A, Page 311*
Manufacturer's technical bulletins will generally give information about the compatibility of aircraft materials with different types of hydraulic fluid.

8424. (C) *EA-ITP-A2, Page 392; EA-ITP-ASG2E, Page 1-50; EA-ITP-AB, Page 376; EA-AC65-15A, Page 311*
5606 hydraulic fluid is processed from petroleum, it is red in color, it is flammable, and it is used with synthetic rubber seals.

8425. (C) *EA-ITP-A2, Page 392; EA-ITP-ASG2E, Page 1-50; EA-ITP-AB, Page 376; EA-AC65-15A, Page 310*
7644 hydraulic fluid is a vegetable base, it is blue in color, it is flammable, and it is used with natural rubber seals.

8426. (B) *EA-ITP-A2, Page 392; EA-ITP-ASG2E, Page 1-51; EA-ITP-AB, Page 376; EA-AC65-15A, Page 311*
When servicing a system using Skydrol® hydraulic fluid, care must be taken to use only seals and hoses having the proper part number. Components which are compatible with mineral or vegetable base fluids will be damaged by the synthetic fluid.

8427. (B) *EA-ITP-A2, Page 392; EA-ITP-ASG2E, Page 1-51; EA-ITP-AB, Page 376*
A hydraulic system that uses mineral base hydraulic fluid can be flushed with naphtha, Varsol®, or Stoddard® solvent.

8428. (A) *EA-ITP-A2, Page 392; EA-ITP-ASG2E, Page 1-51; EA-ITP-AB, Page 376*
A hydraulic system that uses Skydrol® hydraulic fluid can be flushed with trichlorethylene.

8429. (C) *EA-ITP-A2, Page 392; EA-ITP-ASG2E, Page 1-50; EA-AH-1, Page 9*
When servicing a hydraulic system, be sure that only the proper fluid is used. The service manual of the airplane specifies the fluid, and the reservoir should also be marked with the type of fluid required.

8430. (B) *EA-ITP-A2, Page 392; EA-ITP-ASG2E, Page 1-50; EA-ITP-AB, Page 376*
Some compounds will actually attract water from the air. Skydrol is one of these.

8431. (B) *EA-ITP-A2, Page 392; EA-ITP-ASG2E, Page 1-50; EA-AC65-15A, Page 311*
Skydrol® does not appreciably affect common aircraft metals — aluminum, zinc, magnesium, iron, silver, cadmium. Paints which are Skydrol® resistant include epoxies and polyurethanes. Skydrol® is compatible with natural fibers and a number of synthetics, including nylon and polyester.

8432. (B) *EA-ITP-A2, Page 440; EA-ITP-ASG2E, Page 1-56; Aircraft Maintenance and Repair, Page 279*
Quite frequently in hydraulic systems it is necessary to provide alternate or emergency sources of power with which to operate critical parts of the system. In such a case, there must be a means of disconnecting the normal source of hydraulic power and connecting the emergency source of power. This is the function of a shuttle valve.

8433. (B) *EA-ITP-A2, Page 423; EA-ITP-ASG2E, Page 1-55; EA-ITP-AB, Page 398; EA-AC65-15A, Page 328*
An actuating cylinder transforms energy in the form of fluid pressure into mechanical force, or action, to perform work.

8434. (A) *EA-ITP-ASG2E, Page 1-54; Aircraft Maintenance and Repair, Page 281*
Relief valves are used to control maximum system pressure and to control pressure in various parts of the subsystems. For example, a relief valve, called a wing-flap overload valve, is often placed in the down line of the wing-flap subsystem to prevent lowering of the flaps at too high an air speed.

8435. (A) *EA-ITP-A2, Page 423; EA-ITP-ASG2E, Page 1-55; EA-ITP-AB, Page 398; EA-AC65-15A, Page 328*
An actuating cylinder transforms energy in the form of fluid pressure into mechanical force, or action, to perform work. It is used to impart powered linear motion to some movable object or mechanism.

8436. (B) *EA-ITP-ASG2E, Page 1-56; Aircraft Maintenance and Repair, Page 281*
When several relief valves are incorporated in a hydraulic system, they should be adjusted in a sequence which will permit each valve to reach its operating pressure. Thus, the highest pressure valves should be adjusted first, the others in the order of descending pressure valves.

8437. (C) *EA-ITP-A2, Page 421; EA-ITP-ASG2E, Page 1-57; EA-ITP-AB, Pages 395-396; EA-AC65-15A, Page 324*
One function of the accumulator in a hydraulic system is to supply fluid under pressure to compensate for small internal or external leaks which would cause the system to cycle continuously by action of the pressure switches continually "kicking in". If the air preload in the accumulator is low, it won't be able to handle any draw down on the system and the pump will be continually cycling.

8438. (B) *EA-ITP-A2, Page 418; EA-ITP-ASG2E, Page 1-52; EA-ITP-AB, Page 393; EA-AC65-15A, Page 324*
Pressure regulator valves in hydraulic systems sometimes serve as unloading valves as well. Pressure regulators maintain the pressure within a specified range and keep the pump unloaded any time no unit is being actuated.

8439. (C) *EA-AC65-15A, Page 320; EA-ITP-ASG2E, Page 1-52*
Pump drive couplings, used with certain hydraulic pumps, are designed to serve as safety devices. If the pump becomes unusually hard to turn, or becomes jammed, the shear section of the drive coupling will break, preventing damage to the pump or the driving unit.

8440. (C) *EA-ITP-ASG2E, Page 1-52; Aircraft Maintenance and Repair, Page 283*
A thermal relief valve is similar to a regular system relief valve, except that it is installed in parts of the system where fluid pressure is trapped and may need to be relieved because of the increase in pressure caused by high temperatures. Thermal relief valves are adjusted to pressures which are above those required for the operation of the systems.

8441. (B) *EA-ITP-A2, Page 394; EA-ITP-ASG2E, Page 1-51; EA-ITP-AB, Page 368; EA-AC65-15A, Page 315*
A basic hydraulic system needs a reservoir to hold the fluid, a pump to impart pressure to the fluid, a selector valve to direct the fluid, and an actuator to transform the fluid pressure to mechanical force.

8442. (C) *EA-ITP-A2, Page 409; EA-ITP-ASG2E, Page 1-51; EA-ITP-AB, Pages 384-385; EA-AC65-15A, Page 319*
A variable displacement hydraulic pump has a fluid output that is varied to meet the pressure demands of the system by varying its fluid output. The pump output is changed automatically by a pump compensator which is built into the pump.

8443. (C) *EA-ITP-ASG2E, Page 1-51; EA-AC65-15A, Page 320*
Gear-type hydraulic pumps, as well as the other types of hydraulic pumps, have a shear pin or shear section incorporated into their drive shafts. If the load on the pump becomes too great, this portion of the shaft will shear to prevent serious damage to the pump.

8444. (C) *EA-ITP-ASG2E, Page 1-57; Aircraft Maintenance and Repair, Page 272*
The question here talks about a pressure stroke, so the pump in question is a single-acting hand pump. If the output check valve on such a pump is installed backwards, the pump handle can be moved on the intake stroke, but on the output stroke the backwards check valve will prevent fluid from leaving the pump.

8445. (A) *EA-ITP-A2, Page 388; EA-ITP-AB, Page 364; EA-AC65-9A, Page 226; EA-ITP-ASG2E, Page 1-53*
Pressure is a measure of the amount of force that acts on a unit of area. In most American hydraulic systems, pressure is measured in pounds per square inch.

8446. (C) *EA-ITP-A2, Pages 388-390; EA-ITP-GB, Pages 80-81; EA-AC65-9A, Pages 232-233; EA-ITP-ASG2E, Page 1-51*
According to Pascal's Law, the force available at a piston is equal to the cross sectional area of the piston multiplied by the pressure acting on it. The length of stroke of two pistons has no effect on how much force they will exert or the speed at which they will move.

8447. (A) *EA-ITP-A2, Pages 388-390; EA-ITP-GB, Pages 80-81; EA-AC65-9A, Pages 232-233; EA-ITP-ASG2E, Explanation*
According to Pascal's Law, pressure in an enclosed hydraulic system is equal and undiminished in all directions. In this question, the fact that there is 100 PSI in one part of the system means that there is 100 PSI in all parts of the system. This holds true because this is not a complex system making use of pressure reducing devices.

8448. (B) *EA-ITP-A2, Page 407; EA-ITP-ASG2E, Page 1-50*
High pressures and high flow rates will generate heat in the hydraulic system. Many design factors are used to minimize the formation of heat, and heat exchangers may be used to remove heat from the system.

8449. (B) *EA-ITP-A2, Page 399; EA-ITP-ASG2E, Page 1-56; EA-ITP-AB, Page 372; EA-AC65-15A, Page 316*
The design of hydraulic systems is such that they tend to automatically bleed air out of the system when actuators are cycled a few times. If the time it takes for a flap actuator to operate decreases with successive cycles after it is replaced, this is an indication that the air is being worked out of the system.

8450. (C) *EA-ITP-A2, Page 402; EA-ITP-ASG2E, Page 1-57; EA-ITP-AB, Pages 376-377; EA-AC65-15A, Pages 316-317*
When an item such as the wing flaps is operated, whether it is by the main system or the emergency system, the majority of the components which are used are the same. The primary separation between the main and the emergency systems is the supply of fluid and the pump. If the flaps operate normally when lowered by the emergency system, but cannot be lowered by the main system, the probable cause is low fluid level in the main system.

8451. (B) *EA-ITP-A2, Page 402; EA-ITP-ASG2E, Page 1-51; EA-ITP-AB, Pages 376-377; EA-AC65-15A, Page 318*
Some aircraft have emergency hydraulic systems that take over if main systems fail. In many such systems, the pumps of both systems obtain fluid from a single reservoir. The main system draws its fluid through a standpipe located at a higher level in the reservoir, and the emergency system draws its fluid from the bottom of the reservoir.

8452. (B) *EA-ITP-A2, Page 415; EA-ITP-ASG2E, Page 1-55; EA-ITP-AB, Page 390; EA-AC65-15A, Page 349*
Sequence valves are used to allow certain operations to occur in a hydraulic system, in a specific order, so that one operation does not interfere with another. They are especially valuable in a landing gear system to control the operation of the gear and gear doors.

8453. (C) *EA-ITP-A2, Page 418; EA-ITP-ASG2E, Page 1-52; EA-ITP-AB, Page 393; EA-AC65-15A, Page 324*
One purpose of a pressure regulator is to manage the output of the pump to maintain system operating pressure within a predetermined range. The other purpose is to permit the pump to turn without resistance (termed unloading the pump) at times when pressure in the system is within normal operating range.

8454. (C) *EA-ITP-ASG2E, Page 1-57; Aircraft Maintenance and Repair, Page 272*
When the hand pump handle is being moved on the intake stroke, the inport check valve is open to allow reservoir fluid to enter the pump. At this time, the outport check valve is closed, sealing off the pump from the system pressure. If the outport check valve sticks open, the pressure which the system has in it will back up into the hand pump and cause severe kickback.

8455. (C) *EA-ITP-A2, Page 414; EA-ITP-ASG2E, Page 1-53; EA-ITP-AB, Page 389; EA-AC65-15A, Pages 326-327*
The orifice-type in-line check valve is used to allow normal operating speed of a mechanism by providing free flow of fluid in one direction, while allowing limited operating speed through restricted flow of fluid in the opposite direction.

8456. (A) *EA-ITP-A2, Page 395; EA-ITP-ASG2E, Page 1-57; EA-ITP-AB, Page 369*
A power control valve, or pump control valve, is a semi-automatic valve which is manually closed and automatically opened. Its purpose is to allow fluid to circulate back to the reservoir to relieve the pump. This valve must be closed before attempting to adjust the main system pressure relief valve, because pressure will not build in the system with it open.

8457. (B) *EA-ITP-A2, Page 421; EA-ITP-ASG2E, Page 1-53; EA-ITP-AB, Page 395; EA-AC65-15A, Page 324*
The accumulator is a steel sphere divided into two chambers by a synthetic diaphragm. The upper chamber contains system fluid and the lower chamber contains an air pre-charge. Once the hydraulic system is pressurized, the lower pressure is increased by compression until the air pressure equals the system pressure.

8458. (B) *EA-ITP-A2, Page 420; EA-ITP-ASG2E, Page 1-53; EA-ITP-AB, Page 395; EA-AC65-15A, Page 324*
All accumulators consist of a high-strength container divided by some form of movable partition into two sections, or compartments. One compartment is connected to the hydraulic pressure manifold, and the other compartment is filled with either compressed air or with nitrogen.

8459. (C) *EA-ITP-A2, Page 421; EA-ITP-ASG2E, Page 1-53; EA-ITP-AB, Page 372 (Figure 1A-15); EA-AC65-15A, Page 316 (Figure 8-6)*
When an accumulator is installed, right after the installation there is no hydraulic fluid in the fluid side of the accumulator. After the system is operated and hydraulic fluid has had a chance to flow to all components in the system, then the pressure gage will indicate a reading.

8460. (A) *EA-ITP-A2, Page 418; EA-ITP-ASG2E, Page 1-57; EA-ITP-AB, Page 393; EA-AC65-15A, Page 324*
An unloading valve is a type of pressure regulator. Its purpose is to unload the pump when hydraulic pressure is not required for operation of the system components. Before attempting to adjust a main system relief valve, the unloading valve must be held in a closed position so it won't unload the system.

8461. (C) *EA-ITP-A2, Page 392; EA-ITP-ASG2E, Page 1-50; EA-ITP-AB, Page 376; EA-AC65-15A, Page 310*
Natural rubber seals are used with vegetable base hydraulic fluid.

8462. (A) *EA-ITP-A2, Page 433; EA-ITP-ASG2E, Page 1-57*
Pneumatic systems simply exhaust air overboard after it has done the required work.

8463. (A) *EA-ITP-A2, Page 417; EA-ITP-ASG2E, Page 1-55; EA-ITP-AB, Page 390*
Modern aircraft depend on their hydraulic systems not only for raising and lowering the landing gear and flaps, but for control system boosts, thrust reversers, brakes, and many auxiliary systems. If a hydraulic leak were to occur in one of these systems, it could potentially drain the entire hydraulic system. To prevent this, hydraulic fuses are installed which automatically shut off the flow if the volume of fluid passing through the fuse is too great.

8464. (A) *EA-ITP-ASG2E, Page 1-52; EA-AC65-15A, Pages 320-321*
If all the other systems designed to prevent over-pressure were to fail, the last link in the chain is the shear section on the hydraulic pump. The shear section is designed to fail if the pump becomes too difficult to turn, as it would if the pressure in the system were climbing too high.

8465. (A) *EA-ITP-ASG2E, Page 1-57; EA-AC65-15A, Pages 320-321 (Figures 8-13 and 8-16)*
Hydraulic pumps have a drain port located at the bottom of the pump at the shaft end. If the shaft seal is leaking, this will be in evidence by the fluid seen leaking out of the drain port.

8466. (C) *EA-ITP-ASG2E, Page 1-57; EA-ITP-AB, Page 372 (Figure 1A-15); EA-AC65-15A, Page 316 (Figure 8-6)*
If a hydraulic pump is of the correct capacity, it should be able to maintain normal pressure during the operation of a cowl flap actuator, as well as any other actuator in the system. In a typical hydraulic system, the pressure gage is just downstream of the accumulator, and upstream of all the hydraulic actuators and selector valves. A problem with the cowl flap or the selector valve would not affect the system pressure because they are downstream of the pressure gage and pressure relief valve.

8467. (C) *EA-ITP-A2, Page 403; EA-ITP-ASG2E, Page 1-51; EA-ITP-AB, Page 377; EA-AC65-15A, Page 317*
Hydraulic reservoirs are either vented to the atmosphere or closed to the atmosphere and pressurized. If a reservoir is pressurized, the pressure must be released before removing the cap.

8468. (B) *EA-ITP-A2, Pages 407-408; EA-ITP-ASG2E, Page 1-52; EA-ITP-AB, Pages 382-383; EA-AC65-15A, Pages 318-319*
A constant displacement pump is one that puts out a fixed volume of fluid per revolution of the pump. Because the system cannot always use the full volume of displace fluid, there is a pressure regulator in the system which diverts fluid back to the reservoir. This diverting of fluid back to the reservoir happens when a set pressure has been reached. At this time, the pressure in the system is reduced, but the volume output of the pump remains the same.

8469. (C) *EA-ITP-ASG2E, Page 1-53; Aircraft Maintenance and Repair, Page 281*
An accumulator is a high-strength container divided by some form of movable partition. One compartment is connected to hydraulic pressure and the other is filled with compressed air or nitrogen. Using this arrangement we can gain the effect of compressibility for the hydraulic fluid.

8470. (C) *EA-ITP-A2, Pages 407-408; EA-ITP-ASG2E, Page 1-57; EA-ITP-AB, Pages 382-383; EA-AC65-15A, Pages 318-319*
Hydraulic pressure pumps are designed to move liquids, not air. If an air leak exists at the inlet to a hydraulic pump, the pump will draw the air in because it offers the least resistance to flow. If the leak is a small one, the pump will draw a mixture of liquid and air, and will be partially cavitating. The partial cavitating of the pump will be heard as a chattering sound.

8471. (C) *EA-ITP-A2, Page 415; EA-ITP-ASG2E, Page 1-54*
Selections C and D are identical and represent the benefit of using quick disconnect couplings. These valve may be installed to allow the easy removal and replacement of a unit or to disconnect the pump and connect a hydraulic test stand to the system.

8472. (B) *EA-ITP-A2, Page 392; EA-ITP-ASG2E, Page 1-50; Aircraft Maintenance and Repair, Page 291*
Fire-resistant fluids of the phosphate ester type are used with seals made from butyl synthetic rubber of Teflon® fluorocarbon resin.

8473. (C) *EA-ITP-A2, Pages 407-408; EA-ITP-ASG2E, Page 1-51; EA-ITP-AB, Page 382; EA-AC65-15A, Page 318*
A constant displacement pump, regardless of pump RPM, forces a fixed quantity of fluid through the outlet port during each revolution of the pump.

8474. (B) *EA-ITP-A2, Page 425; EA-ITP-ASG2E, Page 1-55; Aircraft Maintenance and Repair, Page 286*
A hydraulic motor is, in essence, a hydraulic pump with the flow of fluid reversed. Instead of rotary motion producing pressure fluid flow as it does in a pump, with the hydraulic motor pressure fluid flow produces rotary motion.

8475. (B) *EA-ITP-ASG2E, Page 1-56; Aircraft Maintenance and Repair, Pages 297-298*
A crossflow valve is used in the landing gear system of some aircraft to connect both sides of the landing gear actuating cylinder. When the landing gear is being lowered, its weight might cause the gear to extend more rapidly than fluid can be supplied to the cylinder. By connecting the two sides of the cylinder, fluid is allowed to flow from one side of the cylinder to the other, ensuring smooth operation. The crossflow valve is also known as a free-fall valve. As a free-fall valve, it allows the retraction side of the cylinder, which has fluid trapped in it under pressure, to be connected with the extension side of the cylinder. This is done if the system malfunctions and an emergency extension of the gear is needed.

8476. (B) *EA-ITP-ASG2E, Page 1-51; EA-AC65-15A, Page 314*
Micronic type filters used in hydraulic systems are made of a specially treated paper. This type of filter should be replaced periodically, according to the applicable instructions, rather than being cleaned.

8477. (C) *EA-ITP-A2, Pages 420-421; EA-ITP-ASG2E, Page 1-51; EA-ITP-AB, Pages 395-396; EA-AC65-15A, Page 324*
In a hydraulic system which is shut down (pump not turning) , there will only be pressure showing on the gage if the accumulator has a preload. If the accumulator has a leaking valve, the preload in the air side of the accumulator will be completely lost. Although the pressure in the system when the pump is running might be normal, the operation of the system with an accumulator which has lost its preload will not be normal.

8478. (B) *EA-ITP-A2, Page 402; EA-ITP-ASG2E, Page 1-51; EA-ITP-AB, Page 376; EA-AC65-15A, Page 316*
In a hydraulic system which uses a nonpressurized reservoir, the cap on the reservoir can be removed while the system is pressurized. This is not the time, however, to service the reservoir; the fluid which it takes to pressurize the system is out of the tank at this time. When the pressure in the system is reduced, fluid will return to the tank and increase its level, possible to the point of being overserviced.

8479. (C) *EA-ITP-A2, Page 402; EA-ITP-ASG2E, Page 1-51; EA-ITP-AB, Page 377; EA-AC65-15A, Page 317*
Pressurizing a reservoir with air is accomplished by forcing air into the reservoir above the level of the fluid. In most cases, the initial source of the air pressure is the aircraft engine from which it is bled. Usually, air coming directly from the engine is at a pressure of approximately 100 PSI. This pressure is reduced to between 5 and 15 PSI, depending on the type of hydraulic system, by using an air pressure regulator.

8480. (B) *EA-ITP-A2, Page 402; EA-ITP-ASG2E, Page 1-51; EA-ITP-AB, Page 377; EA-AC65-15A, Page 317*
In vented reservoirs, atmospheric pressure and gravity are the forces which cause fluid to flow from the reservoir into the pump intake. For some aircraft, however, atmospheric pressure becomes too low at cruise altitudes to adequately force the fluid out of the reservoir and into the pump intake. These aircraft use pressurized reservoirs to supply fluid to the pumps, thereby preventing pump cavitation.

8481. (A) *EA-ITP-A2, Page 390; EA-ITP-GB, Pages 80-81; EA-AC65-9A, Page 233; EA-ITP-ASG2E, Page 1-55*
According to Pascal's Law, the amount of force which a piston will exert is equal to its cross sectional area multiplied by the pressure acting on it. Pressure in a system is equal and undiminished in all directions, so two actuators of different diameter in the same system will operate with the same pressure, but they will exert different amounts of force.

8482. (C) *EA-ITP-A2, Page 402; EA-ITP-ASG2E, Page 1-51; EA-ITP-AB, Page 377; EA-AC65-15A, Page 317*
When the normal supply of fluid and the emergency supply of fluid are coming from the same hydraulic reservoir, the reservoir is fitted with a standpipe. The standpipe extends above the bottom of the reservoir, and it supplies fluid to the normal system power pump. The emergency system draws its fluid from the bottom of the reservoir.

8483. (C) *EA-ITP-A2, Page 402; EA-ITP-ASG2E, Page 1-51; EA-ITP-AB, Page 377; EA-AC65-15A, Page 317*
When hydraulic reservoirs contain a standpipe, the standpipe supplies fluid to the main system pump.

8484. (A) *EA-ITP-A2, Page 421; EA-ITP-ASG2E, Page 1-53; EA-ITP-AB, Page 395; Aircraft Maintenance and Repair, Pages 283-284*
The air charge in the accumulator may be determined by reducing all hydraulic pressure and observe the reading on the accumulator air gage. If the aircraft is not so equipped, the system pressure gage may be used to check the air charge in the accumulator. The procedures for using the system pressure gage will vary with the system, and manufacturer's procedures must be followed.

8485. (C) *EA-ITP-A2, Page 421; EA-ITP-ASG2E, Page 1-53; Aircraft Maintenance and Repair, Pages 283-284*
This question is a repeat of question #8484.

8486. (B) *EA-ITP-A2, Page 392; EA-ITP-ASG2E, Page 1-50; EA-ITP-AB, Page 403; EA-AC65-15A, Page 311*
Synthetic rubber seals and neoprene seals are used with petroleum base hydraulic fluids. The technician must be very careful to only use the proper type seal with each type of hydraulic fluid.

8487. (B) *EA-ITP-ASG2E, Page 1-52; EA-AC65-15A, Page 324; Aircraft Maintenance and Repair, Page 283*
Thermal relief valves are used to relieve excessive pressures that may exist due to thermal expansion of the fluid. They are adjusted to pressures which are above those required for the operation of the systems or the other relief valves, so they do not interfere with normal operation.

8488. (C) *EA-ITP-ASG2E, Page 1-57; Aircraft Maintenance and Repair, Page 324*
The most likely cause of chatter in a hydraulic system is trapped air. Hydraulic systems are designed to handle liquids, not gases. When a hydraulic system has air in it, it causes erratic operation of the system components. This erratic operation is evidenced by a chattering sound.

8489. (C) *EA-ITP-ASG2E, Page 1-57; Aircraft Maintenance and Repair, Page 285*
If hydraulic fluid is found in the air chamber of an accumulator, there is a leak between the two chambers. In such cases, the accumulator must be removed and repaired.

8490. (C) *EA-ITP-G2, Page 303; EA-43.13-1A/3, Page 166; EA-ITP-ASG2E, Page 1-57*
A dent less than 20% of the tube diameter is not objectionable unless it is in the heel of a bend.

8491. (B) *EA-ITP-A2, Pages 420-421; EA-ITP-ASG2E, Page 1-53; EA-ITP-AB, Pages 395-396; EA-AC65-15A, Page 324*
In a hydraulic system which is shut down (pump not turning), there will only be pressure showing on the gage if the accumulator has a preload. Although the pressure in the system when the pump is running might be normal, the operation of the system with an accumulator which has lost its preload will not be normal.

8492. (A) *EA-ITP-A2, Page 414; EA-ITP-ASG2E, Page 1-53; EA-ITP-AB, Page 389; EA-AC65-15A, Pages 327 & 335*
Allows limited operating speed through restricted flow of fluid. Restrictors are a type of control valve used in pneumatic systems. The small outlet port reduces the rate of airflow and the speed of operation of an actuating unit.

8493. (C) *EA-ITP-ASG2E, Page 1-57; EA-AH-1, Page 87*
In addition to slow actuation, fluid movement around the internal seal will cause localized heating by friction. This may help to locate the faulty unit.

8494. (C) *EA-ITP-ASG2E, Page 1-53; EA-AC65-15A, Page 324; EA-AH-1, Page 87*
Supply fluid under pressure to compensate for small internal or external leaks which would cause the system to cycle continuously by action of the pressure switches continually "kicking in".

8495. (A) *EA-ITP-G2, Page 292; EA-43.13-1A/3, Page 166; EA-ITP-ASG2E, Page 1-57*
Teflon® has peculiar characteristics that require extra care in handling. It tends to assume a permanent set when exposed to high temperatures and pressures. Do not attempt to straighten a hose that has been in service.

8496. (C) *EA-ITP-ASG2E, Page 1-58; EA-ITP-AB, Page 409; EA-AC65-15A, Page 339*
The condensation dump valve solenoid is energized and de-energized by the pressure switch. When energized, it prevents the compressor from dumping air overboard; when de-energized, it completely purges the separator's reservoir and lines up to the air compressor.

8497. (A) *EA-ITP-A2, Page 759; EA-ITP-ASG2E, Page 1-59; EA-ITP-AB, Page 497; EA-AC65-15A, Page 546*
With gas turbine engines, the cabin can be pressurized by bleeding air from the engine compressor. Usually the air bled from an engine compressor is sufficiently free from contamination and can be used safely for cabin pressurization.

8498. (A) *EA-ITP-A2, Page 770; EA-ITP-ASG2E, Page 1-64; EA-ITP-AB, Pages 504-505; EA-AC65-15A, Page 565*
An air cycle cooling system consists of an expansion turbine (cooling turbine) , an air-to-air heat exchanger, and various valves which control airflow through the system. High pressure air from the cabin compressor is routed through the turbine section. As the air passes through the turbine, it rotates the turbine and the impeller. When the compressed air performs the work of turning the turbine, it undergoes a pressure and temperature drop.

8499. (B) *EA-ITP-A2, Page 779; EA-ITP-ASG2E, Page 1-65; EA-ITP-AB, Page 514; EA-AC65-15A, Page 577*
When the Freon® gas leaves the compressor in a vapor-cycle air conditioning system, it flows to the condenser. At the condenser, the gas passes through a heat exchanger where outside (ambient) air removes heat from the Freon®. When the heat is removed from the Freon® gas, the gas changes to a liquid and a great amount of heat is given up.

8500. (B) *EA-ITP-A2, Page 768; EA-ITP-ASG2E, Page 1-63; EA-ITP-AB, Page 501; EA-AC65-15A, Page 561*
Regardless of size, every combustion heater needs four things for operation: (1) Fuel to burn; (2) Ignition to ignite the fuel; (3) Combustion air to provide the oxygen required to support the flame; and (4) Ventilating air to carry the heat to the places where it is needed.

8501. (C) *EA-ITP-A2, Page 759; EA-ITP-ASG2E, Page 1-59; EA-ITP-AB, Page 497; EA-AC65-15A, Page 546*
When gas turbine engines are involved in the cabin pressurization process, the air is usually bled from the engine's compressor. This air is generally referred to as bleed air.

8502. (A) *EA-ITP-ASG2E, Page 1-63; EA-AC65-15A, Page 563*
To prevent too much air from entering the heaters as air pressure increases, either a combustion air relief valve or a differential pressure regulator is provided.

8503. (C) *EA-ITP-A2, Page 760; EA-ITP-ASG2E, Page 1-60*
Cabin pressurization is obtained by flowing more air than is needed into the cabin, and controlling the exhaust of excess air.

8504. (A) *EA-ITP-A2, Pages 761-763; EA-ITP-ASG2E, Page 1-60; EA-AC65-15A, Pages 551-552*
What the cabin pressure regulator does, and how it controls the outflow valve, is determined by cabin air pressure.

8505. **(A)** *EA-ITP-A2, Pages 769-770; EA-ITP-ASG2E, Page 1-64; EA-AC65-15A, Page 565*
An air cycle cooling system consists of an expansion turbine (cooling turbine) , an air-to-air heat exchanger, various valves to control airflow, and a source of high pressure air.

8506. **(A)** *EA-ITP-ASG2E, Page 1-60; EA-AC65-15A, Page 555*
A dump valve in a pressurization system might be operated by a switch in the cockpit. The purpose of the dump valve is to release the cabin air pressure to atmosphere.

8507. **(A)** *EA-ITP-A2, Page 782; EA-ITP-ASG2E, Page 1-66; EA-ITP-AB, Page 524*
When servicing a vapor cycle system with Freon®, the Freon® can should be held upright so vapor will come out and not liquid Freon®. If the low-side pressure is below 40 PSIG, the can may be inverted and liquid allowed to enter the system because the liquid will turn into a vapor before it enters the compressor at this low pressure. If the air temperature is below 80°F, the can should never be inverted, because at low temperatures the liquid may not all vaporize. If liquid Freon® is allowed to enter the compressor it will damage the unit.

8508. **(A)** *EA-ITP-A2, Page 786; EA-ITP-ASG2E, Page 1-66; EA-ITP-AB, Page 524; EA-AC65-15A, Page 583*
When charging a vapor cycle system with Freon®, the technician should service with as many pounds of refrigerant as the specifications call for. A full charge will be indicated by the absence of bubbles in the sight glass of the receiver-dryer.

8509. **(A)** *EA-ITP-A2, Page 786; EA-ITP-ASG2E, Page 1-67; EA-ITP-AB, Page 523*
When servicing a vapor cycle system with Freon® after the system has been evacuated, the high-side valve should be opened and the low-side valve watched to see that it starts to come out of a vacuum. If the low-side valve does not come out of a vacuum, it indicates that there is a blockage in the system and that the charge is not being taken.

8510. **(A)** *EA-ITP-A2, Page 775; EA-ITP-ASG2E, Page 1-67; EA-ITP-AB, Page 518; EA-AC65-15A, Page 585*
The most likely culprit if a vapor cycle system will not take a charge is the expansion valve. With just a few drops of moisture in the system, the expansion valve will freeze up due to the extremely cold temperature it operates under.

8511. **(C)** *EA-ITP-A2, Page 783; EA-ITP-ASG2E, Page 1-65; EA-ITP-AB, Page 520*
Evaporator fins should be checked to be sure they are clean and free of any obstruction. If the air over the evaporator should ever be obstructed, the fins will ice up.

8512. **(C)** *EA-ITP-A2, Page 746; EA-ITP-ASG2E, Page 1-62; EA-ITP-AB, Page 483*
Oxygen cylinders must be hydrostatically tested to 5⁄3 of their working pressure by testing with water pressure.

8513. **(A)** *EA-ITP-A2, Page 746; EA-ITP-ASG2E, Page 1-62; Aircraft Maintenance and Repair, Page 445*
Standard weight cylinders are marked DOT 3AA and must be hydrostatically tested every 5 years.

8514. **(A)** *EA-ITP-A2, Page 746; EA-ITP-ASG2E, Page 1-62; EA-ITP-AB, Page 483*
DOT 3HT oxygen cylinders must be hydrostatically tested every three years.

8515. **(B)** *EA-ITP-A2, Page 749; EA-ITP-ASG2E, Page 1-61; EA-ITP-AB, Page 487*
Almost all of the masks used with a continuous flow oxygen system are of the rebreather type.

8516. **(C)** *EA-ITP-A2, Page 759; EA-ITP-ASG2E, Page 1-59; EA-AC65-15A, Page 546*
Included in this group are reciprocating compressors, vane type compressors, and roots blowers. The first two are not suitable for pressurization because of the large quantity of oil present in the air delivered to the cabin.

8517. **(A)** *EA-ITP-ASG2E, Page 1-59; EA-AC65-15A, Page 546*
To achieve an oil free delivery of air, the supercharger bearings are contained in a separate chamber. The use of labyrinth seals permits a small amount of air leakage to ambient.

8518. **(A)** *EA-ITP-A2, Page 754; EA-ITP-ASG2E, Page 1-62; EA-AC65-15A, Pages 589-591*
The solid state oxygen generator is the most efficient system space-wise, also less equipment and maintenance is required.

8519. **(A)** *EA-ITP-A2, Pages 755-756; EA-ITP-ASG2E, Page 1-62; EA-ITP-AB, Page 499; EA-AC65-15A, Page 600*
The main cause of contamination in the system is moisture.

8520. (C) *EA-ITP-A2, Page 770; EA-ITP-ASG2E, Page 1-64; EA-ITP-AB, Page 506; EA-AC65-15A, Page 567*
As the cabin air leaves the secondary heat exchanger, it is routed to the expansion turbine, which is rotated by the air pressure on it. In performing this function, it is further cooled before entering the water separator, then routed to the cabin.

8521. (A) *EA-ITP-A2, Pages 771-773; EA-ITP-ASG2E, Page 1-65; EA-ITP-AB, Page 507; EA-AC65-15A, Pages 576-577*
The refrigerant at a high temperature and pressure flows into the condenser. Here heat flows from the refrigerant to the outside air, condensing the vapor into a liquid.

8522. (B) *EA-ITP-A2, Pages 771-773; EA-ITP-ASG2E, Page 1-65; EA-ITP-AB, Page 507; EA-AC65-15A, Page 580*
The Freon® changes from a liquid to a vapor at the evaporator. The evaporator is designed so that heat is taken from the cabin air.

8523. (B) *EA-ITP-A2, Page 759; EA-ITP-ASG2E, Page 1-60; EA-ITP-AB, Page 499; EA-AC65-15A, Page 550*
Cabin pressurization is usually controlled by a cabin pressure regulator which opens or closes an outflow valve. The opening or closing of the outflow valve maintains the right amount of pressure in the aircraft, according to what is set on the pressurization controller.

8524. (A) *EA-ITP-A2, Page 765; EA-ITP-ASG2E, Page 1-64; EA-ITP-AB, Page 500*
One of the biggest dangers of an aircraft heating system that uses a jacket around the engine exhaust system is the possibility of carbon monoxide being introduced into the cabin. As well as visual inspections of the system, it should be periodically checked for carbon monoxide leakage.

8525. (C) *EA-ITP-ASG2E, Page 1-60; Aircraft Maintenance and Repair, Pages 414-415*
Selecting auxiliary ventilation while in cruising altitude pressurized flight will allow outside ambient air to enter the cabin, which will decrease the cabin pressure, or increase the cabin altitude.

8526. (A) *EA-ITP-A2, Pages 760-761; EA-ITP-ASG2E, Page 1-60; EA-ITP-AB, Pages 499-500; EA-AC65-15A, Page 551*
When the cabin pressurization controller is set for a desired cabin altitude, one of the system components which is affected is the outflow valve. How much the outflow valve will be open, if it is open at all, will be determined by the cabin altitude which is set.

8527. (B) *EA-ITP-A2, Page 776; EA-ITP-ASG2E, Page 1-65; EA-ITP-AB, Page 513; EA-AC65-15A, Page 580*
The purpose of the evaporator in a Freon® cooling system is to cool the air which flows through it. Freon in the evaporator changes from a liquid to a gaseous state, and in so doing absorbs large quantities of heat from the air passing through it.

8528. (A) *EA-ITP-A2, Pages 768-769; EA-ITP-ASG2E, Page 1-65; EA-AC65-15A, Page 574*
A cabin air temperature control regulator, in conjunction with the air temperature selector rheostat and the air duct temperature pickup unit, automatically maintain the temperature of the air entering the cabin at a preselected value. These controls cause a mixing valve to operate, which mixes the proper amounts of cool and hot air to achieve the desired temperature.

8529. (B) *EA-ITP-ASG2E, Page 1-60; EA-AC65-15A, Page 550*
All pressurized aircraft require some form of a negative pressure relief valve. This valve opens when outside air pressure is greater than cabin pressure. The negative pressure relief valve prevents accidentally obtaining a cabin altitude which is higher than the aircraft altitude.

8530. (B) *EA-ITP-A2, Page 784; EA-ITP-ASG2E, Page 1-66; EA-ITP-AB, Page 524; EA-AC65-15A, Page 583*
To check the Freon® level in a vapor cycle system, it is necessary to operate the refrigeration unit for approximately 5 minutes to reach a stable condition. If the system uses a sight glass, the flow of Freon® through the sight glass should be observed. If the Freon® charge is low, bubbles will appear in the sight glass.

8531. (B) *EA-ITP-A2, Pages 761-762; EA-ITP-ASG2E, Page 1-60; EA-AC65-15A, Pages 551-552*
The closing of the outflow valve in a pressurization system causes the cabin to pressurize quicker. If the cabin pressure altitude is not being maintained as the aircraft climbs to higher altitudes, the outflow valve controller needs to be adjusted so the outflow valve will close faster.

8532. (C) *EA-ITP-A2, Page 776; EA-ITP-ASG2E, Page 1-65; EA-ITP-AB, Pages 510-511; EA-AC65-15A, Page 579*
By sensing the temperature and the pressure of the Freon® gas leaving the evaporator, the expansion valve meters the right amount of gas to preclude the possibility of the evaporator becoming flooded and returning liquid refrigerant to the compressor.

8533. (A) *EA-ITP-A2, Page 779; EA-ITP-ASG2E, Page 1-65; EA-ITP-AB, Page 514; EA-AC65-15A, Page 577*
The condenser is the radiator-like component which receives the hot, high-pressure vapors from the compressor and allows cool air to flow over its coils and remove the heat from the Freon®. The condenser cools the Freon® down and turns it back into a liquid.

8534. (C) *EA-ITP-A2, Pages 775-776; EA-ITP-ASG2E, Page 1-65; EA-ITP-AB, Pages 510-511; EA-AC65-15A, Page 579*
The Freon® coming out of the condenser is high-pressure liquid refrigerant. The expansion valve lowers the Freon® pressure and thus lowers the temperature.

8535. (C) *EA-ITP-ASG2E, Page 1-60; EA-AC65-15A, Page 557 (Figure 14-16)*
After leaving the cabin air compressor, pressurized air passes through a combination check valve/shutoff valve on its way to the delivery air ducts. This check valve prevents the air pressure from being lost through an inoperative compressor.

8536. (B) *EA-ITP-A2, Page 784; EA-ITP-ASG2E, Page 1-67; EA-ITP-AB, Page 522; EA-AC65-15A, Pages 583-585*
When adding Freon® to a system, add as much oil as is felt was lost. Anytime the system is opened, it must be evacuated.

8537. (B) *EA-ITP-A2, Page 760; EA-ITP-ASG2E, Page 1-59; EA-ITP-AB, Page 499; EA-AC65-15A, Page 550*
The principal control of the pressurization system is the outflow valve. It is regulated by the altitude controller.

8538. (B) *EA-ITP-ASG2E, Page 1-61; EA-ITP-AB, Page 498; Aircraft Maintenance and Repair, Page 446*
A jet pump flow multiplier can supply cabin pressurization by using engine bleed air.

8539. (A) *EA-ITP-A2, Page 768; EA-ITP-ASG2E, Page 1-64; EA-ITP-AB, Pages 501-503; EA-AC65-15A, Pages 561-562*
Fuel leaks, especially with a combustion heater, can be catastrophic in terms of aircraft and personal safety. Any time the fuel system for a combustion heater is worked on, it should be pressurized and leak checked.

8540. (A) *EA-ITP-A2, Page 768; EA-ITP-ASG2E, Page 1-63; EA-ITP-AB, Page 502; EA-AC65-15A, Page 562*
To increase or decrease the cabin temperature, combustion heaters usually turn fuel on or off automatically by using an amplifier connected to temperature-sensing devices or by cycling switches which open and close the circuit to the fuel solenoid valve. The heater cycles on and off to maintain the temperature selected on a temperature control rheostat located in the cabin.

8541. (C) *EA-ITP-A2, Page 770; EA-ITP-ASG2E, Page 1-65; EA-ITP-AB, Pages 504-505; EA-AC65-15A, Page 565*
In an air-cycle cooling system, compressed air is passed through a turbine which causes the turbine rotor to rotate. When the compressed air performs the work of turning the turbine, it undergoes a pressure and temperature drop. It is this temperature drop which produces the cold air used for air conditioning.

8542. (C) *EA-ITP-A2, Page 772; EA-ITP-ASG2E, Page 1-65; EA-ITP-AB, Page 507; EA-AC65-15A, Page 580*
The next unit in line of cooling flow after the expansion valve is the evaporator.

8543. (C) *EA-ITP-A2, Page 784; EA-ITP-ASG2E, Page 1-66; EA-ITP-AB, Page 510; EA-AC65-15A, Page 579*
The filter/drier in a vapor cycle system usually has a sight glass in it through which the Freon® flow can be observed. A steady flow of Freon® refrigerant observed through the sight glass indicates that sufficient charge is present. If the unit requires additional refrigerant, bubbles will be present in the sight glass.

8544. (A) *EA-ITP-A2, Pages 174-175; EA-ITP-ASG2E, Page 1-59; EA-ITP-AB, Page 76; EA-AC65-15A, Page 543*
When an aircraft is pressurized, it is subjected to a force acting from within which is greater than the force acting from the outside, namely atmospheric pressure. Tension is a primary stress on aircraft that tries to pull an aircraft apart, or stretch a structural member; it is this tensile stress that an aircraft experiences when, under a condition of pressurization, the aircraft tries to force itself apart. One of the problems in designing pressurized aircraft is selecting the proper structural materials which will be able to withstand the great differential in pressure that exists between the inside and outside of a pressurized aircraft.

8545. (B) *EA-ITP-A2, Page 759; EA-ITP-ASG2E, Page 1-59; EA-AC65-15A, Page 545*
Differential pressure is the difference in pressure between the pressure acting on one side of a wall and the pressure acting on the other side of the wall. In aircraft air conditioning and pressurizing systems, it is the difference between cabin pressure and atmospheric pressure.

8546. (C) *EA-ITP-A2, Page 782; EA-ITP-ASG2E, Page 1-67; EA-ITP-AB, Page 522; EA-AC65-15A, Page 585*
Only a few drops of moisture will contaminate and completely block an air conditioning system. If this moisture freezes in the expansion valve, the action stops. Water is removed from the system by evacuation. Anytime the system has been opened, it must be evacuated before recharging.

8547. (B) *EA-ITP-A2, Page 759; EA-ITP-ASG2E, Page 1-60; EA-AC65-15A, Page 553*
By regulating the reference chamber air pressure in the cabin air pressure regulator, the isobaric and differential control systems control the actions of the outflow valve. By controlling the outflow valve, which allows pressurized air to escape to atmosphere or keeps it in the aircraft, the system provides for three modes of operation called unpressurized, isobaric, and differential.

8548. (C) *EA-ITP-A2, Pages 759-760; EA-ITP-ASG2E, Page 1-59; EA-ITP-AB, Pages 497-498; EA-AC65-15A, Page 546*
With aircraft powered by gas turbine engines, the aircraft cabin can be pressurized by bleeding air from the engine's compressor. Usually the air bled from an engine compressor is sufficiently free from contamination and can be used safely for cabin pressurization. This source of air for pressurization is not always the best way to go, however. Another method of obtaining pressurized air is the use of independent cabin compressors. These compressors can be engine driven through the accessory drive gearing, or can be powered by bleed air from a gas turbine engine.

8549. (B) *EA-ITP-A2, Pages 760-761; EA-ITP-ASG2E, Page 1-60; EA-AC65-15A, Pages 550-551*
A pressurization controller looks very much like an altimeter with several added adjustment knobs. Usually there is one adjustment knob for setting the desired cabin altitude. There is a separate knob for setting the existing altimeter setting or sea level barometric pressure. There is a third knob on the controller which adjusts the cabin rate of altitude change.

8550. (A) *EA-ITP-A2, Page 775; EA-ITP-ASG2E, Page 1-65; EA-ITP-AB, Page 510*
The receiver/dryer is the reservoir for a vapor-cycle air conditioning system. It is located in the high side of the system, between the condenser and the expansion valve.

8551. (C) *EA-ITP-A2, Page 779; EA-ITP-ASG2E, Page 1-65; EA-ITP-AB, Page 514; EA-AC65-15A, Page 577*
The condenser is the radiator-like component which receives the hot, high-pressure vapors from the compressor. Cool air flows over its coils and removes the heat from the refrigeration vapors so they will change back into a liquid.

8552. (B) *EA-ITP-A2, Page 773; EA-ITP-ASG2E, Page 1-65; EA-ITP-AB, Page 513; EA-AC65-15A, Page 580*
The actual cooling unit in a vapor-cycle air conditioning system is the evaporator. The evaporator is similar to the condenser in construction, but since it is in the low side of the system, the evaporator is not subject to such high pressures as the condenser. The refrigerant enters the evaporator as a liquid and leaves as a gas.

8553. (C) *EA-ITP-A2, Pages 782-783; EA-ITP-ASG2E, Page 1-67*
No water may be present in a vapor-cycle cooling system. Even a few drops of water can freeze at the thermal expansion valve and block the entire system. To remove water, a vacuum pump is attached to evacuate the system. At the lowered pressure water will boil at temperatures as low as 45°F. The water vapor is then drawn from the system by the vacuum pump.

8554. (B) *EA-ITP-A2, Page 773; EA-ITP-ASG2E, Page 1-65; EA-ITP-AB, Page 513; EA-AC65-15A, Page 580*
The evaporator is in the low pressure side of an air conditioning system. The Freon® enters the evaporator as a liquid, and should use the entire length of the unit when changing from a liquid into a vapor.

8555. (B) *EA-ITP-A2, Page 773; EA-ITP-ASG2E, Page 1-65; EA-ITP-AB, Page 514; EA-AC65-15A, Page 577*
The condenser is the radiator-like component in a vapor-cycle air conditioning system which receives the hot, high-pressure vapors from the compressor. Cool air flows over its coils and removes the heat from the refrigeration vapors so they will change back into a liquid.

8556. (C) *EA-ITP-A2, Page 782; EA-ITP-ASG2E, Page 1-67; EA-ITP-AB, Page 524*
When the Freon® bottle is held vertically with the outlet at the top, gaseous Freon® is introduced to the system. With the outlet at the bottom, however, liquid Freon® is introduced to the system. Liquid Freon® should only be introduced to the system if the low-side pressure is below 40 PSIG and outside temperature above 80°F.

8557. (B) *EA-ITP-A2, Page 785; EA-ITP-ASG2E, Page 1-67; EA-ITP-AB, Page 521*
Any time the Freon® system is opened, all of the refrigerant must be purged. When purging the system, care must be taken not to open the valves to the point where the refrigerant can escape fast enough to blow oil out with the vapor.

8558. (A) *EA-ITP-A2, Page 783; EA-ITP-ASG2E, Page 1-65; EA-ITP-AB, Page 520*
Compressor oil circulates with the refrigeration fluid. When a leak occurs, naturally oil will stain. The refrigeration fluid is colorless.

8559. (B) *EA-ITP-A2, Page 775; EA-ITP-ASG2E, Page 1-66; EA-ITP-AB, Page 510*
When both lines are warm, the expansion valve is not working properly, or not working at all.

8560. (C) *EA-ITP-ASG2E, Page 1-65; EA-AC65-15A, Page 578*
By cooling the refrigerant (in a subcooler) premature vaporization (flash-off) can be prevented.

8561. (B) *EA-ITP-A2, Page 774; EA-ITP-ASG2E, Page 1-67; EA-ITP-AB, Page 510*
Additionally, see question #7588.

8562. (C) *EA-ITP-A2, Page 774; EA-ITP-ASG2E, Page 1-66; EA-ITP-AB, Page 510; EA-AC65-15A, Page 584*
R-12 is not normally toxic, nor is it dangerous when it is breathed. When R-12 passes over an open flame its characteristics change drastically, it becomes deadly phosgene gas.

8563. (B) *EA-ITP-A2, Page 775; EA-ITP-ASG2E, Page 1-65; EA-ITP-AB, Page 521; EA-AC65-15A, Page 585*
The oil level should be maintained at the proper range using oil recommended by the manufacturer.

8564. (C) *EA-ITP-A2, Page 764; EA-ITP-ASG2E, Page 1-62*
Leaks in an oxygen system should be searched out using a special non-oily soap solution.

8565. (C) *EA-ITP-A2, Page 755; EA-ITP-ASG2E, Page 1-62; EA-ITP-AB, Page 494; EA-AC65-15A, Pages 599-600*
One of the precautions to exercise when draining an oxygen system is not to drain it too rapidly, because this will cause condensation within the system. The same problem can be caused by allowing the bottle pressure to drop below a specified minimum.

8566. (A) *EA-ITP-A2, Page 750; EA-ITP-ASG2E, Page 1-61; EA-ITP-AB, Page 490; EA-AC65-15A, Page 587*
In a continuous flow system, oxygen flows from the charged cylinder through the high pressure line to the pressure-reducing valve, which reduces the pressure to that required at the mask outlets. A calibrated orifice in the outlets will control the amount of oxygen delivered to the mask.

8567. (C) *EA-ITP-A2, Page 748; EA-ITP-ASG2E, Page 1-61; EA-ITP-AB, Page 485; EA-AC65-15A, Page 594*
The essential feature of a diluter-demand regulator is a diaphragm-operated valve called the demand valve, which opens by slight suction on the diaphragm during inhalation, and which closes during exhalation.

8568. (B) *EA-ITP-A2, Page 744; EA-ITP-ASG2E, Page 1-61; EA-ITP-AB, Page 481; EA-AC65-15A, Page 598*
Commercial oxygen is used in great quantities for welding and cutting, and for medical use in hospitals and ambulances. Aviators' breathing oxygen is similar to that used for commercial purposes, except that it is additionally processed to remove almost all of the water that could freeze and stop the flow of oxygen.

8569. (A) *EA-ITP-A2, Page 750; EA-ITP-ASG2E, Page 1-61; EA-ITP-AB, Page 488; EA-AC65-15A, Page 594*
In high-pressure oxygen systems, pressure-reducing valves are installed between the supply cylinders and the cockpit and cabin equipment. These valves reduce the pressure down to approximately 300 to 400 PSI.

8570. (B) *EA-ITP-ASG2E, Page 1-61; EA-AC65-15A, Page 594*
A pressure-relief valve is incorporated in the main supply line of a high-pressure system. The relief valve prevents high-pressure oxygen from entering the system downstream of the pressure reducer if the pressure reducer fails.

8571. (C) *EA-ITP-A2, Page 746; EA-ITP-ASG2E, Page 1-61; EA-ITP-AB, Page 483; EA-AC65-15A, Page 598*
Oxygen cylinders for aviation use are identified by their dark green color and the words "Oxygen Aviators' Breathing" stenciled in white letters along the cylinder.

8572. (C) *EA-ITP-AB, Page 509; EA-AC65-15A, Page 599; EA-ITP-ASG2E, Illustration*
First determine what the pressure gage would read if there were no leakage. Because the temperature has gone down 13°, the gage will read differently. The chart supplied may be used to determine the gage reading. Locate the 13° change in temperature on the left side of the chart. Project a line horizontally across the chart to intersect with the diagonal line. From that point drop a vertical line to the bottom of the chart and read the change in pressure due to the change in temperature. This change is about 10 PSI. The instructions on the bottom of the chart tell us to subtract the pressure change value, if the temperature falls. 460 PSI minus 10 PSI will give us a gage reading of 450 PSI. Because a 5 PSI change is allowable, the gage reading must be between 455 and 450 PSI to remain within the allowable limits.

8573. (A) *EA-ITP-A2, Page 743; EA-ITP-ASG2E, Page 1-58; EA-ITP-AB, Page 479; EA-AC65-15A, Page 539*
At 10,000 ft above sea level, oxygen saturation of the blood is about 90%. Long exposure at this altitude will result in headache and fatigue.

8574. (A) *EA-ITP-ASG2E, Page 1-62; EA-AC65-15A, Page 594*
A pressure-relief valve is incorporated in the main supply line of a high-pressure system. The relief valve is vented to a blowout plug in the fuselage skin.

8575. (C) *EA-ITP-A2, Page 758; EA-ITP-ASG2E, Page 1-58; EA-ITP-AB, Page 496; EA-AC65-15A, Page 541*
There are many economic and creature comfort advantages of being able to operate an aircraft at high altitude. High altitude flight means operating above the weather and the majority of the turbulence. To operate at high altitude, however, requires that the cabin be pressurized so the occupants of the aircraft do not suffer from hypoxia.

8576. (A) *EA-ITP-A2, Page 744; EA-ITP-ASG2E, Page 1-61; EA-ITP-AB, Page 481; EA-AC65-15A, Page 598*
Gaseous breathing oxygen used in aircraft is a special type of oxygen containing practically no water vapor, and is at least 99.5% pure. While other types of oxygen (welder, hospital) may be pure enough, they usually contain water, which might freeze and block the oxygen system plumbing.

8577. (A) *EA-ITP-A2, Page 746; EA-ITP-ASG2E, Page 1-61; EA-AC65-15A, Page 587*
In some aircraft a continuous-flow oxygen system is installed for both passengers and crew. The pressure demand system is widely used as a crew system, especially on the larger transport aircraft. Many aircraft have a combination of both systems which may be augmented by portable equipment.

8578. (B) *EA-ITP-A2, Page 748; EA-ITP-ASG2E, Page 1-61; EA-ITP-AB, Page 485; EA-AC65-15A, Page 594*
The purpose of the airflow metering aneroid assembly found in an oxygen diluter demand regulator is to meter the amount of air which will be mixed with oxygen as the aircraft climbs to altitude. As the altitude of the aircraft increases, the amount of oxygen supplied increases, and the air supplied decreases. At approximately 34,000 ft, 100% oxygen is being supplied. The diluter demand position referred to in the answer is better called the normal oxygen position.

8579. (B) *EA-ITP-A2, Page 746; EA-ITP-ASG2E, Page 1-62; EA-ITP-AB, Page 483*
Today, almost all gaseous oxygen is stored in green painted steel cylinders. All cylinders approved for installation in an aircraft must be approved by the Department of Transportation.

8580. (A) *EA-ITP-A2, Page 746; EA-ITP-ASG2E, Page 1-62; EA-ITP-AB, Page 494*
Before filling any aircraft system, the technician must be sure that all of the cylinders are of the approved type, and that they have all been hydrostatically tested within the required time interval.

8581. (A) *EA-ITP-A2, Page 756; EA-ITP-ASG2E, Page 1-62; EA-ITP-AB, Page 495*
If an oxygen system has been opened for servicing, it should be purged of any air that may be in the lines. To purge a continuous-flow system, plug masks into each of the outlets, turn on the oxygen supply valve, and allow the oxygen to flow through the system for about ten minutes.

8582. (B) *EA-ITP-A2, Page 755; EA-ITP-ASG2E, Page 1-62*
A combination shutoff, regulator, and fill valve is usually mounted on the side of a portable high-pressure oxygen cylinder, and usually a pressure gage. This gage indicates the amount of oxygen in the cylinder because the quantity is related to the pressure in the container.

8583. (C) *EA-ITP-ASG2E, Page 1-62; EA-ITP-AB, Page 488*
Oxygen compatible thread lube that meets MIL-T-5542 or Teflon® tape are all that is recommended.

8584. (A) *EA-ITP-A2, Page 746; EA-ITP-ASG2E, Page 1-62; EA-ITP-AB, Page 482*
As air passes through the molecular sieve, nitrogen and other gasses are filtered out. Only oxygen is passed through the sieve.

8585. (A) *EA-ITP-A2, Pages 753-754; EA-ITP-ASG2E, Page 1-62; EA-AC65-15A, Pages 589-591*
A solid state oxygen system utilizes a chemical oxygen "candle" that releases oxygen at a programmed rate. The size and simplicity of the units, and minimal maintenance requirements make them ideal for many applications.

8586. (C) *EA-ITP-A2, Page 616; EA-ITP-ASG2E, Page 1-70; EA-ITP-AB, Page 554; EA-AC65-15A, Page 508*
An aircraft's magnetic compass is swung to correct for magnetic disturbances within the aircraft, called deviation.

8587. (A) *EA-ITP-A2, Pages 589-590; EA-ITP-ASG2E, Page 1-70; EA-ITP-AB, Page 560; EA-AC65-15A, Page 472*
A hydraulic pressure gage generally uses a Bourdon tube and a gear-and-pinion mechanism by which the Bourdon tube's motion is amplified and transferred to the pointer.

8588. (B) *EA-ITP-A2, Page 618; EA-ITP-ASG2E, Page 1-70; EA-ITP-AB, Pages 552-553; EA-AC65-15A, Page 508*
The fixed reference marker attached to the compass bowl of a magnetic compass is called a lubber line.

8589. (A) *EA-ITP-A2, Page 640; EA-ITP-ASG2E, Page 1-68; EA-ITP-AB, Page 568; EA-AC65-15A, Page 470*
Instrument range markings indicate at a glance whether a system or component is operating in a safe range or an unsafe range. See also FAR 23.

8590. (A) *EA-ITP-A2, Pages 616-617; EA-ITP-ASG2E, Page 1-70; EA-ITP-AB, Page 552; EA-AC65-15A, Page 508*
A compensating device containing small permanent magnets is incorporated in the compass to correct for deviations of the compass which result from the magnetic influences of the aircraft structure and electrical system.

8591. (A) *EA-ITP-A2, Pages 591-594; EA-ITP-ASG2E, Page 1-69; EA-ITP-AB, Pages 527 & 530; EA-AC65-15A, Pages 476 & 479*
An aircraft's altimeter and airspeed indicator rely on the static pressure reading to accurately indicate their readings. The altimeter displays altitude as a function of outside (ambient) pressure, a reading which comes from the static system. The airspeed indicator displays a reading by comparing the total pressure in the pitot system to the static pressure reading. If the static pressure value being supplied to both of these systems is coming from a pressurized cabin, it will cause the altimeter and the airspeed indicator to both read low.

8592. (C) *EA-ITP-ASG2E, Page 1-70*
According to FAR 23, the maximum deviation (during level flight) permitted in a magnetic direction indicator installed in an airplane is 10°.

8593. (C) *EA-ITP-ASG2E, Page 1-70; EA-AC65-15A, Page 508*
The magnetic compass consists of a liquid filled bowl containing a pivoted float element to which one or more bar magnets, called needles, are fastened. The liquid in the bowl dampens the oscillations of the float and decreases the friction of the pivot.

8594. (C) *EA-ITP-A2, Pages 593-594; EA-ITP-ASG2E, Page 1-69; EA-ITP-AB, Pages 532-533*
When suction is applied to the static system in an aircraft, the altimeter is made to read a higher altitude (as a function of the decreased pressure). By observing the rate of change (decreasing altitude) in the altimeter, it is possible to check the static system for leakage.

8595. (C) *EA-ITP-ASG2E, Page 1-69*
According to FAR 23, the maximum altitude loss permitted during an unpressurized aircraft instrument static pressure system integrity check is 100 ft in one minute.

8596. (A) *EA-ITP-A2, Page 621; EA-ITP-ASG2E, Page 1-69; EA-ITP-AB, Section 8, Page 24*
Because a dry air pump operates with the carbon vanes providing both the seal and the lubrication, cleanliness is very important. Any solid particles drawn into the pump can damage the carbon vanes and lead to destruction of the pump. To prevent this type of damage the air inlet is covered with a filter.

8597. (A) *EA-ITP-A2, Page 592; EA-ITP-ASG2E, Page 1-68; EA-AC65-9A, Page 242*
When an altimeter is set at 29.92 in. hg. on the ground, the altimeter will read pressure altitude.

8598. (C) *EA-ITP-ASG2E, Page 1-70; EA-AC65-15A, Page 471*
Aircraft mechanics are not allowed to repair instruments. They are, however, allowed to put range markings on the glass, tighten mounting screws, touch up the paint on the case, and tighten a "B" nut.

8599. (A) *EA-ITP-ASG2E, Page 1-70; EA-43.13-1A/3, Pages 303-305*
Items B, C, G, H all require disassembly of the instrument and are beyond the capabilities of an A & P technician. A certified repair station is required to perform these duties.

8600. (B) *EA-ITP-ASG2E, Page 1-70; EA-AC65-15A, Page 471*
The paint being chipped on an instrument case does not require correction. The only difficulty is a cosmetic one.

8601. (A) *EA-ITP-A2, Page 592; EA-ITP-ASG2E, Page 1-68; EA-AC65-9A, Page 242*
When an altimeter's barometric scale is set at 29.92 in. hg., the altimeter indicates the pressure altitude.

8602. (A) *EA-ITP-A2, Page 589; EA-ITP-ASG2E, Page 1-70; EA-ITP-AB, Page 560; EA-AC65-15A, Page 471*
A pointer connected to a Bourdon tube indicates on a dial the temperature of the oil. If an indicator needle is attached to the free end of a Bourdon tube, its reactions to changes in fluid pressure can be observed.

8603. (B) *EA-ITP-A2, Page 612; EA-ITP-ASG2E, Page 1-69; EA-AC65-15A, Page 482*
A turn and bank indicator shows the correct execution of a bank and turn and indicates the lateral attitude of the aircraft in level flight. It also serves as an emergency source of bank information in case the attitude gyro fails.

8604. (A) *EA-ITP-A2, Pages 602-603; EA-ITP-ASG2E, Page 1-70; EA-AC65-15A, Page 494*
Thermocouple leads are designed to provide a definite amount of resistance in the thermocouple circuit. Thus, their length or cross-sectional size cannot be altered unless some compensation is made for the change in total resistance. A typical turbine engine thermocouple system, for example, operates with 8 ohms of resistance.

8605. (C) *EA-ITP-A2, Pages 602-603; EA-ITP-ASG2E, Page 1-70; EA-ITP-P, Pages 302-303; EA-AC65-15A, Pages 483-484*
A selsyn system, which is a synchro-type indicating instrument, consists of a transmitter, an indicator, and connecting wires. An electrical signal is sent from the transmitter to the indicator through wires.

8606. (A) *EA-ITP-A2, Page 629; EA-ITP-ASG2E, Page 1-70; EA-AC65-15A, Page 488*
The operation of the angle-of-attack indicating system is based on detection of differential pressure at a point where the airstream is flowing in a direction that is not parallel to the true angle of attack of the aircraft. This differential pressure is caused by changes in airflow around the probe unit.

8607. (B) *EA-ITP-ASG2E, Page 1-70; EA-ITP-P, Page 292; EA-AC65-15A, Page 495*
The thermocouples used in a turbine engine are made from chromel (a nickel/chromium alloy) and alumel (a nickel/aluminum alloy).

8608. (B) *EA-ITP-A2, Page 627; EA-ITP-ASG2E, Page 1-70; EA-ITP-P, Page 303; EA-AC65-15A, Page 499*
Fuel flow transmitters send an electrical signal to the indicator in the cockpit through wires.

8609. (A) *EA-ITP-A2, Page 616; EA-ITP-ASG2E, Page 1-70; EA-ITP-AB, Page 551; EA-AC65-15A, Page 508*
Magnetic compass deviation can be compensated for by a procedure called "swinging the compass". This is accomplished by a certified mechanic at a location on the airport generally called the "compass rose".

8610. (C) *EA-ITP-ASG2E, Page 1-70; EA-AC65-15A, Page 471*
Instrument repair must be done by an FAA certified instrument repair station. Certified mechanics are not allowed to do instrument repair.

8611. (C) *EA-ITP-A2, Pages 590-591; EA-ITP-ASG2E, Page 1-68; EA-AC65-15A, Page 473*
When an aircraft engine is not running, its manifold pressure gage should read the existing atmospheric pressure.

8612. (C) *EA-ITP-ASG2E, Page 1-69;EA-FAR-M*
FAR Part 43, Appendix E.

8613. (B) *EA-ITP-A2, Page 622; EA-ITP-ASG2E, Page 1-69; EA-ITP-AB, Page 547; EA-AC65-15A, Page 503*
Since the system capacity is more than needed, the adjustable suction relief valve is set for the vacuum desired.

8614. (A) *EA-ITP-A2, Page 642; EA-ITP-ASG2E, Page 1-73*
The signals sent to the various components of these systems are typically linked through a digital data bus. These digital signals consist of short pulses of voltage on or voltage off.

8615. (B) *EA-ITP-A2, Page 641; EA-ITP-ASG2E, Page 1-73*
The CRT (Cathode Ray Tube) is the television-like screen used to display the information processed by the EFIS.

8616. (C) *EA-ITP-A2, Page 642; EA-ITP-ASG2E, Page 1-74*
There may be more than one symbol generator in a complete EFIS system.

8617. (B) *EA-ITP-A2, Page 642; EA-ITP-ASG2E, Page 1-74*

8618. (A) *EA-ITP-A2, Page 593; EA-ITP-ASG2E, Page 1-68*
Do not confuse radar altimeters with encoding altimeters, which are connected to the radar transponder.

8619. (C) *EA-ITP-A2, Page 593; EA-ITP-ASG2E, Page 1-68*
Typically these systems are used at altitudes below 5,000 feet above ground level.

8620. (B) *EA-ITP-A2, Pages 601-603; EA-ITP-ASG2E, Page 1-67*
Wheatstone Bridge type temperature indicators are generally used to measure relatively low temperature ranges. High temperature systems, such as EGT or CHT are usually thermocouple type instruments.

8621. (A) *EA-ITP-ASG2E, Page 1-71*
The flags used on Horizontal Situation Indicators (HSI) tell you that the function indicated on the flag is not operating. This could be because the receiver is not tuned to the proper frequency, the signal being received is unreliable, or the unit is inoperative due to a power interruption.

8622. (B) *EA-ITP-ASG2E, Page 1-68; EA-AC65-15A, Page 471*
The type and number of shock mounts to be used for instrument panels are determined by the weight of the panel.

8623. (C) *EA-ITP-ASG2E, Page 1-69; EA-AC65-15A, Page 471*
Instrument repair, which includes the replacement of an instrument's glass face, is allowed to be done by instrument repair stations only. A certified mechanic would need to replace the instrument.

8624. (C) *EA-ITP-A2, Page 624; EA-ITP-ASG2E, Page 1-68; EA-ITP-AB, Pages 527, 530 and 533; EA-AC65-15A, Page 474*
Vertical speed indicators, altimeters, and airspeed indicators all make use of the aircraft's pitot static system.

8625. (B) *EA-ITP-A2, Pages 640-641; EA-ITP-ASG2E, Page 1-68; EA-ITP-AB, Page 567; EA-ITP-P, Page 311*
Airspeed indicators and cylinder head temperature gages will both have range markings indicated on their faces or cover glass.

8626. (B) *EA-ITP-A2, Page 640; EA-ITP-ASG2E, Page 1-68; EA-ITP-AB, Page 568*
An airspeed indicator is marked with a blue radial line to indicate the best rate-of-climb speed with one engine inoperative.

8627. (B) *EA-ITP-A2, Pages 640-641; EA-ITP-ASG2E, Page 1-68; EA-ITP-P, Page 311; EA-AC65-15A, Page 470*
Green arcs used on aircraft instruments indicate normal or desirable ranges.

8628. (A) *EA-ITP-P2, Page 128; EA-ITP-ASG2E, Page 1-68; EA-ITP-P, Page 310; EA-AC65-15A, Page 470*
White index marks are used on the glass and case of aircraft instruments to indicate whether the glass has slipped.

8629. (B) *EA-ITP-P2, Page 128; EA-ITP-ASG2E, Page 1-68; EA-ITP-P, Page 310; EA-AC65-15A, Page 470*
White index marks are used on the glass and case of aircraft instruments to indicate whether the glass has slipped.

8630. (C) *EA-ITP-A2, Page 640; EA-ITP-ASG2E, Page 1-68; EA-AC65-15A, Page 470*
Instruments should be marked and graduated in accordance with the Aircraft Specifications or Type Certificate Data Sheets and the specific aircraft maintenance or flight manual.

8631. (B) *EA-ITP-A2, Page 639; EA-ITP-ASG2E, Page 1-68; EA-AC65-15A, Page 470*
Instrument panels are usually shock-mounted to absorb low frequency, high amplitude shocks. The mounts are used in sets of two, each secured to separate brackets.

8632. (B) *EA-ITP-ASG2E, Page 1-67; EA-AC65-15A, Page 470*
The method of mounting instruments on their respective panels depends on the design of the instrument case. Depending on their design, they can be mounted from the front or from the rear of the panel.

8633. (B) *EA-ITP-ASG2E, Page 1-68; EA-AC65-15A, Page 470*
An instrument with a flangeless case is mounted from the front of the panel. A special expanding type of clamp, shaped and dimensioned to fit the instrument case, is secured to the rear face of the panel.

8634. (C) *EA-ITP-ASG2E, Page 1-68; EA-AC65-15A, Page 469*
Cases for electrically operated instruments are made of iron or steel because these materials provide a path for stray magnetic force fields that would otherwise interfere with radio and electronic devices.

8635. (B) *EA-ITP-ASG2E, Page 1-68*
According to FAR 43, whenever a mechanic works on an aircraft system, the work must be performed in such a way that the finished product is at least equal to the original or properly altered condition. Ensuring that an instrument is properly marked is the responsibility of the installing person or agency.

8636. (A) *EA-ITP-G2, Pages 525-526; EA-ITP-ASG2E, Page 1-68; EA-AC65-9A, Page 470*
Instruments should be marked and graduated in accordance with the Aircraft Specifications or Type Certificate Data Sheets and the specific aircraft maintenance or flight manual.

8637. (C) *EA-ITP-ASG2E, Page 1-70*
According to FAR 65, a limitation imposed on certified mechanics is that they cannot perform repairs to instruments.

8638. (C) *EA-ITP-P2, Page 128; EA-ITP-ASG2E, Page 1-68; EA-ITP-P, Page 311; EA-AC65-15A, Page 470*
A red radial line on an aircraft instrument is used to indicate maximum or minimum safe operating limits.

8639. (B) *EA-ITP-ASG2E, Page 1-70*
According to FAR 65, a certified mechanic is authorized to perform a 100-hour inspection of an aircraft. This includes the inspection of the instruments.

8640. (B) *EA-ITP-A2, Page 639; EA-ITP-ASG2E, Page 1-68; EA-ITP-AB, Page 567; EA-AC65-15A, Page 470*
An aircraft instrument panel needs to be bonded to the aircraft structure to prevent the buildup of stray electrical charges. This is generally done by using a braided bonding strap.

8641. (A) *EA-ITP-A2, Pages 608-614; EA-ITP-ASG2E, Page 1-69; EA-ITP-AB, Pages 537-545; EA-AC65-15A, Page 499*
Three of the most common flight instruments are controlled by gyroscopes: the attitude indicator, heading indicator, and the turn needle of the turn and bank indicator.

8642. (A) *EA-ITP-A2, Pages 608-609; EA-ITP-ASG2E, Page 1-70; EA-ITP-AB, Page 542*
By using the caging knob on the front of the instrument, the direction the airplane is headed can be adjusted to allow for gyroscope precession.

8643. (A) *EA-ITP-A2, Pages 594-595; EA-ITP-ASG2E, Page 1-68*
The airspeed indicator is connected to both the pitot and static systems, and cabin instruments are not connected to outside references.

8644. (C) *EA-ITP-A2, Page 642; EA-ITP-ASG2E, Page 1-68*
If the static system has been sealed off with the altimeter reading 1000 feet above ground level, the system leakeage should not allow the altimeter to drop more than 100 feet in one minute.

8645. (B) *EA-ITP-A2, Page 597; EA-ITP-ASG2E, Page 1-68*
Start with the portion of the system connected to the instruments and work back toward the static ports.

8646. (C) *EA-ITP-A2, Page 597; EA-ITP-ASG2E, Page 1-68*
Static system leak tests are performed at pressures below atmospheric.

8647. (A) *EA-ITP-ASG2E, Page 1-71; EA-AC65-15A, Page 511*
The purpose of an automatic pilot system is primarily to reduce the work strain, and fatigue of controlling the aircraft during long flights.

8648. (C) *EA-ITP-A2, Page 635; EA-ITP-ASG2E, Page 1-71; EA-AC65-15A, Page 513*
While the automatic pilot system is engaged, the manual operation of the various knobs on the controller allows the pilot to maneuver the aircraft.

8649. (C) *EA-ITP-A2, Pages 633-634; EA-ITP-ASG2E, Page 1-71; EA-ITP-AB, Page 617; EA-AC65-15A, Page 512*
As the aileron surfaces move, a follow-up signal builds up in opposition to the input signal. When the two signals are equal in magnitude, the servo stops moving.

8650. (B) *EA-ITP-A2, Page 631; EA-ITP-ASG2E, Page 1-71; EA-ITP-AB, Page 615; EA-AC65-15A, Page 513*
The directional gyro, turn-and-bank gyro, attitude gyro, and altitude control are the sensing elements in an autopilot system. These units sense the movements of the aircraft, and automatically generate signals to keep these movements under control.

8651. (B) *EA-ITP-A2, Page 631; EA-ITP-ASG2E, Page 1-71; Aircraft Electricity and Electronics, Page 326*
In an autopilot system, attitude is sensed by the relationship between the gyros and their supporting systems.

8652. (C) *EA-ITP-A2, Page 633; EA-ITP-ASG2E, Page 1-71*
If the control surfaces were not returned to neutral until the desired position is reached, the inertia of the aircraft would keep it moving in the same direction until it caused a gyro to sense a new error and send a signal to correct. This overshoot and correct would begin a series of oscillations that would cause the aircraft to rock its wings violently. The follow-up system prevents this from happening.

8653. (A) *EA-ITP-A2, Page 632; EA-ITP-ASG2E, Page 1-71; EA-ITP-AB, Page 616; EA-AC65-15A, Page 511*
In an autopilot system, the servo is the device which is connected to the control surface and converts the electrical signals into mechanical force to move the control surface.

8654. (C) *EA-ITP-A2, Pages 632-633; EA-ITP-ASG2E, Page 1-71; EA-AC65-15A, Pages 514-515*
In an autopilot system, servomotors are the devices which apply the mechanical force to move the control surfaces. Electrical and electro/pneumatic are typical of the servos in use today.

8655. (A) *EA-ITP-ASG2E, Page 1-71; EA-AC65-15A, Pages 512-513*
The elevator channel circuits detect changes in the pitch attitude of the aircraft.

8656. (C) *EA-ITP-A2, Page 633; EA-ITP-ASG2E, Page 1-71; EA-ITP-AB, Page 229; EA-AC65-15A, Page 44*
The elevator of an aircraft provides control about the lateral axis.

8657. (B) *EA-ITP-A2, Page 631; EA-ITP-ASG2E, Page 1-71; EA-AC65-15A, Page 513*
The directional gyro, turn-and-bank gyro, attitude gyro and the altitude control are the sensing elements. They sense the movements of the aircraft and automatically generate signals to keep these movements under control.

8658. (B) *EA-ITP-A2, Page 633; EA-ITP-ASG2E, Page 1-71*
A fully integrated autopilot controls the aircraft around all three axes.

8659. (C) *EA-ITP-A2, Page 635; EA-ITP-ASG2E, Page 1-71; EA-ITP-AB, Page 620*
Dutch roll is a yaw condition. It is sensed by the yaw dampeners and counteracted by the rudder servo system.

8660. (A) *EA-ITP-ASG2E, Page 1-71; EA-AC65-15A, Page 516*
After the aircraft's main power switch has been turned on, allow the gyros to come up to speed and the amplifiers to warm up before engaging the autopilot.

8661. (A) *EA-ITP-A2, Page 700; EA-ITP-ASG2E, Page 1-75; EA-ITP-AB, Page 567; EA-AC65-15A, Page 535*
Radio equipment is sensitive to mechanical shock and vibration and is normally shock mounted to provide protection.

8662. (C) *EA-ITP-A2, Page 701; EA-ITP-ASG2E, Page 1-75; EA-AC43.13-2A, Page 10*
It is advisable to bond radio equipment to the aircraft in order to provide a low impedance ground and to minimize radio interference from static electrical charges. When attaching bonding jumpers, the use of solder should be avoided.

8663. (A) *EA-ITP-ASG2E, Page 1-72; EA-AC65-15A, Page 523*
Prior to transmitting on an aircraft's radio, a person should make sure that the station license is displayed in the aircraft.

8664. (C) *EA-ITP-ASG2E, Page 1-72; EA-AC65-15A, Page 523*
Aircraft station licenses and the operator's permit are both issued by the Federal Communications Commission.

8665. (C) *EA-ITP-A2, Page 679; EA-ITP-ASG2E, Page 1-73*
The ADF system uses the combination of an omnidirectional sense antenna, and a highly directional loop antenna. In the ADF mode the field patterns of the two antennas are combined into what is known as a cardoid pattern.

8666. (C) *EA-ITP-ASG2E, Page 1-75; EA-AC65-15A, Page 538*
When installing coaxial cable (transmission lines), secure the cables firmly along their entire length at intervals of approximately 2 ft. To assure optimum operation, coaxial cables should not be routed or tied to other wire bundles.

8667. (A) *EA-ITP-A2, Page 695; EA-ITP-ASG2E, Page 1-74; EA-ITP-AB, Page 614; EA-AC65-15A, Page 534*
For an emergency locator transmitter, the replacement date for its battery must be marked on the outside of the transmitter. This time is computed from the date of manufacture of the battery.

8668. (B) *EA-ITP-A2, Page 695; EA-ITP-ASG2E, Page 1-74; EA-AC65-15A, Page 534*
The life of the battery is the length of time which the battery may be stored without losing its ability to continuously operate the ELT for 48 hours.

8669. (C) *EA-ITP-A2, Page 695; EA-ITP-ASG2E, Page 1-74; EA-ITP-AB, Page 614; EA-AC65-15A, Page 534*
An emergency locator transmitter may be located anywhere within the aircraft, but the ideal location is as far aft as possible, but just forward of the vertical fin.

8670. (A) *EA-ITP-A2, Page 695; EA-ITP-ASG2E, Page 1-74; AC 91-94A, Page 2*
To meet the "G" force requirements of TSO-C91, automatic fixed-type inertially activated ELTs must activate at any inertial force, parallel to the longitudinal axis of the aircraft of 5 Gs (+2, –0) and greater for a time duration of 11 (+5, –0) milliseconds or longer.

8671. (B) *EA-ITP-A2, Page 695; EA-ITP-ASG2E, Page 1-74; EA-ITP-AB, Page 614; EA-AC65-15A, Page 534*
The battery replacement date for an emergency locator transmitter must be marked on the outside of the transmitter.

8672. (C) *EA-ITP-A2, Page 695; EA-ITP-ASG2E, Page 1-74; EA-AC65-15A, Page 534*
Testing of ELTs should be coordinated with the nearest FAA Tower or Flight Service Station. Tests should be conducted only during the first five minutes of any hour and should be restricted to 3 audio sweeps. The ELT can be checked by tuning a VHF communication receiver to 121.5 MHz and listening for ELT audio sweeps.

8673. (A) *EA-ITP-A2, Page 700; EA-ITP-ASG2E, Page 1-75*
The lower level of discharge current produces an extremely weak magnetic field which is not picked up by radio receivers.

8674. (B) *EA-ITP-A2, Page 696; EA-ITP-ASG2E, Page 1-73*
Airborne LORAN systems choose a master and two or more secondary ground stations.

8675. (A) *EA-356, Page 41; EA-ITP-ASG2E, Page 1-59*
The Passenger Address system may be used to deliver pilot's announcements, attendant's announcements, pre-recorded announcements and boarding music.

8676. (C) *EA-ITP-A2, Pages 651 and 673; EA-ITP-ASG2E, Page 1-72*
VOR stations as well as the vast majority of aeronautical communications operate in the VHF band of frequencies.

8677. (A) *EA-ITP-A2, Page 635; EA-ITP-ASG2E, Page 1-71*
The Flight Management Computer is the main memory unit of a Flight Management System.

8678. (B) *EA-ITP-A2, Page 593; EA-ITP-ASG2E, Page 1-73*
The Ground Proximity Warning System (GPWS) is used to warn the pilot of excessively low altitude.

8679. (C) *EA-ITP-A2, Page 686; EA-ITP-ASG2E, Page 1-74*
The transponder provides reliable information to the air traffic controller as to the location of a specific aircraft.

8680. (C) *EA-ITP-A2, Page 702; EA-ITP-ASG2E, Page 1-75; EA-AC43.13-2A, Page 13*
When attaching an antenna mounting (masts, base receptacles, and/or supporting brackets) it should be done so that the loads imposed are transmitted to the aircraft structure.

8681. (B) *EA-ITP-A2, Pages 679-680; EA-ITP-ASG2E, Page 1-75; EA-AC43.13-2A, Page 21*
After completing the installation of an ADF antenna, it is essential that the loop be calibrated. This can be done in accordance with EA-43.13-1A/3, Chapter 15.

8682. (C) *EA-ITP-ASG2E, Page 1-75; EA-AC65-15A, Page 537*
When installing a typical rigid antenna, a reinforcing doubler of sufficient thickness to reinforce the skin should be used.

8683. (A) *EA-ITP-ASG2E, Page 1-73; EA-AC65-15A, Page 521*
In airborne communications, the same antenna is normally used for both transmission and reception of signals. This is made possible because the two ranges are very close together.

8684. (C) *EA-ITP-ASG2E, Page 1-75; EA-AC43.13-2A, Page 14*
When a rigid antenna is installed on a vertical stabilizer, the technician should evaluate the flutter and vibration characteristics of the installation.

8685. (A) *EA-ITP-A2, Page 703; EA-ITP-ASG2E, Page 1-75; EA-AC65-15A, Page 538*
When installing an antenna, the technician should make sure that the mounting bolts are tightened firmly against the reinforcing doubler and that the mast is drawn tight against the gasket. If a gasket is not used, the technician should seal between the mast and the fuselage with a suitable sealer, such as zinc chromate paste. The gasket or sealant is used to preclude the entry of moisture.

8686. (B) *EA-ITP-ASG2E, Page 1-72; EA-AC43.13-2A, Pages 15-16*
A good location for the VOR localizer receiving antenna on many small airplanes is over the forward part of the cabin. Mount the rigid V-type antenna so that the apex of the "V" points forward and the plane of the "V" is level in normal flight.

8687. (C) *EA-ITP-A2, Page 681; EA-ITP-ASG2E, Page 1-72; EA-ITP-AB, Page 600; EA-AC65-15A, Page 525*
An instrument landing system consists of a runway localizer, a glide slope signal, and marker beacons for position location. The localizer equipment produces a radio course aligned with the center of an airport runway.

8688. (A) *EA-ITP-ASG2E, Page 1-75; EA-AC43.13-2A, Page 14*
The formula given for antenna drag is, in this case:

D = .000327 × .125 × (225) 2

= .000327 × .125 × 50.625

= 2.069 lbs.

8689. (A) *EA-ITP-ASG2E, Page 1-75; EA-AC43.13-2A, Page 14*
Same formula as in #8688, only this time the answer is 3.3804 (Round to 3.387).

8690. (A) *EA-ITP-ASG2E, Page 1-73; EA-AC65-15A, Page 528*
To prevent an interruption in DME operation, the antenna must be located in a position that will not be blanked by the wing when the aircraft is banked.

8691. (A) *EA-ITP-ASG2E, Page 1-75; EA-AC65-15A, Page 538*
When bending coaxial cable, be sure the bend is at least 10 times the size of the cable diameter.

8692. (C) *EA-ITP-ASG2E, Page 1-73*
A DME antenna should be installed so it runs parallel with the center line of the aircraft.

8693. (A) *EA-ITP-G2, Page 271; EA-ITP-ASG2E, Page 1-73; EA-ITP-AB, Page 606; EA-AC65-15A, Page 528*
A typical DME antenna is a blade type.

8694. (A) *EA-ITP-A2, Page 684; EA-ITP-ASG2E, Page 1-73; EA-ITP-AB, Page 604; EA-AC65-15A, Page 526*
A typical glide slope antenna is the one shown in EA-AC65-15A, page 526 at the time that book was written. Currently, a typical glide slope antenna is the "cats whisker" shown in EA-ITP-AB, Page 604.

8695. (A) *EA-ITP-G2, Page 271; EA-ITP-ASG2E, Page 1-75; EA-BEM, Page 7-9*
The addition of any equipment will affect both the empty weight and the aircraft useful load.

8696. (C) *EA-ITP-ASG2E, Page 1-75; EA-AC43.13-2A, Page 7*
See Figure 2.6 — typical underseat installation.

8697. (C) *EA-ITP-A2, Page 683; EA-ITP-ASG2E, Page 1-72; EA-ITP-AB, Page 604; EA-AC65-15A, Page 526*
The glide slope is a radio beam which provides vertical guidance to the pilot, assisting him in making the correct angle of descent to the runway.

8698. (A) *EA-ITP-A2, Pages 717-718; EA-ITP-ASG2E, Page 1-81; EA-AC65-9A, Page 94*
The fuel jettisoning system of an aircraft is usually divided into two separate, independent systems, one for each wing, so that lateral stability can be maintained by jettisoning fuel from the "heavy" wing. Each wing contains either a fixed or an extendable dump chute, depending upon system design.

8699. (A) *EA-ITP-A2, Page 706; EA-ITP-ASG2E, Page 1-81*
If an aircraft's landing weight is less than its take-off weight a means must be provided to jettison fuel to achieve a safe landing weight.

8700. (A) *EA-ITP-ASG2E, Page 1-81*
Both of the statements in the question are true. The fuel jettisoning valve must be designed to allow flight personnel the ability to close it during any phase of the jettisoning operation. In addition, the fuel must discharge clear of any part of the airplane.

8701. (A) *EA-ITP-ASG2E, Page 1-81; EA-AC65-9A, Page 94*
The fuel jettisoning system is usually divided into two separate, independent systems, one for each wing, so that lateral stability can be maintained by jettisoning fuel from the "heavy" wing.

8702. (B) *EA-ITP-A2, Page 706; EA-ITP-ASG2E, Page 1-81; EA-ITP-AB, Page 638; EA-AC65-9A, Page 94*
A fuel jettison system is required for transport category and general aviation aircraft if the maximum takeoff weight exceeds the maximum landing weight.

8703. (A) *EA-ITP-A2, Pages 717-718; EA-ITP-ASG2E, Page 1-81; EA-ITP-AB, Page 640; Aircraft Maintenance and Repair, Page 401*
Systems operation may vary from aircraft to aircraft, but the fuel boost pumps alone, or a combination of boost pumps and gravity will be used to jettison fuel.

8704. (B) *EA-ITP-A2, Page 718; EA-ITP-ASG2E, Page 1-81; EA-ITP-AB, Page 640*
Fuel dump limits are set at a preset level in the fuel tank.

8705. (C) *EA-ITP-ASG2E, Page 1-77; Aircraft Maintenance and Repair, Pages 400-401*
On a sweptback wing airplane, structural limitations usually dictate that the outboard wing tanks be defueled first. If the inboard tanks are defueled first, the weight of the fuel in the outboard tanks will not be balanced out and will create a strain on the wing structure.

8706. (B) *EA-ITP-ASG2E, Page 1-81; EA-AC65-9A, Page 93*
See crossfeed system.

8707. (C) *EA-ITP-A2, Page 715; EA-ITP-ASG2E, Page 1-81*

8708. (A) *EA-ITP-A2, Page 715; EA-ITP-ASG2E, Page 1-81; EA-ITP-AB, Page 629; EA-AC65-9A, Page 93*
The individual engine fuel systems must be interconnected so fuel can be fed from the various tanks to any engine.

8709. (A) *EA-ITP-A2, Page 563; EA-ITP-ASG2E, Page 1-77; EA-ITP-AB, Page 346*
To prevent unwanted potential buildups and to serve as a ground return for shock-mounted equipment, most of the aircraft control surfaces and the electrical and electronic components are bonded. This is especially important on fuel system components where a stray spark could set off an explosion.

8710. (C) *EA-ITP-A2, Pages 715-718; EA-ITP-ASG2E, Page 1-81; EA-ITP-AB, Page 629; EA-AC65-9A, Page 93*
The individual engine fuel systems must be interconnected so fuel can be fed from the various tanks to any engine. In case of engine failure, the fuel supplied to the inoperative engine must be made available to other engines.

8711. (C) *EA-ITP-A2, Page 710; EA-ITP-ASG2E, Page 1-80; EA-ITP-AB, Page 631; EA-AC65-9A, Page 79*
Microbial growth is produced by various forms of micro-organisms that live and multiply in the water interfaces of jet fuels.

8712. (C) *EA-ITP-A2, Page 711; EA-ITP-ASG2E, Page 1-77; EA-ITP-AB, Page 648; EA-AC65-9A, Page 85*
This type of pump is essential, particularly at high altitudes, to keep pressure on the suction side of the engine-driven pump from becoming too low to permit the fuel to boil.

8713. (B) *EA-ITP-A2, Page 708; EA-ITP-ASG2E, Page 1-77; EA-ITP-GB, Page 466; EA-AC65-9A, Page 498*
Care should be taken to identify the aviation fuel and lubricating oil dispensed from each refueling unit before beginning the actual servicing. Aviation technicians should be familiar with the various grades and the aircraft's gasoline requirements so that the appropriate fuel is used.

8714. (A) *EA-ITP-ASG2E, Page 1-77; EA-ITP-GB, Page 471*
When using pressure fueling under the wing, the technician must be sure that the pressure used and the delivery rate are those specified by the manufacturer of the aircraft.

8715. (B) *EA-ITP-ASG2E, Page 1-77; EA-AC65-9A, Page 500*
Advantages of pressure fueling include reduction of aircraft skin damage caused by banging the aircraft with the fueling nozzle, reduced hazards to personnel, and reduced chance of fuel contamination.

8716. (A) *EA-ITP-ASG2E, Page 1-77; EA-AC65-9A, Page 500*
Fueling and defueling procedures are normally placarded on the fueling control panel access door. In addition to knowing these procedures, the fueling operator should possess a thorough knowledge of the aircraft fuel system to recognize malfunction symptoms.

8717. (B) *EA-ITP-A2, Page 718; EA-ITP-ASG2E, Page 1-77*
Pressure fueling is often accomplished through a single connection, often located under the wing.

8718. (C) *EA-ITP-A2, Page 734; EA-ITP-ASG2E, Page 1-81; EA-ITP-AB, Pages 656-657; Aircraft Maintenance and Repair, Page 405*
Riveting in a patch plate is one method that may be used to repair integral fuel tanks. After the riveting is completed, a sealer would typically be used to complete the sealing of the tank.

8719. (C) *EA-ITP-A2, Page 726; EA-ITP-ASG2E, Page 1-77; EA-ITP-AB, Page 648; EA-AC65-9A, Page 85*
Centrifugal-type fuel boost pumps are not positive displacement pumps, meaning they do not necessarily pump a certain amount of fuel per revolution. The amount of pressure they are capable of generating is determined by the clearance between the pump's impeller and the case.

8720. (A) *EA-ITP-A2, Page 718; EA-ITP-ASG2E, Page 1-76; EA-AC65-9A, Page 82*
As fuel is being drawn from a tank, a partial vacuum will try to form because of the void formed as the fuel leaves the tank. To prevent this partial vacuum from forming, fuel tanks are vented so that atmospheric pressure can occupy the space above the fuel.

8721. (A) *EA-ITP-ASG2E, Page 1-81;EA-FAR-M, Part 23*
FAR 23 identifies the test pressures that a fuel tank must be able to withstand, whether new or repaired.

8722. (C) *EA-ITP-A2, Page 737; EA-ITP-ASG2E, Page 1-81; EA-ITP-GB, Page 470; EA-AC65-9A, Page 498*
Aircraft fueling, defueling, or tank purging should be done in a safe place. Do not fuel or defuel an aircraft in a hangar or other enclosed space. Always perform these operations on the ramp.

8723. (C) *EA-ITP-A2, Page 733; EA-ITP-ASG2E, Page 1-81; EA-AC65-9A, Page 96*
Before welding, a fuel tank must be steamed for a minimum of 8 hours. This is done to remove all traces of fuel.

8724. (B) *EA-ITP-A2, Page 72; EA-ITP-ASG2E, Page 1-76; EA-ITP-AB, Page 641; EA-AC65-9A, Page 96*
An aircraft's integral fuel tank is a non-removable part of the aircraft structure.

8725. (B) *EA-ITP-ASG2E, Page 1-81; EA-AC65-9A, Page 96; EA-43.13-1A/3, Page 282*
An aircraft's fuel tank can be purged and made inert by slowly discharging a CO_2 fire extinguisher into the tank.

8726. (A) *EA-ITP-A2, Page 76; EA-ITP-ASG2E, Page 1-76; EA-AC65-9A, Page 84*
The fuel system's main strainer is installed in such a way that all the fuel flows through it before reaching the engine-driven pump. It is located at the lowest point in the fuel system so that it can trap any water that may be present.

8727. (C) *EA-ITP-ASG2E, Page 1-77; EA-AC65-9A, Pages 86-87*
For proper altitude compensation, the vent must be open.

8728. (A) *EA-ITP-G2, Pages 134-135; EA-ITP-ASG2E, Page 1-82; EA-ITP-P, Page 348*
When the mixture control on a normally operating engine is moved to the idle cutoff position, the RPM should increase slightly before dropping off to zero. A rise of 10 to 35 RPM should be seen.

8729. (C) *EA-ITP-A2, Page 710; EA-ITP-ASG2E, Page 1-80; EA-ITP-AB, Page 631*
Because of the viscosity of turbine engine fuels, they are very susceptible to holding water in suspension. To minimize this water problem and what could become an ice problem, most turbine engine fuel is treated with an anti-icing additive that mixes with the water in the fuel, and lowers its freezing point so it will not freeze.

8730. (C) *EA-ITP-A2, Page 734; EA-ITP-ASG2E, Page 1-81*
While leaks in integral tanks do not always constitute a flight hazard, any type of leak that allows fuel vapors to accumulate must be repaired before the aircraft is allowed to fly.

8731. (C) *EA-ITP-P2, Page 232; EA-ITP-ASG2E, Page 1-82; EA-ITP-P, Page 370*
Injector nozzles do not require much service, but every 300 hours of operation they should be removed from the engine and checked. A clogged nozzle can be located by the presence of fuel dye stain on the cylinder head around the nozzle.

8732. (A) *EA-ITP-ASG2E, Page 1-81; EA-AC65-9A, Page 96; EA-43.13-1A/3, Page 282*
Carbon dioxide is the best thing to use to purge and inert an integral fuel tank before attempting to repair it.

8733. (B) *EA-ITP-ASG2E, Page 1-81; EA-43.13-1A/3, Page 282*
A 5 percent solution of nitric or sulfuric acid should be used to remove flux from an aluminum fuel tank after it has been weld repaired. The removal of the flux is especially important because of the possible corrosion it could cause.

8734. (A) *EA-ITP-ASG2E, Page 1-79; EA-43.13-1A/3, Page 283*
Internal leakage of a fuel valve can be checked by placing the appropriate valve in the OFF position, draining the fuel strainer bowl, and observing if fuel continues to flow into it.

8735. (A) *EA-ITP-A2, Page 710; EA-ITP-ASG2E, Page 1-80; EA-ITP-GB, Page 469*
The high viscosity of jet fuel causes it to hold water in suspension much more readily than does aviation gasoline.

8736. (C) *EA-ITP-ASG2E, Page 1-82; EA-43.13-1A/3, Page 281*
On ½″ diameter fuel lines, support clamps or brackets should be spaced no more than 16″ apart.

8737. (C) *EA-ITP-A2, Page 731; EA-ITP-ASG2E, Page 1-81; EA-ITP-AB, Page 562; EA-AC65-9A, Page 89*
Instead of using the float and its attendant mechanical units, the capacitance-type fuel quantity gage uses the dielectric qualities of fuel and air to furnish a measurement of fuel quantity. Essentially, the tank transmitter is a simple electric condenser.

8738. (A) *EA-ITP-A2, Page 731; EA-ITP-ASG2E, Page 1-81; EA-AC65-9A, Page 89*
The electronic-type (capacitance) fuel quantity gage differs from other types in that it has no movable devices in the fuel tank.

8739. (B) *EA-ITP-A2, Page 731; EA-ITP-ASG2E, Page 1-81; EA-ITP-AB, Page 562; EA-AC65-9A, Page 89; EA-AC65-15A, Pages 486-487*
The electronic or capacitance-type fuel quantity indicating system utilizes several probes installed in each fuel tank. By using more than one probe in each tank, the effect which the attitude of the aircraft would have on the fuel quantity reading is compensated for.

8740. (A) *EA-ITP-A2, Pages 730-731; EA-ITP-ASG2E, Page 1-81; EA-ITP-AB, Pages 562-563; EA-AC65-9A, Page 89*
In an electronic-type fuel quantity indicating system, the capacitance of the transmitter is compared to a reference capacitor in a rebalance-type bridge circuit. The unbalanced signal is amplified by the voltage amplifiers that drive a phase discriminating power stage.

8741. (A) *EA-ITP-A2, Page 731; EA-ITP-ASG2E, Page 1-81; EA-AC65-9A, Page 89*
A drip gage is a type of bayonet gage which is used to measure, by manual means, the fuel quantity in the tank.

8742. (A) *EA-ITP-A2, Page 730; EA-ITP-ASG2E, Page 1-81; EA-AC65-9A, Page 89; EA-AC65-15A, Page 485*
The basic components of a capacitor-type (electronic) fuel quantity indicating system are an indicator, a tank probe, a bridge unit, and an amplifier.

8743. (B) *EA-ITP-A2, Page 731; EA-ITP-ASG2E, Page 1-81; EA-ITP-AB, Page 562; EA-AC65-9A, Page 89; EA-AC65-15A, Page 485*
A single probe in a tank, or possibly a series of probes in a tank, is used in a capacitor-type (electronic) fuel quantity indicating system.

8744. (C) *EA-ITP-A2, Page 731; EA-ITP-ASG2E, Page 1-81; EA-ITP-AB, Page 562; EA-AC65-9A, Page 89*
A nice feature of the capacitor-type (electronic) fuel quantity indicating system is that it does not measure fuel quantity by volume, but rather by weight. Engines consume fuel in pounds per hour (indicated on a fuel-flow gage), so it is more valuable to know how many pounds of fuel are in the tanks rather than how many gallons.

8745. (C) *EA-ITP-ASG2E, Page 1-81; EA-AC65-9A, Page 89*
Two important advantages of the electrical and electronic-type fuel quantity indicating systems are that the fuel quantity indicator can be located any distance from the tank, and fuel levels of several tanks can be read on one indicator.

8746. (C) *EA-ITP-ASG2E, Page 1-81; EA-AC65-9A, Page 88*
On some aircraft, one fuel gage, called a totalizer, indicates the total amount of fuel remaining in all the fuel tanks.

8747. (C) *EA-ITP-A2, Page 731; EA-ITP-ASG2E, Page 1-81; EA-ITP-AB, Page 562; EA-AC65-9A, Page 89*
The dielectric (or nonconducting material) of the condenser of a capacitor-type fuel quantity indicating system is fuel and air (vapor) above the fuel in the tank.

8748. (A) *EA-ITP-A2, Page 731; EA-ITP-ASG2E, Page 1-81; EA-ITP-AB, Page 653*
The density of the fuel affects its dielectric constant and thus the capacity of the probe. Because of this the system will show on the fuel quantity indicator the number of pounds of fuel remaining in the aircraft.

8749. (A) *EA-ITP-A2, Pages 730-732; EA-ITP-ASG2E, Page 1-80; EA-AC65-15A, Page 88*
The four types of fuel quantity gages are sight glass, mechanical, electrical and electronic.

8750. (A) *EA-ITP-ASG2E, Page 1-81; EA-AC65-9A, Page 89*
A gallon of gasoline weighs more when it is cold because the volume of fuel changes with the temperature.

8751. (A) *EA-ITP-A2, Pages 730-732; EA-ITP-ASG2E, Page 1-81; EA-AC65-9A, Page 88*
One advantage of the electric and electronic type of fuel quantity gages is they can be located any distance from the tank because they use an electrical wire to carry their information.

8752. (A) *EA-ITP-A2, Page 731; EA-ITP-ASG2E, Page 1-81; EA-AC65-9A, Page 89*
Fuel volume varies with temperature, the higher the temperature the more volume.

8753. (A) *EA-ITP-A2, Page 731; EA-ITP-ASG2E, Page 1-80; EA-AC65-9A, Page 89*
An electrical type fuel quantity gage consists of a float-operated transmitter in the fuel tank and an indicator located in the cockpit.

8754. (A) *EA-ITP-A2, Page 731; EA-ITP-ASG2E, Page 1-80; EA-AC65-9A, Page 89*
The transmitter sends an electrical signal to the indicator.

8755. (A) *EA-ITP-A2, Page 731; EA-ITP-ASG2E, Page 1-80; EA-AC65-9A, Page 89*
The tank transmitter is a simple electric condenser.

8756. (C) *EA-ITP-ASG2E, Page 1-80;EA-FAR-M*
See FAR 23.1337 (b) (1). Each fuel quantity indicator must be calibrated to read zero.

8757. (B) *EA-ITP-A2, Page 715; EA-ITP-ASG2E, Page 1-79; EA-AC65-9A, Page 92*
The fuel pressure warning light is illuminated in the cockpit when the fuel pressure, acting on a pressure-sensitive mechanism, falls below a pre-set value. The value at which the light will illuminate is changed by changing the tension on a spring which is inside the mechanism.

8758. (C) *EA-ITP-P2, Pages 729 and 732; EA-ITP-ASG2E, Page 1-80; EA-ITP-P, Page 397*
A typical means of heating the fuel in a turbine engine powered aircraft is by using an air-to-fuel heat exchanger. This unit makes use of compressor bleed air, which is at an elevated temperature because of compression, and passes it around tubes through which fuel is flowing. The air gives up its heat to the fuel and then is dumped overboard.

8759. (B) *EA-ITP-A2, Page 726; EA-ITP-ASG2E, Page 1-77; EA-AC65-9A, Page 92*
Flapper-type check valves, when used in an integral fuel tank, are designed to prevent the fuel from flowing away from the boost pump when the aircraft is in a high "G" maneuver. Without the check valve, the boost pump could be starved for fuel during the maneuver.

8760. (B) *EA-ITP-A2, Page 715; EA-ITP-ASG2E, Page 1-79; EA-AC65-9A, Page 92*
In a fuel pressure warning system, it is the pressure sensitive mechanism which activates the warning light in the cockpit.

8761. (C) *EA-ITP-A2, Pages 729-730; EA-ITP-ASG2E, Page 1-79; EA-AC65-9A, Pages 92-93*
A means for checking the temperature of the fuel in the tanks and at the engine is provided on some turbine-powered aircraft. During extreme cold, especially at altitude, the gage can be checked to determine when fuel temperatures are approaching those at which there may be danger of ice crystals forming in the fuel.

8762. (A) *EA-ITP-ASG2E, Page 1-79; EA-AC65-9A, Page 92*
In an aircraft with several fuel tanks, there is always the possible danger of allowing the fuel supply in one tank to become exhausted before the selector valve is switched to another tank. To prevent this, pressure warning signal devices are installed in some aircraft.

8763. (C) *EA-ITP-A2, Page 715; EA-ITP-ASG2E, Page 1-79; EA-AC65-9A, Page 92*
The whole purpose of a fuel pressure warning system is to cause a warning light to come on in the cockpit if the fuel pressure drops below specified limits.

8764. (B) *EA-ITP-A2, Page 715; EA-ITP-ASG2E, Page 1-79; EA-AC65-9A, Page 92*
A fuel pressure warning mechanism senses fuel pressure against a diaphragm by balancing the pressure with a spring. If the fuel pressure drops below a specified value, the spring force closes a set of contacts, which causes an electrical signal to be sent to the cockpit to turn on the warning light.

8765. (B) *EA-ITP-A2, Page 716 (Figure 15-8); EA-ITP-ASG2E, Page 1-79; EA-AC65-9A, Page 92 (Figure 4-22)*
The fuel pressure signal sent to the pressure warning mechanism is tapped from the fuel pressure line to the carburetor.

8766. (A) *EA-ITP-ASG2E, Page 1-79; EA-AC65-9A, Pages 94 and 464*
To effectively troubleshoot an aircraft system, the manufacturer's maintenance or service manuals are needed.

8767. (C) *EA-ITP-ASG2E, Page 1-81; EA-AC65-9A, Page 94*
In order to become proficient at the art of troubleshooting, one must be familiar with the complete system which is being worked on. To do this, one can become familiar with the schematics of various portions of the system, the nomenclature of the units, and their particular function within the system by studying aircraft and engine maintenance manuals.

8768. (A) *EA-ITP-ASG2E, Page 1-79; EA-AC65-9A, Page 92*
To prevent fuel starvation caused by one fuel tank becoming empty before fuel tanks are switched with the selector valve, fuel pressure warning systems are installed.

8769. (A) *EA-ITP-A2, Pages 729-730; EA-ITP-ASG2E, Page 1-80; EA-AC65-12A, Page 170*
The function of a fuel heater is to protect the engine fuel system from ice formation. However, should ice form, the heater can also be used to thaw ice on the fuel screen.

8770. (C) *EA-ITP-A2, Pages 729-730; EA-ITP-ASG2E, Page 1-80; EA-ITP-AB, Pages 651-652; EA-AC65-12A, Page 170*
Gas-turbine engine fuel systems are susceptible to the formation of ice because jet aircraft fly at very high altitudes where they are exposed to cold temperatures. The fuel filters are kept from freezing up with heat exchangers.

8771. (B) *EA-ITP-A2, Pages 729-730; EA-ITP-ASG2E, Page 1-80; EA-ITP-P2, Pages 347-348; EA-AC65-12A, Page 170*
A fuel heater can use either engine bleed air or engine lubricating oil as a source of heat.

8772. (B) *EA-ITP-A2, Pages 732-733; EA-ITP-ASG2E, Page 1-79; EA-AC65-9A, Page 90*
The fuel pressure gage is a differential pressure gage used to indicate the pressure of the fuel entering the carburetor.

8773. (A) *EA-ITP-A2, Pages 726-728; EA-ITP-ASG2E, Page 1-77; EA-AC65-9A, Pages 85-86*
Only the positive displacement needs a relief valve. The centrifugal type is not a positive displacement pump.

8774. (B) *EA-ITP-A2, Page 729; EA-ITP-ASG2E, Page 1-76; EA-AC65-9A, Page 82*
A fuel tank sump is the lowest point in the fuel tank, so this is where the water and solid contaminants will collect and can be readily drained.

8775. (C) *EA-ITP-A2, Page 720; EA-ITP-ASG2E, Page 1-76; EA-ITP-AB, Page 641*
Because an integral fuel tank makes use of the existing structure of the airframe, it saves a great deal of weight over the bladder or rigid type fuel tank.

8776. (B) *EA-ITP-A2, Pages 736-737; EA-ITP-ASG2E, Page 1-76; EA-ITP-GB, Pages 468-469; EA-AC65-9A, Page 81*
Even though a fuel tank is serviced with fuel that is known to be free of contamination, contamination of the fuel is still a possibility. Contaminants can come from a leaking fuel cap, a fuel cap left off the tank for a period of time, or corrosion from inside the tank. Periodic checks of the system sumps and strainers are necessary.

8777. (C) *EA-ITP-A2, Page 737; EA-ITP-ASG2E, Page 1-77; EA-ITP-GB, Page 470; EA-AC65-9A, Page 498*
Aircraft should be fueled and defueled in a safe place. Do not fuel or defuel an aircraft in a hangar or other enclosed space. This operation should take place on the ramp.

8778. (C) *EA-ITP-A2, Page 720; EA-ITP-ASG2E, Page 1-76; EA-ITP-AB, Page 641; EA-AC65-9A, Page 83*
Integral fuel tanks are usually built into the structure of the aircraft's wings. By sealing off the inner cavity of the wing, a fuel tank is created.

8779. (A) *EA-ITP-A2, Page 711; EA-ITP-ASG2E, Page 1-77; EA-ITP-AB, Page 629*
According to FAR 23, the airspaces of fuel tanks with interconnected outlets must also be interconnected. This is done so the system can draw a vent through either tank.

8780. (B) *EA-ITP-ASG2E, Page 1-76; EA-AC65-9A, Page 82*
The baffle plate in a fuel tank is designed to resist fuel surging caused by changes in the attitude of the aircraft.

8781. (C) *EA-ITP-ASG2E, Page 1-77*
According to FAR 25, placed on or near the fuel cover of transport category aircraft there must be the word "fuel", the minimum fuel grade or permissible fuel designations, and the maximum permissible fueling and defueling pressure for pressure fueling systems.

8782. (B) *EA-ITP-A2, Page 707; EA-ITP-ASG2E, Page 1-82; EA-AC65-9A, Page 74*
Fuels known as aromatic fuels have a strong solvent and swelling action on some types of hose and other rubber parts of the fuel system.

8783. (A) *EA-ITP-A2, Page 725; EA-ITP-ASG2E, Page 1-77; EA-AC65-9A, Page 85*
The electrically driven centrifugal boost pump, found in many fuel tanks, supplies fuel under pressure to the inlet of the engine-driven fuel pump.

8784. (C) *EA-ITP-A2, Page 726; EA-ITP-ASG2E, Page 1-77; EA-AC65-9A, Page 85*
Flapper valves are used in fuel tanks to prevent fuel from flowing away from the boost pump or tank outlet when the aircraft is in a high "G" maneuver. In this capacity, they serve as check valves.

8785. (B) *EA-ITP-A2, Page 726; EA-ITP-ASG2E, Page 1-77; EA-AC65-9A, Page 85*
A centrifugal-type boost pump is an essential part of the fuel system of high altitude aircraft, because it keeps the pressure on the suction side of the engine-driven pump from becoming low enough to permit the fuel to boil. It also helps keep air out of the system.

8786. (C) *EA-ITP-A2, Page 718; EA-ITP-ASG2E, Page 1-77; EA-AC65-9A, Page 82*
Fuel tanks are vented to the outside air to maintain atmospheric pressure within the tank. Without the vent, the space above the fuel would become a partial vacuum as the fuel was drawn out of the tank. This condition could cause structural damage to the tank, and it could also inhibit the flow of fuel to the engine.

8787. (A) *FAR Part 23.1557(c)(1)(i)*
FAR Part 23.1557 (c)(1)(i) states fuel filter openings for reciprocating engine-powered airplanes must be marked at or near the filler cover with the word "Avgas" and the minimum fuel grade. Therefore, answer A is correct.

8788. (C) *EA-ITP-ASG2E, Page 1-79; EA-AC65-9A, Page 95*
The location of leaks and defects within the internal portions of the fuel system is usually a matter of observation of the pressure gage and operation of the selector valves to determine where the trouble lies. Without the engines running, the pressure gage should read a steady pressure. If the pressure fluctuates when a certain selector valve is opened, there is an indication of a leak in that part of the system.

8789. (B) *EA-ITP-A2, Pages 727-728; EA-ITP-ASG2E, Page 1-77; EA-AC65-9A, Page 85*
Since a sliding vane type of fuel boost pump is a positive displacement type, it must have a relief valve to prevent excessive pressure.

8790. (C) *EA-ITP-A2, Page 725; EA-ITP-ASG2E, Page 1-77; EA-ITP-AB, Page 648; EA-AC65-9A, Page 85*
One of the nice features of a centrifugal-type fuel boost pump is its tendency to separate the air from the fuel as it supplies fuel to the engine-driven fuel pump. As the fuel enters the pump from the tank, a high-speed impeller throws the fuel outward in all directions at a high velocity. The high rotational speed swirls the fuel and produces a centrifuge action that separates air and vapor from the fuel.

8791. (B) *EA-ITP-A2, Page 729; EA-ITP-ASG2E, Page 1-80; EA-AC65-9A, Pages 92-93*
During extreme cold, especially at high altitude, turbine engine fuel is susceptible to the formation of ice crystals. To alert the crew to this potential problem, some fuel tanks are fitted with a temperature probe which powers a temperature gage in the cockpit.

8792. (C) *EA-ITP-ASG2E, Page 1-82; EA-43.13-1A/3, Page 283*
When checking fuel system valves for internal leakage, all valves located downstream of the boost pumps should be checked with the pumps operating. If a leak is present, it will be easier to detect when the valve is under pressure.

8793. (B) *EA-ITP-A2, Page 726; EA-ITP-ASG2E, Page 1-79; EA-ITP-AB, Page 648; EA-AC65-9A, Page 85*
The centrifugal type of boost pump uses a high-speed impeller to throw the fuel outward at a high velocity. The high rotational speed swirls the fuel and produces a centrifugal action that separates air and vapor from the fuel before it enters the fuel line to the carburetor.

8794. (A) *EA-ITP-G2, Page 501; EA-ITP-ASG2E, Page 1-77; EA-AC65-9A, Page 93*
This method of fuel servicing has greatly reduced servicing time on large aircraft.

8795. (A) *EA-ITP-A2, Page 707; EA-ITP-ASG2E, Page 1-82; EA-AC65-9A, Page 74*
If the fuel does not vaporize readily enough it can result in hard starting, slow warmup, poor acceleration, uneven fuel distribution to the cylinders, and excessive crankcase dilution.

8796. (A) *EA-ITP-A2, Page 707; EA-ITP-ASG2E, Page 1-82; EA-AC65-9A, Page 74*
The first statement is not true because when we have too much vapor in a fuel line we have a reduction in fuel flow. But fuel vapor pressure or its tendency to vapor lock is measured by a Reid vapor pressure test apparatus.

8797. (A) *EA-ITP-ASG2E, Page 1-80; EA-AC65-15A, Page 79*
The buildup of microbial growth not only interferes with fuel flow and with fuel quantity indication, but it can also start electrolytic corrosive action in a metal tank.

8798. (A) *EA-ITP-ASG2E, Page 1-82; EA-ITP-AB, Page 631*
Aviation gasoline is required to have a vapor pressure of between 5.5 and 7.0 PSI at 100°F, as opposed to vapor pressures often as high as 14 PSI found in automotive gasolines.

8799. (A) *EA-ITP-A2, Page 710; EA-ITP-ASG2E, Page 1-80; EA-ITP-AB, Page 632*
To eliminate or minimize the microbial growth in an aircraft jet fuel tank, anti-icing and antibacterial additives such as Prist are used.

8800. (A) *EA-ITP-A2, Page 707; EA-ITP-ASG2E, Page 1-82; EA-ITP-AB, Page 631*
Aviation gasoline is required to have a vapor pressure of between 5.5 and 7.0 PSI.

8801. (A) *EA-ITP-A2, Pages 721-722; EA-ITP-ASG2E, Page 1-81; EA-ITP-AB, Page 642*
If it is necessary to leave a bladder-type fuel tank empty, the inside of the tank should be wiped with a rag leaving a film of engine oil on its inside surface.

8802. (A) *EA-ITP-ASG2E, Page 1-82; EA-AC65-9A, Page 75*
A convenient means of improving the anti-knock characteristics of a fuel is to add a knock inhibitor.

8803. (C) *EA-ITP-A2, Page 579; EA-ITP-ASG2E, Page 1-91; EA-ITP-AB, Page 355; EA-AC65-9A, Page 448*
One method for reversing the direction of rotation of electric motors is to employ two field windings wound in opposite directions on the same pole. A single-pole, double-throw switch makes it possible to direct current through either of the two windings.

8804. (C) *EA-ITP-G2, Page 196; EA-ITP-ASG2E, Page 1-87; Aircraft Electricity and Electronics, Page 121*
For testing armatures, a device called a growler is used. This device can be used for shorts, opens, proper grounds, and continuity.

8805. (C) *EA-ITP-A2, Page 566; EA-ITP-ASG2E, Page 1-84*
The ring-type terminal is less likely to fail than slide-on type terminals.

8806. (A) *EA-ITP-A2, Page 576; EA-ITP-ASG2E, Page 1-91; EA-ITP-AB, Page 355; EA-AC65-9A, Page 446*
If high starting torque is needed under heavy load conditions, the series wound motor is the best one to use.

8807. (A) *EA-ITP-ASG2E, Page 1-88; EA-AC65-9A, Page 399*
When an added load is applied to a generator using a vibrator-type voltage regulator, the field rheostat is shorted for a longer period of time, which causes the solenoid armature to vibrate more slowly. The vibrating solenoid armature is what opens and closes the points.

8808. (C) *EA-ITP-ASG2E, Page 1-88; EA-AC65-9A, Page 406*
Excessive pressure on generator brushes will cause rapid wear, but too little pressure will cause "bouncing" of the brushes which results in arcing and burned and pitted surfaces.

8809. (B) *EA-ITP-A2, Page 549; EA-ITP-ASG2E, Page 1-89; EA-AC65-9A, Pages 430-431*
Synchronizing, or paralleling, of AC generators (alternators) is somewhat similar to paralleling DC generators, except that there are more steps with the alternators. In order to synchronize (parallel) two or more alternators to the same bus, they must have the same phase sequence as well as equal voltages and frequencies.

8810. (C) *EA-ITP-A2, Page 576; EA-ITP-ASG2E, Page 1-91; EA-ITP-AB, Page 355; EA-AC65-9A, Page 446*
Because of the low resistance in the windings, the series motor is able to draw a large current in starting. This starting current, in passing through the field and armature windings, produces a high starting torque.

8811. (B) *EA-ITP-A2, Page 581; EA-ITP-ASG2E, Page 1-86; EA-ITP-AB, Page 356; Aircraft Electricity and Electronics, Pages 430-431*
When a motor needs to have a precise point at which it will stop, an armature brake is utilized. A landing light retraction motor is an application where a precise stopping point is needed.

8812. (A) *EA-ITP-G2, Pages 189-191; EA-ITP-ASG2E, Page 1-87; EA-AC65-9A, Page 397; Aircraft Electricity and Electronics, Page 107*
An interpole is a pole placed between the main poles of a generator. Interpoles may be used to counteract the effects of armature reaction.

8813. (A) *EA-ITP-A2, Page 529; EA-ITP-ASG2E, Page 1-88; EA-ITP-AB, Page 312; EA-AC65-9A, Page 398*
Efficient operation of electrical equipment in an aircraft depends on a constant voltage supply from the generator. Among the factors which determine the voltage output of a generator, only one, the strength of the field current, can be conveniently controlled.

8814. (B) *EA-ITP-G2, Page 184; EA-ITP-ASG2E, Page 1-86; EA-AC65-9A, Page 390*
In small generators, the frame is made of one piece of iron, while in large generators it is usually made up of two parts bolted together. The frame has high magnetic properties and, together with the pole pieces, forms the major part of the magnetic circuit. The frame and the pole pieces are part of the field assembly.

8815. (B) *EA-ITP-G2, Page 206; EA-ITP-ASG2E, Page 1-89; EA-AC65-9A, Page 417*
The frequency of the alternator voltage depends upon the speed of rotation of the rotor and the number of poles. When a rotor has rotated through an angle so that two adjacent rotor poles (a north and a south pole) have passed one winding, the voltage induced in that winding will have varied through one complete cycle. Two poles produce one cycle, so six poles will produce three cycles.

8816. (C) *EA-ITP-A2, Page 531; EA-ITP-ASG2E, Page 1-89; EA-ITP-AB, Pages 314-315; EA-AC65-9A, Pages 400-401*
The reverse-current cutout is the relay in the voltage regulator which disconnects the battery from the generator. If the battery is not disconnected, it will discharge through the generator armature when the generator voltage falls below that of the battery, thus driving the generator as a motor. When this happens, the current flows opposite the normal direction through the generator armature and in the normal direction through the shunt field.

8817. (C) *EA-ITP-A2, Page 581; EA-ITP-ASG2E, Page 1-86; EA-ITP-AB, Page 356; Aircraft Electricity and Electronics, Page 164*
An armature magnetic brake consists of a drum mounted on the armature shaft and internal brake shoes, controlled by a magnetizing coil. The coil is placed inside the brake shoes, and when the motor current is turned off, the coil is de-energized and the brake shoes are forced against the drum by spring pressure. Conversely, when the power is turned on, the coil pulls the brake shoes away from the drum.

8818. (A) *EA-ITP-ASG2E, Page 1-88; EA-AC65-9A, Page 406*
The purpose of the pigtail on the generator's brushes is to conduct current, rather than subjecting the brush spring to currents which would alter its spring action by overheating. The pigtails also eliminate any possible sparking to the brush guides caused by the movement of the brushes within the holder, thus minimizing side wear of the brush.

8819. (C) *EA-ITP-A2, Page 576; EA-ITP-ASG2E, Page 1-91; EA-ITP-AB, Page 355; EA-AC65-9A, Page 446*
The load on a series wound motor is highest when the motor is at low RPM. The high load at low RPM causes a high current flow.

8820. (A) *EA-ITP-A2, Page 566; EA-ITP-ASG2E, Page 1-82*
Ring-type terminals are used because they are less likely to fail, causing a circuit interruption.

8821. (C) *EA-ITP-A2, Page 565; EA-ITP-ASG2E, Page 1-83; EA-AC65-15A, Page 447*
Aluminum wire must be stripped very carefully, using extreme care, because individual strands will break very easily after being nicked.

8822. (B) *EA-ITP-A2, Page 529; EA-ITP-ASG2E, Page 1-86*
As the generator rotates, an alternating current is generated within the coils and is converted into DC as it is picked up by the brushes riding on the copper segments of the commutator.

8823. (C) *EA-ITP-ASG2E, Page 1-90; EA-AC65-9A, Page 400 (Figure 9-26)*
The function of an ammeter is to indicate current flow. In a battery charging circuit, the ammeter shows the amount of current being used to charge the battery.

8824. (A) *EA-ITP-G2, Pages 189-190; EA-ITP-ASG2E, Page 1-87; EA-AC65-9A, Page 397*
Interpoles in a generator are used to help reduce the effect of armature reaction. They do this by counteracting the field distortion caused by the electromagnetic fields in the windings. Interpoles also improve the efficiency, output, and service life of the brushes. By counteracting field distortion, interpoles effectively increase the field strength.

8825. (A) *EA-ITP-G2, Page 195; EA-ITP-ASG2E, Page 1-87; Aircraft Electricity and Electronics, Page 145*
When an armature is placed in a growler and the unit is turned on, a potential for current flow exists between segments of the commutator and the armature. A tests lamp is connected between a segment of the commutator and the armature and will light if no "open" exists.

8826. (B) *EA-ITP-A2, Page 582; EA-ITP-ASG2E, Page 1-86; EA-AC65-15A, Page 460*
Navigation lights for civil airplanes are located on the forward tip of the left and right wing tips, and on the tail.

8827. (C) *EA-ITP-G2, Page 197; EA-ITP-ASG2E, Page 1-87; EA-AC65-9A, Page 406*
After a commutator under repair has been turned on a lathe, it is necessary to undercut the mica insulation between the segments to a depth of approximately the width of the mica.

8828. (C) *EA-ITP-A2, Page 529; EA-ITP-ASG2E, Page 1-88; EA-ITP-AB, Page 312; EA-AC65-9A, Page 398*
Among the factors which determine the voltage output of a generator, only one, the strength of the field current, can be conveniently controlled.

8829. (B) *EA-ITP-G2, Page 186; EA-ITP-ASG2E, Page 1-87; EA-AC65-9A, Page 394*
Series wound generators have very poor voltage regulation qualities, so they are never used as an aircraft generator. In this type of generator, as the load goes up, the voltage goes up. A series wound generator is what would be used with an arc welder.

8830. (A) *EA-ITP-G2, Page 173; EA-ITP-ASG2E, Page 1-86; EA-ITP-GB, Page 213*
For precise frequency measurement, integrated circuit chips having clock circuits are used to actually count the cycles in a given time period and display the frequency as a digital display.

8831. (B) *EA-ITP-ASG2E, Page 1-?; Aircraft Electricity and Electronics, Page 136*
Cooling air for generators may be supplied, on the ground, by a fan mounted on the generator shaft or by bleed air from the turbine engine compressor.

8832. (B) *EA-ITP-G2, Page 151; EA-ITP-ASG2E, Page 1-89; EA-ITP-GB, Page 181; EA-AC65-9A, Page 374*
A rectifier is a device which transforms alternating current into direct current by limiting or regulating the direction of current flow.

8833. (A) *EA-ITP-G2, Page 165; EA-ITP-ASG2E, Page 1-86; EA-ITP-GB, Page 210; EA-AC65-9A, Page 331*
The megger, or megohmmeter, is a high-range ohmmeter containing a hand-operated generator. It is used to measure insulation resistance and other high resistance values.

8834. (C) *EA-ITP-G2, Page 176; EA-ITP-ASG2E, Page 1-86; EA-ITP-GB, Page 217*
A semiconductor diode is an electrical check valve and, as such, it allows electron flow in one direction but not in the other. To check for a shorted or an open diode, disconnect it from the circuit and measure its resistance with an ohmmeter set on the R × 1 scale.

8835. (C) *EA-ITP-A2, Page 585; EA-ITP-ASG2E, Page 1-86*
Capacitors store electricity in an electrostatic field. Large capacitors are capable of storing a charge large enough to injure or kill a person. They should be fully discharged prior to maintenance operations on these systems.

8836. (B) *EA-ITP-G2, Page 100; EA-ITP-ASG2E, Page 1-55*
Excessive rates of current flow in a nickel-cadmium battery can cause high temperatures. These temperatures can cause a breakdown of the cellophane-like material that separates the plates in the battery cell. This breakdown creates a short circuit, and current flow increases. The increased current flow will generate additional heat, and cause further breakdown of the separator material. This process is known as vicious-cycling or thermal runaway.

8837. (C) *EA-ITP-G2, Page 176; EA-ITP-ASG2E, Page 1-90; EA-ITP-GB, Page 218*
If is often difficult to find a shorted transformer winding by measuring its resistance. It is easier to determine whether or not any coils are shorted by feeling the transformer after it has been operating for a while. If any of the windings are shorted, the transformer will be hot.

8838. (A) *EA-ITP-G2, Page 210; EA-ITP-ASG2E, Page 1-91; EA-AC65-9A, Page 444*
The major parts in a practical motor are the armature assembly, field assembly, brush assembly and end frame.

8839. (C) *EA-ITP-G2, Page 211; EA-ITP-ASG2E, Page 1-91; EA-AC65-9A, Page 445*
The three basic types of DC motors are series motors, shunt motors, and compound motors. The basic series motor has field windings consisting of a relatively few turns of heavy wire, connected in series with the armature winding.

8840. (A) *EA-ITP-A2, Page 566; EA-ITP-ASG2E, Page 1-84*
A properly installed crimp-on terminal will be at least as strong as the wire itself.

8841. (A) *EA-ITP-A2, Pages 557-561; EA-ITP-ASG2E, Page 1-83; EA-AC65-15A, Pages 433 and 435*
Several factors must be considered in selecting the size of wire for transmitting and distributing electrical power. They are the allowable power loss, permissible voltage drop in the line, the current-carrying ability of the conductor, and whether or not the load is continuous or intermittent.

8842. (A) *EA-ITP-A2, Page 563; EA-ITP-ASG2E, Page 1-83; EA-AC65-15A, Page 454*
Hardware used to make bonding or grounding connections should be selected on the basis of mechanical strength, the current to be carried, and ease of installation.

8843. (A) *EA-ITP-A2, Page 567; EA-ITP-ASG2E, Page 1-83; EA-43.13-1A/3, Page 444*
Stagger splices so the bundle doesn't become too large.

8844. (A) *EA-ITP-ASG2E, Page 1-83; EA-43.13-1A/3, Page 444*
Avoid a bending radius of less than 10 times the outer diameter of the bundle.

8845. (A) *EA-ITP-A2, Page 567, EA-AC43.13-1A/3, Page 240; EA-ITP-ASG2E, Page 1-83*
Splices should be kept to a minimum and avoided entirely in areas subject to extreme vibrations.

8846. (B) *EA-ITP-A2, Page 568; EA-ITP-ASG2E, Page 1-84*
AN (Air Force-Navy) and MS (Military Standards) are both hardware identification systems originating in the military.

8847. (A) *EA-ITP-A2, Page 568; EA-ITP-ASG2E, Page 1-84*
Earlier versions of AN/MS connectors used solder to connect each wire to its pin or socket. Newer connectors used crimp-type terminations.

8848. (B) *EA-ITP-A2, Page 568; EA-ITP-ASG2E, Page 1-84*
To eliminate possible short circuits to ground, the end of the connector that carries the power uses socket connectors.

8849. (B) *EA-ITP-A2, Page 533; EA-ITP-ASG2E, Page 1-89; EA-AC65-9A, Page 417*
The output voltage of an alternator is best controlled by regulating the voltage output of the DC exciter, which supplies current to the alternator rotor field.

8850. (B) *EA-ITP-ASG2E, Page 1-84; EA-AC65-15A, Pages 442-443*
Dusting of electrical cable with powdered soapstone prior to routing it through rigid conduit is advisable because the cable will have less tendency to bind or hang up. This will also reduce chafing or abrasion.

8851. (C) *EA-ITP-ASG2E, Page 1-84; EA-AC65-15A, Page 452*
Grounding is the electrical connecting of a conducting object to the primary structure for a return path for current. One of the reasons for grounding is to prevent the development of radio frequency potentials.

8852. (B) *EA-ITP-ASG2E, Page 1-84; EA-43.13-1A/3, Page 188; EA-AC65-15A, Page 454*
Aluminum alloy jumpers are recommended for most bonding connections. Copper jumpers, however, are used to bond together parts made of stainless steel, cadmium-plated steel, copper, brass, or bronze.

8853. (C) *EA-ITP-A2, Page 574; EA-ITP-ASG2E, Page 1-85; EA-ITP-GB, Page 112; EA-AC65-9A, Page 317*
Fuses are designed to break the connection to a circuit if the current flow is too great. For this reason, fuse capacity is rated in amps.

8854. (B) *EA-ITP-G2, Page 81; EA-ITP-ASG2E, Page 1-86; EA-ITP-GB, Page 115; EA-AC65-9A, Page 296*
A rheostat is a variable resistor used to vary the amount of current flowing in a circuit. In order to control brightness of a light, it would need to be connected in series with the light.

8855. (A) *EA-ITP-A2, Page 571; EA-ITP-ASG2E, Page 1-85; EA-43.13-1A/3, Page 178; EA-AC65-15A, Page 459*
One way to prevent a switch from being accidentally turned on or off is to install a guard over it. This forces the operator to move the guard before being able to turn the switch on or off, which would ensure that it is not done accidentally.

8856. (C) *EA-ITP-A2, Page 582; EA-ITP-ASG2E, Page 1-86; EA-AC65-15A, Page 461*
Lights in a circuit are connected parallel to each other, so the voltage available to each light will be the same. If one switch is going to control all the lights in the circuit, the switch will need to be in series with the parallel circuit of the lights.

8857. (B) *EA-ITP-A2, Page 568; EA-ITP-ASG2E, Page 1-84*
Junction boxes should be fabricated of such material, and installed in such a manner as to prevent oil-canning.

8858. (A) *EA-ITP-A2, Pages 561-562; EA-ITP-ASG2E, Page 1-83; EA-43.13-1A/3, Page 186*
When wire is being run in bundles, to simplify maintenance and to minimize the damage that may result from a single fault, the number of wires in the run should be limited.

8859. (B) *EA-ITP-ASG2E, Page 1-84; EA-AC65-15A, Page 451*
Terminal lugs should be installed on terminal blocks so that they are locked against movement in the direction of loosening.

8860. (B) *EA-ITP-A2, Page 562; EA-ITP-ASG2E, Page 1-84; EA-ITP-AB, Page 345; EA-AC65-15A, Page 457*
Conduit is used in aircraft installations for the mechanical protection of wires and cables.

8861. (C) *EA-ITP-ASG2E, Page 1-84; EA-43.13-1A/3, Page 201; EA-AC65-15A, Page 457*
When installing conduit, be sure not to locate it where it can be used as a handhold or footstep.

8862. (C) *EA-AC65-9A, Pages 328-329; EA-ITP-ASG2E, Page 1-86*
The connection of the voltmeter described in this question is backwards, so the voltmeter will read backwards. This means the needle will peg out below the "zero" value.

8863. (A) *EA-ITP-GB, Pages 107-108; EA-AC65-9A, Pages 288-289; EA-ITP-ASG2E, Page 1-83*
One method of determining the resistance in a series circuit is to determine the voltage drop at each of the resistors. Because current is constant in the circuit, if the voltage drop at each resistor is known, the resistance can be calculated by using Ohm's Law.

8864. (C) *EA-ITP-A2, Page 571; EA-ITP-ASG2E, Page 1-85; EA-43.13-1A/3, Page 177; EA-AC65-15A, Page 459*
The nominal current rating of the conventional aircraft switch is usually stamped on the switch housing and represents the continuous current rating with the contacts closed.

8865. (C) *EA-ITP-ASG2E, Page 1-84; EA-43.13-1A/3, Page 174*
When fire-proofing is necessary, a junction box made out of stainless steel is recommended.

8866. (A) *EA-ITP-ASG2E, Page 1-84; EA-43.13-1A/3, Page 201*
Flexible brass conduit conforming to Specification MIL-C-7931 is available and normally used instead of flexible aluminum where necessary to minimize radio interference.

8867. (A) *EA-ITP-A2, Pages 557-561; EA-ITP-ASG2E, Page 1-83; EA-ITP-AB, Pages 336-337; EA-AC65-15A, Page 435*
When selecting the size of electric cable to use for a job, the things to consider are the allowable power loss in the line, the permissible voltage drop in the line, and the current carrying ability of the conductor.

8868. (C) *EA-ITP-GB, Page 111; EA-AC65-9A, Page 318; EA-ITP-ASG2E, Page 1-86*
Electric switches which control a single circuit, and have only an ON and an OFF position, are referred to as single-pole, single-throw, two-position switches. Single-pole means that the switch has only one pole or circuit energized when it is moved to the ON position. Single-throw means that from any position, the switch has only one throw or change of position available to it. Two-position means that the switch has two stopping points — it is either on or off.

8869. (C) *EA-ITP-A2, Page 574; EA-ITP-ASG2E, Page 1-85; EA-ITP-GB, Page 112; EA-AC65-9A, Page 317*
Fuses are current sensitive devices. If the current in a circuit becomes too great, the fuse protecting the circuit will be subjected to the excessive current and it will blow. With excessive current comes excessive heat. When a fuse flows, what it actually does is melt because of the heat.

8870. (C) *EA-ITP-A2, Page 703; EA-ITP-ASG2E, Page 1-84; EA-AC65-9A, Page 442*
All wiring should be installed so that it is mechanically and electrically sound and neat in appearance. Most electrical wiring should be routed parallel with, or at right angles to, aircraft structural members like stringers or ribs. An exception to this method of routing is coaxial cable, which is routed as directly as possible.

8871. (B) *EA-ITP-A2, Page 558; EA-ITP-ASG2E, Page 1-83; EA-ITP-AB, Page 337; EA-43.13-1A/3, Page 180*
Aluminum wire can be substituted for copper wire if the size of the wire is increased. Two sizes larger should be used when the substitution is made, but aluminum wire smaller than a number 6 should not be used. It must be remembered that, as the number gets smaller, the wire size gets larger.

8872. (C) *EA-ITP-ASG2E, Page 1-90; EA-43.13-1A/3, Page 175*
In installations where the ammeter is in the generator or alternator lead, and the regulator system does not limit the maximum current that the generator or alternator can deliver, the ammeter can be redlined at 100 percent of the generator or alternator rating. If the ammeter reading is never allowed to exceed the red line, except for short intermittent loads, the generator or alternator will not be overloaded.

8873. (B) *EA-ITP-ASG2E, Page 1-83; EA-43.13-1A/3, Page 188*
Terminals are attached to the ends of electric wires to facilitate connection of the wires to terminal strips or items of equipment. The tensile strength of the wire to terminal joint should be at least equivalent to the tensile strength of the wire itself, and its resistance negligible relative to the normal resistance of the wire.

8874. (A) *EA-ITP-A2, Page 563; EA-ITP-ASG2E, Page 1-85; EA-43.13-1A/3, Page 188*
When a bonding jumper is being made, it should be as short as practicable. It should be installed in such a manner that the resistance of each connection does not exceed .003 ohm.

8875. (A) *EA-ITP-A2, Page 572; EA-AC65-9A, Page 318; EA-ITP-ASG2E, Page 1-85*
A switch installed in a single wire circuit that must be held in the ON position would be a single-pole, single-throw, two-position normally open switch. A switch which stays off, except when held in the ON position, is a normally open switch.

8876. (A) *EA-ITP-A2, Page 574; EA-ITP-ASG2E, Page 1-85; EA-ITP-AB, Pages 349-350; EA-AC65-15A, Page 458*
Conductors should be protected with circuit breakers or fuses located as close to the electrical power source bus as possible. Normally, the manufacturer of the electrical equipment specifies the fuse or circuit breaker to be used when installing equipment.

8877. (B) *EA-ITP-G2, Page 175; EA-ITP-ASG2E, Page 1-86; EA-AC65-9A, Page 328*
Voltage measuring instruments are always connected across (in parallel with) the circuit. Care should be taken to ensure that the positive terminal of the voltmeter is connected to the positive terminal of the source. If the meter is hooked up backwards, the needle will be driven below zero.

8878. (B) *EA-ITP-A2, Page 558; EA-ITP-ASG2E, Page 1-83; EA-ITP-AB, Page 337; EA-43.13-1A/3, Page 180*
The explanation for this question is the same as that for question #7891.

8879. (B) *EA-ITP-ASG2E, Page 1-84; EA-AC65-15A, Page 455*
Electrical connectors are particularly vulnerable to corrosion due to the condensation which can collect in the shell. Special connectors with waterproof features have been developed which can be used to replace non-waterproof plugs in areas where moisture causes a problem.

8880. (B) *EA-ITP-ASG2E, Page 1-85; EA-AC65-9A, Page 317*
Fuses are installed in two types of fuse holders in aircraft. "Plug-in holders" are used for small type and low capacity fuses. "Clip" holders are used for high capacity fuses and current limiters.

8881. (C) *EA-ITP-A2, Page 562; EA-ITP-ASG2E, Page 1-84; EA-43.13-1A/3, Page 204; EA-AC65-15A, Page 443*
It is essential when installing wiring bundles in aircraft that they be routed so that contact with moving parts is avoided. If it is not possible to avoid this contact, such as wiring attached to a retractable landing gear, the wire should be installed in conduit (probably of the flexible type).

8882. (C) *EA-ITP-A2, Page 556; EA-ITP-ASG2E, Page 1-82; EA-ITP-AB, Page 335; EA-AC65-15A, Page 433*
In the American Wire Gage system, electrical wire size numbers relate to the cross-sectional area of the wire. The way the system works, however, the larger the wire size, the smaller the number size. For example, a number 8 wire is larger than a number 12 wire.

8883. (C) *EA-ITP-A2, Page 558; EA-ITP-ASG2E, Page 1-83; EA-43.13-1A/3, Page 179*
The values given in this question are plugged into the voltage drop formula as follows:
Step 1: Resistance per foot (.00644) × 50 feet of wire = .322.
Step 2: .322 × the current flow of 12.5 = 4.025 voltage drop.

8884. (C) *EA-ITP-A2, Page 571; EA-ITP-ASG2E, Page 1-85; EA-ITP-AB, Page 349; EA-AC65-15A, Page 459*
Although a switch might be rated for a continuous current load of 15 amps, if the switch is derated and used in a circuit with a current load of only 10 amps, the switch efficiency and service life will be increased.

8885. (C) *EA-ITP-A2, Page 574; EA-ITP-ASG2E, Page 1-85; EA-ITP-GB, Page 113; EA-AC65-9A, Page 317*
When a fuse being used to protect an electrical circuit blows, the fuse must be replaced. A big advantage of using circuit breakers instead of fuses is the breaker's ability to be reset and reused. The circuit breaker is, however, considerably more expensive than the fuse.

8886. (C) *EA-ITP-A2, Page 574; EA-ITP-ASG2E, Page 1-85; EA-ITP-AB, Page 350; EA-AC65-9A, Page 317*
Most fuses have a strip of metal (which is designed to melt at a set temperature) made of a tin alloy and bismuth. Some fuses have a metal strip which is made of copper, and they are referred to as current limiters. Current limiters will stand a considerable overload for a short period of time before they will blow.

8887. (B) *EA-ITP-A2, Page 564; EA-ITP-ASG2E, Page 1-84; EA-ITP-AB, Page 344; EA-AC65-15A, Page 442*
Wires and wire groups should be protected against chafing or abrasion in those locations where contact with sharp surfaces or other wires would damage the insulation. Damage to the insulation can cause short circuits, malfunction, or inadvertent operation of equipment. When wires are being passed through holes, it may be necessary to install a rubber grommet to protect the wire from any sharp edges.

8888. (A) *EA-ITP-A2, Page 574; EA-ITP-ASG2E, Page 1-85; EA-ITP-GB, Page 113; EA-43.13-1A/3, Page 176*
In an aircraft electrical system, automatic reset circuit breakers should not be used. If a fault exists in a circuit which causes a circuit breaker to trip, the fault needs to be identified before the circuit is powered again. To have a circuit breaker which would continue to reset and then trip could cause serious problems.

8889. (A) *EA-ITP-A2, Page 572; EA-ITP-GB, Page 111; EA-AC65-9A, Page 318; EA-ITP-ASG2E, Page 1-85*
When a switch is described as being a double-throw switch, what is being identified is the number of circuits which can be completed through the switch.

8890. **(B)** *EA-ITP-A2, Page 557; EA-ITP-ASG2E, Page 1-83; EA-ITP-AB, Pages 336-337; EA-AC65-15A, Page 435*
This question is the same as #6887. The size of electric cable to use is determined by the amount of current it will be asked to carry and the amount of voltage drop which will be allowable.

8891. **(C)** *EA-ITP-A2, Page 574; EA-ITP-ASG2E, Page 1-85; EA-AC65-9A, Page 317*
Each aircraft electrical circuit has a requirement in terms of the current flow it will draw. Based on the current flow in the circuit, the device which is installed to protect the circuit (fuse or circuit breaker) will need to be of sufficient capacity.

8892. **(B)** *EA-ITP-A2, Page 574; EA-ITP-ASG2E, Page 1-85; EA-AC65-9A, Page 315*
Circuit breakers are designed to protect an electrical system from too much current flow. In an instrument lighting system, it is the wiring which is protected from excessive current.

8893. **(B)** *EA-ITP-A2, Page 522; EA-ITP-ASG2E, Page 1-89; EA-ITP-GB, Page 137; EA-AC65-9A, Page 338*
Alternating current has largely replaced direct current in commercial power systems for a number of reasons. It can be transmitted over long distances more readily and more economically than direct current, and it can be increased or decreased quite easily by the use of transformers.

8894. **(C)** *EA-ITP-ASG2E, Page 1-86; Aircraft Electricity and Electronics, Pages 31-32*
The primary windings of an induction coil consist of relatively few turns of large wire and will carry sufficient current to set up a strong magnetic field through the soft iron core. The core consists of thin iron laminations or insulated iron wires bound together in the shape and size desired. This type of construction reduces the effects of the eddy currents induced in the core.

8895. **(C)** *EA-ITP-A2, Page 547; EA-ITP-ASG2E, Page 1-89; EA-ITP-GB, Page 181; EA-AC65-9A, Page 374*
Aircraft which have AC electrical systems, and utilize an AC producing alternator, still have batteries which put out direct current. The batteries on these aircraft are kept charged by rectifying the AC coming off the alternator, which converts it to direct current. The device that does this is called a rectifier.

8896. **(A)** *EA-ITP-G2, Page 120; EA-ITP-GB, Page 155; EA-AC65-9A, Page 359; EA-ITP-ASG2E, Page 1-90*
Transformers, whether AC or DC, cannot produce power. Their output power, if they are 100% efficient, will be the same as the input power. Since power is the product of the voltage and the current, if a transformer is stepping up the voltage, it is stepping down the current. A transformer secondary with more loops than the primary steps up the voltage, and steps down the current.

8897. **(B)** *EA-ITP-ASG2E, Page 1-89; Aircraft Electricity and Electronics, Page 120*
When residual voltage shows up on a test voltmeter, it indicates that the generator is operating but that there is no field current. This means that the field circuit is open because of a broken or loose wire or that the voltage regulator is defective.

8898. **(A)** *EA-ITP-G2, Page 194; EA-ITP-ASG2E, Page 1-86; EA-AC65-9A, Page 407*
To restore residual magnetism to a generator, the field needs to be flashed by removing the regulator and connecting a jumper from the positive bus to the generator field. This is done while the engine is running at cruise RPM.

8899. **(C)** *EA-ITP-G2, Pages 104-105; EA-ITP-ASG2E, Page 1-89; EA-ITP-GB, Page 137; EA-AC65-9A, Page 338*
Transformers are used to step up or step down the voltage in a system. It is advantageous to transmit AC at a high voltage and then step it down to a lower voltage by using a transformer.

8900. **(A)** *EA-ITP-A2, Pages 582-583; EA-ITP-ASG2E, Page 1-86; EA-AC43.13-2A, Page 26*
According to EA-AC43.13-2A, when an anticollision light is installed, a switch needs to be used which is independent of the position light system switch.

8901. **(C)** *EA-ITP-ASG2E, Page 1-88; EA-AC65-9A, Page 436*
Inductor-type rotary inverters control their output voltage by controlling the current supplied to the DC stator field.

8902. **(C)** *EA-ITP-G2, Page 196; EA-ITP-ASG2E, Page 1-87; Aircraft Electricity and Electronics, Page 122*
To test the field coil of a generator for continuity, the leads of an ohmmeter are connected to the terminals of the coil. The shunt field coil should show low resistance, approximately 2 to 30 ohms. A series field coil should show practically no resistance because it carries the entire load to the generator and the internal resistance of the generator must be as low as possible.

8903. (B) *EA-ITP-G2, Page 69; EA-ITP-ASG2E, Page 1-86; EA-ITP-GB, Pages 102-103; EA-AC65-9A, Page 305*
With a fixed core material, the strength of an electromagnet will increase in proportion to the number of turns of wire around the core and the amount of current passing though the wire. As current passes through a wire, a magnetic field is created around the wire. For this reason, the greater the number of turns of wire or the greater the current flow, the stronger the electromagnet.

8904. (C) *EA-ITP-A2, Page 529; EA-ITP-ASG2E, Page 1-88; EA-ITP-AB, Page 312; EA-AC65-9A, Page 398*
If the current supplied to the field circuit is increased, the voltage output of a generator is increased. By varying the resistance in the field circuit, the voltage regulator controls the current flow and consequently the voltage output.

8905. (A) *EA-ITP-A2, Page 538; EA-ITP-ASG2E, Page 1-88; EA-AC65-9A, Page 403*
As the name implies, the overvoltage control protects the system when excessive voltage exists. The overvoltage relay is closed when the generator output reaches 32 volts and completes a circuit to the trip coil of the field control relay. The closing of the field control relay trip circuit opens the shunt field circuit and completes it through a resistor, causing the generator voltage to drop.

8906. (C) *EA-ITP-A2, Pages 544-545; EA-ITP-ASG2E, Page 1-89; EA-ITP-AB, Pages 326-327; EA-AC65-9A, Page 403*
The generator equalizing system depends on the voltage drop in some calibrated resistors to determine what it will do. If all the generators are supplying the same current, the voltage drop in all the ground leads will be the same. If the current supplied by the generators is unequal, there will be a greater voltage drop in the ground lead of the generator which is supplying more current. The equalizing system senses this and acts to aid the voltage coil in the regulator of the weak generator and oppose the voltage coil in the regulator of the strong generator. This causes the output of the weak generator to increase and the output of the strong generator to decrease.

8907. (B) *EA-ITP-ASG2E, Page 1-86; Aircraft Electricity and Electronics, Page 173*
Intermittent duty circuits are those which are operated for only a few seconds or minutes and then are turned off. Examples include the landing gear, flaps, trim motors, landing lights, and emergency hydraulic pumps.

8908. (A) *EA-ITP-A2, Page 529; EA-ITP-ASG2E, Page 1-87; EA-AC65-9A, Page 395 (Figure 9-17)*
A compound generator is one that has series field coils and shunt field coils. Regardless of the type of generator, voltage output is always regulated by varying the current flowing through the shunt field coils.

8909. (C) *EA-AC65-9A, Page 306; EA-ITP-ASG2E, Illustration*
Storage batteries can be connected either in series to add voltage, or in parallel to increase capacity, but not as combined in drawing 3.

8910. (C) *EA-ITP-A2, Page 540; EA-ITP-AB, Page 322; EA-ITP-ASG2E, Illustration*
The red warning light in this circuit gets its power through wire #8, the up limit switch, wire #19, and circuit breaker #5. The only one of these items that could cause the light to be on when it shouldn't be is the up limit switch.

8911. (A) *EA-ITP-ASG2E, Page 1-89; EA-AC65-9A, Page 409*
See low voltage isolation procedure.

8912. (B) *EA-ITP-A2, Page 579; EA-ITP-ASG2E, Page 1-91; EA-ITP-AB, Page 355; EA-AC65-9A, Page 448*
The direction a motor will turn is determined by the direction of current flow in either the armature or the field windings. By reversing either of these current flows, the direction of rotation will be reversed.

8913. (A) *EA-ITP-A2, Page 547; EA-ITP-ASG2E, Page 1-90; EA-AC65-9A, Page 374*
Because batteries produce direct current and therefore need direct current to be charged, aircraft which operate with AC producing alternators generally use a stepdown transformer and a rectifier to produce a suitable direct current for charging the batteries.

8914. (C) *EA-ITP-A2, Page 582; EA-ITP-ASG2E, Page 1-86; EA-AC43.13-2A, Page 26*
See question #7922.

8915. (A) *EA-ITP-ASG2E, Page 1-90; EA-43.13-1A/3, Page 174*
When adjustments are being made to regulators, contacts, and inverters, the adjustment procedures outlined by the equipment manufacturer should be followed.

8916. (B) *EA-ITP-ASG2E, Page 1-89; EA-AC65-9A, Page 435*
An inverter is a device which converts a portion of the aircraft's direct current to alternating current. This AC is used mainly for instruments, radio, radar, and lighting.

8917. (A) *EA-ITP-G2, Page 78; EA-ITP-GB, Page 112; EA-AC65-9A, Page 319; EA-ITP-ASG2E, Page 1-89*
A relay switch consists of a coil, or solenoid, an iron core, and both fixed and movable contacts. When the solenoid is energized (powered) , it creates a magnetic force which pulls the movable contact down until it touches the fixed contact, which completes the circuit.

8918. (C) *EA-ITP-G2, Page 138; EA-ITP-ASG2E, Page 1-89; EA-ITP-GB, Page 181; EA-AC65-9A, Page 374*
The purpose of a rectifier is to change alternating current to direct current.

8919. (A) *EA-ITP-G2, Page 120; EA-ITP-GB, Pages 155-156; EA-AC65-9A, Pages 358-359; EA-ITP-ASG2E, Page 1-89*
In a transformer, the ratio between the number of turns in the primary and the voltage in the primary is directly proportional to the ratio between the number of turns in the secondary and voltage in the secondary. If the secondary has three times as much voltage as the primary it will have three times as many turns.

8920. (C) *EA-ITP-A2, Page 525; EA-ITP-GB, Pages 142-143; EA-AC65-9A, Pages 357-358; EA-ITP-ASG2E, Page 1-90*
In an alternating current circuit, the true or real power is seldom equal to the apparent power, because the voltage and the current are not always in phase with each other. If they are in phase, however, the true power will equal the apparent power.

8921. (B) *EA-ITP-G2, Page 191; EA-ITP-ASG2E, Page 1-87; EA-AC65-9A, Page 398*
A generator is rated in power output. Since a generator is designed to operate at a specified voltage, the rating is usually given as the number of amperes the generator can safely supply at its rated voltage.

8922. (B) *EA-ITP-G2, Page 187; EA-ITP-ASG2E, Page 1-87; EA-AC65-9A, Pages 394-395*
In a shunt-wound DC generator the fields are in parallel with (shunted across) the armature.

8923. (C) *EA-ITP-ASG2E, Page 1-86; EA-AC65-9A, Pages 390-391*
The field poles of a generator are bolted to the inside of the frame and form a core on which the field coil windings are mounted. The poles are usually laminated to reduce eddy current losses and serve the same purpose as the iron core of an electromagnet; that is, they concentrate the lines of force produced by the field coils.

8924. (B) *EA-ITP-A2, Page 534; EA-ITP-ASG2E, Page 1-89; EA-AC65-9A, Page 417*
The frequency of the alternator voltage depends upon the speed of rotation of the rotor and the number of poles. The faster the speed, the higher the frequency will be; the lower the speed, the lower the frequency becomes. The more poles on the rotor, the higher the frequency will be for a given speed.

8925. (B) *EA-ITP-G2, Page 191; EA-ITP-ASG2E, Page 1-86; EA-AC65-9A, Page 398*
Generator rating and performance data are stamped on the name plate attached to the generator. When replacing a generator, it is important to choose one of the proper rating.

8926. (B) *EA-ITP-G2, Page 194; EA-ITP-ASG2E, Page 1-86; Aircraft Electricity and Electronics, Page 120*
The two most likely indications of generator system trouble are (1) no voltage and (2) residual voltage. Residual voltage is the result of residual magnetism in the field poles. When residual voltage shows on the voltmeter, it indicates that the generator is operating but that there is no field current.

8927. (C) *EA-ITP-G2, Page 175; EA-ITP-GB, Page 215; EA-AC65-9A, Page 352; EA-ITP-ASG2E, Page 1-87*
The combined effect of resistance, inductive reactance, and capacitive reactance makes up the total opposition to current flow in an AC circuit.

8928. (A) *EA-ITP-A2, Page 549; EA-ITP-ASG2E, Page 1-89*
The Constant Speed Drive (CSD) unit has its own lubrication system, and the flight engineer has a low oil pressure warning light and an oil temperature gauge for the unit.

8929. (C) *EA-ITP-A2, Page 549; EA-ITP-ASG2E, Page 1-89*
A CSD uses a hydraulic constant-speed drive unit to turn the alternator.

8930. (C) *EA-ITP-G2, Page 206; EA-ITP-ASG2E, Page 1-90*
The Integrated Drive Generator (IDG) contains both the constant speed drive unit and a brushless alternator in one unit.

8931. (B) *EA-ITP-A2, Page 552; EA-ITP-ASG2E, Page 1-74*
The BITE system employes an LED display which can be viewed by maintenance personnel during troubleshooting. The technician then refers to a code book to determine the defective component and necessary repair.

8932. (A) *EA-ITP-A2, Page 550; EA-ITP-ASG2E, Page 1-89*
CSD units can be disconnected in flight, in the event of a malfunction, but they can be connected only on the ground.

8933. (C) *EA-ITP-A2, Page 550; EA-ITP-ASG2E, Page 1-89*
CSD units can be disconnected in flight, in the event of a malfunction, but they can be connected only on the ground

8934. (C) *EA-ITP-A2, Page 480; EA-ITP-ASG2E, Page 1-92; EA-ITP-AB, Pages 455; EA-AC65-15A, Page 400*
As it rotates, it develops a voltage and current signal. Signal strength indicates wheel rotation speed.

8935. (B) *EA-ITP-A2, Page 481; EA-ITP-ASG2E, Page 1-92; EA-ITP-AB, Page 456; EA-AC65-15A, Page 400*
In an antiskid brake system, the skid control valves are located downstream of the brake control valves. When the skid control generator senses an impending wheel skid, it sends a signal to the skid control box which in turn sends a signal to the skid control valve. The skid control valve takes the brake system pressure supplied to it and bypasses some of it back to the system return line. The amount of pressure the skid control valve allows to go to the brakes should not allow a wheel skid to occur.

8936. (B) *EA-ITP-A2, Page 479; EA-ITP-ASG2E, Page 1-92; EA-ITP-AB, Pages 455-456; EA-AC65-15A, Page 400*
An antiskid system makes use of electrical components, such as the skid control generator and the skid control box, as well as hydraulic components, such as the skid control valves. This combination of electrical and hydraulic components makes the antiskid an electrohydraulic system.

8937. (B) *EA-ITP-A2, Page 484; EA-ITP-ASG2E, Page 1-92; EA-ITP-AB, Page 454; EA-AC65-15A, Page 400*
The pilot can turn off the operation of the antiskid system by a switch in the cockpit.

8938. (B) *EA-ITP-ASG2E, Page 1-93*
Take-off Warning Systems will monitor the take-off position of critical control surfaces and systems.

8939. (C) *EA-ITP-ASG2E, Page 1-93; EA-AC65-15A, Page 517*
Aircraft with retractable landing gear use an aural warning system to alert the crew to an unsafe condition. A bell will sound if the throttle is retracted and the landing gear is not in a down and locked condition.

8940. (C) *EA-ITP-A2, Page 594; EA-ITP-ASG2E, Page 1-68; EA-ITP-AB, Page 527; EA-AC65-15A, Page 479*
Airspeed indicators are sensitive instruments which measure the difference between the pitot and static pressures.

8941. (B) *EA-ITP-A2, Page 629; EA-ITP-ASG2E, Page 1-70; EA-AC65-15A, Page 488*
The operation of the angle-of-attack indicating system is based on detection of differential pressure at a point where the airstream is flowing in a direction that is not parallel to the true angle of attack of the aircraft.

8942. (C) *EA-ITP-A2, Page 480; EA-ITP-ASG2E, Pages 1-92; EA-ITP-AB, Pages 454-455*
At speeds below approximately 20 MPH the antiskid system automatically gives the pilot full control of the brakes for maneuvering and parking. The antiskid system consists basically of three components: the wheel speed sensors, the control boy, and the control valves.

8943. (A) *EA-ITP-A2, Page 479; EA-ITP-ASG2E, Page 1-93; EA-ITP-AB, Page 455*
In an antiskid system the wheel skid is detected by an electrical sensor of either an AC or DC type.

8944. (B) *EA-ITP-A2, Pages 482-483; EA-ITP-ASG2E, Page 1-92; EA-AC65-15A, Page 399*
The skid control system performs four functions: normal skid control, locked wheel skid control, touchdown protection and fail safe protection.

8945. (A) *EA-ITP-A2, Page 479; EA-ITP-ASG2E, Page 1-92; EA-AC65-15A, Page 399*
If a wheel goes into a skid, its braking value is greatly reduced.

8946. (A) *EA-ITP-A2, Page 479; EA-ITP-ASG2E, Page 1-92; EA-ITP-AB, Page 453; EA-AC65-15A, Page 399*
Normal skid control comes into play when wheel rotation slows down but has not come to a stop.

8947. (C) *EA-ITP-A2, Pages 483-484; EA-ITP-ASG2E, Page 1-92; EA-AC65-15A, Page 400*
The antiskid system is designed to apply enough force to operate the wheel's speed just below the skid point because this gives the most effective braking. A warning lamp lights when the system is turned off or if there is a system failure.

8948. (C) *EA-ITP-ASG2E, Page 1-93*

8949. (C) *EA-ITP-A2, Page 629; EA-ITP-ASG2E, Page 1-93*
As the angle of attack increases, the low pressure air traveling over the wind moves into the area where the reed inlet is located, causing it to sound.

8950. (A) *EA-ITP-A2, Page 628; EA-ITP-ASG2E, Page 1-93*
The stall warning system tells the pilot that the aircraft is reaching an angle of attack that is approaching a stall.

8951. (A) *EA-ITP-A2, Pages 540-541; EA-ITP-AB, Pages 322-325; EA-ITP-ASG2E, Illustration*
The described condition would cause the warning horn to sound and the red position light to be extinguished. The easiest way to determine the condition of the red position light is to start at the light and trace the circuit back to the power source. The only way the red light can receive power without engaging the press to test function is through wire #8, the up limit switch, wire #19, the 5 amp breaker, and the bus. Since the contacts in the up limit switch change position only when the gear is in the fully retracted position, the red light only receives power when the gear is retracted. Since the landing gear described in this question is in transit, the red position light is extinguished.

8952. (C) *EA-ITP-A2, Pages 540-541; EA-ITP-AB, Pages 322-325; EA-ITP-ASG2E, Illustration*
When the gear switch is placed in the UP position, power will supplied from the 20 amp breaker, through the switch to the relay. In order for power to flow through the relay to the gear motor, the relay must be activated. A break in wire No. 12 would prevent the relay from closing.

8953. (C) *EA-ITP-A2, Pages 540-541; EA-ITP-AB, Pages 322-325; EA-AC65-15A, Page 517; EA-ITP-ASG2E, Page 1-92*
Aircraft with retractable landing gear use an aural warning system to alert the crew to an unsafe condition. A bell will sound if the throttle is retracted and the landing gear is not in a down and locked condition.

8954. (C) *EA-ITP-A2, Pages 540-541; EA-ITP-AB, Pages 322-325; EA-ITP-ASG2E, Illustration*
The amber indicator light is in the "gear full retracted indicator system" part of the circuit. The light is illuminated when all three gear are in the full retracted position. The light obtains its ground through the gear full retract switches.

8955. (C) *EA-ITP-A2, Pages 540-541; EA-ITP-AB, Page 322; EA-AC65-15A, Page 353; EA-ITP-ASG2E, Illustration*
The horn blows when both throttles are retarded and landing gear is in any position other than down and locked.

8956. (A) *EA-ITP-A2, Page 540; EA-ITP-ASG2E, Page 1-92; EA-ITP-AB, Page 322; EA-AC65-15A, Page 351*
Landing gear safety switches are usually located on the main gear shock struts.

8957. (C) *EA-ITP-A2, Page 540; EA-ITP-ASG2E, Page 1-92; EA-ITP-AB, Pages 322-325; EA-AC65-15A, Page 351*
Airplanes that have retractable landing gear usually have a ground safety switch attached to one or more of their gear struts. When the airplane is on the ground and the gear struts are compressed, the ground safety switch breaks the circuit that would allow the gear to retract. When the airplane takes off and the gear struts extend, the switch completes the circuit and the gear is allowed to retract.

8958. (C) *EA-ITP-A2, Page 540; EA-ITP-ASG2E, Page 1-93; EA-ITP-AB, Pages 322-325; EA-43.13-1A/3, Page 145*
Of the choices given in this question, the only repair that would need a retraction test to check out is the gear downlock microswitch. If the microswitch is repaired or replaced, its adjustment will need to be checked. Since this is the microswitch which turns on the green light in the cockpit to identify the gear as being down and locked, it is essential that its adjustment be correct.

8959. (C) *EA-ITP-A2, Page 540; EA-ITP-ASG2E, Page 1-92; EA-ITP-AB, Page 322; EA-AC65-15A, Page 353*
Most gear warning systems provide a red light for an unsafe gear, a green light for the gear down, and no indication for gear up.

8960. (A) *EA-ITP-A2, Page 415; EA-ITP-ASG2E, Page 1-55; EA-ITP-AB, Page 425; EA-AC65-15A, Page 351*
When the landing gear of an airplane is being retracted or lowered, sequence valves are used to control the order of events.

8961. (C) *EA-ITP-A2, Page 540; EA-ITP-ASG2E, Page 1-92; EA-AC65-15A, Page 353*
An airplane which has retractable landing gear is required, by FAR 23, to have an aural and a visual warning system to let the pilot know if the gear is not down and locked.

8962. (B) *EA-ITP-A2, Page 482; EA-ITP-ASG2E, Page 1-92; EA-AC65-15A, Page 400*
When an airplane takes off and the ground safety switch closes, it completes the circuit for the retraction of the gear. Depending on the airplane, it may also deactivate the antiskid system. What this actually does is deactivate the brakes so they cannot be locked up upon landing.

8963. (A) *EA-ITP-A2, Page 604; EA-ITP-ASG2E, Page 1-92; EA-ITP-AB, Page 621*
In an Autosyn® indicating system the rotor is an electromagnet.

8964. (A) *EA-ITP-A2, Page 604; EA-ITP-ASG2E, Page 1-92; EA-ITP-AB, Page 622*
The Autosyn® system uses an electromagnet for its rotor while the Magnesyn® system utilizes a permanent magnet.

8965. (A) *EA-ITP-A2, Page 604; EA-ITP-ASG2E, Page 1-92; EA-AC65-15A, Page 484*
See the explanation for question #7977.

8966. (B) *EA-ITP-A2, Page 572; EA-ITP-ASG2E, Page 1-93; EA-AC65-9A, Page 319*
Microswitches are used primarily as limit switches to provide automatic control of electrical units.

8967. (A) *EA-ITP-A2, Pages 603-604; EA-ITP-ASG2E, Page 1-91; EA-AC65-15A, Page 483*
The DC Selsyn® system can be used to indicate position of retractable landing gear, cowl flaps, oil cooler doors, or wing flaps, as well as fuel quantity with some types of fuel quantity gages.

8968. (C) *EA-ITP-A2, Page 603-604; EA-ITP-ASG2E, Page 1-91; EA-AC65-15A, Page 483*
A syncro system is an electrical system used for transmitting information from one point to another. The DC selsyn system is a widely used electrical method of indicating a remote mechanical condition.

8969. (A) *EA-ITP-A2, Pages 379-380; EA-ITP-ASG2E, Page 1-94; EA-ITP-AB, Page 669; EA-AC65-15A, Page 287*
Before installing pneumatic surface-bonded deicer boots, the installation area must be thoroughly cleaned and all paint removed before the glue is applied.

8970. (C) *EA-ITP-ASG2E, Page 1-95; EA-AC65-15A, Page 300*
Laminated windshields which have integral heaters incorporate autotransformers, heat control relays, heat control toggle switches, and indicating lights. The heating elements are powered by the 115 volt AC bus.

8971. (A) *EA-ITP-A2, Page 373; EA-ITP-ASG2E, Page 1-95; EA-AC65-15A, Page 303*
To check if a pitot/static tube heater is operating properly, turn the heater on and observe the ammeter reading. The needle on the gage should move to indicate current flow.

8972. (C) *EA-ITP-A2, Page 376; EA-ITP-ASG2E, Page 1-94; EA-ITP-AB, Page 665; EA-AC65-15A, Page 291*
In a pneumatic deicer boot system, the distributor valve accepts the air from the compressor and directs it to the boots in the proper sequence.

8973. (A) *EA-ITP-A2, Page 376; EA-ITP-ASG2E, Page 1-93; EA-ITP-AB, Page 666; EA-AC65-15A, Page 288*
On a typical reciprocating engine airplane, the deicer boots are powered by a rotary, four-vane positive-displacement pump. The pressure side of the pump inflates the boots, and the suction side of the pump holds the boots down when they are not inflated.

8974. (C) *EA-ITP-A2, Page 376; EA-ITP-ASG2E, Page 1-94; EA-AC65-15A, Page 291*
The suction side of the pressure pump, regulated by a suction relief valve, holds the deicing boots deflated when they are not in use.

8975. (C) *EA-ITP-A2, Page 381; EA-ITP-ASG2E, Page 1-94; EA-ITP-AB, Page 669; EA-AC65-15A, Page 292*
Deicer boots should be cleaned using mild soap and water solution.

8976. (C) *EA-ITP-ASG2E, Page 1-93; EA-ITP-P, Page 430; EA-AC65-15A, Page 293*
Anti-icing systems which heat the leading edges of wings and intake ducts are generally operated when icing conditions are encountered or when they are expected to occur.

8977. (B) *EA-ITP-ASG2E, Page 1-94; EA-AC65-15A, Page 291*
With the deicer system controls in the proper positions, the suction and pressure gages for the pneumatic deicer system can be checked. The pressure gage will fluctuate as the deicer tubes inflate and deflate. A relatively steady reading should be maintained on the vacuum gage.

8978. (A) *EA-ITP-ASG2E, Page 1-93; EA-AC65-15A, Page 294*
An anti-icing system using surface combustion heaters is automatically controlled by overheat switches, thermal cycling switches, a balance control, and a duct pressure safety switch. The overheat and cycling switches allow the heaters to operate at periodic intervals, and they also stop the heater operation completely if overheating occurs.

8979. (B) *EA-ITP-A2, Pages 377-378; EA-ITP-ASG2E, Page 1-94; EA-ITP-AB, Page 665; EA-AC65-15A, Page 291*
In a pneumatic deicer boot system, the distributor valve sequences the inflating and deflating of the different boots in the system.

8980. (A) *EA-ITP-A2, Page 376; EA-ITP-ASG2E, Page 1-93; EA-AC65-15A, Page 290*
In a pneumatic deicer boot system, an oil separator is provided for each wet-type air pump. Each separator has an air inlet port, an air outlet port, and an oil drain line which is routed back to the engine oil sump. Since the air pump is internally lubricated, it is necessary to provide a means of separating oil from the pressurized air. If this were not done, the oil could damage the deicer boots.

8981. (A) *EA-ITP-ASG2E, Page 1-95; EA-AC65-15A, Page 300*
On most electrically heated windshields, the thermistor-type heat sensors are imbedded in the glass.

8982. (B) *EA-ITP-A2, Page 371; EA-ITP-ASG2E, Page 1-93; EA-AC65-15A, Page 293*
Wing thermal anti-icing systems may get their hot air from a turbine engine, (bleed air from the compressor), from a fuel/air combustion heater, or from augmentor tubes which draw heat from a reciprocating engine's exhaust.

8983. (B) *EA-ITP-ASG2E, Page 1-95; EA-AC65-15A, Page 300; Aircraft Maintenance and Repair, Page 445*
Windshield temperature control in an electrically heated windshield is normally maintained by thermistors. A thermistor is a device in which the resistance changes with the temperature. When placed in series in an electrical circuit, the thermistor will cause the current flow in the circuit to change as temperature changes.

8984. (C) *EA-ITP-ASG2E, Page 1-95; EA-AC65-15A, Page 302*
Arcing in a windshield panel usually indicates that there is a breakdown in the conductive coating. Where arcing exists, there is invariably a certain amount of local overheating which, depending upon its severity and location, can cause further damage to the panel.

8985. (C) *EA-ITP-A2, Page 376; EA-ITP-ASG2E, Page 1-94; EA-ITP-AB, Page 665; EA-AC65-15A, Page 291*
The distributor valve in a pneumatic deicer boot system directs the pressure air and the suction to the deicer boot, depending on whether the boot is being inflated or deflated.

8986. (B) *EA-ITP-A2, Page 376; EA-ITP-ASG2E, Page 1-93; EA-ITP-AB, Page 665; EA-AC65-15A, Page 286*
The deicing boots of a pneumatic system are composed of a series of inflatable tubes. During operation, the tubes are inflated with pressurized air, and deflated in an alternating cycle. This inflation and deflation causes the ice to crack and break off.

8987. (A) *EA-ITP-A2, Page 377; EA-ITP-ASG2E, Page 1-93; EA-ITP-AB, Page 665; EA-AC65-15A, Page 286*
Deicer boots are installed in sections along the wing with the different sections operating alternately and symmetrically about the fuselage. This is done so that any disturbance to airflow caused by an inflated tube will be kept to a minimum by inflating only short sections on each wing at a time.

8988. (A) *EA-ITP-P2, Page 172; EA-ITP-ASG2E, Page 1-95; EA-ITP-P, Pages 407-409; EA-AC65-15A, Pages 286 & 302*
Carburetor icing may be eliminated by heating the induction air or by using an alcohol spray. The most common method in use today is heating of the induction air.

8989. (B) *EA-ITP-A2, Page 384; EA-ITP-ASG2E, Page 1-96; EA-ITP-AB, Page 672; EA-AC65-15A, Page 308*
The rain repellent system should not be operated on dry windows because heavy undiluted repellent will restrict window visibility. Should the system be operated inadvertently, do not operate the windshield wipers or rain clearing system as this tends to increase the smearing.

8990. (B) *EA-ITP-A2, Page 384; EA-ITP-ASG2E, Page 1-96; EA-ITP-AB, Page 672; EA-AC65-15A, Page 307*
With the advent of turbine powered aircraft, the pneumatic rain removal system became feasible. This method uses high pressure, high temperature engine compressor bleed air which is blown across the windshields. This air blast forms a barrier that prevents raindrops from striking the windshield surface.

8991. (A) *EA-ITP-A2, Page 382; EA-ITP-ASG2E, Page 1-94; EA-AC65-15A, Page 299*
A mixture of ethylene glycol and isopropyl alcohol is used for aircraft deicing.

8992. (A) *EA-ITP-A2, Page 382; EA-ITP-ASG2E, Page 1-94; EA-AC65-15A, Page 299*
Wet snow should be removed with a brush or a squeegee.

8993. (A) *EA-ITP-A2, Pages 373-375; EA-ITP-ASG2E, Page 1-95; EA-AC65-15A, Pages 300-302*
Windshield wipers, circulated hot air and electric elements can be used as methods of anti-icing an aircraft windshield.

8994. (C) *EA-ITP-P2, Page 172; EA-ITP-ASG2E, Page 1-95; EA-AC65-15A, Page 285*
Ice may be expected to form whenever there is visible moisture in the air and the temperature is near or below freezing. An exception is carburetor icing which can occur during warm weather with no visible moisture present.

8995. (C) *EA-ITP-ASG2E, Page 1-95; EA-AC65-15A, Page 300*
If the compressor cannot be turned by hand because of the pressure of ice or snow, hot air should be blown through the engine until the rotating parts are free.

8996. (B) *EA-ITP-ASG2E, Page 1-95; EA-AC65-15A, Page 300*
The temperature sensing element in an electrically heated windshield is a thermistor.

8997. (B) *EA-ITP-A2, Page 507; EA-ITP-ASG2E, Page 1-99; EA-AC65-15A, Page 430*
Carbon monoxide detectors, which detect concentrations of carbon monoxide gas are rarely used to monitor cargo and baggage compartments. However, they have gained widespread use in conducting tests for the presence of carbon monoxide gas in aircraft cabins and cockpits.

8998. (B) *EA-ITP-A2, Pages 507-508; EA-ITP-ASG2E, Page 1-99; EA-AC65-15A, Page 431*
On a few aircraft, visual smoke detectors provide the only means of smoke detection. When smoke is present, a lamp with the indicator is illuminated automatically by the smoke detector.

8999. (A) *EA-ITP-A2, Page 510; EA-ITP-ASG2E, Page 1-97; EA-ITP-AB, Page 678; EA-AC65-15A, Page 429*
Fire protection for the aircraft interior is usually provided by hand-held extinguishers. Four types of fire extinguishers are available for extinguishing interior fires: (1) water (2) carbon dioxide (3) dry chemical, and (4) halogenated hydrocarbons.

9000. (B) *EA-ITP-A2, Page 507; EA-ITP-ASG2E, Page 1-99; EA-AC65-15A, Pages 430-431*
There are several types of portable carbon monoxide testers in use. One type has a replaceable indicator tube which contains a yellow silica gel, impregnated with a complex silico-molybdate compound and is catalyzed using palladium sulfate. When a sample of air is drawn through the tube, if carbon monoxide is present, the gel will turn green.

9001. (C) *EA-ITP-A2, Pages 507-508; EA-ITP-ASG2E, Page 1-98; EA-AC65-15A, Page 430*
Smoke detectors are classified by their method of detection.

9002. (B) *EA-ITP-A2, Page 507; EA-ITP-ASG2E, Page 1-98; EA-AC65-15A, Page 431*
Photoelectric smoke detectors consist of a photoelectric cell, a beacon lamp, a test lamp, and a light trap. An accumulation of 10% smoke in the air causes the photoelectric cell to conduct electric current. This current feeds an amplifier, which activates a warning light and bell.

9003. (B) *EA-ITP-A2, Page 507; EA-ITP-ASG2E, Page 1-99; EA-AC65-15A, Page 430*
The portable carbon monoxide tester which has the yellow silica gel is returned to service after contamination by replacing the tube which contains the gel.

9004. **(C)** *EA-ITP-A2, Page 503; EA-ITP-ASG2E, Page 1-97; EA-ITP-AB, Page 675; EA-AC65-15A, Page 408*
The thermocouple fire warning system operates on an entirely different principle than the thermal switch system. A thermocouple depends upon the rate of temperature rise and will not give a warning when an engine slowly overheats or a short circuit develops.

9005. **(A)** *EA-ITP-A2, Page 510; EA-ITP-ASG2E, Page 1-97; EA-AC65-15A, Page 429*
A carbon dioxide hand held fire extinguisher can be used on an electrical fire, provided the discharge horn is made from a nonmetallic material. A metallic horn would tend to transfer an electrical charge back to the fire extinguisher and to ground through the person holding the extinguisher.

9006. **(C)** *EA-ITP-A2, Page 510; EA-ITP-ASG2E, Page 1-97; EA-ITP-GB, Pages 450-452; EA-AC65-9A, Pages 500-503*
Aircraft brake systems may utilize components made of magnesium, a highly flammable metal. Burning magnesium is considered a Class D fire. While the cause may be a Class A, B, or C fire, flammable metals must be extinguished with a special Dry Powder material. This material is not to be confused with the common Dry Chemical extinguisher.

9007. **(B)** *EA-ITP-A2, Page 507; EA-ITP-ASG2E, Page 1-98; EA-AC65-15A, Page 431*
Photoelectric smoke detectors are the ones most commonly used in aircraft cargo and baggage compartments.

9008. **(B)** *EA-ITP-A2, Page 507; EA-ITP-ASG2E, Page 1-99*
This type of detector draws air samples through the unit. A beacon lamp shines through the labyrinth. An accumulation of 10% smoke in the air will cause enough light to fall on the photocell to energize the warning circuit.

9009. **(C)** *EA-ITP-A2, Pages 502-503; EA-ITP-ASG2E, Page 1-96; EA-AC65-15A, Page 408*
The Fenwal fire detection system is wired in parallel between two separate circuits so that a short in either leg of the system will not cause a false fire warning. The system is wired so that one leg of the circuit supplies potential to the detectors and the other leg serves as a path to ground. If the ground leg of the circuit has a short in it, a false fire warning will not occur because this portion of the circuit is already grounded. If the powered leg of the circuit has a short in it, the current flow will be great enough to trip the relay in the circuit. The tripped relay causes the powered leg to become the grounded part of the system and the grounded leg to become powered. In either case, the short has not caused a false fire warning.

9010. **(B)** *EA-ITP-A2, Page 517; EA-ITP-ASG2E, Page 1-97; EA-ITP-AB, Page 679; EA-AC65-15A, Page 427*
Fire extinguisher containers are checked for their state of charge by periodically weighing them. By knowing the weight of the container and its contents, the amount of agent in the container can be calculated.

9011. **(A)** *EA-ITP-G2, Page 302; EA-ITP-ASG2E, Page 1-97; EA-AC65-9A, Page 102 (Figure 5-3)*
The color code for fire extinguisher lines is brown.

9012. **(C)** *EA-ITP-ASG2E, Page 1-97; EA-AC65-15A, Page 425*
In a continuous-loop fire detection system, a single wire or two wires (depending on manufacturer) are running in an inconel tube. The wires and the tube are separated at all points by a special ceramic material. Continuity between the single wire and the tube, or between the two wires, is what causes a fire warning to be sounded. If the tube becomes dented, kinked, or crushed, the circuit can be completed even though a fire is not present.

9013. **(A)** *EA-ITP-A2, Pages 503-504; EA-ITP-ASG2E, Page 1-97; EA-ITP-AB, Page 675; EA-AC65-15A, Page 408*
A thermocouple in a fire detection system makes use of two dissimilar metals, such as chromel and constantan, which when heated produce a small current flow. It is by the production of this small current flow that a fire may be sensed and a fire warning generated in the cockpit.

9014. **(C)** *EA-ITP-A2, Pages 503-504; EA-ITP-ASG2E, Page 1-97; EA-ITP-AB, Page 675; EA-AC65-15A, Page 408*
A thermocouple fire warning system operates on an entirely different principle than the thermal switch system. A thermocouple depends upon the rate of temperature rise and will not give a warning when an engine slowly overheats or a short circuit develops.

9015. **(A)** *EA-ITP-A2, Page 502; EA-ITP-ASG2E, Page 1-97; EA-ITP-AB, Page 674; EA-AC65-15A, Page 408*
In a fire detection system, thermal switches are generally wired in parallel with each other and in series with the warning light. By wiring the switches in parallel, an open anywhere in the parallel circuit will not affect the operation of the system. Because there is a path to ground from either of the parallel legs, the entire system is still functional.

9016. (B) *EA-ITP-A2, Page 512; EA-ITP-ASG2E, Page 1-97; EA-AC65-15A, Page 427*
Halogenated hydrocarbons (Freon®) are one of the most common extinguishing agents used in aircraft fire extinguishing systems. The Freon® bottles are charged with nitrogen to a pressure sufficient to ensure the complete discharge of the agent. Nitrogen is used as a pressurizing gas because it is inert and noncorrosive.

9017. (B) *EA-ITP-ASG2E, Page 1-97; Aircraft Powerplants, Page 331*
Because fire extinguisher discharge cartridges contain explosive material, extreme care must be taken to prevent inadvertent discharge. The terminal should be shorted any time that the unit is not in place, and there should be no voltage at the terminal connections when the cartridge is reinstalled.

9018. (B) *EA-ITP-A2, Pages 511-512; EA-ITP-ASG2E, Page 1-97; EA-AC65-15A, Page 418*
Carbon dioxide bottles are generally equipped with a red thermosafety discharge indicator disk set to rupture at or above a pressure of 2,650 PSI. Discharge overboard will occur at temperatures above 74°C.

9019. (B) *EA-ITP-A2, Page 502; EA-ITP-ASG2E, Page 1-97; EA-ITP-AB, Page 674; EA-AC65-15A, Page 408*
Bimetallic thermal switches in fire detection systems are usually wired in parallel with each other and in series with the warning light. By wiring the switches in parallel, the system can withstand an open circuit in the thermal switch loop without affecting the ability of the switches to indicate a fire. A short circuit in the system, however, will trigger a false fire warning.

9020. (C) *EA-ITP-A2, Page 515 (Figure 22) ; EA-ITP-ASG2E, Illustration*
Due to the narrow range of the choices between the answers available, we believe the answer to be 45° to 73°F.

9021. (A) *EA-ITP-A2, Page 515; EA-AC65-15A, Page 426; EA-ITP-ASG2E, Illustration*
To answer this question, it is necessary to interpret the chart. For a temperature of 70°F, the allowable pressure is 319 to 405. At 80°F, the allowable pressure is 356 to 443. The difference between 319 and 356 is 37, and the difference between 405 and 443 is 38. Because our temperature is 77°F, our allowable pressure will be $\frac{7}{10}$ of the way between the 70 and 80° pressures. Seven-tenths of 37 is 26, and this added to 319 gives us a lower limit of 345 PSIG. Seven-tenths of 38 is 26.6 and this added to 405 gives us an upper limit of 431.6 PSIG.

9022. (B) *EA-ITP-A2, Page 515; EA-ITP-ASG2E, Page 1-97; EA-AC65-15A, Page 426*
If pressure does not fall between graph limits, the extinguisher container should be replaced.

9023. (C) *EA-ITP-A2, Page 512; EA-ITP-ASG2E, Page 1-97?; EA-AC65-15A, Page 418; EA-ITP-P, Page 321*
The yellow disk indicates which bank of bottles has been emptied by normal discharge.

9024. (B) *EA-ITP-A2, Page 515; EA-ITP-ASG2E, Page 1-97; EA-AC65-15A, Page 426*
If a cartridge is removed from a discharge valve for any reason, it should not be used in another discharge valve assembly, since the distance the contact point protrudes may vary with each unit.

9025. (C) *EA-ITP-ASG2E, Page 1-62; EA-AC65-15A, Page 600*
When working on an oxygen system it is essential that the warnings and precautions given in the aircraft maintenance manual be carefully observed. In general, before any work is attempted on an oxygen system, the following fire precautions should be taken: 1. Provide adequate fire-fighting equipment. 2. Display "NO SMOKING" placards. 3. Avoid checking aircraft radio or electrical systems. 4. Keep all tools and oxygen servicing equipment free from oil or grease.

9026. (B) *EA-ITP-A2, Page 509; EA-ITP-ASG2E, Page 1-97; EA-AC65-9A, Page 502*
See Figure 11-10, toxicity table.

9027. (C) *EA-ITP-A2, Page 513; EA-ITP-ASG2E, Page 1-97; EA-AC65-15A, Pages 423-424*
When performing maintenance on fire detection systems, damaged sensing elements are replaced and not repaired. Continuous loop elements come in pre-made lengths, so they are not altered in any way. Fire detection sensing elements are not calibrated by the maintenance technician. When sensing elements are found to be damaged, they can be replaced by the maintenance technician.

9028. (B) *EA-ITP-A2, Page 512; EA-ITP-ASG2E, Page 1-98; Aircraft Maintenance and Repair, Page 452*
In the explosive cartridge which causes the extinguishing agent to be released, the squib is the explosive charge which drives the slug through the disk.